Climate Change and Its Impact on Fertility

Khursheed Ahmad Wani
Government Degree College, Bijbehara, India

Nibedita Naha
ICMR, National Institute of Occupational Health, Ahmedabad, India

IGI Global
PUBLISHER of TIMELY KNOWLEDGE

A volume in the Advances in Medical Diagnosis, Treatment, and Care (AMDTC) Book Series

Published in the United States of America by
IGI Global
Medical Information Science Reference (an imprint of IGI Global)
701 E. Chocolate Avenue
Hershey PA, USA 17033
Tel: 717-533-8845
Fax: 717-533-8661
E-mail: cust@igi-global.com
Web site: http://www.igi-global.com

Library of Congress Cataloging-in-Publication Data

Names: Wani, Khursheed Ahmad, editor. | Naha, Nibedita, 1972- editor.
Title: Climate change and its impact on fertility / Khursheed Ahmad Wani and Nibedita Naha, editors.
Description: Hershey, PA : Medical Information Science Reference, [2021] | Includes bibliographical references and index. | Summary: "This book studies climate change as the planet's biggest threat affecting fertility of mammals across the globe through its potential effects on heat stress, food and water security, extreme weather events, vulnerable shelter and population migration. It highlights the important and timely issues and concerns to address the latest impact of climate change, and mitigation strategies for enhancing early embryo survival and uterine potential"-- Provided by publisher.
Identifiers: LCCN 2020020130 (print) | LCCN 2020020131 (ebook) | ISBN 9781799844808 (hardcover) | ISBN 9781799844815 (ebook)
Subjects: MESH: Fertilization | Embryo Implantation | Climate Change | Mammals
Classification: LCC RG135 (print) | LCC RG135 (ebook) | NLM WQ 205 | DDC 618.1/780599--dc23
LC record available at https://lccn.loc.gov/2020020130
LC ebook record available at https://lccn.loc.gov/2020020131

This book is published in the IGI Global book series Advances in Medical Diagnosis, Treatment, and Care (AMDTC) (ISSN: 2475-6628; eISSN: 2475-6636)

British Cataloguing in Publication Data
A Cataloguing in Publication record for this book is available from the British Library.

Advances in Medical Diagnosis, Treatment, and Care (AMDTC) Book Series

ISSN:2475-6628
EISSN:2475-6636

Mission

Advancements in medicine have prolonged the life expectancy of individuals all over the world. Once life-threatening conditions have become significantly easier to treat and even cure in many cases. Continued research in the medical field will further improve the quality of life, longevity, and wellbeing of individuals.

The **Advances in Medical Diagnosis, Treatment, and Care (AMDTC)** book series seeks to highlight publications on innovative treatment methodologies, diagnosis tools and techniques, and best practices for patient care. Comprised of comprehensive resources aimed to assist professionals in the medical field apply the latest innovations in the identification and management of medical conditions as well as patient care and interaction, the books within the AMDTC series are relevant to the research and practical needs of medical practitioners, researchers, students, and hospital administrators.

Coverage

- Disease Prevention
- Critical Care
- Alternative Medicine
- Internal Medicine
- Medical Procedures
- Chronic Conditions
- Experimental Medicine
- Medical Testing
- Emergency Medicine
- Patient Interaction

IGI Global is currently accepting manuscripts for publication within this series. To submit a proposal for a volume in this series, please contact our Acquisition Editors at Acquisitions@igi-global.com or visit: http://www.igi-global.com/publish/.

The Advances in Medical Diagnosis, Treatment, and Care (AMDTC) Book Series (ISSN 2475-6628) is published by IGI Global, 701 E. Chocolate Avenue, Hershey, PA 17033-1240, USA, www.igi-global.com. This series is composed of titles available for purchase individually; each title is edited to be contextually exclusive from any other title within the series. For pricing and ordering information please visit http://www.igi-global.com/book-series/advances-medical-diagnosis-treatment-care/129618. Postmaster: Send all address changes to above address.

Titles in this Series

For a list of additional titles in this series, please visit:
http://www.igi-global.com/book-series/advances-medical-diagnosis-treatment-care/129618

Handbook of Research on Advancements in Cancer Therapeutics
Sumit Kumar (National Institute of Pathology, New Delhi, India) Moshahid Alam Rizvi (Jamia Millia Islamia, India) and Saurabh Verma (National Institute of Pathology, India)
Medical Information Science Reference • ©2021 • 913pp • H/C (ISBN: 9781799865308) • US $465.00

Cases on Medical Nutrition Therapy for Gastrointestinal Disorders
Elham Abbas Aljaaly (King Abdulaziz University, Saudi Arabia) Nahlaa Abdelwahab Khalifa (King Abdulaziz University, Saudi Arabia) Rouba Khalil Naaman (King Abdulaziz University, Saudi Arabia) and Manar Abduljalil Bakhsh (King Abdulaziz University, Saudi Arabia)
Medical Information Science Reference • ©2021 • 254pp • H/C (ISBN: 9781799838029) • US $245.00

Handbook of Research on Disease Prediction Through Data Analytics and Machine Learning
Geeta Rani (Manipal University Jaipur, India) and Pradeep Kumar Tiwari (Manipal University Jaipur, India)
Medical Information Science Reference • ©2021 • 586pp • H/C (ISBN: 9781799827429) • US $395.00

Handbook of Research on Evidence-Based Perspectives on the Psychophysiology of Yoga and Its Applications
Shirley Telles (Patanjali Research Foundation, India) and Ram Kumar Gupta (Patanjali Research Foundation, India)
Medical Information Science Reference • ©2021 • 577pp • H/C (ISBN: 9781799832546) • US $345.00

New Developments in Diagnosing, Assessing, and Treating ADHD
Rejani Thudalikunnil Gopalan (Mahatma Gandhi Medical College and Hospital, India)
Medical Information Science Reference • ©2021 • 419pp • H/C (ISBN: 9781799854951) • US $265.00

Handbook of Research on Oncological and Endoscopical Dilemmas in Modern Gynecological Clinical Practice
Konstantinos Dinas (2nd Department of Obstetrics and Gynaecology, Aristotle University of Thessaloniki, Greece) Stamatios Petousis (2nd Department of Obstetrics and Gynaecology, Aristotle University of Thessaloniki, Greece) Matthias Kalder (Department of Obstetrics and Gynaecology, Philipps-University of Marburg, Germany) and George Mavromatidis (2nd Department of Obstetrics and Gynaecology, Aristotle University of Thessaloniki, Greece)
Medical Information Science Reference • ©2021 • 460pp • H/C (ISBN: 9781799842132) • US $385.00

Diagnostic Techniques and Therapeutic Strategies for Parotid Gland Disorders
Mahmoud Sakr (Alexandria University, Egypt)
Medical Information Science Reference • ©2021 • 362pp • H/C (ISBN: 9781799856030) • US $295.00

IGI Global
PUBLISHER of TIMELY KNOWLEDGE

701 East Chocolate Avenue, Hershey, PA 17033, USA
Tel: 717-533-8845 x100 • Fax: 717-533-8661
E-Mail: cust@igi-global.com • www.igi-global.com

Editorial Advisory Board

Table of Contents

Detailed Table of Contents

Climate change can disturb the characteristics of the soil either indirectly, or directly, or both. The direct effects include changes in the soil properties and composition by organic carbon transformation, precipitation, temperature, and % of moisture. The indirect impact includes nutrient cycling, improved soil erosion rate due to an increase in rate and intensity of rainfall, irrigation facility, changes in the crop rotation, and spadework practices. The presence of soil organic carbon (SOC) greatly influenced by the climatic condition. The crop cultivation depends upon the exchange of carbon between the troposphere and soil, which also vastly influences the fertility of the soil. The use, development, and management of soil depend upon soil structure, soil texture, soil stability, water holding capacity of the soil, availability of the nutrients, and erosion of the soil. Hence, deterioration of soil fertility by climatic change may affect the several soil lifeforms like fauna and flora either directly, or indirectly through nutritional value of the soil.

It has been shown that reproductive activity is responsive to changes in the physical, psychosocial, and chemical environments. Occupational exposure has been identified to a wide range of putative hazards and adverse reproductive outcomes, or fertility. Such risks include the physical environment such as VDT and noise, as well as psychosocial stress and chemical agents. Some of these may have a direct or indirect effect on IVF outcomes. Psychosocial factors such as inadequate coping mechanisms, anxiety and/or depression are related to a lower incidence of IVF pregnancy. In this chapter the authors focus on the difficulties involved in linking exposure to putative hazardous substances to adverse reproductive outcomes in environmental and occupational settings, particularly the effectiveness of IVF procedures.

Nibedita Naha, ICMR-National Institute of Occupational Health (NIOH), Ahmedabad, India
Gokul Manickavachagam, Biochemistry Division, ICMR-National Institute of Occupational Health (NIOH), India

Textile industry is one of the primarily concerned industries in the world generating huge revenues where both the skilled and unskilled labours are employed. Raw materials are mainly synthetic chemicals rather than the natural one. Several bleaching agents, additives are also used to get sufficiently high and uniform degree of whiteness in the textile materials. Environmental contamination is also common by the textile effluents. All these substances are mutagenic and carcinogenic, and cause enormous health hazards to humans, animals, and aquatic lives. Infertility, implantation failure, and miscarriage due to exposure to dyes, dye intermediates, and other raw materials are major concern worldwide as integrity of both the male and female gametes are massively affected by them. However, little or no attention has been paid in real life scenario, mainly in the developing and under developed countries including adaptation of successful and advanced mitigation strategies. Therefore, the chapter highlights the common issues and possible remedial measures of the textile industry exposure with respect to fertility potential and pregnancy outcome.

Chandana Haldar, Department of Zoology, Banaras Hindu University, India
Jayita Pal Chowdhury, Department of Zoology, Banaras Hindu University, India

Entrainment of 24 hrs light/dark cycle is nowadays getting altered in corporate sector and even at home due to over exposure to artificial light at night (ALAN) and is commonly denoted as chronodisruption. Chronodisruption interferes with the daily physiology that is regulated by SCN/pineal-melatonin/Hypothalamo-Pituitary axis. In females, reduction in sleep due to shift work desynchronizes HPG-HPA axis induces a stress like condition, accompanied by increased free radical generation in ovary leading to polycystic ovaries, depletion of ovarian reserve, luteal phase defects, endometriosis, implantation failure, etc. Evidence suggest that there is remedial rescue by management of circulatory melatonin due to its chronobiotic/antioxidant/anti-inflamatory/antiapoptotic potentials and its role in ovarian folliculogenesis and steroidogenesis. Advancement in melatonin research suggests that it could be beneficial for commercial use, that is, (1) as an endogenous synchronizer, (2) to improve the quality of oocyte, and (3) for promoting success in the frequency of IVF.

Khursheed Ahmad Wani, Government Degree College, Bijbehara, India
Jamila Irfan, Government Womens College, Srinagar, India
Junaid Ahmad Malik, Government Degree College, Bijbehara, India

One of the expected effects of mammalian developmental defects is a rise in air temperature. Heat stress adversely affects embryo oogenesis, oocyte maturation, fertilization, and implantation. The number of defects caused by heat stress in all mammals is almost identical, but each species has its own particular sensitivity to specific defects. It suggests that genotype may have a significant effect on the type of defect, its occurrence, and its extent. By heat output and loss, the body temperature is usually preserved in a restricted range, but the equilibrium can be disrupted by illness, inadequate nutrition, and severe environmental temperature. Elevated maternal temperature during pregnancy, either by fever or any other means of heating, may result in embryo death, retarded development, abortion, and many embryonic defects such as cell proliferation, migration, differentiation and apoptosis or programmed cell death, structural and functional defects, and changes in maternal physiology. Maternal heat stress also reduces the levels of placental hormones. This chapter deals with the heat stress effect on reduction of reproductive function, implantation defects, etc. of different animals and humans.

Chapter 6
Impact of Heat Stress and Nutritional Stress on Early Embryo Development 113
Golden Gokhale, Devi Ahilya Vishwavidyalaya, India
Pir Mohammad Ishfaq, Dr. Harisingh Gour Central University, Sagar, India
Gurudutt Sharma, Devi Ahilya Vishwavidyalaya, India

The nutritional, physiological, and reproductive function has detrimental effects on heat stress, which is found in many species of mammals. High ambient temperature in mammals cause a decrease in the length and intensity of estrus by disturbing ovarian function as well as decreasing pregnancy rate after artificial insemination. The effects of nutritional stress on developing oocytes in the ovarian follicle and in the reproductive tract on early embryos are because of the environment where a breeding female lives before conception and at the early stages of pregnancy. Maturity of oocyte, blastocyst yield, prenatal survival, and the number of offspring born alive are affected by change in consumption and quantity of the food taken during the pre-mating period. To improve reproductive efficiency and offspring quality, it is necessary to detect and evaluate the deteriorating effects of heat stress on reproductive organs and cells and to plan nutrition related strategies.

Chapter 7
Thermotolerance for Physiological and Endocrine Regulation of Embryo-Uterine Development 135
Sameena Khanday, Devi Ahilya Vishwavidyalaya, Indore, India
Rayees Ahmad, Jamia Millia Islamia, New Delhi, India
Guru Dutt Sharma, Devi Ahilya Vishwavidyalaya, Indore, India

Hyperthermia affects most aspects of reproductive performance in mammals by compromising the physiology of reproductive tract, through hormonal imbalance, disrupting the development and maturation of oocyte, causing embryonic mortality, abortion, growth retardation, and major developmental defects. Heat stress reduces the steroidogenic capacity of its theca and granulosa cells by altering the efficiency of follicular selection resulting in drop of luteinizing hormone and estradiol secretions from the dominant follicle in the plasma, reduced intensity, and duration of estrus expression. The mechanism for the developmental stage-dependent change in heat tolerance is considered to be the accumulation of antioxidants in embryos in response to heat-inducible production of reactive oxygen species. Morula or blastocysts can repair heat-induced misfolded or unfolded proteins or facilitate DNA damage induced apoptosis. Therefore, embryo transfer (ET) that can bypass the heat-sensitive stage could be a good solution to improve the conception rate under heat stress. However, further research is required to improve the reduction in pregnancy rates due to summer heat stress.

Foreword

Climate change will have an impact on the economy of countries together with COVID-19 pandemic during the coming decade. These will affect the human social life, in particular, fertility. The climate change will affect both the agricultural as well as non-agricultural sectors. The changes may affect fertility patterns following the structural transformation process. Temperature alone will have long-term effects on fertility via reproductive health. It may also lead to large societal disruptions following impact on fertility patterns. The climate change may affect fertility through child mortality, longevity, and the gender wage gap. Now-a-days parents prefer to have fewer children with greater health and education investment or more children with less investment on them. In general, the climate change leads to higher fertility and lower educational attainment. We know that at high latitudes, the demographic effects of climate change are reversed, as lower fertility and greater education attainment. The impacts also differ in rich and poor countries depending on the income of a family. In addition, the cultural norms top the list regarding the consequences for fertility decisions. The families decide whether to use the resources to have more children, or invest more in the future of each child. All this depends on the total amount of resources spent on bringing up children. If the relative benefit of having another child decreases the same then fertility goes down. The effects on fertility decisions also depend on the relative cost-benefit interactions of having children and investing in the well-being of each child. If climate damages decrease income, consumers spend much of their income on agricultural goods. Economic theory and evidence suggest that families must adapt to these changes by spending lesser resources on educating children and more on increasing fertility. However, the gender wage gap also affects fertility through structural changes. In many parts of the world women have to undertake the responsibility for raising children. If labour opportunities for women increase, the cost of raising children increases. Generally, females have a comparative advantage in education-intensive work, while males have comparative advantage in other production modes. The opportunity cost for women for raising children increases when economy reallocates towards agriculture. The fertility also depends on access to birth control. The gender wage gap has been an important factor in long-run demographic changes even in highly industrialised countries. The effects of climate change on fertility through structural transformation depend on the assumption of its differential impacts across sectors. Such a differential effect exists because crop yields are weather-dependent and highly sensitive to the changes in climate. The changes in temperature directly affect productivity. There is enough evidence suggesting that increase in temperatures can lower worker productivity, or decrease attendance. The change can at the same time affect long-term worker health, increase the potential for extreme weather, and lead to coastal erosion, all connected to the agricultural production. There is a vast difference in temperature around the globe. Increasing temperatures have differential effects on the suitability of local conditions for growing any particular crop. As for the impact across all crops, this process will lead to differential effects on overall agricultural productivity. There is a potential link between climate and fertility, and any change will affect mortality. The increases in life expectancy are associated with greater human capital accumulation and lower fertility according to the theory as longer life expectancy increases the benefits from early investment in human capital.

Any long-run changes in fertility can also be an important driver of change in child mortality. The parents try to increase fertility in case of high and uncertain mortality rates. The parents want to ensure that they achieve some minimum desired level of fertility.There is theoretical relationship between mortality and children number however, statistical evidence in this connection is mixed. In the case of linking higher mortality with decrease in fertility, higher costs of having healthy children and increased investments in the health of each living child lead to an uncertain and context-dependent overall effect. There are simple causal effects from increasing income to decreasing fertility. It is not that richer families try to have few children, as it is unlikely because raising children is resource-intensive. A greater return to education in industrialised countries suggests that the quantity-quality trade-off would suggest a negative correlation between fertility and income.Many mechanisms determine the full strength of climate change effect on fertility. If climate change lowers the productivity of agrodiversity, food scarcity will lead to an increase in relative food prices that will increase the relative wages for agricultural workers and lower the incentive for parents to invest in child quality, leading to an increase in the fertility but educational attainment decreases. New findings can highlight the potential for international development assistance to increase climate change adaptation. The future work needs to consider non-economic mechanisms as, cultural factors, large-scale disruptions like natural disasters and civil wars, should be attached to the climate change, resulting in significant effects on fertility. Many least developed countries on the equator are facing climate change with broader negative effects on agriculture. Any scarcity in agricultural products end up with higher agricultural prices and wages. This leads to labour reallocation in the sector because agriculture makes less use of skilled labour. The climate damages decrease the return to skilled jobs, compelling parents to invest less resources in the education of children and to increase fertility. At higher latitudes these patterns are reversed suggesting that climate change may exacerbate inequities by reducing fertility and increasing education in rich nations, while increasing fertility and reducing education in poor tropical nations. The working paper published in July 2020 has emphasized the fact on globally comprehensive and empirically grounded estimates of mortality risk due to future temperature increases related to climate change. Investigations undertaken in 40 countries have revealed that a U-shaped relationship is found in the countries where extreme cold and hot temperatures increase mortality rates, especially for the elderly, it is flattened by higher incomes and adaptations to local climate. The mean increase in mortality risk is valued at roughly 3.2% of global GDP in 2100, with today's cold locations benefiting and damages being especially large in today's poor and/or hot locations. The data stresses the fact that climate change will indeed affect fertility and there is definitely need for future work on this topic. In this book, editors have pooled up the papers presenting the potential for human and animal adaptation to climate change concerning the long-run fertility patterns and highlight this important issue and spur future work in this area.

This edited volume on *Climate Change and Its Impact on Fertility* by Dr. Khursheed A. Wani and Dr. Nibedita Naha has 14 chapters. The Chapter 1 deals with the climate change and its impact on soil fertility and life forms. It argues that such a change can disturb the characteristics of soil, may change the soil properties and composition by organic carbon transformation, precipitation, temperature, and moisture as well as indirectly through nutrient cycling, improve soil erosion rate because of high rainfall, irrigation facility, changes in the crop rotation and spadework practices, resulting in detrimental effects on food security and health. The topic of the effect of environmental and occupational factors on implantation failure presented in Chapter 2, which discusses the reproductive activity is responsive to changes in the physical, psychosocial, and chemical environment. Authors in this chapter have identified occupational exposure to a wide range of putative hazards and adverse reproductive outcomes, and fertility. They present the difficulties involved in linking exposure to putative hazardous substances to adverse reproductive outcomes in environmental and occupational settings, particularly the effectiveness of IVF procedures. Chapter 3 highlights the issues of textile industry and health hazards related to climate change and fertility. It presents the possible remedies with respect to exposure in such industries as these are the ones primarily concerned industries in the world that generate huge revenue, where both skilled and unskilled labours are employed, moreover raw materials are mainly synthetic chemicals rather than the natural ones. In Chapter 4 the topic of chronodisruption and loss of female reproductive potential has been evaluated. An entrainment of 24 hours light/dark cycle is changing in corporate sector and even at homes due to over exposure to artificial light at night. It is commonly denoted as chronodisruption which interferes with the daily physiology regulated by SCN/pineal-melatonin/hypothalamo-pituitary axis. The advancement in melatonin research suggests that it could be beneficial for commercial use both as an endogenous synchronizer as well as improve the quality of oocyte and for promoting success in the frequency of IVF. An increase in average temperature is one of the predicted consequences of embryonic defects in mammals as such, heat stress adversely affects oogenesis, oocyte maturation, fertilization and implantation of embryo; the genotypes can have a strong influence on the type of defect, its incidence and its severity. All these findings have been presented in Chapter 5 titled as impact of heat stress on embryonic implantation. This chapter is followed by Chapter 6 with title as impact of heat stress and nutritional stress on early embryo development. It argues nutritional, physiological and reproductive functions under heat stress face detrimental effects as recorded in many mammal species. High ambient temperature in mammals decreases the length and intensity of estrus by disturbing ovarian function as well as decreasing pregnancy rate after artificial insemination. Authors stress that to improve reproductive efficiency and offspring quality it is necessary to detect and evaluate the deteriorating effects of heat stress on reproductive organs and cells and to plan nutrition related strategies. In Chapter 7 on thermotolerance for physiological and endocrine regulation of embryo-uterine development, authors present the mechanism for the developmental stage-dependent change which is considered to be due to the accumulation of antioxidants in embryos in response to heat-inducible production of reactive oxygen species. They propose that embryo transfer can bypass the heat-sensitive stage and could be a good solution to improve the pregnancy rate under heat stress. The neuroendocrine cross-talk and axial regulation of embryonic implantation and the role of climate change issues have been summarised in the Chapter 8, whereas an effect of temperature stress on sperm motility and viability is discussed in Chapter 9. The Chapter 10 presents details on the fertility and reproductive behavior due to environmental contaminants in extreme conditions. Authors discuss the association between environmental contaminants and reproductive disorders with particular emphasis on infertility. The impact of climate change on animal fertility is given in Chapter 11. It stresses that the fertility and reproductive efficiency of male/female animals is very sensitive to climatic disturbances, particularly hyperthermia. The heat stress decreases the quantity-quality of sperm production in males and compromised fertility in females. Authors have separately and thoroughly elaborated the effects of climate change on male and female reproduction separately in this chapter. The Chapter 12 deals with the climate change and livestock fertility because it is a global threat to the sector including several species and ecosystems at global level. According to the authors climate change, heat stress and nutritional stress are the major intriguing factors responsible for reduced fertility in farm animals in tropical countries. Adaptation to these changes so as to lower its negative effect by alteration of animal micro-environment using different essential technologies are discussed as the main mitigation strategies to recover heat stress damage in this connection. In Chapter 13 titled as climate change effects on fertility and reproductive processes of livestock and amelioration strategies, the author has summarised the data on climate change-induced heat stress disruptions in reproductive and fertility systems in livestock, which can be reversed through genetic modifications, nutritive supplementation, physical cooling mechanisms and hormonal therapies. The author stress that a successful implementation of these strategies is pegged on improved research and their combined administration. The Final Chapter (Chapter 14) has pooled up information on the multidimensional impact of climate change on human reproduction and fertility; and presenting information on the medical perspective on changing dynamics in this connection as it is of great interest in human reproduction and fertility. According to the author in the real-world scenario, humans face threat of concurrent exposure to multiple, adverse climatic parameters, which may potentiate, augment negative effects of each other, causing cumulative ultimate, measurable damage to reproduction and fertility. The evidence although foreseeable is limited in this connection.

Overall, in the book the authors show that reproductive activity is responsive to physical, psychosocial, and environmental changes. Occupational exposure can also lead to a wide range of putative hazards leading to adverse reproductive outcomes or infertility. Authors also have focussed on the difficulties involved in linking exposure to different hazardous substances to adverse reproductive outcomes in environmental and occupational settings. Further, authors collectively tried to reveal the hidden cross-talk and regulatory mechanisms to unfold the coordinating role of climate change issues on embryonic implantation and offspring development in animals, and humans; and provided a new insight for future research on the subject which is becoming a major public health issue of reproductive wastages due to adverse climatic conditions.

Best Wishes,

Münir Öztürk
Islamic World Academy of Sciences, Turkey & Centre for Environmental Studies, Ege University, Turkey & Universiti Putra Malaysia, Malaysia

Preface

World Health Organization (WHO) has recognized infertility as a worldwide public health problem. Infertility is a major issue to the residents of the developing and underdeveloped countries, and worldwide nearly 186 million people have suffered with this disorder, where climate change is playing a crucial role to this direction. Several organisms including microbes, plants, animals etc. are influenced by the prolong environment changes along with humans; thus number of factors has been identified in which altered climatic conditions can affect both the general and reproductive health hazards. Climate change variables have potential to mediate their ultimate impact on human as well as animal reproduction through a variety of complexes, and interlinked pathways including physiological, biochemical, molecular, diurnal and psychological factors along with adverse occupational and economic consequences. Therefore, how these climate change variables ultimately affect the overall aspect of the reproductive health and fertility of the living organisms (whether single cell, or multi-cellular), will be a great importance of research need of the day from the point of public health view.

The reproductive processes of both the males and females, especially in animals and humans, are very sensitive to disturbance by hyperthermia, with the most evident consequences being decreased quantity and quality of sperm production in the males, and compromised fertility in the females. The embryo-maternal dialogue that begins very early stage in the life of the embryo, is critical for its own implantation. Successful embryo implantation involves an appropriate interaction between the receptive endometrium, and an implantation-competent blastocyst, which occurs in a limited time period, known as “the window of implantation”. The role of the circadian clock is indispensable for the regulation of female reproductive physiology and its de-synchrony is a major cause of several pathological conditions. The phenomenon of chronodisruption in literature is commonly studied in photo-periodic animals while exploring potential mechanisms behind it. Important gaps in the present knowledge explaining the major impact of light, shift work, heat stress etc. on the physiology, and/or pathophysiology of reproduction that may serve as intentions for future investigation.

Further, the sustainable use of the land surface can restore the major ecosystem and will promote ecological balance. There is a link in between different segments of environment and these segments influence each other in one, or the other way; for example, deficiency of food and safe drinking water, poor sanitation, population migration, changing disease pattern and morbidity, more recurrent great weather events, lack of shelter, adverse occupational outcome, and so on, that can affect human and animal heaths through altered climatic conditions over the years. Nutrition plays a critical role during embryonic development, as nutrients have the efficiency to modify oocyte and embryo micro-environment, making it more favorable to fertilization as well as pre-implantation growth. Globally, another important area of research related to climate change issue is livestock sector although optimization of production capacities is hampered by several environmental conditions, which directly and indirectly affect the demand for animal products; and thus the overall economy of the country. Since climate change has multidimensional effects on fertility potential and there is dearth of literature in this direction, a need was felt to compile the relevant literature at one single place.

This edited book included 14 chapters providing in-depth coverage of the wide spectrum climate change issues and impacts on fertility potential such as, soil fertility; environmental and occupational factors on implantation failure; role of textile industry on climate change and fertility potential; chronodisruption in female reproductive potential due to shift work; heat stress and nutritional stress on implantation and embryonic development, as well as sperm motility and viability; thermotolerance for physiological and endocrine regulation of the embryo-uterine cross-talk; neuroendocrine and axial regulation during implantation; fertility and reproductive behavior during extreme conditions; animal fertility; livestock fertility including amelioration strategies; and multidimensional impact on human reproduction from medical perspective.

The Chapter 1 argued that the climate change can disturb the characteristics of the soil either directly, or indirectly, or both, and the impact on different life forms through nutritional value of the soil and thus food security. The Chapter 2 focused on the difficulties involved in linking exposure to putative hazardous substances to adverse reproductive outcome in environmental and occupational settings, in particular, the effectiveness of IVF procedures. The Chapter 3 highlighted the climate change issues, impacts and possible remedial measures of male and female fertility potential including carcinogenesis and mutagenesis with respect to exposure in the textile industry. Reduction in sleep due to shift work desynchronizes HPG-HPA axis, induces a stress-like condition, accompanied by increased free radical generation in ovary leading to polycystic ovaries, depletion of ovarian reserve, defects in luteal phase, endometriosis, and failure of implantation, are well debated in the Chapter 4 of the book. The Chapter 5 argued that an increase in average temperature is one of the predicted consequences of embryonic defects in the mammals. Heat stress adversely impacts oogenesis, oocyte maturation, fertilization and implantation of the embryo. The Chapter 6 advocated that reproductive efficiency and offspring quality are necessary to detect and evaluate the deteriorating effects of heat and nutritional stress on reproductive organs and cells, and to plan nutrition-related strategies. The mechanism for the developmental stage-dependent changes in thermotolerance is considered to be the accumulation of antioxidants in embryos in response to heat-inducible production of ROS, which are explained in the Chapter 7 of the book. The Chapter 8 focused to reveal the hidden cross-talk and regulatory mechanisms to unfold the coordinating role of neuroendocrine axial system in the embryonic implantation, and also to explore the impact of climate change in this regard. The impact of temperature on the motility and viability of spermatozoa in some fish species like, salmonids, cyprinids and sturgeons, and in humans also have been fundamentally surveyed and abridged in the Chapter 9. The goal of the Chapter 10 is to discuss the association between the environmental contaminants in extreme conditions and the reproductive behavioral disorders with particular emphasis on infertility. The Chapter 11 discussed that the fertility and reproductive efficiency of both the male and female animals are very sensitive to climatic disturbances, in particular, hyperthermia. Adaptation to climatic changes and lowering the negative effects by alteration of the animals especially, livestock's micro-environment using different essential technologies are the main mitigation strategies to recover heat stress-induce damage, as explained by the authors in the Chapter 12. The Chapter 13 especially dealt with female livestock, and argued that heat stress decreases the production of gonadotrophins, resulting in hormonal imbalance, deterioration of oocyte quality and lengthens the estrus period, leading to infertility. The Chapter 14 finally pooled up information on the multidimensional impact of the acute or chronic changes in the climate change variables on human reproduction and fertility, and the threat of concurrent exposure to multiple, adverse climatic parameters in reality that may potentiate/augment negative effects of each other, thus causing cumulative but ultimate, measurable damage to reproduction and fertility outcome.

Overall, this edited book brings diverse topics together, representing a novel resource for researchers to address the *'climate change and fertility potential issues'* in both the male and female animals and humans. This book is intended for researchers, academicians, scientists, environmentalist, reproductive biologists, experimental toxicologists, mammalian cell biologists, embryologists, fertility centers personnel, medical students and doctors, industrialist, NGOs, health and safety agencies, regulatory authorities, public health officials and policy makers, who are working in this field. The vast information for all researchers to continue their research beyond the topics covered, and the research gaps as identified through different chapters, will open new avenues for future research. We hope this volume will meet the demands of the researchers all over the world, and we welcome the suggestions from all the readers to be incorporated in our next volume/edition.

Thanks to all.

Khursheed A. Wani

Nibedita Naha

Acknowledgment

Disclaimer

This edited book embodies the study on climate change issues and the impacts on different aspects of fertility. The authors have taken care to ensure that information compiled in the book chapters are correct. The editors and publishers however does not assume and hereby disclaim any liability to any part for completeness of information, errors of omission, commission, sequence, due to an inadvertent negligence, accident, or any cause.

Drs. KA Wani and N Naha would like to make special recognition of Prof. Münir Öztürk for his contribution in forwarding the book *Climate Change and Its Impact on Fertility* in spite of his busy schedule. Eminent editors provided editorial assistance both via continuing interactions with each of the authors as well as expert reviewers of the subjects. The suggestions of the Editorial advisory board and timely service of IGI-Global are also highly acknowledged.

This edited book embodies the study on climate change issues and the impacts on different aspects of fertility. The authors have taken care to ensure that information compiled in the book chapters are correct. The editors and publishers however does not assume and hereby disclaim any liability to any part for completeness of information, errors of omission, commission, sequence, due to an inadvertent negligence, accident, or any cause.

Chapter 1
Climate Change and Its Impact on Soil Fertility and Life Forms

Trinath Biswal
VSS University of Technology, India

ABSTRACT

Climate change can disturb the characteristics of the soil either indirectly, or directly, or both. The direct effects include changes in the soil properties and composition by organic carbon transformation, precipitation, temperature, and % of moisture. The indirect impact includes nutrient cycling, improved soil erosion rate due to an increase in rate and intensity of rainfall, irrigation facility, changes in the crop rotation, and spadework practices. The presence of soil organic carbon (SOC) greatly influenced by the climatic condition. The crop cultivation depends upon the exchange of carbon between the troposphere and soil, which also vastly influences the fertility of the soil. The use, development, and management of soil depend upon soil structure, soil texture, soil stability, water holding capacity of the soil, availability of the nutrients, and erosion of the soil. Hence, deterioration of soil fertility by climatic change may affect the several soil lifeforms like fauna and flora either directly, or indirectly through nutritional value of the soil.

INTRODUCTION

Soil is an important and also most ignored material over the surface of the earth, although it is indirectly responsible for the food source of the plants and animals. Carbon is an essential element for the existence of all living being on the globe and is the major constituent of both plants and animals. The ocean is the largest store of carbon and after that the second largest sink or store of carbon is soil. Due to climate change the carbon balance is disturbed and more carbon is stored in plants and soil because of agricultural productivity, and in accordance to that more carbon also released to the atmosphere, disturbing the equilibrium condition of the atmosphere(Henrysonet al., 2018;PatiandLamnganbi, 2018). The sustainable use of the land surface can restore the major ecosystem, which may probably alleviate and familiarize to the climate change. There is an interlink in between atmospheric carbon and soil carbon. During the process of photosynthesis, plant takes carbon from the atmosphere, but the carbon present

DOI: 10.4018/978-1-7998-4480-8.ch001

at the atmosphere also affects the soil, because the percentage (%) of carbon, which is not only used for the growth of plantsabove the ground surface but also normally distributed into the roots of the plants and this carbon deposits in soil. If this carbon is not distributed properly, then it becomes stable and is locked for many years. Climate change something occurs within the atmosphere because of different climatic factors and healthy standard soil can able to mitigate climate change properly (Frazer, 2009). The largest pool of carbon, sulphur and nitrogen in terrestrial ecosystem is soil and it very closely involved the interlinking of human being in atmosphere and land. Soil is also the primary source in which human beings and animals depend upon their food, shelter dress materials, fuels etc. Hence the use of land in a sustainable manner is a vital factor that accountable for climate change and is the cause of major impact to the human population. The carbon storage capacity of all soils are not equal, it was observed that peat lands are the highest carbon containing soil in the universe and this kind of soil are mostly found in the UK, Ireland and some parts of northern Europe(Lori et al., 2017). The soil containing grasslands also contains higher % of carbon per hectare of land. The soil in dry and hot areas contains a less amount of carbon and these kinds of soils are specially found in the southern European countries. Due to rapid urbanization and industrial development, the amount of CO_2 and other greenhouse gases are rapidly increasing within the troposphere, resulting the change in climate which causes an unfavourable situation of our society and in the terrestrial ecosystem. The unplanned management of soil by the human being is the cause of loss of bio-genic carbon from the biomass, which is one of a major cause of addition of $CO_2$to the atmosphere(Pareek, 2017).The organic material present in the soil is the major nutrient source of the soil and it is the cause of increase in the water holding capacity of the soil, if the organic matter present in the soil decreases the quality of the soil and the agricultural productivity also accordingly decrease.

It was closely observed that the earth's atmosphere is continuously warming and it was found that in recent years the temperature of the earth increased to an unexpected level, since from past 50 years. There is strong evidence found that the warming effect observed in the last 50 years is due to human activity, especially,the burning of fossil fuels and deforestation(Smith et al., 2017) The climate change impacts stress on land surface and cause of changing the quality, properties and composition of the soil. This climate change not only increases the temperature on the surface of the earth, but also causes many hazards such as, frequent droughts, floods, other natural disasters, earlier flowering of plants and the pole ward changing in the distribution of numerous species. The main forcing variables of soil forming are directly due to the global change which includes supply of organic matter and organic carbon from biomass, management of soil temperature, soil hydrology and the change in the potential evaporation. It was observed that in the last 30 years the global climate becomes unpredicted mainly because of industrialization, urbanization and agricultural origin (Karmakaret al.,2016).The International Geosphere Biosphere Program (IGBP) started in order to study and analyse these changes and proposed possible solutions towards the impact on the human population and its influence towards the environment. The organic matter content and production of the cultivated soil depends upon a number of factors such as, climatic situation, soil features and organic matter present in the soil. It was reported that in different climatic conditions the % of organic matter in different soils are different, the increase in productivity also causes increase in plant residues, and these plant residues on biodegradation also cause an increase in organic carbon in the soil(Sherbininet al.,2007). Hence, the high yield of crops with appropriate management of crop residue is important to enhance the increase in soil organic carbon (SOC), especially, for agricultural land soil. Soil erosion is an important factor, which is greatly influenced by climate change; however, the climate change is significantly affected by the hydrological cycle, nitrogen cycle and carbon sequestration which influences the soil nutrients (N and P) and level of the organic matter present in the soil(Yu et al.,2019).Hence, more research is needed to know about the mechanism of soil erosion in connection with climate change. The organic matter present in the soil can alter two important cycles such as, carbon cycle and nitrogen cycle and cause of significant contribution towards the global food security. Hence the interaction of soil to the climate is an important matter for the study of soil, food security and agricultural productivity, which directly, or indirectly affect the life forms(Rachelleet al.,2018).

IMPACT OF CLIMATIC FACTORS ON SOIL FERTILITY

Impact of High Concentration of CO_2

- CO_2 is an important greenhouse gas in the atmosphere. The burning of fossil fuel such as, coal and petroleum products for the generation of electricity is one of the major sources of CO_2 especially in the countries like China and India.
- The transportation is the second principal source of CO_2. The international travel, including ships in the ocean is another cause of enhancing CO_2 in the troposphere.
- The industrial sectors, mainly steel, cement, and other manufacturing units are the third largest source of the production of CO_2.
- Another important source of the generation of CO_2 is deforestation. About 20% of the production of CO_2 is due to the burning of wood and forest materials as fuel (Stewartet al.,2020).

The higher is the % of CO_2, more is the rate of growth and water absorptive capacity of the natural crops, but it is limited to some extent. For some plants, at elevated temperature the increase in the concentration of CO_2 causes adverse effect mainly in the respiration in night time. For a particular species the reduced growth of the cycle due to more CO_2 and higher temperature can be remunerated by natural vegetation in the alterations of dominance or the composition of the species. The short growth cycle of the major crops is greatly affected due to climate change and the unproductive periods of the cropping pattern might be raised. More agricultural litters, residues of crops and root mass are the cause of increase in the concentration of the mycorrhiza colonization and the activity and function of other soil micro-organisms resulting in the increase in nitrogen fixation and is considered as a positive effect on supplying nitrogen to the plants or crops. This increase in the root and microbial activity is the cause of the increase in CO_2 both in the soil and air, therefore the release of plant nutrients and micronutrients drastically increases(Francisco et al.,2019).The activity due to mycorrhiza is also a cause of improved phosphate uptake. Hence, higher is the microbial activity, more is the cycling of the plant nutrients and again the population of the soil micro-organism depends upon the various parameters of the local climate. More is the root materials and crop residues, greater is the soil organic matter and fertility of the soil. It is found that in the high concentration of CO_2, the C/N ratio in the crop litters will be higher, which is the cause of slower remobilization and slower decomposition of the plant nutrients from the agricultural waste litters and is then more time is taken for the insertion into the soil by different kinds of soil micro-organisms such as, termites, earthworms, etc. If the soil temperature is more, then it can able to counteract the soil organic matter and made it stable, but the microbial activity of the soil is stimulated. The enhancement of organic matter and soil micro-organisms present in the soil is due to more CO_2 and air, and may be the reason of competition of plant nutrients temporally. This temporary

Table 1. Chemical composition of the soil.

Sl. No.	Characteristics of soil	Composition
1	pH of the soil	6.5
2	Available Phosphorous	16.1 ppm
2	Total organic Carbon	1.4%
4	Total Nitrogen	0.15%
5	Exchangeable form of Carbon	1.89 meq/ 100g
6	Exchangeable form of Potassium	0.4 meq/ 100g
7	Exchangeable form of Magnesium	0.66 meq/ 100g
8	Soil Colour	Dark Brown

effect may be the cause of negative impact on the plant and its productivity. Due to the increase in the% of CO_2 and temperature, the activity of micro-organisms also increases, leading to the production of more polysaccharides and soil stabilizers. Hence the root mass agricultural litters, residues of crops, and other soil organic matter are the cause of stimulating the function of soil macro fauna such as earthworms, bacteria, actinomycetes, fungi, algae and protozoa, which causes the increase in the infiltration rate of the soil and increases the flexibility of the soil and soil fertility.

The increase in the partial pressure of the CO_2 within the soil air is the cause of decreasing oxygen content in the soil, leading to damage of the function of roots, which cannot be materialized; but if the exchange of the gas in between the soil and atmosphere is increased, then it results the formation of more stable biopores which maintains the equilibrium in between the CO_2 and O_2 within the soil either naturally, or artificially as in the form of well-drained soils. The crops cultivated in wetland areas, mainly rice, jute possess their own gas exchange mechanisms, and therefore the vegetation and % of yield cannot be affected. The major greenhouse gas in the troposphere is CO_2,which contributes to an average of about 55% towards global warming and major source of it, is combustion of oil, coal and gasoline. Soil is a huge pull of organic carbon in the terrestrial eco-system which counteracts the high level of CO_2 and hence possibly playing an important role in controlling climate change and maintaining the equilibrium condition of the troposphere. Carbon has accumulated in the soil in the process of photosynthesis through numerous years and is lost from the soil by the soil micro-organisms (bacteria and fungi). These micro-organisms convert the soil carbon to CO_2 and finally released to the troposphere. The equilibrium of both of these two processes is indeterminate to each other, but the climate change due to CO_2 is the model that shows how much % of carbon converted into CO_2 by soil micro-organisms and how much persist in the soil. The higher % of CO_2 in the atmosphere is the combined effect of fossil fuel burning and release due to conversion of organic carbon to CO_2 by soil micro-organism (Anjaliet al., 2019; Salehiet al., 2017; Onwukaand Mang, 2018). The average composition of the soil is shown below.

Impact of Rainfall Pattern and Temperature

The organic material existing in the soil also depends upon the variation of average temperature, rainfall pattern and precipitation. All these factors, not only affect the organic matter, but also affect some other important properties such as, soil stability, aggregate formation, infiltration capacity, cation exchange capacity, and water holding capacity and nutrient content present in the soil. The climate change and soil properties are complex process and proceeds slowly because the soil does not only affect directly by climate change, but also it affects indirectly in many ways. The change in temperature in the atmosphere is the cause of variation in the decomposition of the organic matter, change in % of moisture and can affect the evapotranspiration which are partially accountable for the climate change. In addition to the climate change, the use and function of the soil are more dependent on socio-economic factors of the local environment. In many parts of the world, the rainfall gradient is more associated with factors of soil fertility, but it was observed that the seasonal drought may affect the soil fertility. This loss of CO_2 is the cause of weakening soil structure, soil stability, water holding capacity of the top soil, humus matter present at the surface of the soil, soil erosion and nutrient availability. The soil carbon is destroyed due to the increase in climatic temperature, but the decrease in the soil carbon is counterbalanced by the increase in the release of nutrients, which leads to the increase in plant productivity and subsequently increases the plant litters. The release of CH_4 gas and formation of peat increases by an increase in rainfall rates, whereas the decrease in the rainfall rate is the cause of deficiency of the moisture, especially, in shallow soils, increase in CO_2 and peat formation. In case of forest soils, the rainfall pattern affects the survivability and reproduction of the soil organisms, foraging patterns, food web, food chain and pathogens of natural plants. The decrease in rainfall rate is the cause of droughts and other related natural hazards, and increases the probability of shrink-swell in the clay type soils. It is the cause of the increase in temperature in soil, the foundation of the building, increase in leachate formation and generation of more landfill contaminant gases. Otherwise the increase in rainfall rate is the cause of increasing the atmospheric nitrogen and the deposition of nitrogen in the soil and stimulates disturbances in the soil, cause of flooding, which results in the change of habitants of waterlogged and wetland areas and also increases erosion of the soil causing the increase in silt and clay particles in the surface water bodies (Brevik, 2013; Issoufouet al.,2020).

Additionally, there is some indirect impact of climate change in the fertility and texture of the soil. The integrated impact of change in the climate may generally expect the increase in the % of crop yield, such as in winter, sugar beet, sunflower, wheat etc., which probably due to the combined impact of CO_2 fertilization, the efficiency of the use of radiation and the growing season is comparatively long. All these factors are mostly applicable to the C_3 photosynthetic pathway and not necessarily applicable to the C_4 pathway. The increase in the CO_2 is resulted the increase in the dry weight and size of the plants, trees and their components, especially in C_3 plants, which is due to more photosynthesis in the leaves of the crops and plants, and gradually tends to decrease the yield. The increase in the % yield is generally expected in the sunflower plants, whereas less increase in the yield, or probably the decrease in the yield is observed in the crops like oilseed rape, potatoes and high quality of horticultural crops if these crops grow under water deficit light textured soils. Hence the soil organic carbon is the cause of increase in soil organic matter and impacts greatly the productivity of the soil. Soil is simultaneously interacting with a number of factors such as climatic factors (wind, temperature, pH, moisture, etc.), parent soil materials, soil micro-organisms and takes thousands and thousands years of its complete formation. In many areas of our globe soil is still in the developing stage. Hence climate is considered as the most valuable factor in the development and formation of soil, its use, proper management with the implication to the soil stability, structure, water holding capacity of the top soil, availability of nutrient and erosion of the soil. Indirectly the climatic factors are the cause of the change of growth rate of plants, efficiencies of using water, raising on sea level, cause of increase or decrease in the vegetated areas, etc. Hence we can conclude that the greater the productivity, the higher is the organic carbon in the soil and thus resulting increase in the organics. In the monsoon climates and humid tropical areas, the intensity of rainfall is more, therefore causing high amount of leachate formation in the areas of well-drained soils having more infiltration rate, which is the cause of water saturation and temporary flooding in these areas. Hence it is the reason of reducing the decomposition of organic matter in the soil. In some local areas, there may cause of mudflows or landslides in some spongy sedimentary materials. In the semi-arid and subtropical regions, there is more efficiency of the use of the water which increases the productivity. This is mostly due to higher concentration of $CO_{2,}$ which may cause the increase in the ground cover and balances the impact of increase in temperature. If the intensity of rainfall in the local regions is less, then it leads to the formation of less dry-mass and cause of lowering the nutrient and organic contents in the soil (Kyei-Mensahet al., 2019; Várallyay, 2010 ;Kumarand Das, 2014)

Impact of pH

In many soils the climatic change does not affect the pH of the soil, but there might be an exception in some soils, which includes the soils having rich in acid sulphate, coastal areas and estuary areas. Every soil has its buffering capacity and due to this property, it can resist the change of pH, but such kind of soils are mostly clays having moderate or higher power of cation exchange capacity and the quantities of the acid generated in such kinds of soils is mainly due to oxidation, which normally exceed the buffering capacity of such kind of soils. Hence the pH value of such kinds of soil may be temporarily falling in a range of 2.5 to 3.5 and some portion of the clay may undergo decomposition at that range of pH. The soil generally buffers in the pH range of 3.5 and 4 in the long time basis. In the calcareous soils the soil reaction generally carried out in the range of the pH of about 8.5 to7.0, which depends upon the partial pressure exerted by the CO_2 into the soil and this range of pHis maintained against the leaching of basic cations by the various soil processes as long as a some % of finely dispersed lime remains in the soils. But the buffering in the non-calcareous soils are comparatively less, which depends upon the cation exchange capacity of the soil and pH. In most of the soils, the decay of organic matter is the cause of maintaining concentration of CO_2 in the soil air, which is far lower than the atmospheric CO_2concentration and the solubility of $CaCO_3$ is estimated by the partial pressure of CO_2 present in the soil air and also its activity in soil water but not in the atmosphere. Hence the leaching rate of lime is positively related to the rate of decomposition of organic matter, negatively impact to the diffusion rate of the gas and also positively impacts the amount of water that infiltrating to the soil. Hence the rate of leaching is accelerated due to the change in climatic condition, if the soil undergoes rapid acidification for a longer period of time, then the leaching rate is also affected. The soil may be steadily and continuously depleted due to basic cations but the change in pH either starts or sometimes becomes comparatively more rapid once certain buffering pools are exhausted or nearly exhausted. The pH of the soil is a direct function of the weathering time, a function of the parent materials, vegetation products and climatic factors is treated as important indicators of soil properties and fertility. Although most of the soil resists at the change in pH due to the change in climate (temperature change, CO_2 fertilization, deposition of the atmospheric N, variation in the precipitation); but it can affect the status of the organic matter present in the soil, nutrient cycling, carbon content, water available for plants and also plant productivity. The soil pH becomes lower with increasing height because of, due to leaching of the basic cations in a higher rate, more precipitation and an increase in the rate of nitrification. Hence it is concluded that with the increase in height, the soil pH decreases and vice-versa (Subediet al., 2019; Chatterjee et al., 2019; Asabere et al.,2018).

Impact of Rising of sea Level and Climate in Coastal Areas

The characteristics of the soil in sea level mostly depend upon the type of estuaries and coastal areas. The level of sea water raises normally depends upon the various external and local factors with the forest area and the interrelationship in between them. The rise in the level of the sea water is the cause of the erosion of the soil in the coastal areas and the existing coastal line forwarded towards the land surface. The extend of the movement of the sea, mainly depends upon the elevation in the sea level, resistance due to the materials in the coastal areas, sediments due to river flow, drift of the long seashore and the strength of the long seashores, the tidal waves and the human interaction in the coastal areas which probably either accelerate or prevent the erosion of seashore areas. The sediments generated from the different major deltas such as, the deltas of Brahmaputra–Ganga River and the major Chinese rivers in the estuarine areas are sufficient to counterbalance the impacts of the rise of the sea level. The aggregations of such kinds of deltas are reduced under the following circumstances(Mimura, 2013; Praveen et al.,2016).

- The inland of human interventions which includes the large dams, water reservoirs or victorious soil conservation programs that substantially prevents the supply of sediments into the delta areas. The construction of huge numbers of dams may cause the erosion of the soil and frequent flooding.
- The embankment construction within the premises of delta valley prevents the supply of sediments to the nearby back swamps and the cause of the increase in the level of the sea water.
- The land sinking generally occurs because of the abstraction of oil, natural gas and water, which is already happening in the northern part area of Netherlands and in the Bangkok. The lowland areas of coastal regions are not sufficiently protected by the embankments, or sediment supply, or the tidal flooding by the saline water and it penetrates into the inland area, which is extended by seasonally saline soils.
- The presence of Phragmites (*Rhizophora* mangrove) areas over many decades are causes the formation of the soil rich in sulphate acids. The drainage of water from the land surface by the higher sea level, by the congruently higher levels of the connecting estuarine areas of the rivers and their embankments may extend the areas continuously, or seasonally reducing soil coverage and increasing normal flood depths, and also durations in the river and estuary area basins and also on the embankment back slopes (Willy et al.,2019).
- The regions where the soil becomes continuously wet, the % of organic matter in the soil will more and eventually are the cause of the formation of peat in that region, but otherwise where the erosion in the soil of coastal areas able to remove the present barrier of mineral soils, or the mangrove forests and hence causes the formation of more frequent formation of storms, hurricanes which results the increase in the level of the sea water. It causes the destruction of the peat swamps in the coastal estuarine areas. But otherwise, the coastal erosion is prevented by the existing preventive barrier of minerals present in soils or mangrove forest, higher storm surges associated with a rise in sea level could allow sea-water saltwater lagoons to the freshwater (Sidibé et al.,2018).
- There are some coastal areas where the level of the lands in the estuarine areas are subsided due to the extraction of oil and natural gases, abstraction of water etc., which causes the change in the level of sea water. Although the water rise in sea impacts in most of the coastal areas, but the effect is different in different areas depending upon the type of soil. It was found that some island states are completely disappearing even if modest increase in sea level. About 33.5% population of the

world is living within a 100 meters of sea level (vertically) and almost 23% of the population of the world is living within 100km from the sea area(Gelybó et al., 2018).

- There are different physical impacts of increasing the level of the seawater are as follows;
    - Decrease in the land areas, or even complete vanishing of some of the low-lying islands.
    - Increase and submergence of flooding in the estuarine and coastal land areas.
    - Intrusion of the saltwater of the subsurface and surface waters.
    - Increase in the erosion of the soil.
    - Destruction of the habitats of the species in the coastal areas.
- The warming of the global climate is also partly due to the increase in sea level. In the past, people had no idea regarding the rise in sea level and at that age people are not affected by the climate change. From continuous marine research it was found that the level of the water of various oceans is continuously increasing since last 100 years.
- The change in the level of the sea water depends upon the nature of the land and the different land areas are of a different property. The relative change in the sea level in several regions is different, but the actual change in the sea level in specific places on the globe is the sum of the global change in sea level value and landfall or raise value. There are some large deltas in the globe like the Yellow River deltas, Yangtze River deltas where the rate of land sinking, is mainly due to the change in the rate of sea level in the estuarine areas. Recently the scientist of different countries has been focusing on the increase in the sea level and the interlinking in between sea level and altitude of the land because the rise in the sea level greatly impacts the human society and its habitats.
- Although salt is an essential ingredient of cooking of food, but higher % of salt in the soil are the cause of devastation of crops and loses the fertility of the soil. The rise in the level of the sea water in the low-lying estuarine and coastal areas is the cause of temporary flooding with sea water, which results the increase in the concentration of the salt present in the soil and the soil becomes infertile. Due to flooding of sea water, the land surface is covered by a layer of salt and that salt layer is dissipated due to rainwater. The climate change is not only the cause of a rise in the level of the sea water, but also resulted even if extreme weather conditions and rise in the frequency of heat waves and droughts, the major causes of excessive use of the groundwater for irrigation, drinking and other human consumption leading to the depletion of the ground water level. Due to spreading of salt water, more salt has leached into the soil and made it infertile and affects the agricultural production(Williams, 2013;Kulpand Strauss, 2019).

Climate Change and its Impact on Soil Erosion

The abrasion of topsoil due to various environmental factors and climate change is normally called as soil erosion. The topmost layer of the soil is the most fertile layer because this layer is rich in nutrient and organic materials. Hence farmers always want to protect the top layer of the soil for cultivation and grass growing for cattle feeding. The cause of soil erosion is mostly due to rainfall, wind and flood; hence the variation in the frequency of the intensity of the rainfall greatly affects the erosion process of the soil. In addition to that the climate change has a major effect on the chemical and physical properties of the soil, which subsequently influences the infiltration rate and erosion process of the soil. When anything erodes it undergoes destruction because some external force acting on it and in the similar manner it was observed that the continuous beating forces of tidal waves and wind are the cause of the erosion of the rocky structures nearby many coastal lines, leaving behind a number of fascinating caves, cliffs and structures of different shapes. The soil erosion normally occurs due to a number of forces such as, hydrostatic force due to water, wind speed, and usual farming (Hussain and Javadi, 2016).The formation of soil is an extremely slow process that takes year after year. The soil is naturally formed by mixing the weathered rocks with minerals and organic matter produced in it due to the biodegradation of dead bodies of plants and animals, although soil during formation is continuously disrupted by both the man-made and natural forces.

The sensitivity of the erosion in soil is due to the change in climate which is greatly influenced by the farmer's management. In the inland area of the Pacific Northwest (PNW) the change in the climate is the cause of enhancing the greenhouse gases and the cause of an increase in temperature, winter precipitation for which a great portion of precipitated water dwindling as rain water rather than snow fall and leads to an increase in the intensity of precipitation in extreme condition of the events. This change in the atmosphere may alter the interrelationship between the atmosphere and land. Therefore, the agricultural sector becomes stable, profitable and predictable, which may increase the economic growth of the farmers. Although any projected changes in the climatic conditions are beneficial for certain sectors, but some group of farmers may not be able to change the pace of the global climate change through the mitigation, they may adapt the change in the climate and plan the management practices, which are tougher to the impacts of the climate. There is a great relationship between the climate change and soil erosion and the broad range of the change in the climate is due to precipitation and warming condition. The rate of the soil erosion varies with the variety of crop management features and the development of the landscape and some other climatic factors. Hence all these factors are analysed by the model known as Water Erosion Prediction Project (WEPP) model, which predicts the loss of soil, biomass change, scenarios of winter erosion and evapotranspiration due to climatic and environmental variables.

The soil fertility rate is larger in the month of December and January in the warming experiments because of the intense reduction in the snowpack on the ground surface and causes an increase in both rainfall and melting of snow during these months. The climate change related to the soil fertility is the correlation in between runoff, precipitation, temperature, plant vegetation and soil erosion. Due to the change of the climate both precipitation and temperature changes, but the availability of the moisture balances in between evapotranspiration and precipitation. If temperature and precipitation both are increasing the moisture available either decreases or increases that depends upon the relative increase in the precipitation and evapotranspiration. The increase in the available moisture is also otherwise termed as effective precipitation, which tends to promote both the soil erosion and runoff simultaneously, whereas the vegetation coverage influences "competing" processes of these factors. Hence, there is a good interrelationship exist among the soil erosion, vegetation, runoff under a specified condition and changing of the effective precipitation. The vegetation decreases the soil erosion and vice-versa whereas the erosion of the soil is also tending to increase the surface runoff, because of a decrease in the soil moisture storage capacity and exposure to the low-permeability in the subsoil. Therefore, soil is ultimately responsible for sustainable development because of their multifunctional role, which can providing services that includes the biomass production such as, fiber materials, food, fuel, feedstuff, habitats for living organisms, gene pools, biodiversity, cleaning of air and water, extenuation of the emissions of greenhouse gas, contributions to the sequestration of carbon, buffering of the precipitation in excesses and provisions to recreation, culture and human health assets. So, finally it can be said that the change in the climate is interrelated with temperature and extreme weather events, mainly intense rainfall, storms, frosts, droughts and raising the level of the sea water in the coastal areas. All these effects are the cause of increasing threats to the soil erosion, soil fertility reduction, soil compaction, decrease in the agricultural productivity, and ultimately the cause of deterioration of environmental sustainability and food security.

The climate change can impact the soil functions both directly and indirectly. The direct effect is the transformation of the organic carbon, nutrient cycling through the changed moisture, temperature management of the soil and increasing the rate of soil erosion due to the increase in the intensity and frequency of the rainfall. It was found that in the soil of higher organic content, the increase in temperature and a decrease in the soil moisture are linked with the impact of warming, drought, which subsequently reduces the carbon storage capacity and soil fertility. There is also some indirect effect of climate change on soil functions. The effect of climate change can be mitigated by agricultural management. For example, due to the increase in the soil organic carbon, the carbon sequestration, and the farmers can implement adaptations as a result of the multiple tangled driving forces, change in market price, the implication of new technologies and improvement of knowledge in combination with climate change. The global warming effect can lead to a more intense and rapid hydrological cycle, which results the increase in more frequency and intensity of the rainfall events and according to the models of the climate change this trend is expected to be continued in an increased rate in the 21st century also. This change in rainfall impacts the changes in temperature, concentration of atmospheric CO and CO_2, solar radiation and results more significant impacts on the erosion rate of the soil. Moreover, the process of climate change and its impact on the erosion of the soil due to water flow is a more complex thing which involves the changes in the intensities and amounts of rainfall, number of days of precipitation, production of plant biomass, ratio of snow to rain, the decomposition rate of plant residues, microbial activity in soil, the rate of evapotranspiration, and shifting of land use, which is required to accommodate in the new climatic regime (Hamidov and Helming, 2018;Klik and Eitzinger, 2010).

Impact of Climate Change on Soil Functions

The impact of the climate change in the properties of the soil is a very slow and complex process because the soil is greatly affected by the climatic condition both directly and indirectly. Example-The decomposition of the soil organic matter directly depends upon temperature, whereas the variation of the soil moisture indirectly influences due to change in evapotranspiration of the plants and trees and in addition to that it is accountable for the creation of the greenhouse gases, which is significantly responsible for the change in the climate. The function and use of the soil provides more attention in the socio-economic factor rather than environmental factors. But however the interaction in between the different processes of soil formation and development, especially biological processes makes it problematic to quantitative measurement of the changes (Novaraa and Pisciottaa, 2018).

- ***Direct impact on soil property***

The model of the 'Soil-climate' predicts a continuous increase of the fertility of the soil due to vegetation. This also predicts the probable changes in the temperature, evaporation, evapotranspiration and precipitation that are connected with the increase in organic matter and facilitating the more release of CO_2 in the minerals and from the organic soils. The reduction of the organic carbon in the soil affects some other soil properties such as soil structure, water holding capacity of the top layer soil, availability of nutrients, soil stability and soil erosion. The carbon reduction in the soil is increased by the rise in the temperature of the soil or atmosphere, but this effect is counterbalanced by the release of more nutrients because of the increase in the plant productivity. The increase in the intensity of the rainfall is the cause of increase in the formation of peat and more release of CH_4 gas whereas the decrease in the rainfall concentration causes loss of CO_2, cause of increase in the deficiency of the moisture especially in the arable crops, decrease in the formation of peat and in case of forest soil it affects the reproduction capacity and survivability of the soil invertebrates, foraging patterns, food chain, food web, ecological balance and some natural plant pathogens. The decrease in the rainfall is the cause of chance of drought formation which results the probability of shrink-swell in the clay type of soils and disturb to the foundations of the building, increases soil temperature and also cause of intensifying the chemical attack to the foundations of engineered structures (clay caps).The contaminated landfill site is the cause of generation of more leachate and emission of toxic contaminant gases. The increase in the temperature is the cause of the increase in the perseverance of pesticides. The diseases in the crops or plants and the types of pests are changing with the change of the climate and hence there is a complex relationship exists in between the climate change and its impact towards the environment. The increase in rainfall due to climate change is the cause of depositing more amounts of nitrogen into the soil and disturbs the soil composition resulting in equilibrium in food production, change in the habitants of water logging areas, an increase in the rate of soil erosion and more potentially cause of potential increase in surface water pollution (Vries and Griffiths, 2018).

- ***Indirect impact on soil property***

The combined impact of changes in climate is generally assumed to be the increase in the crop yields such as, in the winter season the yield of some crops such as, curly flower, cabbage, beans, wheatsugar, beet wheat etc. This is due to the combined impact of radiation efficiency, CO_2 fertilization, which is mainly applicable to the pathway of C_3 photosynthesis and not in the species of C4 pathway. The rise in the % composition of the CO_2 causes the increase in the dry weight and size of the plant components of most of the C_3 plants, mainly petioles and stems during the vegetative development in order to support the leaves. The harvesting index tends to be decrease with the increase in the concentration of CO_2 and temperature. Although the increase in the production is predictable for sunflower, but the slight increase or even decrease is observed for the crops like oilseed rape, potatoes and some high quality of horticultural crops which grows especially in water deficit light textured soils(Soares et al., 2019; Classen and Sundqvist, 2015). Fig 1 shows a number of ways of depletion of soil organic matter, the prime cause of soil infertility.

Figure 1. Effects of depletion of soil organic matter.

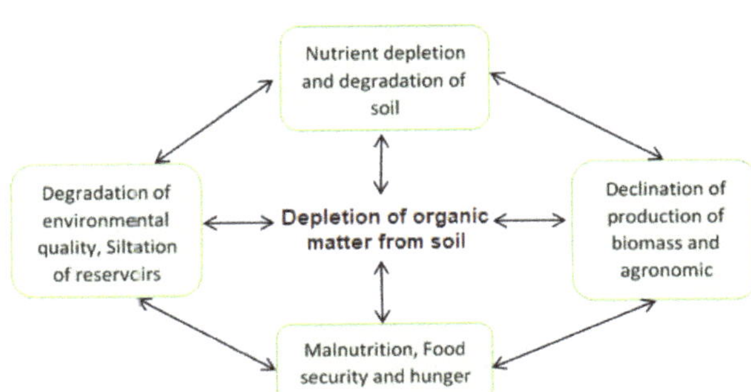

Impact of Variation in Land use due to Climate Change

The climate change forced the human beings to modify the land use pattern. The possible change of the land use includes a decrease in the arable land change of the grassland to the arable afforestation, change in the soil properties by changing the properties of the soil like the alternation of the soil biota which may cause the modification of the structural stability of the soil biodiversity of the soil, interactions in between plants and soil and also nutrient cycling. The continuous change in the vegetation also brings about the nutrient loss, changes the runoff and the % of soil organic matter. But actually the socio-economic factors play a significant role in the distribution of the land use patterns. It was reported by many scientists that afforestation is the cause of decrease in respiration of the soil and hence decreases the emission of soil CO_2, although the details for this are not still clear as there is only an important relationship with the temperature. The respiration of the soil has been found to be ambitious by some specific climatic parameters which includes precipitation, air temperature and status of soil water. The wide range of the chemical, physical and biological characteristics that affect the formation of soil and modifying the soil properties will also replay to the climate change due to the variabilityof timescales (Qambrani et al., 2017; Le et al., 2019).

IMPACT OF CLIMATE CHANGE ON SOIL PARAMETERS

The different properties of the soil responsible for climate change, are represented below, which reflecting the different properties of soil (Fig. 2).

Figure 2. Different properties of the soil.

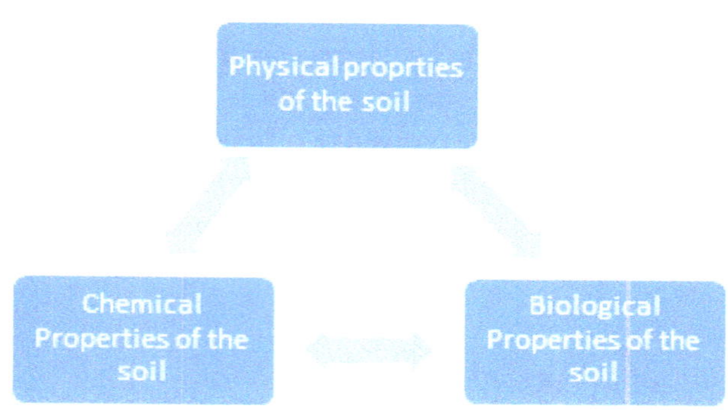

Physical Parameters

- ***Soil water***

The soil water changes by a number of climatic factors, i.e. precipitation is the cause of a great change in the soil water, the increase in temperature of the atmosphere causes more evapotranspiration, which results more water loss from the soil and depend upon the kind of land use. There is an integrated relationship of climate-vegetation-hydrology-land utilization, variations of which are replicated by the soil moisture and the arena of water balance. The lower precipitation rate or restricted water storage capacity is the cause of decrease in the water storage (S),infiltration (I),groundwater recharge (G) and surface runoff (R) specially attended by less precipitation, moderate rate of the hostile hydrological consequences of the frost and quick melting of snow, resulting in waterlogging hazard and providing more chance for penetration of water. The decrease in the atmospheric precipitation will be the major cause of decrease in the water storage (S),infiltration of water (I), surface runoff (R),also decreases the water supply to the soil and subsequently to the plants and trees especially in the hilly areas having swelling surfaces results hazards effect of water erosion. But for dry surface there is an increasing risk of wind prevention, losses due to filtration, recharge of groundwater (G)and hence causes the increase in the evaporation losses. The transpiration rate in the plants increases if the crop cover area or the area of vegetation has not declined due to deficiency of water, drought sensitivity, environmental and ecological consequences with its physiological activity. The direct effect of soil fertility can be modified due the impact of the properties and characteristics of vegetation. It directly depends upon the types of the vegetation, density of soil, species composition, dynamic factors, litters, biomass production and root characteristics. The human interference like the land use, agro- technics, cropping pattern, improvement of wind and controlling water erosion, irrigation, chemical recovery, drainage and some other activities that can drastically modify the field water balance and also its mechanisms. There are a number of soil forming procedures such as, turnover of organic matter, weathering, clay translocation, podsolization, which are strongly influenced by the moisture the of the soil. The structural formation of the soil affects the developmental of runoff, percolation into soil, infiltration, drainage and the distribution of the water across the region's landscape (Bonfante et al.,2019; Gogoi et al., 2019).

- ***Soil temperature***

The variation of soil temperature is the most vital indicator of the climate change. There is a very good relationship in between the soil temperature and atmospheric temperature and if the temperate of the air increases by any cause then there will be prominently lead to be an increase in the soil temperature. The management and control of temperature of the soil is mainly affected by the loss and gain of radiation at the surface of the soil, the evaporation process, conduction of heat by the process of soil profile and convective transmission of heat on the basis of the movement of the water and gas. There is a substantial decreasing trend of heat in the depth of snow cover is observed in winter and spring season and was related to increasing the air temperatures. The soil temperature and soil moisture are the chief mover in most of the soil processes. More soil temperature will increase the soil processes, cause of quick decomposition of the organic matter, enhancement of microbial activity, faster release of nutrients, increase in the rate of nitrification and more commonly accentuate to chemical weathering of the minerals. But however the soil temperatures also affected by the nature and type of vegetation cultivated in the land surface which may also change itself as a result of the change in climate or any adaptation management (Dong et al., 2019).

- ***Texture differentiation and soil structure***

The soil structure means the type and nature of combination of soil particles together and is an important property of the soil. The soil structure is the cause of movement of soil gases, soil water, different contaminates, nutrients, seepage, maintaining the quality of the water, soil fauna and appearance of crops and also the foundation of the buildings. The quality and characteristics of soil structure is greatly influenced by the quality and amount of organic matter, ingredients of inorganic material present in the soil matrix, method of cultivation, freeze-melt performance and some natural processes such as the shrink-swell, which generally occurs in the soils having a higher of clay, mainly in semiotic mineralogy. The soil organic material plays a significant role in impacting the fertility and soil properties. The declination of the organic matter present in the soil causes the decrease in the stability of the soil aggregate, rate of infiltration and increases the surface run-off, vulnerability to compaction and susceptible to soil erosion. In some regions flash flooding may happen due to change in soil structure resulting more cracking in the land surface which affects the soil fertility (Arévalo-Gardini et al.,2015).

- ***Soil colour***

The colour of the soil is treated as the indicator of the different developments that can be going to happen in the soil depending upon the category and amount of minerals present in the soil. Example: If the soil is red colour it indicates the higher of iron oxide in the oxidizing condition present in the soil. The dark colour of the soil is due to a higher of the decomposed organic matter. The yellow colour of the soil is mainly due to the presence of hydroxides and hydrated iron oxides in the soil. The black nodes of the soil are due to the existence of MgO. The mottling and gluing of the soil are mainly due to poor drainage facility and higher water table. The plentiful pale yellow mottles coupled with low pH are the indication of acid sulphate rich soil. The colour of the soil is described by different parameters termed as value, hue and chroma. Hue signifies the colour or major wavelength of the light, value generally refers to the lightness of the colour, chroma generally refers the relative purity or strength of the colour. On the basis of these parameters the colour of the soil can be rapidly determined by the comparison with a sample with a standard set of colour chips equestrian in a note-book termed as Munsell soil colour charts. In the soil chart the right hand side top corner signifies the hue, the vertical axis shows the value and the horizontal axis shows the chroma (Awaland Safeeq, 2019).

- ***Soil permeability***

The ability or capacity of the soil to transmit air and water is called as soil permeability. The soil, which is more impermeable, is considered as a good quality of the soil for aquaculture because the loss of water due to infiltration or seepage is very less. Since the different layers of the soil or horizons are different in the characteristics, therefore the permeability of the soil also differs from layer to layer. The different factors including soil texture, Pore size, and structure of the soil and the presence of impermeable layers such as clay pan determine the rate of permeability of a soil. The Clayey soils having platy structures are generally exhibits low permeability. The Permeability of the soil is measured as in terms of the coefficient of permeability or permeability rate (cm per sec, cm per hour, cm per day). Both these parameters are determined in the laboratory by calculating the rate of flow of water from a particular water head over a column of the soil at definite moisture concentrations and some other conditions. The coefficient of permeability of the soils appropriate for aquaculture should be less than 5×10^{-6} m/s. Hence the permeability of the soil has a direct influence to the climatic condition (Raza et al.,2019).

Biological Parameters

- ***Soil organic matter***

The organic material present in the soil is definitely one of the most important components of the soil because it determines the quality of the soil in terms of water holding ability, soil structure, nutrient storage, oxygen-holding capacity and soil stability. The organic matter present in the soil specifically useful for a number and variety of soil microflora and fauna that is responsible for the productivity and soil health. The organic matter of the soil is extremely susceptible for the change of the land management and land use, soil moisture and soil temperature. It was seen that in some regions due to proper management and change of the land use pattern the level of the organic matter present in the soil gradually decreases which subsequently increases the susceptibility to the erosion of the soil. The organic matter in the soil is functioning as both sink and source of carbon in the environment causing the climate change. It was calculated that globally the quantity of carbon stored in soils is almost about three times that present in the troposphere. It was found that the increase in the CO_2 and global warming in the troposphere favours the plant growth and productivity and can deliver more organic matter to the soil. The increase in the temperature of the atmosphere and also the soil is the cause of soil decomposition and decrease in the organic matter present in the soil. The increase in the efficiency of the use of water in the higher CO_2 situation results higher productivity mainly in the regions of limited water supply, but it depends upon other limiting factors mainly soil nitrogen and phosphorus. Peat soils are especially susceptible to the climate change. This group of soils is particularly rich in the organic matter throughout the whole depth(Schoonover and Crim, 2015).

- ***Soil life forms: fauna and flora***

There are thousands of species generally found in a square meter of all most all soils called as soil flora and fauna, which serves a substantial role in different processes of soil such as breakdown, retention and incorporation of plant residues, nutrient cycling and their impact on the structure of the soil and porosity. Global warming has almost no direct impact on the ecological structure because the soil flora and fauna have a quite broad optimum temperature, although the changes in ecosystems and the immigration of vegetation zones may greatly affect by less migratory soil flora and fauna due to the increase in temperature and change in the rainfall pattern. There is also another significant impact to climate change because of soil flora and fauna due to highCO_2 in the troposphere, which leads to an increase in plant growth, productivity and allocation of more carbon below the ground level that enhances the microbial population and increases the rate of nitrogen fixation in the soil. The denitrification and immobilization of nitrogen are the cause of the increase in the mycorrhizal population and soil aggregation and finally increased weathering process of minerals. However, always there is equilibrium maintained in between the decomposition of the organic matter in the soil, plant productivity and also affects the climate change. The change of the climate affects either indirect or direct or both on the activities and functions of the soil micro-organisms which is one of the feeding source of the greenhouse gases to our atmosphere and partly responsible for global warming. The direct impact includes the soil micro-organism and its influence on the production of greenhouse gases, temperature change, change in the rate of precipitation and also extreme climatic phenomenon.

The indirect impact includes climate-driven changes in the productivity of plants and change in physicochemical conditions of the soil, the supply of carbon to the soil, the activity, function, structure and of microbial populations involved in the decomposition processes of the organic matter and release of carbon from the soil. The decomposition of the organic matter and global warming are the cause of increase in the rate of the activities of heterotrophic microbial communities and affects the % of CO_2 in the atmosphere and transfers dissolved organic carbon by the process of hydrologic leaching. The indirect response to the climate-microbe interaction operated in a number of ways with prediction of different mechanisms, which are responsible for climate change in a variety of ways, but broadly it may be divided into two types. The first one deals with the indirect effects of increasing the atmosphericCO_2concentrations on soil microbes, through the process of increased plant photosynthesis and transfer of the photosynthetic carbon to fine roots of the plants and mycorrhizal fungi. It is the indication of a higher % of CO_2inthe photosynthesis process and growth of the plants increases mainly in the nutrient-rich soil, which causes the increase in the flux of the carbon to the roots, symbionts of those plants, heterotrophic micro-organisms, organic acids, sugars, and amino acids. There is a major outcome due to soil micro-organisms and carbon exchange, which includes the enhancement of the loss of soil carbon because of respiration and the drainage of water as in the form of dissolved organic carbon due to microbial activity in accordance with the climatic changes (Robinson et al., 2018).

Chemical Parameters

The most important and rapid procedure of mineralogical or chemical changes is due to the change in the external conditions which causes the loss of cations from soil nutrient, salts in soil and leads to an increase in the leaching process and salinization where the movement of water occurs upward due to increase in evapotranspiration rate or decrease in irrigation water supply or rainfall intensity. The composition of the clay mineral per mineralogy of the coarser portions is generally unchangeable or little changeable even for a longer period of time, but the exception is the transformation of halloysite formation in the moisture rich environment to the periodic drying or of the steady dehydration of the goethite to haematite at elevated temperatures or of condition of severe drying or both. The change in the surface characteristics of clay fraction is always less than the movement of the salts present in the soil, which happens in a much faster rate than the change of the crystal structure or bulk composition. The surface change has a great influence on the chemical and physical properties of the soil. The change in the bulk composition or mineral surfaces of the clay portion of the soils are carried outby using a number of few transformation processes or it can be inhibited due to some variations in the external climatic conditions due to the global climate change. For example, the water hydrolysis containing CO_2 can able to remove silica particles and some basic cations and also be accelerated by the increase in the rate of leaching.

Cheluviation is a process which can remove or dissolves the iron and aluminium from the soil due to chelating the organic acids and can be accelerated by increasing the rate of leaching. Ferrolysis is a cyclic process of dissolution and transformation of clay with the alternating oxidation and reduction of iron which causes the decrease in the cation exchange ability due to the interlayering of aluminium because of swelling of clay minerals and it generally happened in the soil subjected to alternating the process of reduction and leaching with oxidation. This may be occurring large areas in the warmer climate, especially in monsoon climates and high latitudes. The dissolution of the minerals of the clay due to strong mineral acids, acidic aluminium salts and amorphous structure of silica where sulphidic resources of materials in the coastal areas are normally oxidized with the development of drainage system, but however the increase in the level of the sea water is the cause of reduction of the probability of this process in natural manner. Reverse weathering is a process of formation of clay and transformation of it under strongly alkaline or neutral medium and create the clays like montmorillonite, palygorskite, or analcime, these areas are however are dried continuously in global warming(Elhakim, 2016; Lutfalla et al., 2017).

Salinization, Acidification and Solidity Problem in Soil

Presently it was forecasted that in most of the regions of the world, temperature increases due to global warming, but however there is a less certainty observed regarding the changes in precipitation. The drastic increase in the intensity of rainfall is the cause of the increase in leaching, loss of the soil nutrients, increasing the acidification of the soil, which depends upon the buffer capacity and buffer index of that soil. The direction of the increase in the rate of leaching and the increase in the rate of evaporation mainly depends upon the extent of the rainfall, change in the temperature, and change in the use of the land pattern and land management. In any of the cases there occurs a change in soil properties and health. The increase in the alkalization and salinization generally observed in the areas of high rate of evaporation and lower rate of rainfall. The transient salinity generally increases with the raise in capillary action which transporting salts with water into the root areas of sodic soils. Due to sealing of the land surface, leaching occurs during the episodic rainfall events observed to be limited. The increase in the drying of the subsoil is the cause of increase in the concentration of the salts in the aqueous solution of the soil. But conversely the intensity of the saline scalds is mainly because of secondary salinization, which may be the cause of the declination of the level of groundwater with the reduction in the rate of rainfall. It causes great impacts on the large areas of semi-arid zones. In the regions where salinity is the cause of recharge processes, the salinization increases if the upstream recharging rainfall continuously increased. The increase in the concentration of the CO_2 in the atmosphere is the cause of reducing the impact of salinity on the plant growth.

The climate change and its impact are anticipated to be more significant in the summer. The increase in the intensity of the rainfall, the chance of an increase in the frequency of the drought in many areas, increase in the frequency of the tropical cyclones, hurricanes, raising the level of the sea water, storm surges are the major cause of climate change. All these phenomena and its anticipated impacts have a direct significance to the coastal acid sulphate soil landscapes, through either intensifying sulphide oxidation due to drought, re-instating reductive geochemical processes or altering the mobilization and export of contaminants. The interaction of some particular land management factors such as drainage system made by human beings also plays an important role in predicting how the impact of the climate change affects the landscape areas. The impacts of the climate change, especially in the lowland acid sulphate soils of coastal areas are particularly given more attention because these lands have a major utility for urban communities and agricultural activities (Lartey, 2006; Penn and Camberato, 2019).

Fertility of Soil and Acquisition of Nutrient

The change of climate may function and interacts differently, such as weaker or stronger, periodical or permanent, direct (primary) or indirect (secondary), unfavourable or favourable (in some cases catastrophic in nature) impact significantly on soil processes. Among all these processes, the regime of soil moisture plays an important and distinguished role. It regulates the influence of heat regimes, air circulation, water supply to the plants, different biological functions and presence of nutrient status in the soil. But in many cases it controls the agro-ecological potential, production of biomass in different agro and natural ecosystems and also the toxic impact of soil or water contamination. The deficit of the soil moisture directly affects the soil fertility and crop productivity and also decreases the % of yield because its impact on the transport of soil nutrients and its availability. Drought is the cause of increase in the susceptibility to nutrient losses from the root zones due to erosion of the soil because the nutrients present in the soil are taken from the roots of the plants through water. The deficit of the soil moisture may retard the diffusion of nutrients over a short distance and the mass flow rate of water-soluble nutrients such as calcium, magnesium, sulphates, nitrates and silicon over a longer distance. The roots of the plants can extend its length, increases the area of its coverage, changes their architecture, and so that it can give more effort to gain the mobility of nutrients mainly nitrogen and phosphorus from the soil. But in the drought condition the water scarcity is maximum which prevents the growth of the roots and destroys the function of the roots effectively, which subsequently retards the nutrient acquisition ability of the root systems. In the drought condition the fluxes of both oxygen and nitrogen are reduced for which the accumulation of nitrogen in the root nodules prevents the nitrogen fixation especially in legume type crops. The drought is also the cause of changing the number, composition and function of the microbial communities in the soil such as reduction of nitrifying bacteria in soil.

In many developing countries the excessive precipitation is the cause of a major source of the loss of soil nutrients mainly due to leaching. The agricultural lands of less facility of drainage with the frequent and intense rainfall produce waterlogged soils that soil is hypoxic in nature. The alternation of the redox status of the soil may cause the toxicity of many essential elements present in the soil such as manganese, boron, iron and aluminium, and decreases the % of crop yield with the formation of more phytotoxic organic solutes which can impair growth rate of root and its function. In the absence of oxygen, a substantial nitrogen loss is observed under the hypoxic conditions through the process of nitrification and denitrification because $NO_3^{3-}$is the alternative electron accept or by using soil microorganisms. The increase in the soil temperature subsequently increases capturing capacity of soil nutrients from 100-300%due to increases in the surface area of the roots and also increases the rate of diffusion of the nutrients and also water influx. Since with the increase in temperature the rate of transpiration increases, therefore the water soluble form of the nutrients such as, nitrites, calcium, magnesium, sulphates etc. can be easily flows towards the roots through the process of transpiration(Manuel et al., 2017; Schjoerrin et al., 2019; Shrivastava and Kumar, 2015).

IMPACT OF SOIL FERTILITY DUE TO CLIMATE CHANGE ON HUMAN AND OTHER LIFE FORMS

Soil is considered as one of the major necessities to maintain life on the surface of the globe. Soil is responsible for the growth of plants and formation of forest, which provides not only food to our biological community, but also maintains the equilibrium and sustainability of our ecosystem. The soil provides us medicines, clothes, construction materials either directly or indirectly in addition to food for marinating healthy and comfortable life. Oxygen required for our breathing originated from plants, therefore besides the plant growth, soil plays a crucial role for marinating the equilibrium in climatic condition and sustaining the life on the surface of the earth (plants, animals, human beings and microorganisms). Soil is responsible for the purification of water sources through drains into ground, oceans, rivers, lakes, reservoirs. The soil decomposition also supports the recycling of the nutrients because of disintegration of plants and animals. The nutrients obtained in this decomposition are the cause of growth of plants. In addition to the plant nutrients, soil is considered as a shelter house for a wide variety of organisms from single cell walled to small kinds of mammals. The land management can influence the nutrient concentration and contamination level of the soil. Besides deterioration of soil quality due to climatic change for a prolong period, any man-made activity that exposes soil to rain and wind, may also cause soil erosion and soil contamination, leading to negative impact on food security, and hence, detrimental to development of any life forms throughout the world (Steffan et al., 2017; Yang et al., 2020;Ye et al., 2020).Since soil is a stock resource, therefore, its contamination and degradation is almost irreversible in the average life span of a human being. It was found that almost 95% of the food either indirectly or directly produced due to soil. Healthy and uncontaminated soil is the crucial element of our sustainable development and biodiversity. The contamination of the soil is a shocker to our biological community on globe; unplanned agricultural activity is the cause of decreasing soil organic matter and promotes the transfer of contaminants into the food chain and food web to reach plants, animals and humans (Rohila et al., 2017; Thrall et al., 2011; Bünemann et al., 2018).

It was estimated that at every year about700, 000 deaths in the worldwide are due to antimicrobial resistant bacteria. It was predicted that by 2950, if this situation is not tackled, then mortality rate is more of it than cancer. It was predicted that the global population increased to be 9 billion by 2050, therefore our future food security is now a threat due to extensive soil contamination by human activities. Hence soil contamination impacts negatively to our crop production, leading to the uncertainty in the global food security. Further, the accumulation of massive toxic chemicals may cause permanent infertility of the soil or production of very low quality of crops. Farmers and people working on the extensive agricultural land are continually exposed to harmful and deadly chemicals leading to the threat to the health of human being. The domestic animals also exposed to these contaminated lands and suffered by many deadly diseases. The surface run-off of the contaminated soil carries the toxic chemicals and heavy metals; which during heavy rain mixed with the nearest sources of surface water leading to contamination of the underground water by chemical and metal leaching, resulting in many diseases of both animals and humans. Hence it is very important to aware and educates people regarding the negative effect of soil contamination, which can help in preventing soil contamination; and we can able to stop the devastation caused to plants, human beings, microorganisms and animal life (Yao et al., 2012; Zwolak et al. 2019, Martinho, 2020). So following may be the solution for controlling soil contamination, and indirect protection of our health such as, adopting proper farming techniques, crop rotation and land management; proper recycling of wastes specially soil wastes before adding to the land surface; suitable disposal technique for both domestic and industrial wastes; uses of natural bio-fertilizer and bio-insecticides instead of toxic chemical fertilizer and synthetic insecticides; proper awareness and community education; proper laws, implementation and penalties for violation of laws; and suitable maintenance and management of sewage system (Yang et al., 2019 ;Sodango et al., 2018)

CONCLUSION

The most significant change in the soil properties is expected to be a steady increase in the physical conditions and soil fertility, especially in the sub-humid and humid climatic environment. The change of the climate is well interlinked in the mineral composition of the soil, % of soil organic matter, structural stability of the soil and soil temperature. The climate change and its impact on the soil properties and fertility can be easily monitored in different agro-ecological regions continuously. The climate change and degradation of lands are very closely interlinked with each other and the farming conservation can potentially minimize the land degradation. The soil fertility and the necessary physical properties are increasing due to increase in the atmospheric CO_2.The increase in the productivity and % of the use of water for crops and vegetation, higher rainfall rate cannot be completely countered by higher rates of evapotranspiration. One of the major changes in the soil properties is also due to the biological activity of soil micro-organisms and also the continuous degradation of the soil. In the unprotected lowland coastal areas, the continuous infringement of the *Phragmites* or *Rhizophora* mangroves produces more extensive brackish-water and easily forms a layer of acid sulphate soil after a number of decades. Frequent and long duration of flooding causes formation of more organic matter over the soil surface and causes peat formation in local areas which increases the soil fertility and productivity. There are two factors such as precipitation and temperature, which are mainly because of climate change are predicted to be comparatively a well buffered due to the composition of the minerals, % of organic matter, structural stability of the soil and nutrient content. In some of the fragile soils the characteristics of the soil formation process can be varied with the increase or decrease in the intensity of the seasonal rainfall. It was observed that in many cases and situations the change in properties and structure of the soil by the direct human impact of both on-site and off-site are much more than the direct impact due to climate change. The constant productive capacity of the soil in a sustainable way can be optimized by designing adequate soil management and we can able to counteract any degradation of the agricultural land due to climate change. The soil having less management like semi-natural forest covers used for extraction of wood and some other forest products and these forests are less protected from the climate change, but such kind of soils are also less threatened by the natural factors, but more by human activities such as acid deposition, excessive extraction of the nutrients from the lower input agricultural land. The land management and human activity has to be more studied because most of the lands of the whole world are used and managed in different degrees by human beings rather than the natural climatic condition.

AUTHOR CONTRIBUTION

Conceived, designed and wrote the book chapter along with literature survey.

FUND DISCLAIMER

No grant received from any funding agency in public, commercial, or not-for-profit sector.

CORRESPONDING AUTHOR

Trinath Biswal

REFERENCES

Anjali, M. C., & Dhananjaya, B. C. (2019). Effect of Climate Change on Soil Chemical and Biological Properties-A Review. *International Journal of Current Microbiology and Applied Sciences*, *8*(2), 1502–1512. doi:10.20546/ijcmas.2019.802.174

Arévalo-Gardini, E., Canto, M., Alegre, J., Loli, O., Julca, A., & Baligar, V. (2015). Changes in Soil Physical and Chemical Properties in Long Term Improved Natural and Traditional Agroforestry Management Systems of Cacao Genotypes in Peruvian Amazon. *PLoS One*, *10*(7), 1–18. doi:10.1371/journal.pone.0132147 PMID:26181053

Asabere, S. B., Zeppenfeld, T., Nketia, K. A., & Sauer, D. (2018). Urbanization Leads to Increases in pH, Carbonate, and Soil Organic Matter Stocks of Arable Soils of Kumasi, Ghana (West Africa). *Frontiers in Environmental Science*, *6*, 119–136. doi:10.3389/fenvs.2018.00119

Awal, R., Safeeq, M., Abbas, F., Fares, S., Deb, S. K., Ahmad, A., & Fares, A. (2019). Soil Physical Properties Spatial Variability under Long-Term No-Tillage Corn. *Agronomy (Basel)*, *9*(11), 750–767. doi:10.3390/agronomy9110750

Bonfante, A., Terribile, F., & Bouma, J. (2019). Refining physical aspects of soil quality and soil health when exploring the effects of soil degradation and climate change on biomass production: An Italian case study. *Soil (Göttingen)*, *5*(1), 1–14. doi:10.5194oil-5-1-2019

Brevik, E. C. (2013). The Potential Impact of Climate Change on Soil Properties and Processes and Corresponding Influence on Food Security. *Agriculture*, *3*(3), 398–417. doi:10.3390/agriculture3030398

Bünemann, E. K., Bongiorno, G., Bai, Z., Creamer, R. E., De Deyn, G., de Goede, R., & Brussaard, L. (2018). Soil quality – A critical review. *Soil Biology & Biochemistry*, *120*, 105–125. doi:10.1016/j.soilbio.2018.01.030

Chatterjee, D., Kuots, R., Ao, M., Saha, S., Ray, S. K., & Ngachan, S. V. (2019). Does rise in temperature adversely affect soil fertility, carbon fractions, microbial biomass and enzyme activities under different land uses? *Current Science*, *116*(12), 2044–2054. doi:10.18520/cs/v116/i12/2044-2054

Classen, A. T., Sundqvist, M. K., Henning, J. A., Newman, G. S., Moore, J. A. M., Cregger, M. A., Moorhead, L. C., & Patterson, C. M. (2015). Direct and indirect effects of climate change on soil microbial and soil microbial-plant interactions: What lies ahead? *Ecosphere*, *6*(8), 1–24. doi:10.1890/ES15-00217.1

de Sherbinin, A., Carr, D., Cassels, S., & Jiang, L. (2007). Population and Environment. *Annual Review of Environment and Resources*, *32*(1), 345–373. doi:10.1146/annurev.energy.32.041306.100243 PMID:20011237

Dong, N., Liu, Z., Luo, M., Fang, C., & Lin, H. (2019). The effects of anthropogenic land use changes on climate in China driven by global socioeconomic and emission scenarios. *Earth's Future*, *7*(7), 784–804. doi:10.1029/2018EF000932

Elhakim, A. F. (2016). Estimation of soil permeability. *Alexandria Engineering Journal*, *55*(3), 2631–2638. doi:10.1016/j.aej.2016.07.034

Francisco, I., Pugnaire, Morillo, J. A., Peñuelas, J., Reich, P. B., & Bardgett, R. D. (2019). Climate change effects on plant-soil feedbacks and consequences for biodiversity and functioning of terrestrial ecosystems. *Science Advances*, *5*(11), 1–12. PMID:31807715

Frazer, L. (2009). Climate change: Will warmer soil be as fertile? *Environmental Health Perspectives*, *117*(2), 58–61. doi:10.1289/ehp.117-a59 PMID:19270772

Gelybó, G., Tóth, E., Farkas, C., Horel, Á., Kása, I., & Bakacsi, Z. (2018). Potential impacts of climate change on soil properties. *Agrokémia és Talajtan*, *67*(1), 121–141. doi:10.1556/0088.2018.67.1.9

Gogoi, P. P., Vinoj, V., Swain, D., Roberts, G., Dash, J., & Tripathy, S. (2019). Land use and land cover change effect on surface temperature over Eastern India. *Scientific Reports*, *9*(1), 8859–8869. doi:10.103841598-019-45213-z PMID:31222135

Hamidov, A., Helming, K., Bellocchi, G., Bojar, W., Dalgaard, T., Ghaley, B. B., Hoffmann, C., Holman, I., Holzkämper, A., Krzeminska, D., Kvaernø, S. H., Lehtonen, H., Niedrist, G., Øygarden, L., Reidsma, P., Roggero, P. P., Rusu, T., Santos, C., Seddaiu, G., ... Schönhart, M. (2018). Impacts of climate change adaptation options on soil functions: A review of European case-studies. *Land Degradation & Development*, *29*(8), 2378–2389. doi:10.1002/ldr.3006 PMID:30393451

Henryson, K., Sundberg, C., Kätterer, T., & Hansson, P. A. (2018). Accounting for long-term soil fertility effects when assessing the climate impact of crop cultivation. *Agricultural Systems*, *164*, 185–192. doi:10.1016/j.agsy.2018.03.001

Hussain, M. S., & Javadi, A. A. (2016). Assessing impacts of sea level rise on seawater intrusion in a coastal aquifer with sloped shoreline boundary. *Journal of Hydro-environment Research*, *11*, 29–41. doi:10.1016/j.jher.2016.01.003

Issoufou, A. A., Soumana, I., Maman, G., Konate, S., & Mahamane, A. (2020). Dynamic relationship of traditional soil restoration practices and climate change adaptation in semi-arid Niger. *Heliyon*, *6*(1), 1–7. doi:10.1016/j.heliyon.2020.e03265 PMID:32042980

Karmakar, R., Das, I., Dutta, D., & Rakshit, A. (2016). Potential Effects of Climate Change on Soil Properties: A Review. *Science International*, *4*(2), 51–73. doi:10.17311ciintl.2016.51.73

Klik, A., & Eitzinger, J. (2010). Impact of climate change on soil erosion and the efficiency of soil conservation practices in Austria. *The Journal of Agricultural Science*, *148*(05), 529–541. doi:10.1017/S0021859610000158

Kulp, S. A., & Strauss, B. H. (2019). New elevation data triple estimates of global vulnerability to sea-level rise and coastal flooding. *Nature Communications*, *10*(1), 4844–4856. doi:10.103841467-019-12808-z PMID:31664024

Kumar, R., & Das, A. J. (2014). Climate Change and its Impact on Land Degradation: Imperative Need to Focus. *Journal of Climatology and Weather Forecasting*, *2*(1), 1–3. doi:10.4172/2332-2594.1000108

Kyei-Mensah, C., Kyerematen, R., & Adu-Acheampong, S. (2019). Impact of Rainfall Variability on Crop Production within the Worobong Ecological Area of Fanteakwa District. Ghana, *Advances in Agriculture. Article ID*, *7930127*, 1–7.

Lartey, R. T.Robert T. Lartey. (2006). Dynamics of Soil Flora and Fauna in Biological Control of Soil Inhabiting Plant Pathogens. *Plant Pathology Journal*, *5*(2), 125–142. doi:10.3923/ppj.2006.125.142

Le, P. V. V., Kumar, P., Ruiz, M. O., Mbogo, C., & Muturi, E. J. (2019). Predicting the direct and indirect impacts of climate change on malaria in coastal Kenya. *PLoS One*, *14*(2), 1–18. doi:10.1371/journal.pone.0211258 PMID:30726279

Lori, M., Symnaczik, S., Mäder, P., De Deyn, G., & Gattinger, A. (2017). Organic farming enhances soil microbial abundance and activity—A meta-analysis and meta-regression. *PLoS One*, *12*(7), 1–14. doi:10.1371/journal.pone.0180442 PMID:28700609

Lutfalla, P. S. S., Riley, W. J., Torn, M. S., Schmidt, M. W. I., & Soussana, J.-F. (2017). The changing faces of soil organic matter research. *European Journal of Soil Science*, *69*(1), 23–30.

Manuel, R., Machado, A., & Serralheiro, R. P. (2017). Soil Salinity: Effect on Vegetable Crop Growth. Management Practices to Prevent and Mitigate Soil Salinizatio. In *Soil Science: An Introduction*, (pp.353-384). Indian Society of Soil Science.

Martinho, V. J. P. D. (2020). Exploring the Topics of Soil Pollution and Agricultural Economics: Highlighting Good Practices. *Agriculture*, *10*(1), 24–43.

Mimura, N. (2013). Sea-level rise caused by climate change and its implications for society. *Proceedings of the Japan Academy. Series B, Physical and Biological Sciences*, *89*(7), 281–301. doi:10.2183/pjab.89.281 PMID:23883609

Novaraa, A., & Pisciottaa, A. (2018). The impact of soil erosion on soil fertility and vine vigor. A multidisciplinary approach based on field, laboratory and remote sensing approaches. *The Science of the Total Environment*, *622–623*, 474–480. doi:10.1016/j.scitotenv.2017.11.272 PMID:29223076

Onwuka, B., & Mang, B. (2018). Effects of soil temperature on some soil properties and plant growth. *Advances in Plants & Agriculture Research*, *8*(1), 34–37. doi:10.15406/apar.2018.08.00288

Pareek, N. (2017). Climate change impact on soils: Adaptation and mitigation. *Shengtai Huanjing Xuebao*, *2*(3), 136–139. doi:10.15406/mojes.2017.02.00026

Pati, A., & Lamnganbi, M. (2018). Impct of climate change on soil health: A review. *International Journal of Chemical Studies*, *6*(3), 2399–2404.

Penn, C. J., & Camberato, J. J. (2019). A Critical Review on Soil Chemical Processes that Control How Soil pH Affects Phosphorus Availability to Plants. *Agriculture*, *9*(6), 120–138. doi:10.3390/agriculture9060120

Praveen, D., Andimuthu, R., & Palanivelu, K. (2016). The urgent call for land degradation vulnerability assessment for conserving land quality in the purview of climate change: Perspective from South Indian Coast. *AIMS Agriculture and Food*, *1*(3), 330–341. doi:10.3934/agrfood.2016.3.330

Qambrani, N. A., Rahman, M. M., Won, S., Shim, S., & Ra, C. (2017). Biochar properties and eco-friendly applications for climate change mitigation, waste management, and wastewater treatment: A review. *Renewable & Sustainable Energy Reviews*, *79*, 255–273. doi:10.1016/j.rser.2017.05.057

Rachelle, S., Brendan, M., & Cullena, R. (2018). Potential impacts of climate change on soil organic carbon and productivity in pastures of south eastern Australia. *Agricultural Systems*, *167*, 34–46. doi:10.1016/j.agsy.2018.08.010

Raza, A., Razzaq, A., Mehmood, S., Zou, X., Zhang, X., Lv, Y., & Xu, J. (2019). Impact of Climate Change on Crops Adaptation and Strategies to Tackle Its Outcome: A Review. *Plants*, *8*(2), 34–63. doi:10.3390/plants8020034 PMID:30704089

Robinson, S. I., McLaughlin, Ó. B., Marteinsdóttir, B., & O'Gorman, E. J. (2018). Soil temperature effects on the structure and diversity of plant and invertebrate communities in a natural warming experiment. *Journal of Animal Ecology*, *87*(3), 634–646. doi:10.1111/1365-2656.12798 PMID:29368345

Rohila, A. K., Duhan, A., Mann, D., & Kumar, A. (2017). Impact of agricultural practices on environment. Asian Journal of Microbiology. *Biotechnology and Environmental Sciences*, *19*(2), 381–384.

Salehi, A., Fallah, S., & Sourki, A. A. (2017). Organic and inorganic fertilizer effect on soil CO2 flux, microbial biomass, and growth of Nigella sativa. *International J. of Agrophysics*, *31*(1), 103–116. doi:10.1515/intag-2016-0032

Schjoerrin, J. K., Cakmak, I., & White, P. J. (2019). Plant nutrition and soil fertility: Synergies for acquiring global green growth and sustainable development. *Plant and Soil*, *434*(1-2), 1–6. doi:10.100711104-018-03898-7

Schoonover, J. E., & Crim, J. F. (2015). An Introduction to Soil Concepts and the Role of Soils in Watershed Management,*Journal of Contemporary Water Research andamp. Education*, *154*(1), 21–47.

Shrivastava, P., & Kumar, R. (2015). Soil salinity: A serious environmental issue and plant growth promoting bacteria as one of the tools for its alleviation. *Saudi Journal of Biological Sciences*, *22*(2), 123–131. doi:10.1016/j.sjbs.2014.12.001 PMID:25737642

Sidibé, Y., Foudi, S., Pascual, U., & Termansen, M. (2018). Adaptation to Climate Change in Rainfed Agriculture in the Global South: Soil Biodiversity as Natural Insurance. *Ecological Economics*, *146*, 588–596. doi:10.1016/j.ecolecon.2017.12.017

Smith, P., Lutfalla, S., Riley, W. J., Torn, M. S., Schmidt, M. W. I., & Soussana, J.-F. (2017). The changing faces of soil organic matter research. *European Journal of Soil Science*, *69*(1), 23–30. doi:10.1111/ejss.12500

Soares, J. C., Santos, C. S., Carvalho, S. M. P., Pintado, M. M., & Vasconcelos, M. W. (2019). Preserving the nutritional quality of crop plants under a changing climate: Importance and strategies. *Plant and Soil*, *443*(1-2), 1–26. doi:10.100711104-019-04229-0

Sodango, T. H., Li, X., Sha, J., & Bao, Z. (2018). Review of the Spatial Distribution, Source and Extent of Heavy Metal Pollution of Soil in China: Impacts and Mitigation Approaches. *Journal of Health and Pollution*, *8*(17), 53–70. doi:10.5696/2156-9614-8.17.53 PMID:30524849

Steffan, J. J., Brevik, E. C., Burgess, L. C., & Cerdà, A. (2017). The effect of soil on human health: An overview. *European Journal of Soil Science*, *69*(1), 159–171. doi:10.1111/ejss.12451 PMID:29430209

Stewart, Z. P., Pierzynski, G. M., Middendorf, B. J., & Vara Prasad, P. V. (2020). Approaches to improve soil fertility in sub-Saharan Africa. *Journal of Experimental Botany*, *71*(2), 632–641. doi:10.1093/jxb/erz446 PMID:31586430

Subedi, R., Bhatta, L. D., Udas, E., Agrawal, N. K., Joshi, K. D., & Panday, D. (2019). Climate-smart practices for improvement of crop yields in mid-hills of Nepal. *Cogent Food & Agriculture*, *5*(1), 1–15. doi:10.1080/23311932.2019.1631026

Thrall, P. H., Oakeshott, J. G., Fitt, G., Southerton, S., Burdon, J. J., Sheppard, A., & Ford Denison, R. (2011). Evolution in agriculture: The application of evolutionary approaches to the management of biotic interactions in agro-ecosystems. *Evolutionary Applications*, *4*(2), 200–215. doi:10.1111/j.1752-4571.2010.00179.x PMID:25567968

Várallyay, G. (2010). The impact of climate change on soils and on their water management. *Agronomy Research*, *8*, 385–396.

Vries, F. (2018). Impacts of Climate Change on Soil Microbial Communities and Their Functioning. *Developments in Soil Science*, *35*, 111–129. doi:10.1016/B978-0-444-63865-6.00005-3

Williams, S. J. (2013). Sea-Level Rise Implications for Coastal Regions. *Journal of Coastal Research*, *63*(Special Issue), 184–196. doi:10.2112/SI63-015.1

Willy, D. K., Muyanga, M., Mbuvi, J., & Jayned, T. (2019). The effect of land use change on soil fertility parameters in densely populated areas of Kenya. *Geoderma*, *343*, 254–262. doi:10.1016/j.geoderma.2019.02.033 PMID:31160827

Yang, T., Siddique, K. H. M., & Liu, K. (2020). Cropping systems in agriculture and their impact on soil health-A review. *Global Ecology and Conservation*, *23*, 1–13. doi:10.1016/j.gecco.2020.e01118

Yang, X., Zhang, S., Ju, M., & Liu, L. (2019). Preparation and Modification of Biochar Materials and their Application in Soil Remediation. *Applied Sciences (Basel, Switzerland)*, *9*(7), 1365–1390. doi:10.3390/app9071365

Yao, Z., Li, J., Xie, H., & Yu, C. (2012). Review on Remediation Technologies of Soil Contaminated by Heavy Metals. *Procedia Environmental Sciences*, *16*, 722–729. doi:10.1016/j.proenv.2012.10.099

Ye, L., Zhao, X., Bao, E., Li, J., Zou, Z., & Cao, K. (2020). Bio-organic fertilizer with reduced rates of chemical fertilization improves soil fertility and enhances tomato yield and quality. *Scientific Reports*, *10*(1), 177–188. doi:10.103841598-019-56954-2 PMID:31932626

Yu, M., Wang, Y., Jiang, J., Wang, C., Zhou, G., & Yan, J. (2019). Soil Organic Carbon Stabilization in the Three Subtropical Forests: Importance of Clay and Metal Oxides. *Journal of Geophysical Research. Biogeosciences*, *124*(10), 2976–2990. doi:10.1029/2018JG004995

Zwolak, A., Sarzyńska, M., Szpyrka, E., & Stawarczyk, K. (2019). Sources of Soil Pollution by Heavy Metals and Their Accumulation in Vegetables: A Review. *Water, Air, and Soil Pollution*, *230*(7), 164–173. doi:10.100711270-019-4221-y

KEY TERMS AND DEFINITIONS

Active Carbon (AC): The fraction of the total soil organic carbon, which can be relatively metabolized easily or can be used effectively by microorganisms. The 50% of the active carbon generally range from a few weeks to a few years. Active carbons are mostly glucose and simple polysaccharides, which are equivalent to reducing sugars, proteins, amino acids, extractable carbon, and also biomass of microbial carbon, etc.

Aerobic: The term aerobic means the presence of oxygen or plants growing in the existence of oxygen. The aerobic soils possess huge amounts of oxygenated air for oxidative reactions mainly of nutrient cycling and decomposition of soil organic matter.

Anaerobic: The absence of oxygen or the process operating without oxygen. The clay materials Soils compacted soil, wet or flooded soil tend to be anaerobic because these categories of soils are less oxygenated to do oxidative reactions.

Bacteria: A group of large number of microorganisms of single-celled with absence of prokaryotic and chlorophyll are called as bacteria having few micrometers in length. Bacteria have a wide range of shapes (rods and spirals). These species are more significant for the functioning of biochemical properties. These organisms are the cause of the recycling of nutrients. Decomposing of the organic matter, and detoxification of contaminants.

Carbon Sequestration Index (C_{index}): It is a new term which explains the simple and integrated index that recognizes the rate of the accumulation of carbon in the soil. The value of this index is a sensitive indicator of carbon or accumulation of soil organic matter and helps in explaining the complex nature of global carbon cycling.

Carbon-Nitrogen Ratio (C/N): The ratio of the mass of the soil organic carbon to the mass of total nitrogen present in the soil, plants, or any other organic compounds called as C/N ration. Soil C/N is generally the ratio of 10:1for effective ecological functions.

Soil Aggregates: The soil particles, including silt, sand and clay are always found together in a single cluster, such as block, crumb, clod or prism by using the organic matter as cementing material. Soil aggregates are usually greater than ten millimeters in diameter and formed by different natural forces and organic materials that are derived from root exudates, roots, soil animals and by-products of microbial communities.

Chapter 2
Environmental and Occupational Factors on Implantation Failure

Khursheed Ahmad Wani
https://orcid.org/0000-0001-5577-3455
Government Degree College, Bijbehara, India

Javid Ahmad Lone
https://orcid.org/0000-0002-2821-206X
Department of Environmental Science, JJT University, Rajasthan, India

Waseem Yaseen
Department of Environmental Science, Governmentt Degree College, Bijbehara, India

Junaid Ahmad Malik
https://orcid.org/0000-0003-4411-2015
Department of Zoology, Government Degree College, Bijbehara, India

ABSTRACT

It has been shown that reproductive activity is responsive to changes in the physical, psychosocial, and chemical environments. Occupational exposure has been identified to a wide range of putative hazards and adverse reproductive outcomes, or fertility. Such risks include the physical environment such as VDT and noise, as well as psychosocial stress and chemical agents. Some of these may have a direct or indirect effect on IVF outcomes. Psychosocial factors such as inadequate coping mechanisms, anxiety and/or depression are related to a lower incidence of IVF pregnancy. In this chapter the authors focus on the difficulties involved in linking exposure to putative hazardous substances to adverse reproductive outcomes in environmental and occupational settings, particularly the effectiveness of IVF procedures.

DOI: 10.4018/978-1-7998-4480-8.ch002

INTRODUCTION

Although the focus of this chapter is on the success of *in vitro* fertilization (IVF) after exposure to hazardous substances in the workplace, the outcome of any attempt to attain a live birth depends on several factors, primarily the union of putatively normal sperm with oocytes. From 1979 to 2004, the Medline and PubMed databases are searched for papers that link IVF achievements and infertility to occupational exposure to hazardous substances. There is, however, a lack of data which links human exposure to hazardous substances and IVF success. Hence, in relation to reproductive outcomes, we will address the available exposure data and explain their importance to IVF.

Infertility has frequently been described as failure to attain pregnancy within one year of unprotected intercourse. Delays in the time needed to become pregnant or pregnancy loss before term is regarded as proof of subfertility. The understanding of better infertility prevalence along with rising knowledge of the toxicity of environmental contaminants to reproductive function in wildlife and laboratory animals has resulted in increased research focus on infertility aetiology. The physical environment, behavioral and socioeconomic factors, as well as environmental pollutants are among the factors thought to influence human fertility. Schull (1984) stated that the evidence as to whether adverse conditions at the workplace present a risk to human fertility is not conclusive. Nine years later, Baranski (1993) examined the then recent literature on the adverse effects of occupational influences on fertility and related reproductive outcomes, and concluded that the current findings are not adequate to support the hypothesis of female infertility triggered by occupational exposure. Lindbohm (1999) indicated that psychological stress at work is becoming a significant infertility cause. Work in this field is still considered young, though (Bonde, 1999).McElgunn (1998) claimed that critical periods prior to conception and during pregnancy are significant periods of adverse impact on fertility and the outcome of pregnancy, and that ambient tobacco smoke and exposures from video display terminals (VDT) and indoor air quality are the most common concerns of women at their workplaces. Any of these environmental conditions may have an effect on IVF tests.

In addition to physical and psychosocial factors thought to have an impact on human reproductive capacity and fertility, there is concern that the exposure to chemical pollutants has adversely affected human fertility. The impulse for enlightening the causal link between chemicals released in the atmosphere and human health came from Silent Spring, a book by Carson (1962). She suggested that there is a correlation between changes observed in wildlife ecology and human cancers and suggested that these results are the consequences of releasing countless chemicals into the atmosphere through industrial processes and agricultural processes.It took two more decades before Colborn advanced the now generally accepted environmental endocrine hypothesis (Colborn, 1991, 1995; Colborn et al., 1993, Colborn et al., 1996). The theory suggests that environmental endocrine-disrupting toxins have greater control over the production of exposed adult offspring than the genes inherited by the child, or the training they undergo.

Endocrine disrupting chemicals (EDCs) are synthetic and naturally occurring chemicals that cannot be labeled with any particular physical or chemical properties but are distinguished by their ability to replicate the effects of endogenous hormones.These features of EDC have caused concern that exposure to such compounds could be associated with adverse effects on human health (Damstra et al., 2002). Endocrine disrupters have deleterious effects in wildlife and fish species (Damstra et al., 2002), but there is no strong evidence of adverse health effects in humans. To date, epidemiological studies have failed to support a correlation between exposure to endocrine disrupters and infertility, or decreased fecundity (Foster and Holloway, 2003).However, quantification of endocrine toxicants in human ovarian follicular fluid and their association with IVF outcomes (Younglai et al., 2002), together with reported adverse effects in animals and in vitro studies (Gray et al., 2001), support concerns that exposure to endocrine toxicants may adversely affect the functioning of human ovaries.

Many studies over the past few years also have concentrated on environmental factors as detrimental as possible to fertility effects (Chapin et al., 1996; Daston et al., 1997; Harrison et al., 1997;Van Oostdam et al., 1999; Cocco, 2002; Joffe, 2003), with anemphasis either on the male (Twombly, 1995; DeCelis et al.,1996; Spira and Multigner, 1998; Cooper et al., 1999; Fischet al., 2000; Foster et al., 2001; Guillette and Gunderson, 2001;Oliva et al., 2001; Sharpe, 2001; Damgaard et al., 2002; Myerset al., 2003; Fisher, 2004; Myers, 2004), or female (Feichtinger,1991; Kamrin et al., 1994; Sharara et al., 1998; Nicolopoulou-Stamati and Pitsos, 2001; Foster and Holloway, 2003).

OCCUPATIONAL EXPOSURES AND IMPLANTATION OUTCOME

Occupational exposure has been identified to a wide range of putative hazards and adverse reproductive effects, or fertility. Such risks include the physical environment such as VDT and noise, as well as psychosocial stress and chemical agents, which may have a direct or indirect effect on IVF outcomes.

Physical Environment

Goldhaber et al. (1988) stated in a case-control study of 1583 pregnant women in Northern California that the odds ratio (OR) for miscarriage is 1.8, 95% confidence interval (CI) 1.2– 2.8 for working women using VDT for.20 h per week. Lindbohm et al. (1992) found in a Finnish study of bank clerks and clerical employees that the OR for miscarriages in employees using a VDT with a high level of extremely low frequency magnetic fields (0.9 mT) is 3.4 (95% CI 1.4–8.6) compared to workers using a VDT with a low level of such magnetic fields (0.4 mT).On the other hand, in an infertile population of 177 people attending the Yale New Haven infertility clinic (Lundsberg et al., 1995), or in 57 Danish couples planning their first child, there is no correlation between occupationally linked categories of magnetic field exposure and male subfertility (Hjollund et al., 1999).

Seasonal variation is identified in fertility (Rojansky et al., 1992; Lam and Miron, 1994). However, the data in support of this view has been considered inconclusive (Bronson, 2004). It has been suggested that while some individuals are photoresponsive, others are not, and the incongruities are due to individual variability. In the output of IVF some seasonality is observed. Weigert et al. (2001) recorded that in 8184 IVF cycles of two European centers the lowest pregnancy rate is 25.7% in July compared to 35.5% in December. Further research found that the most important anomalous moon time is (Weigert et al., 2002). This is contrary to research by British (Fleming et al., 1994) and Canadian (Yie et al., 1995), in which no major seasonal variability in IVF performance is found.

Frequent shifts in time zones in the workplace as encountered by flight attendants seem to have a marginally higher probability of spontaneous abortion (SA) (OR 1.3, CI 0.9–1.8) but this is not consistent between the two time periods analyzed (Aspholm et al., 1999), as there is a significant correlation in the one-time frame, but none in another. In a Danish population, male heat exposure and female noise exposure are associated with infertility (Rachootin and Olsen, 1983). Babies with low birth weight (OR 5.38, CI 1.42–20.46) are substantially correlated with work in the electronics industry (Lipscomb et al., 1991). These tests, howeverare not definitive, and no related adverse effects are reported on IVF outcomes; and as with other possible fertility threats, the evidence is both incomplete and incoherent.

Chemical Environment

Several studies have linked reduced fertility of both males and females to occupational exposure to several chemical agents like, solvents, metals, pharmaceuticals, and agricultural products etc., discussed below.

- ***Solvents***

Female dental surgeons who are regularly exposed to mercury, chloroform, and benzene may not have trouble getting pregnant (Dahl et al., 1999), but exposure rates may be too small to cause an impact. Using SA as a fertility marker, maternal exposures to toluene, xylene and formalin (Taskinen et al., 1994), chloroform (Wennborg et al., 2000), and ethylene glycol ethers (Correa et al., 1996) have been shown to decrease fertility. No increased risk is found when males are exposed (Correa et al., 1996). McMartin et al. (1998) concluded, in a meta-analysis of research papers from 1966–1994, that there is a trend (OR 1.25, CI 0.99–1.58) towards an increased risk to SA with maternal exposure to organic solvent. The sensitivity to first trimester solvents is substantially correlated with spontaneous abortion (SA) (Lindbohm et al., 1990; Lipscomb et al., 1991) (OR 5.38, CI 1.42–20.46). The inference has been verified by other female workers (Taskinen et al., 1986; Khattak et al., 1999).In case of males, using the TTP endpoint (i.e., time to pregnancy, length of time to get pregnant), Sallmen et al. (1998) found limited support for the hypothesis that paternal exposure to organic solvents could be associated with reduced fertility. There is also a trend towards increased SA when only the husbands are exposed to organic solvents (Taskinen et al. 1989; Lindbohm et al., 1991).Nonetheless, the expert panel of the NTP-CERHR (enter for the Evaluation of Risks to Human Reproduction) recently concluded that there is a negligible danger to human reproduction of exposure to ethylene glycol or propylene glycol, two solvents that are released into the atmosphere omnipresent (NTP-CERHR, 2004a,b).

- ***Metals***

Metal industry employees (mostly males) are exposed to numerous dangerous substances including lead and steel. In a study of 400 Danish battery staff (Bonde and Kolstad, 1997), blood lead levels are in the region of 35.9 mg/dl but there is no reduction in birth rate (OR 0.983, CI 0.88–1.1.3) over the period studied. Occupational lead exposure often doesn't affect pregnancy time (Joffe et al., 2003). Welders are continuously exposed to metals and solder vapours, and the consistency of their semen is equated with fertility. Such men are more at risk of low sperm content (Mortensen, 1988) and less fertility (OR 0.89, CI 0.83–0.97) (Bonde et al., 1990).In a Canadian study (Bigelow et al., 1998), the poor quality of the sperm is confirmed, but the quality of the sperm is not correlated with reduced fertility. Having SA as a marker for fertility, Hjollund et al. (1995) stated that the proportion of SA for pregnancies at risk from stainless steel welding is not increased in 2520 pregnancies of 1715 married metal workers. Later, however, the same group (Hjollund et al., 2000) stated that stainless steel welding is associated with an increased risk of SA. It is not clear if IVF-conceived embryos with sperm suffer the same fate from welders. The opposing results of the same party, however, show the difficulty in interpreting data from these studies.

- ***Pharmaceuticals***

Occupational exposure to antineoplastics can affect both gametes and the developing fetus. In a study of 7094 pregnancies among 2976 pharmacy and nursing workers (Valanis et al., 1999), mom's exposure to or handling of antineoplastic agents during pregnancy is associated with substantially increased SA risk (OR 1.5, CI 1.2–1.8). In biomedical research facilities, similar threats to fertility are seen for staff (Wennborg et al., 2002).

- ***Agricultural agents***

Farmers and agricultural employees are exposed to a number of potentially harmful chemicals. Previously, there had been no connection between chemical exposure and infertility (Gerber et al.,1988). By comparison, occupational exposure to pesticides in fruit growers in the Netherlands (DeCock et al., 1994) is found to have an adverse effect on TTT (OR 0.46, CI 0.28–0.77) and higher prevalence of male factor infertility compared to normal controls is found among farm workers in Austria pursuing IVF (Strohmer et al., 1993).Additionally, reduced fertility is documented in farm workers exposed to pesticides (DeCock et al., 1995; Fuortes et al., 1997; Tielemans et al., 1999b). In comparison, there has been no correlation between male pesticide exposure and fertility among French and Danish men (Larsen et al., 1998; Thonneau et al., 1999a).However, Arbuckle and Sever (1998) stressed that many epidemiological studies are suffering from methodological issues, and while the literature reviewed indicated an increased risk of fetal death associated with pesticides in general and maternal jobs in the agricultural industry, they cautioned that work must concentrate on particular products and improve exposure assessment.Conclusions that vary depending on the endpoint used, e.g. when TTP is measured using questionnaires in the Ontario Farm Family Health Study (OFFHS), no trend of exposure to pesticides is identified (Curtis et al., 1999), but when SA is used as an index at 1,12 weeks, there is an increased risk (Arbuckle et al., 2001). Similarly, in France and Denmark, Thonneau et al. (1999b) found no correlation between TTP and male pesticide exposure. However, with SA as the endpoint, phenoxy herbicide exposure has been found to be hazardous for pregnancy (OR 2.5, CI 1.0–6.4) (Arbuckle et al., 1999).

The picture is more confusing with IVF population. Using data from a common questionnaire and the laboratory findings from in vitro insemination of oocytes, Tielemans et al. (1999a) found a substantially lower fertilization rate among 16 Dutch couples where the male partner had possible exposure to pesticides; OR ranged from 0.22 to 0.54 depending on the alleged extent of exposure. The seven fruit farmers with high exposure to pesticides had, however, an increased rate of implantation within the same study group (Tielemans et al., 2000).This is contrary to a Danish study in which 128 couples had partners with documented exposure to pesticides and growth retardants, from a pool of 5879 people who had IVF treatment (Hjollund et al., 2004b). Surprisingly, fewer SA occurred in the exposed group of 192 pregnancies (19.7–22.2%) compared with 2925 pregnancies (28.4%) in the control group. Therefore it cannot be inferred that exposure to pesticides can lead to adverse IVF outcomes.

Psychosocial Environment (Stress)

Psychological stress in the workplace may affect fertility (Negro-Vilar, 1993), and this aspect seems to affect IVF outcomes (Eugster and Vingerhoets, 1999). Psychosocial factors such as inadequate coping mechanisms, anxiety and/or depression are related to a lower incidence of IVF pregnancy. 8% females are pregnant in a group of 192 patients waiting for IVF (Jarrell et al., 1993a). The live birth rate in the IVF population is 29% in the IVF-treated group compared to 1% in the untreated group in a more recent multicenter, randomized controlled trial of 139 couples randomized to receive IVF or no care for 90 days (Hughes et al., 2004). Whether the untreated party sought counseling has not been reported. In Austria, where psychological therapy is legally necessary before IVF, 1156 such patients had a cumulative pregnancy rate of 56.4% (Poehl et al., 1999). While the reduction of stress by relaxation therapy or therapeutic treatment increases the levels of pregnancy well before IVF is performed. Acute and chronic stress led to adverse outcomes in a prospective study of seven clinics involving 151 patients (Klonoff-Cohen et al., 2001a). The adverse findings are due to the elevation of activated T cells in chronic stress environments, leading to a decrease in the rate of implantation among those women who have IVF (Gallinelli et al., 2001). The study of Milad et al. (1998), which found no difference in hormonal markers of stress and adverse outcome of pregnancy in 40 patients suffering from IVF, did not support those conclusions. This sample size may be argued as being too small. However, using SA as a fertility measure, Fenster et al. (1995) found no correlation with the effectiveness of stressful labor and IVFs.

Further in males, the negative outcome of stress caused by IVF may have resulted from fertilization with low quality spermatozoa. Bigelow et al. (1998) found that there is a important dose-response relationship between perceived level of stress at work and poor quality of sperm. Using a stress assessment questionnaire, Hjollund et al. (2004a) found no effect on sperm production and only those people with low sperm concentration and high stress rates had significantly reduced fecundity. On the other hand, even with subsidization, the emotional burden of ineffective repeat IVF procedures may lead to cycle discontinuation (Hammarberg et al., 2001; Olivius et al., 2004).

ENVIRONMENTAL EXPOSURE (HEAT STRESS) AND BIRTH OUTCOME

Owing to elevated ambient temperatures the effects of direct heat exposure suggest the possible health effects of increased core body temperature. Physical heat exchange between the human body and the surrounding air is necessary to maintain a steady core temperature of the body (37°C). If the air is colder than the body, so heat is applied to the body. Depending on the clothing or shading facilities, strong heat radiation that reaches the skin (e.g. via sun rays or from hot equipment) also adds heat to the body (Cunningham et al., 2010).The ability of a person to remove excessive heat by sweat evaporation to maintain core body temperature is highly affected by the temperature, humidity, wind and clothing around him. The rate of sweat evaporation is decreased and increased by air movement with not more than 80% humidity. Hence, measures of heat stress include humidity, wind speed and temperature.If core body temperature for many hours reaches 38°C, there will be heat exhaustion and a decrease in psychometric and motor power. An unconscious condition can occur if the core body temperature continues to rise >39°C (Kjellstrom, 2009). People can adjust to ambient temperatures in many ways during the leisure time, for example by spending time in the shade and indoors, drinking water, and swimming. However, many people in developed countries are unable to make such changes during working hours, since they do hard physical labor in the sun or work in a factory without air conditioning (Kjellstrom, 2009).

With rising temperatures due to climate change, the possibility of heat shock in already hot countries and for people with physically demanding work would be highest. Many cases of severe chronic kidney disease are observed among young people working in the sugar industry in El Salvador and Nicaragua (Kjellstrom, 2009). For several months the staff spends 6–8 hours outside each day. They also experience dehydration, which is possibly causing kidney disease. During the hot periods, decreased productivity in factories is observed in Southern India and Northern Vietnam because the temperature may be above 40°C, as air conditioning is not available in many manufacturing units (Kjellstrom, 2009). Despite the hormonal condition at all stages of pregnancy, pregnant women are at particular risk of 'overheating' (i.e., too high core body temperature) (Cunningham et al., 2010). This raises the risk to mom and fetus for safety (Cunningham et al., 2010). Because of his limited temperature regulation ability, the newborn is especially vulnerable to too high or too low temperature in the atmosphere (Poursafa and Kelishadi, 2011).Strand et al. (2011) explored seasonal patterns of birth outcomes in relation to ambient temperature, such as low birth weight, preterm birth, and stillbirth. The authors analyzed the epidemiological evidence on the seasonality of birth outcomes and the effect on birth outcomes of prenatal exposure to ambient temperature. Most research find increases of premature birth, stillbirth, and low birth weight in winter, summer, or both, suggesting temperature extremes may be a risk factor for poor birth outcome. Sheffield and Landrigan (2011) took a regional view on evaluating the health and climate change on children. They concluded that the heat-related health effects include decreased school results, increased levels of complications from pregnancy, and renal effects. The magnitude of such results may differ based on geographic area and socioeconomic status, raising health inequality even further.

CONCLUSION

This chapter discussed the occupational and environmental factors on implantation failure, including environmental pollutants, and the correlation with IVF performance, the issues associated with the analysis of such data sets, and the need for more well-designed studies. There is no continuity between the results of the study in each area discussed in this chapter. Moreover, the scarcity of literature renders it impossible to draw any definite conclusions other than to indicate that, amid increasing alarm, the evidence connecting environmental influences to impaired human fertility is poor. Several measures must be taken to advance this field of research and to create a connection between exposure to environmental pollutants and effects on fertility. These involve, but are not limited to: (I) final exposure results for allergic reactions accompanied by validation in laboratory studies; (II) proof that quantities found in the atmosphere as mixtures which have effects; (III) regional, gender and ethnic variations in response to exposures. Such research may provide evidence to suggest that workplace vulnerability to environmental hazards presents a real danger to adverse IVF outcomes. Also with increasing knowledge and better-designed exposure testing tests, the increasingly changing technical field and the development of new pharmaceuticals would bring further obstacles. There is a clear need for novel Impact Markers to advance our knowledge of toxic impacts on human reproductive system. We also assume that more attention should be given to the possibilities of IVF result variables.

AUTHOR CONTRIBUTION

TB conceived, designed and wrote the book chapter along with literature survey.

FUND DISCLAIMER

No grant received from any funding agency in public, commercial, or not-forprofit sector.

CORRESPONDING AUTHORS

Khursheed Ahmad Wani

REFERENCES

Arbuckle, T. E., Lin, Z., & Mery, L. S. (2001). An exploratory analysis of the effect of pesticide exposure on the risk of spontaneous abortion in an Ontario farm population. *Environmental Health Perspectives*, *109*(8), 851–857. doi:10.1289/ehp.01109851 PMID:11564623

Arbuckle, T.E., Savitz, D.A., Mery, L.S., & Curtis, K.M. (1999). Exposure to phenoxyherbicides and the risk of spontaneous abortion. *Epidemiology, 10*, 752–760.

Arbuckle, T. E., & Sever, L. E. (1998). Pesticide exposures and fetal death: A review of the epidemiologic literature. *Critical Reviews in Toxicology*, *28*(3), 229–270. doi:10.1080/10408449891344218 PMID:9631282

Aspholm, R., Lindbohm, M. L., Paakkulainen, H., Taskinen, H., Nurminen, T., & Tiitinen, A. (1999). Spontaneous abortions among Finnish flight attendants. *Journal of Occupational and Environmental Medicine*, *41*(6), 486–491. doi:10.1097/00043764-199906000-00015 PMID:10390700

Baranski, B. (1993). Effects of the workplace on fertility and related reproductiveoutcomes. *Environmental Health Perspectives*, *101*(Suppl 2), 81–90. PMID:8243410

Bigelow, P. L., Jarrell, J., Young, M. R., Keefe, T. J., & Love, E. J. (1998). Associationof semen quality and occupational factors: Comparison of casecontrolanalysis and analysis of continuous variables. *Fertility and Sterility*, *69*(1), 11–18. doi:10.1016/S0015-0282(97)00437-8 PMID:9457925

Bonde, J. P. (1999). Environmental fertility research at the turn of the century. *Scandinavian Journal of Work, Environment & Health*, *25*(6), 529–536. doi:10.5271jweh.476 PMID:10884149

Bonde, J. P., & Kolstad, H. (1997). Fertility of Danish battery workers exposedto lead. *International Journal of Epidemiology*, *26*(6), 1281–1288. doi:10.1093/ije/26.6.1281 PMID:9447408

Bronson, F. H. (2004). Are humans seasonally photoperiodic? *Journal of Biological Rhythms*, *19*(3), 180–192. doi:10.1177/0748730404264658 PMID:15155003

Carson, R. (1962). Silent Spring. Fawcett Press, New York, USA.Chapin RE, Stevens JT, Hughes CL, Kele WR, Hess RA and Daston GP(1996) Endocrine modulation of reproduction. *Fundamental and Applied Toxicology*, *29*, 1–17.

Chapin, R. E., Stevens, J. T., Hughes, C. L., Kele, W. R., Hess, R. A., & Daston, G. P. (1996). Endocrine modulation of reproduction. *Fundamental and Applied Toxicology*, *29*(1), 1–17. doi:10.1006/faat.1996.0001 PMID:8838635

Cocco, P. (2002). On the rumors about the silent spring. Review of thescientific evidence linking occupational and environmental pesticideexposure to endocrine disruption health effects. *Cadernos de Saude Publica*, *18*(2), 39–402. doi:10.1590/S0102-311X2002000200003 PMID:11923880

Colborn, T. (1991) Nontraditional evaluation of risk from fish contaminants. In *Proceedings of a Symposium on Issues in Seafood Safety*. National Academy of Sciences, Institute of Medicine, Food and Nutrition Board.

Colborn, T. (1995). Pesticides—how research has succeeded and failed to translate science into policy: Endocrinological effects on wildlife. *Environmental Health Perspectives*, *103*(Suppl 6), 81–85. PMID:8549498

Colborn, T., Dumanoski, D., & Meyers, J. P. (1996). *Our Stolen Future*. Dutton.

Colborn, T., Vom Saal, F. S., & Soto, A. M. (1993). Developmental effects ofendocrine disrupting chemicals in wildlife and humans. *Environmental Health Perspectives*, *101*(5), 378–384. doi:10.1289/ehp.93101378 PMID:8080506

Cooper, R. L., Goldman, J. M., & Stoker, T. E. (1999). Neuroendocrine and reproductive effects of contemporary-use pesticides. *Toxicology and Industrial Health*, *15*, 26–36.

Cooper, R L., Goldman, J. M., & Stoker, T. E. (1999). Neuroendocrine and reproductiveeffects of contemporary-use pesticides. *Toxicology and Industrial Health*, *15*(1), 26–36. doi:10.1191/074823399678846727 PMID:10188189

Correa, A., Gray, R. H., Cohen, R., Rothman, N., Shah, F., Seacat, H., & Corn, M. (1996). Ethylene glycol ethers and risks of spontaneous abortion and subfertility. *American Journal of Epidemiology*, *143*(7), 707–717. doi:10.1093/oxfordjournals.aje.a008804 PMID:8651233

Cunningham, F. G., Leveno, K. J., Bloom, S. L., Hauth, J. C., Rouse, D. J., & Spong, C. Y. (2010). Overview of obstetrics. In F. G. Cunningham, K. J. Leveno, S. L. Bloom, J. C. Hauth, D. J. Rouse, & C. Y. Spong (Eds.), *Williams obstetrics* (23rd ed.). McGraw-Hill.

Curtis, K. M., Savitz, D. A., Weinberg, C. R., & Arbuckle, T. E. (1999). The effect ofpesticide exposure on time to pregnancy. *Epidemiology (Cambridge, Mass.)*, *10*(2), 112–117. doi:10.1097/00001648-199903000-00005 PMID:10069244

Dahl, J. E., Sunbdy, J., Hensten-Pettersen, A., & Jacobsen, N. (1999). Dental workplace exposure and effect on fertility. *Scandinavian Journal of Work, Environment & Health*, *25*(3), 285–290. doi:10.5271jweh.436 PMID:10450781

Damgaard, I. N., Main, K. M., Toppari, J., & Skakkebaek, N. E. (2002). Impact ofexposure to endocrine disrupters in utero and in childhood and adultreproduction. *Best Practice & Research. Clinical Endocrinology & Metabolism*, *16*(2), 289–309. doi:10.1053/beem.2002.0205 PMID:12064894

Damstra, T., Barlow, S., Bergman, A., Kavlock, R., & Van der Kraak, G. (2002). *Global assessment on the state-of-the-science of endocrine disruptors. WHO publication no. WHO/PCS/EDC/02.2*. World Health Organization.

Daston, G. P., Gooch, J. W., Shuey, D. L., Nikiforov, A. I., Fico, T. A., & Gorsuch, J. W. (1997). Environmental estrogens and reproductive health: A discussion ofthe human, environmental data. *Reproductive Toxicology (Elmsford, N.Y.)*, *11*(4), 465–481. doi:10.1016/S0890-6238(97)00014-2 PMID:9241667

DeCelis, R., Pedron-Nuevo, N., & Feria-Velasco, A. (1996). Toxicology ofmale reproduction in animals and humans. *Archives ofAndrology*, *37*(3), 201–218. doi:10.3109/01485019608988523 PMID:8939299

DeCock, J., Heederik, D., Tielemans, E., te Velde, E., & Van Kooij, R. (1995). Offspring sex ratio as an indicator of reproductive hazards associated with pesticides. *Occupational and Environmental Medicine*, *52*, 429–430. PMID:7627322

Eugster, A., & Vingerhoets, A. J. J. M. (1999). Psychological aspects of in vitro fertilization: A review. *Social Science & Medicine*, *48*(5), 575–589. doi:10.1016/S0277-9536(98)00386-4 PMID:10080360

Feichtinger, W. (1991). Environmental factors and fertility. *Human Reproduction (Oxford, England)*, *6*, 1170–1175. doi:10.1093/oxfordjournals.humrep.a137505 PMID:1806580

Fenster, L., Schaefer, C., Mathur, A., Hiatt, R. A., Pieper, C., Hubbard, A. E., von Behren, J., & Swan, S. H. (1995). Psychologic stress in the workplaceand spontaneous abortion. *American Journal of Epidemiology*, *142*(11), 1176–1183. doi:10.1093/oxfordjournals.aje.a117576 PMID:7485064

Fisch, H., Hyun, G., & Golden, R. (2000). The possible effects of environmentalestrogen disrupters on reproductive health. *Current Urology Reports*, *1*(4), 253–261. doi:10.100711934-000-0004-6 PMID:12084301

Fisher, J.S. (2004) Environmental anti-androgens and male reproductive health: Focus on Phthalates and testicular dysgenesis syndrome. *Reproduction, 127*, 305–315.

Fleming, C., Nice, L., Hughes, A. O., & Hull, M. G. R. (1994). Apparent lack ofseasonal variation in implantation rates and in-vitro fertilization. *Human Reproduction (Oxford, England)*, *9*, 2164–2166. doi:10.1093/oxfordjournals.humrep.a138411 PMID:7868692

Foster, P. M. D., Mylchreest, E., Gaido, K. W., & Sar, M. (2001). Effects of phthalateesters on the developing reproductive tract of male rats. *Human Reproduction Update*, *7*(3), 231–235. doi:10.1093/humupd/7.3.231 PMID:11392369

Foster, W. G., & Holloway, A. C. (2003). Do environmental contaminantsadversely affect human reproductive physiology? *Journal of Obstetrics and Gynaecology Canada*, *25*(1), 33–44. doi:10.1016/S1701-2163(16)31080-5 PMID:12548323

Fuortes, L., Clark, M. K., Kirchner, H. L., & Smith, E. M. (1997). Associationbetween female infertility and agricultural work history. *American Journal of Industrial Medicine*, *31*(4), 445–451. doi:10.1002/(SICI)1097-0274(199704)31:4<445::AID-AJIM11>3.0.CO;2-# PMID:9093660

Gallinelli, A., Roncaglia, R., Matteo, M. L., Ciaccio, I., Volpe, A., & Facchinetti, F. (2001). Immunological changes and stress are associated with differentimplantation rates in patients undergoing in vitro fertilization-embryotransfer. *Fertility and Sterility*, *76*(1), 85–91. doi:10.1016/S0015-0282(01)01826-X PMID:11438324

Gerber, W.L., De la Pena, V.E., & Mobley, W.C. (1988). Infertility, chemical lexposure, and farming in Iowa: Absence of an association. *Urology, 31*, 46–50.

Goldhaber, M. K., Polen, M. R., & Hiatt, R. A. (1988). The risk of miscarriage andbirth defects among women who use visual display terminals duringpregnancy. *American Journal of Industrial Medicine*, *13*(6), 695–706. doi:10.1002/ajim.4700130608 PMID:3389364

Gray, L. E., Ostby, J., Furr, J., Wolf, C. J., Lambright, C., Parks, L., Veeramachaneni, D. N., Wilson, V., Price, M., & Hotchkiss, A. (2001). Effects of environmentalantiandrogens on reproductive development in experimental animals. *Human Reproduction Update*, *7*(3), 248–264. doi:10.1093/humupd/7.3.248 PMID:11392371

Guillette, L. J. Jr, & Gunderson, M. P. (2001). Alterations in development ofreproductive and endocrine system of wildlife populations exposed toendocrine-disrupting contaminants. *Reproduction (Cambridge, England)*, *122*, 857–864. doi:10.1530/rep.0.1220857 PMID:11732981

Hammarberg, K., Astbury, J., & Baker, H. W. G. (2001). Women's experience ofIVF: A follow-up study. *Human Reproduction (Oxford, England)*, *16*(2), 374–383. doi:10.1093/humrep/16.2.374 PMID:11157838

Harrison, P. T., Holmes, P., & Humfrey, C. D. (1997). Reproductive health inhumans and wildlife: Are adverse trends associated with environmentalchemical exposure? *The Science of the Total Environment*, *205*(2-3), 97–106. doi:10.1016/S0048-9697(97)00212-X PMID:9372623

Hjollund, N. H., Bonde, J. P., Ernst, E., Lindenberg, S., Andersen, A. N., & Olsen, J. (2004b). Pesticide exposure in male farmers and survival of in vitro fertilizedpregnancies. *Human Reproduction (Oxford, England)*, *19*(6), 1331–1337. doi:10.1093/humrep/deh203 PMID:15117907

Hjollund, N. H., Bonde, J. P., & Hansen, K. S. (1995). Male-mediated risk of spontaneousabortion with reference to stainless steel welding. *Scandinavian Journal of Work, Environment & Health*, *21*(4), 272–276. doi:10.5271jweh.37 PMID:8553001

Hjollund, N. H., Skotte, J. H., Kolstad, H. A., & Bonde, J. P. (1999). Extremely lowfrequency magnetic fields and fertility: A follow up study of couplesplanning first pregnancies. The Danish First Pregnancy Planner StudyTeam. *Occupational and Environmental Medicine*, *56*, 253–255. doi:10.1136/oem.56.4.253 PMID:10450242

Hjollund, N. H. I., Bonde, J. P. E., Henriksen, T. B., Giwercman, A., & Olsen, J. (2004). Reproductive effects of male psychologic stress. *Epidemiology (Cambridge, Mass.)*, *15*(1), 21–27. doi:10.1097/01.ede.0000100289.82156.8b PMID:14712143

Hughes, E. G., Beecroft, M. L., Wilkie, V., Burville, L., Claman, P., Tummon, I., Greenblatt, E., Fluker, M., & Thorpe, K. (2004). A multicentre randomizedcontrolled trial of expectant management versus IVF in women with fallopiantube patency. *Human Reproduction (Oxford, England)*, *19*(5), 1105–1109. doi:10.1093/humrep/deh209 PMID:15044399

Jarrell, J.F., Labelle, R., Goeree, R., Milner, R., & Collins, J. (1993a). In vitro fertilizationand embryo transfer: A randomized controlled trial. *Journal of Current Clinical Trials, 73*.

Joffe, M. (2003). Infertility and environmental pollutants. *British Medical Bulletin*, *68*(1), 47–70. doi:10.1093/bmb/ldg025 PMID:14757709

Kamrin, M. A., Carney, E. W., Chou, K., Cummings, A., Dostal, L. A., Harris, C., Dostal, L. A., Harris, C., Henck, J. W., & Loch-Caruso, R. (1994). Femalereproductive and developmental toxicology: Overview and currentapproaches. *Toxicology Letters*, *74*(2), 99–119. doi:10.1016/0378-4274(94)90089-2 PMID:7940600

Khattak,, S., Moghtader, G., McMartin, K., Barrera, M., Kennedy, D., & Koren, G. (1999). Pregnancy outcome following gestational exposure to organicsolvents: A prospective controlled study. *Journal of the American Medical Association*, *281*, 1106–1109. doi:10.1001/jama.281.12.1106 PMID:10188661

Kjellstrom, T. (2009). Climate change, direct heat exposure, health and well-being in low and middle-income countries. *Global Health Action*, 2. PMID:20027264

Klonoff-Cohen, H., Chu, E., Natarajan, L., & Sieber, W. (2001a). A prospectivestudy of stress among women undergoing in vitro fertilization or gameteintrafallopian transfer. *Fertility and Sterility*, *76*(4), 675–687. doi:10.1016/S0015-0282(01)02008-8 PMID:11591398

Lam, D. A., & Miron, J. A. (1994). Global patterns of seasonal variation inhuman fertility. *Annals of the New York Academy of Sciences*, *709*(1), 9–28. doi:10.1111/j.1749-6632.1994.tb30385.x PMID:8154738

Larsen, S. B., Joffe, M., & Bonde, J. P.ASCLEPIOS Study Group. (1998). Time to pregnancy and exposure topesticides in Danish farmers. *Occupational and Environmental Medicine*, *55*(4), 278–283. doi:10.1136/oem.55.4.278 PMID:9624283

Lindbohm, M. L. (1999). Women's reproductive health: Some recent developmentsin occupational epidemiology. *American Journal of Industrial Medicine*, *36*(1), 18–24. doi:10.1002/(SICI)1097-0274(199907)36:1<18::AID-AJIM3>3.0.CO;2-N PMID:10361582

Lindbohm, M. L., Hemminki, K., Bonhomme, M. G., Anttila, A., Rantala, K., Heikkila, P., & Rosenberg, M. J. (1991). Effects of paternal occupationalexposure on spontaneous abortions. *American Journal of Public Health*, *81*(8), 1029–1033. doi:10.2105/AJPH.81.8.1029 PMID:1853994

Lindbohm, M. L., Hietanen, M., Kyyronen, P., Sallmen, M., von Nandelstadh, P., Taskinen, H., Pekkarinen, M., Ylikoski, M., & Hemminki, K. (1992). Magneticfields of video display terminals and spontaneous abortion. *American Journal of Epidemiology*, *136*(9), 1041–1051. doi:10.1093/oxfordjournals.aje.a116569 PMID:1462964

Lindbohm, M. L., Taskinen, H., Sallmen, M., & Hemminki, K. (1990). Spontaneousabortions among women exposed to organic solvents. *American Journal of Industrial Medicine*, *17*(4), 449–463. doi:10.1002/ajim.4700170404 PMID:2327413

Lipscomb, J. A., Fenster, L., Wrensch, M., Shusterman, D., & Swan, S. (1991). Pregnancy outcomes in women potentially exposed to occupational solventsand women working in the electronics industry. *Journal of Occupational Medicine.*, *33*, 597–604. PMID:1870012

Lundsberg, L. S., Bracken, M. B., & Belanger, K. (1995). Occupationally relatedmagnetic field exposure and male subfertility. *Fertility and Sterility*, *63*(2), 384–391. doi:10.1016/S0015-0282(16)57373-7 PMID:7843447

McElgunn, B. (1998). Reproductive and developmental hazards in the workplace. *Clinical Excellence for Nurse Practitioners*, *2*, 140–145. PMID:12675082

McMartin, K. I., Chu, M., Kopecky, E., Einarson, T. R., & Koren, G. (1998). Pregnancyoutcome following maternal organic solvent exposure: A meta-analysisof epidemiologic studies. *American Journal of Industrial Medicine*, *34*(3), 288–292. doi:10.1002/(SICI)1097-0274(199809)34:3<288::AID-AJIM12>3.0.CO;2-Q PMID:9698999

Milad, M. P., Klock, S. C., Moses, S., & Chatterton, R. (1998). Stress and anxietydo not result in pregnancy stage. *Human Reproduction (Oxford, England)*, *13*(8), 296–300. doi:10.1093/humrep/13.8.2296 PMID:9756314

Mortensen, J. T. (1988). Risk for reduced sperm quality among metal workerswith special reference to welders. *Scandinavian Journal of Work, Environment & Health*, *14*(1), 27–30. doi:10.5271jweh.1954 PMID:3353693

Myers, J. P., Dumanoski, D., & Colborn, T. (2006). Our Stolen Future: A Decade Later. *San Francisco Medicine*, *79*, 32–33.

Myers, J. P., Guillette, L. J., Palanza, P., Parmigiani, S., Swan, S. H., & Vom Saal, F. S. (2003). The emerging science of endocrine disruption. *International Seminar on Nuclear War and Planetary Emergencies*, *28*, 1–13.

Negro-Vilar, A. (1993). Stress and other environmental factors affecting fertilityin men and women: Overview. *Environmental Health Perspectives*, *101*(Suppl 2), 59–64. PMID:8243408

Nicolopoulou-Stamati, P., & Pitsos, M. A. (2001). The impact of endocrinedisrupters on female reproductive system. *Human Reproduction Update*, *7*(3), 323–330. doi:10.1093/humupd/7.3.323 PMID:11392379

NTP-CERHR expert panel report on the reproductive and developmental toxicity of ethylene glycol. (2004a). *Reproductive Toxicology, 18*, 457–532.

NTP-CERHR expert panel report on the reproductive and developmental toxicityof ethylene glycol. (2004b). *Reproductive Toxicology, 18*, 533–579.

Oliva, A., Spira, A., & Multigner, L. (2001). Contribution of environmental factorsto the risk of male infertility. *Human Reproduction (Oxford, England)*, *16*(8), 1768–1776. doi:10.1093/humrep/16.8.1768 PMID:11473980

Olivius, C., Friden, B., Borg, G., & Bergh, C. (2004). Why do couples discontinuein vitro fertilization treatment? A cohort study. *Fertility and Sterility*, *81*(2), 258–261. doi:10.1016/j.fertnstert.2003.06.029 PMID:14967352

Poehl, M., Bichler, K., Wicke, V., Dorner, V., & Feichtinger, W. (1999). Psychotherapeuticcounseling and pregnancy rates in *in vitro* fertilization. *Journal of Assisted Reproduction and Genetics*, *16*(6), 302–305. doi:10.1023/A:1020449831333 PMID:10394525

Poursafa, P., & Kelishadi, R. (2011). What health professionals should know about the health effects of air pollution and climate change on children and pregnant mothers. *Iranian Journal of Nursing and Midwifery Research, 16*, 257–264. PMID:22224116

Rachootin, P., & Olsen, J. (1983). The risk of infertility and delayed conceptionassociated with exposures in the Danish workplace. *Journal of Occupational Medicine., 25*, 394–402. PMID:6854429

Rojansky, N., Brzezinski, A., & Schenker, J. G. (1992). Seasonality in humanreproduction. *Human Reproduction (Oxford, England), 7*(6), 735–745. doi:10.1093/oxfordjournals.humrep.a137729 PMID:1323571

Sallmen, M., Lindbohm, M. L., Anttila, A., Kyyronen, P., Taskinen, H., & Nykyri Eand Hemminki, K. (1998). Time to pregnancy among the wives of menexposed to organic solvents. *Occupational and Environmental Medicine, 55*(1), 24–30. doi:10.1136/oem.55.1.24 PMID:9536159

Sharara, F. I., Seifer, D. B., & Flaws, J. A. (1998). Environmental toxicants andfemale reproduction. *Fertility and Sterility, 70*, 613–622. doi:10.1016/S0015-0282(98)00253-2 PMID:9797086

Sharpe, R. M. (2001). Hormones and testis development and the possibleadverse effects of environmental chemicals. *Toxicology Letters, 120*(1-3), 221–232. doi:10.1016/S0378-4274(01)00298-3 PMID:11323180

Sheffield, P. E., & Landrigan, P. J. (2011). Global climate change andchildren's health: Threats and strategies for prevention. *Environmental Health Perspectives, 119*(3), 291–298. doi:10.1289/ehp.1002233 PMID:20947468

Spira, A., & Multigner, L. (1998). Environmental factors and male infertility. *Human Reproduction (Oxford, England), 13*, 2041–2042. doi:10.1093/humrep/13.8.2041 PMID:9756262

Strand, L. B., Barnett, A. G., & Tong, S. (2011). The influence of season andambient temperature on birth outcomes: A review of theepidemiological literature. *Environmental Research, 111*(3), 451–462. doi:10.1016/j.envres.2011.01.023 PMID:21333980

Taskinen, H., Anttila, A., Lindbohm, M. L., Sallme'n, M., & Hemminki, K. (1989). Spontaneous abortions and congenital malformations among the wivesof men occupationally exposed to organic solvents. *Scandinavian Journal of Work, Environment & Health, 15*(5), 345–352. doi:10.5271jweh.1839 PMID:2799322

Taskinen, H., Kyyronen, P., Hemminki, K., Hoikkala, M., Lajunen, K., & Lindbohm, M. L. (1994). Laboratory work and pregnancy outcome. *Journal of Occupational and Environmental Medicine, 36*, 311–319. doi:10.1097/00043764-199403000-00008 PMID:8195901

Taskinen, H., Lindbohm, M. L., & Hemminki, K. (1986). Spontaneous abortionsamong women working in the pharmaceutical industry. *British Journal of Industrial Medicine, 43*, 199–205. PMID:3947584

Thonneau, P., Abell, A., Larsen, S. B., Bonde, J. P., Joffe, M., Clavert, A., Ducot, B., Multigner, L., & Danscher, G. (1999a). Effects of pesticide exposure ontime to pregnancy: Results of a multicenter study in France andDenmark. ASCLEPIOS Study Group. *American Journal of Epidemiology, 150*(2), 157–163. doi:10.1093/oxfordjournals.aje.a009975 PMID:10412960

Thonneau, P., Larsen, S. B., Abell, A., Clavert, A., Bonde, J. P., Ducot, B., & Multinger, L. (1999b). Time to pregnancy and paternal exposure to pesticidesin preliminary results from Danish and French studies. Asclepios. *Scandinavian Journal of Work, Environment & Health, 25*(Suppl 1), 49–61. PMID:10235412

Tielemans, E., Burdorf, A., te Velde, E. R., Weber, R. F. A., Van Kooji, R. J., Veulemans, H., & Heederik, D. J. J. (1999b). Occupationally relatedexposures and reduced semen quality: A case-control study. *Fertility and Sterility*, *71*(4), 690–696. doi:10.1016/S0015-0282(98)00542-1 PMID:10202880

Tielemans, E., Van Kooij, R., Looman, C., Burdorf, A., te Velde, E., & Heederik, D. (2000). Paternal occupational exposures and embryo implantation ratesafter IVF. *Fertility and Sterility*, *74*(4), 690–695. doi:10.1016/S0015-0282(00)00720-2 PMID:11020508

Tielemans, E., Van Kooij, R., te Velde, E.R., Burdorf, A., & Heederik, D. (1999a). Pesticide exposure and decreased fertilisation rates in vitro. *Lancet, 354*, 484–485.

Twombly, R. (1995). Assault on the male. *Environmental Health Perspectives*, *103*(9), 802–805. doi:10.1289/ehp.95103802 PMID:7498090

Valanis, B., Vollmer, W. M., & Steele, P. (1999). Occupational exposure to antineoplasticagents: Self-reported miscarriages and stillbirths among nursesand pharmacists. *Journal of Occupational and Environmental Medicine*, *41*(8), 632–638. doi:10.1097/00043764-199908000-00004 PMID:10457505

Van Oostdam, J., Gilman, A., Dewailly, E., Usher, P., Wheatley, B., Kuhnlein, S., Walker, J., Tracy, B., Feeley, M., & Jerome, V. (1999). Human healthimplications of environmental contaminants in Canada: A review. *The Science of the Total Environment*, *230*(1-3), 1–82. doi:10.1016/S0048-9697(99)00036-4 PMID:10466227

Weigert, M., Feichtinger, W., Kulin, S., Kaali, S. G., Dorau, P., & Bauer, P. (2001). Seasonal influences on in vitro fertilization and embryo transfer. *Journal of Assisted Reproduction and Genetics*, *18*(11), 598–602. doi:10.1023/A:1013160905497 PMID:11804428

Wennborg, H., Bodin, L., Vainio, H., & Axelsson, G. (2000). Pregnancy outcomeof personnel in Swedish biomedical research laboratories. *Journal of Occupational and Environmental Medicine*, *42*(4), 438–446. doi:10.1097/00043764-200004000-00022 PMID:10774513

Wennborg, H., Bonde, J. P., Stenbeck, M., & Olsen, J. (2002). Adverse reproduction outcomes among employees working in biomedical research laboratories. *Scandinavian Journal of Work, Environment & Health*, *28*(1), 5–11. doi:10.5271jweh.640 PMID:11871853

Yie, S.-M., Brown, G. M., Liu, G.-Y., Collins, J. A., Daya, S., Hughes, E. G., Foster, W., & Gand Young-lai, E. V. (1995). Melatonin and steroids in human pre-ovulatoryfollicular fluid: Seasonal variations, granulosa cell steroid production. *Human Reproduction (Oxford, England)*, *10*(1), 50–55. doi:10.1093/humrep/10.1.50 PMID:7745070

Younglai, E. V., Foster, W. G., Hughes, E. G., Trim, K., & Jarrell, J. F. (2002). Levelsof environmental contaminants in human follicular fluid, serum andseminal plasma of couples undergoing in vitro fertilization. *Archives of Environmental Contamination and Toxicology*, *43*(1), 121–126. doi:10.100700244-001-0048-8 PMID:12045882

Chapter 3
Textile Industry and Health Hazards: Impact of Climate Change Issues and Fertility Potential

Nibedita Naha
ICMR-National Institute of Occupational Health (NIOH), Ahmedabad, India

Gokul Manickavachagam
Biochemistry Division, ICMR-National Institute of Occupational Health (NIOH), India

ABSTRACT

Textile industry is one of the primarily concerned industries in the world generating huge revenues where both the skilled and unskilled labours are employed. Raw materials are mainly synthetic chemicals rather than the natural one. Several bleaching agents, additives are also used to get sufficiently high and uniform degree of whiteness in the textile materials. Environmental contamination is also common by the textile effluents. All these substances are mutagenic and carcinogenic, and cause enormous health hazards to humans, animals, and aquatic lives. Infertility, implantation failure, and miscarriage due to exposure to dyes, dye intermediates, and other raw materials are major concern worldwide as integrity of both the male and female gametes are massively affected by them. However, little or no attention has been paid in real life scenario, mainly in the developing and under developed countries including adaptation of successful and advanced mitigation strategies. Therefore, the chapter highlights the common issues and possible remedial measures of the textile industry exposure with respect to fertility potential and pregnancy outcome.

DOI: 10.4018/978-1-7998-4480-8.ch003

INTRODUCTION

The industry which benefited most from the industrial revolution was the textile industry that started its journey as small as cottage industry prior to 1700s. People at that time produced and sold their textiles, mainly, cotton and wool to local communities as the long distance trade was not familiar. However, due to industrialization, different kind of textiles are produced rapidly in a central place on a large scale with the help of new technologies and marketed from one country to other, even to continent; thereby a huge number of skilled and unskilled man-power are employed day-by-day. Textile is a flexible network of natural (i.e., wool, silk from animals; and cotton, flax, jute, bamboo from plants) and artificial (nylon, polyester, acrylic, rayon) fibers, produced by spinning, weaving, knitting, crocheting, knotting, felting, or braiding. Textile is also made up of interacting fibers such as, carpeting and geotextiles.

Today, clothing, or garment of vibrant colors rules the types of textile production and shelf-life of the textiles, along with market as well, which involve several kinds of chemicals, or dyes (mostly azo dyes), resins, fiber additives, minerals etc., causing deterioration of the ecological balance. More than 10000 types of dyes are used these days in textile manufacturing; of which almost 70% are synthetic azo dyes because of their cost-effectiveness and having complex structures (HSE, 2016; Hassan et al., 2017). The world-wide production of dyes is about 800,000 tons per year; 10–15% of which are lost during different processes in the textile industry (HSE, 2016; Hassan et al., 2017); thereby releases harmful colors into environment due to incomplete exhaustion of dyes, leading to health hazards. Further, textile dyeing requires a huge amount of water i.e., 200 tons of water per ton production of textiles (IARC monograph, 1998); hence, effluents are the main source of water pollution as well as a biggest threat to aquatic lives. Also, consumers play a crucial role in the pollution chain of the textile industry through several apparels.

Worldwide India is the 3rd leading textile exporters with export values of 17 billion USD after European Union (EU, 69 billion USD≅23% of global market share) and China (110 billion USD≅37.2% of global market share) (Statista, 2019). India is 1st in jute production, and 2nd in silk and cotton production in 2017–18 alone. However, lowest production cost, huge number of workers, reduction of commercial barriers and availability of materials remain behind the advantages of China as compared to India (Statista, 2019). EU, Italy, Germany, Spain, and France are the prime countries comprising of more than 170 thousand textile and clothing companies in 2017; 70% of them are clothing companies, 30% are textile companies, while man-made fiber companies account for less than 1% in contrast to China and India (Statista, 2019). Pakistan, Bangladesh and Ethiopia are also the leading countries in this regard.

As per ILO (2015), about 60–75 million people are employed in the textile and clothing industries worldwide; of them about three quarters are females. According to OECD (2017), the global middle class could reach 3.2 billion by 2020 and 4.9 billion by 2030 that ensure hundreds of millions of new consumers come mostly from the developing and underdeveloped countries; hence, overall demands of the skilled workforce and challenges for decent work are likely to be increased remarkably. In India, the textile industry employs 40 million workers directly and 60 million workers indirectly (https://www.ibef.org/archives/detail/) in contrast to 1.7 million workers in the EU textile industry (https://ec.europa.eu/growth/sectors/fashion/textiles-clothing/eu_en). As per 2006 official figures, out of total 760 million workforces, around 20 million migrant workers are employed in the textile and garment sectors (Wong, 2008), who come from the unorganized sector of the Afro-Asian countries, and are suffering from 'flexible/no' contracts as well as no minimum health care benefit, or protection.

TEXTILE DYES AND TEXTILE DYEING PROCESSES

The world's annual production of dyes accounts to more than 7×10^5 tones (Moosvi et al., 2007), and the textile industries accounts for the largest consumption of dyes at nearly 70% (Mathur et al., 2012). Dyes are the crystal structures of the colored substances that can adhere to compatible surface of fabric by solution, or by covalent bond with salt and metals, or by physical adsorption, or by mechanical retention (Kirk-Othmer, 2004; Bafana et al., 2011). They are classified according to application (i.e., acid, basic, direct, disperse, mordant, reactive, sulfur dye, pigment, VAT dye, azo insoluble) and chemical structure (i.e., nitro, azo, carotenoid, diphenylmethane, xanthene, acridine, quinoline, indamine, sulfur, amino- and hydroxyl-ketone, anthraquinone, indigoid, phthalocyanine, inorganic pigment) (Christie, 2001). Except colorant precursors like, oxidation bases and sulfur dyes, almost 2/3rd of all the organic dyes are azo dyes (R1–N=N–R2). Several starting materials, or intermediates are also included in the list such as, aniline, chloroanilines, naphthylamines, methylanilines, benzidines, phenylenediamines, and so on.

Overall, based on the general structure, textile dyes are fall into three categories:- anionic (direct, acid and reactive dyes), nonionic (disperse dyes that not ionize in an aquatic medium) and cationic (azo basic, anthraquinone disperse and reactive dyes) (Christie, 2001). Some disperse dyes having bioaccumulation property, whereas azo and nitro compounds are prefer to sediment. Most of these dyes have a high chemical, photolytic, bleaching and temperature stability; thus the conventional textile effluent treatment does not degrade these dyes; hence, a large quantity is found in the water bodies (Couto, 2009). Also, antimicrobial agents in textile manufacturing mainly, cotton, are resistant to biodegradation, whereas synthetic dyes with aromatic structure are more recalcitrant to biodegradation (Couto, 2009; O'Neill et al., 1994). Further, due to extensive use of cotton since past century, use of azo dyes are increased in the textile industry as cotton fibers are mainly dyed by the azo compounds (Akarslan and Demiralay, 2015).

Successful dyeing process earns enormous revenue, but at the same time causes environmental pollution and occupational health hazards. Although basic dyeing process does not change much with time, modern technologies are introduced according to the nature and properties of the fibres, dyes and pigments. Dyeing process involves three steps targeting the expected end-users as revealed by Moore and Ausley (2004): (1) preparation (i.e., removal of unwanted impurities from fabrics using aqueous alkaline substances and detergents, or enzymes; bleaching with hydrogen peroxide, or chlorinated compounds; or optical brightening agents:- depend on the demand of the fabric color); (2) Dyeing (i.e., aqueous application of synthetic organic dyes at elevated temperatures and pressures to the textile fabrics along with chemical aids such as, surfactants, acids, alkali/bases, electrolytes, carriers, levelling agents, promoting agents, chelating agents, emulsifying oils, softening agents etc., as well as pigments that differ from dyes by absence of physical/chemical affinity towards fibres, together with binders i.e., polymers that fix the pigments with the fibres); and (3) finishing with chemical treatments to improve the fabric quality (i.e., permanent press treatments, water proofing, softening, antistatic protection, soil resistance, stain release, microbial, or fungal protection etc.).

Liposome-based technology is widely popular to develop ecofriendly textile processing such as, dyeing and finishing where natural surfactant phospholipids i.e., phosphatidyl choline, plosphatidyl serine and phosphatidyl ethanolamine are used as liposome vesicles for auxiliary materials by interaction of their hydrophilic head groups with water molecule (Lesoin et al., 2011; Barani and Montazer, 2008), resulting in reduction in dyeing temperature and contamination load with low environmental impact. Further, ultrasonic energy-based textile dyeing is another technique operating under less extreme conditions such as, low temperature and chemical concentration. The idea is based on the enhanced mass transfer by reducing the stagnant cores in the yarns through ultrasound frequencies above 17kHz, which are attributed to cavitation phenomena and other resultant physical effects like, dye dispersion, or breaking of the aggregates with high molecular mass, degassing, or increasing the trapped air in the fibre capillaries, strong agitation of liquid at the fibre-liquid boundary layer, and increasing the diffusion rate of dye inside the fibre (Vajnhandl et al., 2005; Ferrero and Periolatto, 2012); hence popular in wet textile process as an alternative to harmful synthetic dyes, i.e., laundering, desizing, scouring, bleaching, mercerization of cotton fabrics, enzymatic treatment, dyeing and leather processing, together with the decoloration/mineralization of textile dyes in wastewater.

Moreover, irrespective of the characteristics of dyes, the final process involves washing in baths to remove the excesses of dyes, which is not fixed to the fibre. During this process 10–15% of the dyes are lost as effluents, and contaminating environment with about 1 million tons of these compounds due to activated sludge treatment in the effluent treatment plants (Umbuzeiro et al., 2005). The reduction of azo dyes by sodium hydrosulfite and successive chlorination with hypochlorous acid form 2-benzotriazoles fenilbenzotriazol (PBTA) derivatives, which is highly mutagenic aromatic amines than the original dye forms (Shiozawa et al., 1999). In aquatic environment, reduction of dye can occur in two phases:- (1) application of reducing agents to newly-dyed fibres to remove the excess unbound dye, leading to 'bleeding' of the textile fabrics during washing, and (2) use of reducing agents during bleaching process in order to make the 'color' less effluent to avoid legislation (Abou-Okeil et al., 2010), which ultimately produce PBTA in sewerage system and cause enormous impact on the environment; and thus, impose a threat to humans and aquatic lives due to climatic change effects (Guaratini and Zanoni, 2000).

Additionally, the textile industry came to focus due to large volume of water use in the dyeing process that produces huge amount of effluents, for example, 200 Litre of water is required for per kg of cotton production (Carneiro et al., 2004). Also, heavy metals and pesticides are detected in the effluents along with high organic load (PBTA derivatives and others), and low dissolved oxygen concentration and biodegradability, resulting in alteration of photosynthesis and oxygenation process of the aquatic biota (Pereira and Freire, 2005). Azo dyes and by-products are carcinogenic and mutagenic (Chequer et al., 2011; Ferraz et al., 2011), and resistant to biodegradation to meet the durability of colors in the fabric (Firminoet al., 2010), consequently endangering humans, animals and aquatic health, as the wastewater treatment plants are ineffective in removing their color and mutagenic properties (Oliveira et al., 2010; Konstantinou and Albanis, 2004).

CLIMATE CHANGE ISSUES AND HEALTH HAZARDS

Although various primary, secondary and tertiary treatments are applied to the textile industry effluents such as, flocculation, trickling filters, electrodialysis etc., textile dyes remain in environment for an extended period of time due to their high thermal- and photo-stability to resist biodegradation (Eswaramoorthi et al., 2008; Hassaan et al., 2016); which apart from aesthetic issue, creates a huge impact on environmental and ecological imbalance along with several hazards to plants, animals and humans. Use of the textile effluents for irrigating agricultural land is a worldwide problem; thus textile hub facing soil contamination issues, mostly in the developing countries. Sulphide contamination in surface water imposes soil infertility in Tamilnadu, India, the country's biggest knitwear centre; hence urgent need opts for zero liquid discharge (ZLD) (https://www.business-standard.com/article/current-affairs/tirupur-textile-hub-facing-soil-contamination-114120901345_1.html).

With respect to both the quantity and quality, textile processing wastewater is considered as the most polluted sources among all the industrial sectors (Chang et al., 2009; Rusan et al., 2007) that adversely affect the growth of several crop plants due to toxicity, and raise concern on the sustainability of soil environment especially, the physicochemical properties of soil contaminated with the toxic effluent. Study also reported that salts like, chloride in the dyeing process are leached to the nearby soil and affected the pH towards alkaline side, manifested by the yields of the crops (Mojiri, 2011; Joshi and Santani, 2012). Also, irrigation with wastewater increases organic carbon content and electrical conductivity of the soil (Mojiri, 2011; Joshi and Santani, 2012). High concentration of sodium in the contaminated soil irrigated with the textile effluents are contributed to excessive minerals in the wastewater (Khai et al., 2008). High amounts of sodium ions can result in precipitation of calcium and magnesium ions from the soil; thus, affecting their effectiveness in enhancing physical internal drainage (Ishaya et al., 2011). High level of metals such as, lead and copper further affected the physico-chemical property of soil, resulting in negative effect on the yield of the seasonal crops as well as selection of the tolerant crops (Mehta and Yadav, 2013) as shown in Fig. 1. Moreover, the yearly estimates of soil pollutants in the textile industry are about 2,00,000 tons of salt-like substances, 50,000 tons impurities of natural fibres, 80,000–1,00,000 tons of polyvinyl alcohol and carboxymethyl cellulose, and 15,000–20,000 tons of carboxylic acids (EWA, 2005).

The textile industry requires huge quantity of water for various processes like, sizing, scouring, bleaching, dyeing, printing and other finishing processes. Further, wastewater mainly due to dyeing and finishing contains several dyes, dye intermediates and other chemicals that poses additional environmental challenge. Globally, almost 10,000 different dyes with yearly production of 7.105 metric tons are produced commercially; 30% of which are used in quantum of 1000 tons per annum, and 90% of the textile products are used in the range of 100 tons per annum, or less (Baban et al., 2010). Approximately, 10–25% of textile dyes are misplaced during the dying process and 2–20% is directly released in the form of water effluents such as, benzidine, naphthalene and other aromatic compounds during several processes as discussed above (Suteu et al., 2011; Zaharia et al., 2009), inadequate treatment of which further increases their existence in the environment for longer period of time; for example, the half-life of hydrolyzed reactive blue 19 is as high as 46 years at pH 7.0 and 25°C temperature (Anjaneyulu et al., 2005).

Additionally, wastewater contains high quantity of several dye intermediates and heavy metals (Bisschops et al., 2003; Bornick and Schmidt, 2006) that increases high salt contents, total dissolved solids (TDS), total suspended solids (TSS), biological oxygen demand (BOD) and chemical oxygen demand (COD) (EWA, 2005; Lellis et al., 2019) into rivers and other water bodies of the nearby factories. However, low willingness of the industry to meet the environmental requirements, inadequate monitoring and enforcement of the government authorities, high installation cost and poor maintenance are playing crucial role for non-compliance of controlling water pollution in the developing countries (Sakamoto et al., 2019); hence, exert toxicity to fish, aquatic flora, sewage bacteria and plants as well as human health through food chain (Fig. 1).

Figure 1. Worldwide impact of textile industry effluents and heat stress on environment, plants, microbes, animals and human healths by contaminated soil, water and air. TDS, Total dissolved solids; TSS, Total suspended solids; BOD, Biological oxygen demand; COD, Chemical oxygen demand; HR, Heart rate.

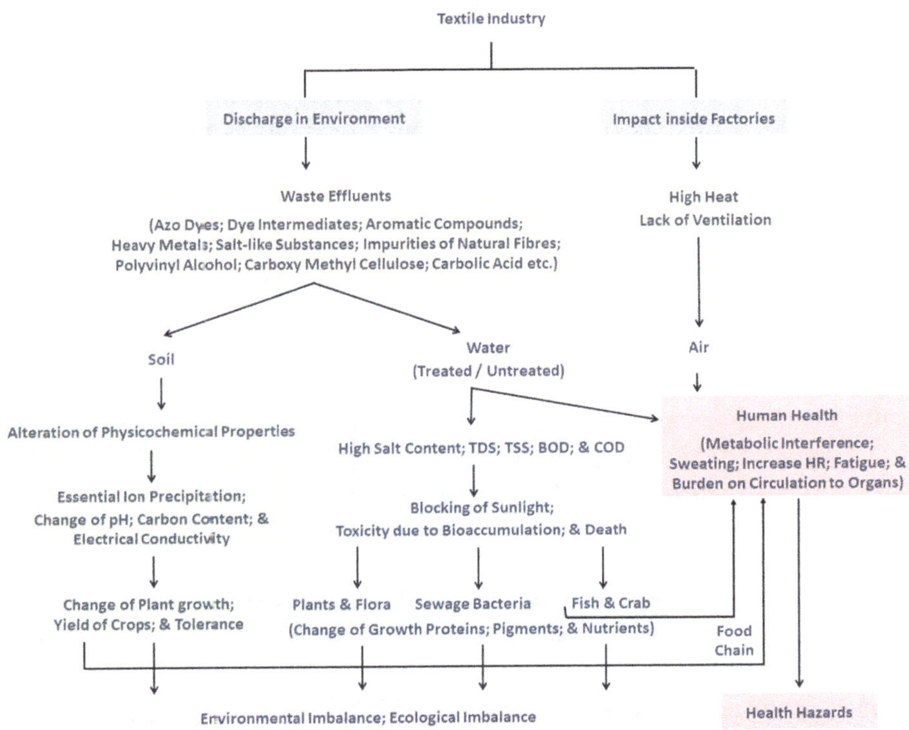

Textile dyes in both the treated and untreated effluents significantly compromise the quality of water bodies through BOD and COD values; provide recalcitrance and bioaccumulation, thereby promoting toxicity. Suspended solids in the textile effluents either killing, or reducing the growth of fish by clogging at the gills (Tufekci et al., 2007), erratic swimming, hyperexcitation, rapid opercular movement, thick mucus covering, enlarging primary gill bar, detaching secondary gill bar, disintegrating intestinal villi and infiltrating haemocytes (Selvaraj et al., 2015), high mortality and cytotoxicity on RBC including low count and poikilocytosis (Soni et al., 2006), alteration of ionic regulations in tissues like, liver, kidney, muscles and other organs (Karthikeyan et al., 2006), and lowering of feeding rate, absorption and food conversion (Sripriya et al., 2014). Further, hyperemia, necrosis and degeneration in liver, loss of nutritive value such as, protein, carbohydrate and lipids in fresh water female crab *Spiralothelphussa hydrodroma* are also reported as a consequence of water pollution due to textile dyes and synthetic agents (Sekar et al., 2009).

Moreover, solid wastes of the textile industry reduce oxygen concentration in water bodies due to presence of hydrosulphides and block the passage of sunlight; thereby decrease food and oxygen production ability of algae through diminishes photosynthetic activity, which is detrimental to water ecosystem (Tufekci et al., 2007; Khan and Malik, 2013). Depletion of dissolved oxygen further hinders the self-purification process of water. Sulphates in normal level are not considered toxic to any live forms but at high concentration cause prolong illness (Tufekci et al., 2007). High dye content with low biodegradability (i.e., half-life time of several years) in water bodies affect growth protein, pigment and other nutrients, as well as biomass production in algae (Dwevedi, 2013; Chia and Musa, 2014), probably due to their 50% more sensitivity towards contaminants in aquatic environment (Klaine and Lewis, 1995) along with ecological imbalance (Sousa et al., 2012). Heavy metals in effluents also accumulate in primary organs, and over time, begin to fester, leading to various symptoms of toxicity in both the aquatic and human lives through influences of the food chain (Fig. 1).

Chemicals and heavy metals from the textile industry entre into organism through respiration from air, or absorption through skin, or ingestion through food; thereby causing allergic reaction to eyes and skin (contact dermatitis), chemosis, lacrimation, exophthalmos, permanent blindness, rhabdomyolysis, and acute tubular necrosis (LGC report, 1999). Aromatic amines, bleaching agents, formaldehyde-based resins, fibre additives, ammonia, acetic acid, shrink-resist chemicals, optical whiteners, soda ash and caustic soda are mobilised by water, or sweat, which encourage their absorption through skin and other exposed areas like, mouth, and causes skin irritation, itchy, or blocked nose and sore eyes (HSE, 2016; Akarslan and Demiralay, 2015). Epoxy resins, based on the monomer diglycidyl ether of bisphenol A (DGEBA), also provoke sensitization after a single exposure in about 50% of the working people (Akarslan and Demiralay, 2015). Fabric containing larger diameter fibers like, wool causes more itching sensation than the fabric with smaller diameter fibers i.e., cotton (Akarslan and Demiralay, 2015). Trace metals like, chromium, copper, zinc and arsenic in the textile effluents also cause haemorrhage, skin ulcer, irritation, dermatitis and nausea. Absorption by ingestion is faster; thus potentially more dangerous as more dye is absorbed in a lesser time frame (Puvaneswari et al., 2006). Water soluble azo dyes become dangerous when metabolized by liver enzymes, and cause vomiting, gastritis, hypertension, and even vertigo and edema of face, neck, pharynx, tongue and larynx upon ingestion (LGC Report, 1999).

Moreover, textile mills generate huge amount of heat in the environment (Setyawan et al., 2017) by various textile machines like, jet, stenters, agers, jigger, drum, drier, printing machine, boiler etc. as well as lack of ventilation inside the factories during day and night hours as measured by heat stress indices (HSI) that causes detrimental effects on health. The average HSI is 84.81 during day time while the value is 83.07 at night, resulting in heat cramps among the working population if continuously exposed to $\geq$8 hr, as well as heat exhaustion (Sheth and Marvania, 2017). The study also showed that the weaving section of Indonesian textile hub generate high-temperature environment among all the production processes (Setyawan et al., 2018), which impact health and productivity of the workers, might be due to metabolic interference, sweating, high heart rate, fatigue, and so on. Heat exposure causes additional burden on blood circulation to different organs in the body. Inhalation of dust of the textile industry causes respiratory burden like, byssinosis, asthma, coughing, and wheezing (Hassaan et al., 2017) as well as hearing impairment due to noise generated through machine work especially, from drying process (Shaikh, 2009). Overall, health hazards directly affect the workers and later entre in the life cycle of consumers through the use of the products. Also, long half-life of dyes and intermediates, bleaching agents etc. are the source of continuous bioaccumulation in the organs due to repeated daily exposure of these toxic substances.

MALE REPRODUCTION AND INFERTILITY

Reproductive system is the soft target of slow but prolong exposure to textile dyes, bleaching agents and dye intermediates, although mostly studies are concentrated on the experimental models. Amin et al. (2016) found that the effects are relatively mild amongst animals that receive treated textile dye wastewater vs. those exposed to untreated one; untreated effluents cause complete sterility in mice. Congo red, diamine blue and cholrazol black E are attributed to low testicular weight, external malformation and consequent damages including capacitation, the pre-requisite of fertilization (Amin et al., 2016). Suryavathi et al. (2005) revealed a decrease of body weight (7–25%) and reproductive organ weights (2–48%) in both the rats and mice exposed to untreated azo dye wastewater along with a lowering of total protein (14–70%), cholesterol (14–91%) and total lipid (10–30%) in sperms and reproductive organs as well as fructose (18–44%) in seminal vesicle. The whole event is manifested by low sperm count, alteration of sperm motility, liter size and body weight gain of both the F1 and F2 generations of the rats and mice, although magnitudes of the effects are higher among rats (Suryavathi et al., 2005); however, the species discrepancy needs further in-depth confirmatory study.

The waste effluents from textile dyes and printing industries also inhibit spermatogenesis in rats, might be due to heavy metals such as, lead, cadmium, chromium, copper, zinc, nickel and iron more than their respective permissible limits (Gupta et al., 2003, resulting in lowering of spermatids (>70%), spermatogonia (>67%), preleptotene spermatocytes (>71%) and secondary spermatocytes (>73%) (Gupta et al., 2003). Sertoli cells and Leydig cells as well as several biochemical parameters such as, proteins, sialic acid and glycogen are also affected (Gupta et al., 2003); thereby suggesting interference of several enzymatic reactions especially, oxidative phosphorylation with constituents of the textile industry wastewater that might be responsible for overall low sperm density and presence of non-motile spermatozoa in the semen. Benoff et al. (2000) reported male infertility of human exposed to metal aerosols in environment that accumulate in the reproductive organs; thereby play a crucial role in fertilization as metal aerosols enter into mature sperms through calcium and potassium channels, and decrease acrosome reaction (Bedford, 1983).

Further, exposure to ethylene glycol ethers, a constituent of organic solvents related to the textile industry, for example, 2-methoxyethanol (2-ME) and 2-ethoxyethanol (2-EE), or their acetates are estimated by urinary metabolites methoxyacetic acid (MAA) and ethoxyacetic acid (EAA), respectively, which cause a considerable toxicity in both the male and female reproduction as compared to the patent molecules (Hardin, 1983; Foster et al., 1986; Oudiz and Zenick, 1986; Moss et al., 1985).Workers associated with preparatory phases that contain organic solvents, also report infertile, or subfertile upon ethylene glycol ethers exposure (Veulemans et al., 1993). Disperse Red 1 (i.e., N-ethyl-N-(2-hydroxyethyl)-4-(4-nitrophenylazo) aniline), like other azo dyes of the textile industry, exerts cytotoxicity and genotoxicity by inducing DNA damage in type A spermatogonia, resulting in abnormal sperm morphology and low DNA integrity (Fernandes et al., 2015); however, potential harmful effects and actual caution level on human reproduction are yet to be explored. Cryptorchidism has a relation with significant reduction of the surface area of Sertoli cells during wastewater treatment among experimental animals (Ronis et al., 1998). Moreover, sperm motility inhibition test also indicates higher cytotoxicity of reactive dyes and dyed fabrics especially cotton, with blue dyes as compared to yellow dyes (Klemola et al., 2007). Table 1 provides an overview of the male reproductive impairments due to harmful agents used in the textile industry and related discharge/effluents as described above.

Female Reproduction and Miscarriage

Female reproductive system is equally affected by textile dyes, bleaching agents and their intermediates as well as the textile industry wastewater (Amin et al., 2016; Suryavathi et al., 2005), leading to implantation failure due to impairment of embryo-uterine cross-talk and miscarriage; thus raise worldwide concern. The equilibrium between cell proliferation and cell death through apoptosis is a key factor of embryogenesis that reflects the dynamic aspects of neural tube development. Manimaran et al. (2018) recently observed a dose-dependent developmental defects (i.e., teratogenicity) in zebrafish embryo induced by apoptotic cell death, and even cause lethal at higher dose when exposed to reactive black 5, a most commonly used synthetic dye in the textile industry. Further, the rate of hatching is significantly affected in the zebrafish, might be due to interference of the cleavage enzymes produce by hatching glands of chorionic sac upon exposure, resulting in shortage of oxygen supply, an essential pre-requisite for embryonic development (Manimaran et al., 2018). Embryo toxicity is also evaluated by reactive blue 160 (Barathi et al., 2020). Azo dyes like, congo red significantly reduces the number of germ cells in both the male and female animals (Gray and Ostby, 1993), but only female offspring reports reduction in fertility potential (Gray et al., 1992), which requires an in-depth molecular study.

Several epidemiological studies report the increase number of women involvement even during pregnancy; hence, adverse effects of occupational risk factors on fertility outcome have immense clinical importance. For example, occupational exposure to synthetic fibres, fibre dust, solvents and a mixture of synthetic and natural fibres etc. in the textile industry causes miscarriages as reported by Wong et al. (2009). Carbon disulfide as a major ingredient of rayon manufacturing, induces adverse effects on menstruation and pregnancy outcome in the female rayon workers (Zhou et al., 1988; Hemminki et al., 1980). A very recent study of Egypt reported a significant association of adverse pregnancy outcome with the duration of exposure to the textile industry pollution (Abo El Gheit et al., 2020). Endotoxin, a lipopolysaccharide component of the cell walls of gram-negative bacteria induces miscarriages in the cotton textile industry workers (Wang et al., 1998; Foley and Schlafer, 1994), although not in accordance with the findings of Wong et al. (2009), might be due to the one-child Govt. policy in China. However, the role of specific agents of the synthetic fibres on the elevated risk of pregnancy outcome needs to be studied further.

Occupational exposure to VAT dyes as used by textile dyers further reveals high prevalence of reproductive impairments in female workers as compared to the non-exposed group, namely, infertility, miscarriage and low birth weight, although absence of significant association might be due to limitation by recall bias of such kind of retrospective study (Soyinka et al, 2020). Moreover, a study in Denmark reported a five-fold increase in the risk of infertility among the female textile workers exposed to cutting oils, dry cleaning chemicals, and heavy metals such as, lead, cadmium, or mercury when data are adjusted for age, education, residence and parity (US Congress, 1985). Previous study also showed a 12% high rate of spontaneous abortion in the Iranian women of average 30 years' age and more than 15 years' exposure to the textile industry as compared to the non-working group (US Congress, 1985). In Finland, similar trends of high spontaneous abortion (12.2%) are noted in the women employees of the textile industry (US Congress, 1985). In Sweden, the women textile workers and the wives of the men at rayon factories are also experienced high rate of spontaneous abortion (US Congress, 1985). Hence, all these data strongly suggest that occupational exposure due to textile processes impose highest risk to female infertility including termination of pregnancy, mostly due to implantation failure and miscarriages.

Moreover, earlier studies with synthetic textile dyes not only attribute to embryonic stage but also early developmental stages of life by inducing DNA damage and hindering DNA repair through generation of reactive oxygen species (ROS) and inhibition of DNA ligase (Liu et al., 2007; De Oliveira et al., 2015). These effects are either direct impact of original compounds, or their intermediates/by-products due to cleavage of the azo bond by microorganisms (Ayadi et al., 2016). The reduction of textile azo dyes may also bind with DNA binding motifs (Levine, 1991; Stiborova et al., 1992) and cause multiple toxic effects, manifested by implantation failure, miscarriage, cancer, and so on. The overall female reproductive impairments upon exposure to the textile industry discharge is mentioned in Table 1. As most of these compounds/intermediates are proven carcinogen and mutagen; hence, a separate section highlights their specific functions in this regard.

Table 1. Overview of the textile industry exposure on male and female reproductive impairments and fertility outcome

Male Reproductive Potential				*Female Reproductive Potential*			
Source	*Agents*	*Species*	*Observation*	*Source*	*Agents*	*Species*	*Observation*
Hardin, 1983; Foster et al., 1986	Ethylene glycol ethers, or metabolites	Rats Rabbits	§ Teratogenicity § Testicular atrophy § Infertility	Hardin, 1983	Ethylene glycol ethers	Rats Rabbits	§ Embryonic and fetal death § Teratogenicity
Gupta et al., 2003	Lead Cadmium Chromium Copper Zinc Nickel Iron	Rats	§ Low germ cell counts § Low sperm density § Fall in total proteins, sialic acid, glycogen § Low sperm motility	Gray et al., 1992	Congo red	Mice	§ Reduce germ cell number § Low fertility potential
Suryavathi et al., 2005	Untreated azo dye (wastewater)	Mice and progeny	§ Low testis weight § Decrease total protein, cholesterol, total lipid in sperms § Low seminal fructose § Low sperm count	De Oliveira et al., 2015	Direct black 38 Reactive blue 15 Reactive orange 16 VAT green 3	Zebrafish	§ Delay in hatching § Embryonic malformation § Delay in larval development and malformation
Klemola et al., 2007	Reactive dyes Dyed fabrics (cotton)	Mice	§ Cytotoxicity § Reduce sperm motility	Manimaran et al., 2018	Reactive black 5	Zebrafish	§ Teratogenicity § Low hatching rate in embryo
Fernandes et al., 2015	Disperse red 1	Mice	§ Cytotoxicity § Genotoxicity, germ cell DNA damage § Abnormal sperm morphology	Barathi et al., 2020	Reactive blue 160	Zebrafish	§ Decrease survival, hatching and heart rate in embryo

continued on following page

Table 1.Continued

Male Reproductive Potential				Female Reproductive Potential			
Source	*Agents*	*Species*	*Observation*	*Source*	*Agents*	*Species*	*Observation*
Amine et al., 2016	Congo red Diamine blue Cholrazol black E	Mice and progeny	§ Reduce testis weight and malformation § Capacitation defect	Ayadi et al., 2016	Synthetic azo compounds, or metabolites	Aquatic life Animal Human	§ DNA damage § Hinder DNA repair § Endocrine disruptors
Veulemans et al., 1993	Ethylene glycol ethers Dyes metabolites	Human (worker)	§ Infertility, or subfertility	Saletta et al., 2007	Arylamine derivatives (arylnitrenium ions form)	Human	§ Bind with DNA to form DNA adduct in mammary glands
				Zhou et al., 1988; Hemminki et al., 1980	Carbon disulfide	Human (rayon, chemical workers)	§ Menstrual irregularity, abnormal duration and quantity § Spontaneous abortion
				Wang et al., 1998; Foley and Schlafer, 1994	Endotoxin	Human (cotton workers)	§ Miscarriage
				Wong et al., 2009	Synthetic fibres Fibre dust Solvent impurities (natural fibres)	Human (workers)	§ Miscarriage
				Soyinka et al., 2020; US Congress, 1985	VAT dyes Cutting oils Dry cleaning chemicals Lead Cadmium Mercury	Human (workers, users)	§ Infertility § Implantation failure § Spontaneous abortion § Low birth weight

Carcinogenic and Mutagenic Effects

Depending on the extent of exposure, duration and concentration of the substances, chemicals, heavy metals, dyes and dye intermediates cause irritation of the mucous membrane, which might be due to interaction with amino ($–NH_2$), or sulphydryl (–SH) groups of the target proteins in the body, irrespective of age, and even before birth (Mirghani et al., 2008). Alteration of normal cell physiology and biochemical mechanisms results in impairment of reproductive functions, and even cancer and mortality (Khan and Malik, 2013; Akarslan and Demiralay, 2015). Development of bladder cancer is found in the painters upon long-time exposure to azo dyes by generation of the aromatic amines due to metabolism at intestinal wall and liver, which interfere with several normal biochemical and physiological processes. These harmful agents through chemical and biological assimilations also cause eutrophication, consume dissolved oxygen, prevent re-oxygenation, and sequester metal ions to accelerate mutagenicity while targeting DNA, and carcinogenicity on reductive cleavage.

International Agency for Research on Cancer (IARC) of World Health Organization (WHO) classified azo dyes and its metabolic product benzidine as category 1 carcinogen (IARC, 2010). Benzidine congeners such as, 3,3′-dimethylbenzidine (*o*-tolidine), 3,3′-dimethoxylbenzidine (*o*-dianisidine) and *p*-phenylenediamine as well as their N-acetylated and N,N′-diacetylated derivatives are also potent carcinogen and mutagen in animals and humans (Chung, 2015; Chung and Cerniglia, 1992). Benzidine- and benzidine congener-derived dyes undergo reductive metabolism before exerting their mutagenic property. Arylamines, an azo dye product, induce genitourinary cancer in humans (Chung, 2015). Thus to protect human health, USA and many European countries restrict/prohibit the use of more than 20 aromatic amines in the consumer goods (Puntener and Pages, 2013). In European countries, azo dyes and dye intermediates (i.e., amines) are significantly reduced due to strict national policies (Claxton et al., 2001). Public awareness regarding arylamines and their regulation may also contribute a lot in this regard, although till today these agents are widely used in many developing countries including India.

Azo dyes and their breakdown products possess highest toxicity and cause huge body burden through circulation in animals and humans. In the textile factory due to high heat and lack of ventilation, body temperature of the workers (directly affected) remain warm and skin pores become open for perspiration, which facilitate absorption of these dyes and intermediates, heavy metals etc. through skin followed by accumulation in organs and negatively affect their functions (Khan and Malik, 2013). Studies also revealed that the electron-donating substituent in *ortho*, or *para* position increase carcinogenic potential of these textile dyes (Carmen and Daniela 2012). The genotoxic potency of aromatic amines and arylamines depends on the hydrophobicity, and electronic (reactivity and propensity to be metabolically transformed) and steric properties (Benigni and Passerini, 2002) as well as functional groups like, sulphonate (SO_3), halogen, or nitro (NO_2) groups (Chung and Cerniglia, 1992).

Further, arylamine derivatives (i.e., hydroxyl, acetoxy, or glucuronides derivatives) are converted to highly reactive arylnitrenium ions that bind with DNA to form DNA adducts in mammary glands (Saletta et al., 2007). Electrophilic metabolites of aromatic amines and amides are carcinogens such as, reactive esters N-acetoxyarylamines, the precursors of short lived nitrenium ions (NH_2^+), are responsible for arylamine-derived DNA adducts formation and cancer (Josephy and Novak, 2013). N-hydroxyl and nitroso forms of arylamine (4-aminobiphenyl and 2-naphthylamine)-induce DNA damage are mediated by DNA adduct 7,8-dihydro-8-oxo-2′deoxyguanosine (8-oxo-dG) lesions and mutations via ROS and hydrogen peroxide (H_2O_2) (Murata and Kawanish, 2011). This process is also induced by 2-naphthylamine (2-NA), 4-aminobiphenyl (4-ABP), and benzidine (Ohnishi et al., 2001; Makena and Chung, 2007). ABP-induced frame shift, base-pair and complex mutations are proved by DNA sequencing analysis (Mathur et al., 2012). Several other factors such as, promotion, tissue growth, differentiation, genetic susceptibility, metabolic pathways, expression, or suppression of oncogenes and suppressor genes, etc. are important determinants of arylamine-induced carcinogenesis (Menashe et al., 2012).

Nicotinamide adenine dinucleotide hydrogen (NADH)-dependent, NADH-enhanced, or superoxide dismutase (SOD)/manganese-enhanced redox reactions are also observed in metal-mediated oxidative DNA damage by hydroxy and nitroso derivatives of the arylamines as auto-oxidation of *o*-phenylenediamine and other arylamine metabolites are accelerated by SOD or manganese (Murata and Kawanish, 2011). Correlation of DNA adduct and mutation is clearly supported by the fact that DNA adducts are pro-mutagenic lesions, and mutation causes activation of oncogenes. Also mutagenicity due to reduction of dyes/dye intermediates through oxidative mechanism is comparatively higher than that of the parent congener amines, which provide the proof of mutagenicity of chromophores (Mathur et al., 2012). Although oxidative DNA damage by arylamines contributes to chemical carcinogenesis (Murata and

Kawanish, 2011), the in-depth molecular mechanism of genotoxic and epigenotoxic potentials of these agents for tumorigenesis is yet to be investigated, which would certainly be beneficial for the development of translational strategy. Also, in contrast to USA and European countries, as most of these dyes are sold in the market of the developing countries including India without any information about chemical properties, purity, or possible mutagenicity; hence, unlimited and uncontrolled use of such agents in reality causes huge consequences in human and animal healths, and ecological imbalance.

MITIGATION STRATEGIES

Wastewater Treatment and Optimizing Dyeing Process

There are several mitigation strategies pertaining to optimizing dyeing process and wastewater treatment of the textile effluents but having certain drawbacks like, incineration that produces dioxins and furans due to incomplete combustion, absorption onto solid matrices based on phase transfer of contaminants without any destruction (Babel and Kurniawan, 2003), and so on. Lint filters of the dyeing machines also play crucial role to keep the lint out from heat exchangers and clothes so that total suspended solids (TSS) levels are low in raw textile dyeing wastewater (Tufekci et al., 2007). Further, use of filtration and separation drawn attention as common mitigation strategies (Walsh, 2001). Augugliaro et al. (2000) reported the importance of '*Langmuir Hinshelwood model*' for degradation of aromatic ring of the amine. But as all these processes involves only phase transfer and generate huge sludge in the tank, and exhibit low efficiency of color removal and reduction of organic load; hence, new technology targeting minimizing volume and toxicity of textile waste is of high demand. The existing remedial processes for most of the textile industries are originally developed for other wastes; therefore, colored effluents with higher organic load significantly contribute to health hazards by accumulation in the surface water (Pereira, 2004).

Temperature, pH etc. are the important determinants for mineralization during treatment of the textile wastewater (Cantle et al., 2005; Chu et al., 2007). Advanced oxidation processes (AOPs) is comparatively functioning to accelerates oxidation process of a wide spectrum organic contaminants present in the textile effluents (Glaze et al., 1987; Finklea, 1998) through generation of hydroxyl free radicals ($OH^{\cdot}$) by several oxidants under different combinations, which overcome the barrier of conventional oxidation processes (Kdasi et al., 2004; Arslan et al., 1999; Kestioglu et al., 2005). Ozone, ultra violet (UV), titanium dioxide (TiO_2), hydrogen peroxides (H_2O_2), fenton (i.e., H_2O_2 with ferrous sulphate, $FeSO_4$), photo-fenton (i.e., H_2O_2 and UV radiation with $FeSO_4$) and ultrasonic are potent agents in this respect that use chemical, electrical, mechanical, or radiation energy, either alone, or in different combinations (Chung et al., 1992; Pramauro et al., 1995; Fukushima et al., 2010; Augugliaro et al., 2000). Other advantages of AOPs are mineralization of organic contaminants in wastewater during oxidation (Kdasi et al., 2004), and destruction of complex structure of dyes/dye intermediates as low and non-waste generation technology through short-lived chemical species having high oxidation power (Kestioglu et al., 2005; Chung et al., 1992).

Silver nanoparticles due to their excellent surface properties and chemical reactivity emerge as a better option for dyes/dye intermediates by adsorption, photocatalytic degradation, or combination of both; but generate toxicity to environment as they are released to water stream, sludge, soil as well as underground water (Marimuthu et al., 2020). Raman and Kanmani (2016) reported the application of nano zero valent iron (nZVI) on textile dye degradation as a better option due to high reactivity towards pollutant, less toxicity and cost-effectiveness. nZVI quickly aggregate with organic dyes and dye intermediates but inclusion of resin, nickel, zinc, bentonite, biopolymer, kaolin, rectorite, nickel-montmorillonite, bamboo, cellulose, biochar, graphene and clinoptilolite increases the stability of iron, which ensures integration of nZVI with other suitable techniques for complete and rapid degradation of organic materials found in the textile industry waste and imposes less toxicity in the environment (Raman and Kanmani, 2016).

Further, BOD and COD are high in slashing, fabric formation and wet processing (Tufekci et al., 2007); thus become more important mitigation targets although depends on the production facilities, toxicity of the textile industry wastewater varies. Also treatment plants using biological procedure show preferences due to less excess sludge production, lower operational costs and better COD removal as compared to the chemical processes (Tufekci et al., 2007). Further, prior to biological treatment, use of hydrochloric acid or carbon dioxide, and sodium chloride instead of sulphuric acid in the neutralization

Figure 2. Multidimensional strategies including traditional and newer techniques for management of the textile industry effluents and solid wastes. AOPs, Advanced oxidation processes; nZVI, nano zero valent iron.

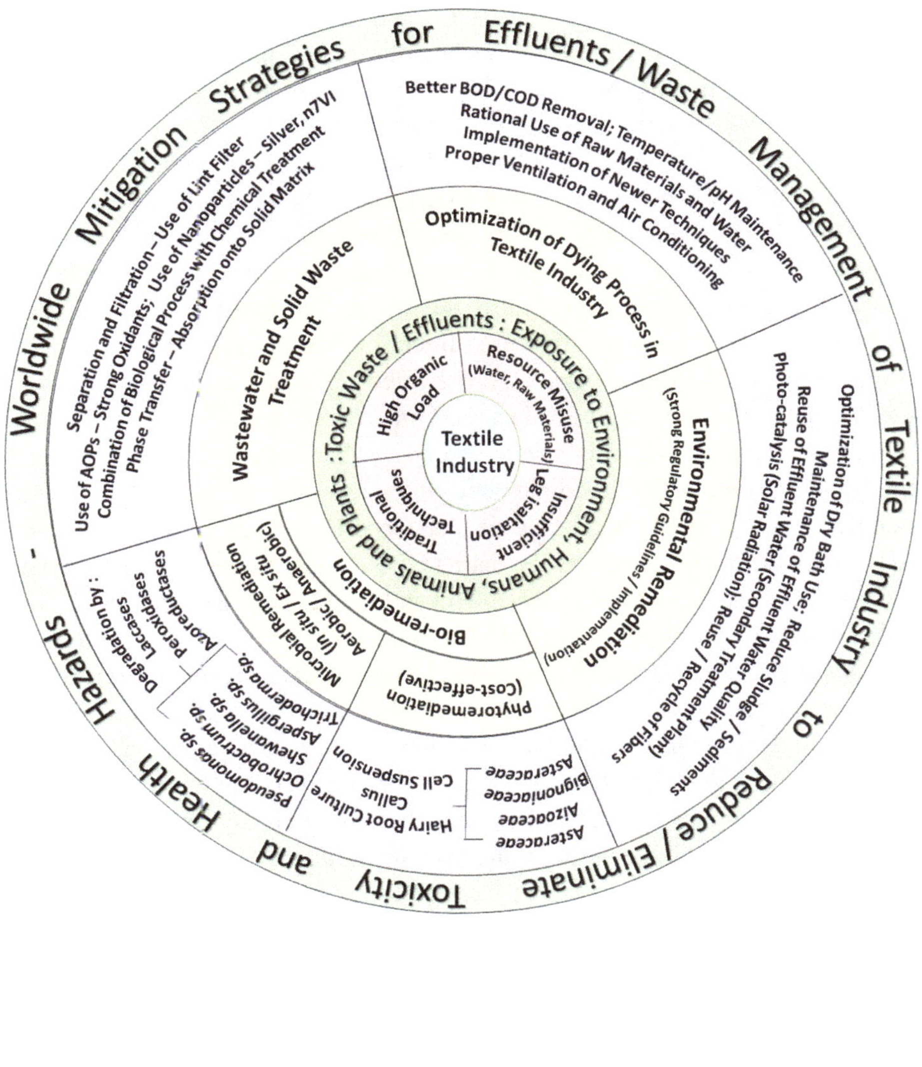

process and sodium sulphate in the dyeing process, respectively, mitigate the excess amount of sulphate ($-SO_4$) present in the textile effluents (Tufekci et al., 2007). Proper ventilation and implementation of air conditioning system are also important to mitigate heat stress that generate in the factories due to textile processing. The overall mitigation strategies related to the wastewater treatment and optimizing dyeing process are mentioned below (Fig. 2).

Environmental- and Bio-remediation

Development of new methods to promote the treatment of the textile industry effluents (i.e., decolorization, and/or removal of organic load) with maximum efficiency are encouraged, although trigger further damage to environment at the same time. Also environmental issues with used dye baths are related to wide spectrum components that add to the dye bath in high concentrations; hence, the textile industry faces a reusing problem of incoming fresh water as traditional methods are insufficient for obtaining the required water quality. So a better approach to solve the increase demand of fresh water in the textile industry is to elevate the water quality of the effluent to high standard for reuse using a secondary wastewater treatment plant, although little attention has been paid in reality (Lin and Chen, 1997). Studies also found that ultrasonic energy as a renewable energy, is ecofriendly in the dyeing process at 60°C that minimizes the damage of the mechanical properties of the fibres especially wool, which is much lesser than the common industrial temperature (98°C), and has a potential for environmental remediation (Ferrero and Periolatto, 2012). Photocatalysis by solar radiation as well as artificial radiation are the alternate waste treatment process but not widely used in the textile sectors (Pramauro et al., 1995; Augugliaro et al., 2000).

Reuse and recycling of the fibres (57%), polymer/oligomer (37%), monomer (29%) and fabric (14%) reduce environmental impacts as compared to the physical and chemical treatments as discussed above (Sandin and Peters, 2018); hence consider one of the best environmental remediation strategies. Moreover, reuse is more favourable than recycling under certain conditions where customer plays a crucial role (Sandin and Peters, 2018). An exploratory study in Sweden revealed charity also an important determinant for reuse of the products through a semi-structured interview based questionnaire on propositions and model for different drivers of the reuse-based clothing value chain such as, business system, legislation, awareness, design, material quality, brand values and product prices as well as the consumer/donor's attitude (Paras et al., 2018). However, this may not be applicable in the developing and underdeveloped countries due to their different socio-economic conditions.

Phytoremediation of a number of textile dyes viz., basic red 46, red 198, poly R-478, reactive red M5B, emazol red, red HE8B, reactive red 2, red HE 7B, scarlet GDR, golden yellow HER, methyl orange, orange M2RL, navy blue HE2R, reactive green 19A-HE4BD and malachite green are noted where secondary metabolites and enzymes play crucial roles (Lellis et al., 2019). Several plants such as, *Tagetespatula* (Asteraceae), *Sesuvium portulacastrum* (Aizoaceae), *Tecoma stans* (Bignoniaceae), *Blumea malcolmii* (Asteraceae) etc. through the process of rhizofiltration, phytoextraction, or phytoaccumulation, phytodegradation and rhizodegradation, become the most popular mitigation agents as revealed by cultures of hairy roots, callus and cell suspension (Lellis et al., 2019). Overall, phytoremediation of the complex textile dyes and dye intermediates is ecofriendly, economic, aesthetically satisfactory and require small quantity of nutrients due to their large autotrophic biomass system; hence, has a promising future.

Microbial remediation of textile dyes under aerobic, or anaerobic conditions in waste effluents are another way of mitigation strategy with the aim of increasing the yield of degradation. For example, methanogenic bacteria *Acetoclastic* under anaerobic conditions mineralize various compounds of the textile effluents with a good yield especially, in case of nitroaromatic and azo dyes (Flores et al., 1997). Aerobic microorganisms also degrade aromatic amines as observed in the sediments of river Elbe in Germany (Bornick, 2001). Wang et al. (2007) reported that PN1001, a member of *Pseudomonus sp*., induces degradation of pentylamine and aniline by 82% and 78% respectively, from petrochemical wastewater. Microbial bioremediation of reactive blue 160 that causes embryo toxicity, shows positive impact on environment and aquatic lives (Barathi et al., 2020). The *in situ* bioremediation process involves biostimulation (i.e., introduce nutrients to favor local microorganisms), or bioaugmentation (i.e., inoculation of exogenous microorganisms) (Kasai, 2011; Lellis et al., 2019), whereas *ex situ* bioremediation consists of composting and land farming by bioreactors, hybrid crops and genetic improvement of the lineages (Das and Dash, 2014).

Not only bacteria, but also algae, filamentous fungi and yeasts in soil, water, sludge, waste or effluent actively participate in bioremediation process (Das and Dash, 2017; Lellis et al., 2019). *Escherichia coli, Pseudomonas luteola, Bacillus subtilis, Aeromonas punctate, Shewanella putrefaciens, Pseudomonas aeruginosa, Aeromonas hydrophila, Providencia vermicola, Shewanella aquimarina, Scheffersomyces spartinae, Pichia occidentalis, Aspergillus niger, Aspergillus flavus, Aspergillus fumigatus, Cladosporium cladosporioides* as well as *Ochrobactrum sp. And Trichoderma sp* are some of the important names of the long list where laccases, peroxidases, azoreductases etc. cause cleavage of the azo bonds (–N=N–) of the textile dyes (Lellis et al., 2019). However, although the use of microorganisms in the treatment of textile wastes containing aromatic amines draw attention, in reality far from the practical use due to low investment, high maintenance costs, long treatment times and huge deposition of sludge at the bottom of the treatment plants; hence, more extensive research is needed to overcome these drawbacks from commercial point of view. Fig. 2 represents the worldwide view of the environmental- and bio-remediation strategies targeted to the textile industry effluents and solid wastes.

CONCLUSION

It is concluded that the textile dyes and bleaching agents represent a large group of complex organic compounds that have undesirable effects on environment as well as impose a great threat to humans, animals and aquatic lives. The higher organic load and difficulty in wastewater treatment of the textile industry lead to a constant search for new and efficient methods that at the same time are economically viable too. But unfortunately, till date, no such efficient methods have been invented for complete removal of colors as well as toxic properties of the substances like, dyes and dye intermediates, bleaching agents, additives, carriers, levelling agents, promoting agents, chelating agents, emulsifying oils, softening agents, and many more as used in the textile industry.

The use of nanoparticles and bioremediation are quite promising, although most of the nanoparcles like, nZVI require green supports and additives as mentioned above. Further, scale-up and commercialization of the environmental- and bio-remediation processes are yet to be done. It is highly recommended to perform more bench- and pilot-scale studies to apply both these new techniques to treat the contaminated textile industry waste, so that the effluents quality become improve to such an extent that could be reusable without creating scarcity of fresh water and health hazards to humans, animals and aquatic lives. It also reduces, or nullifies environmental pollution and pay huge attention to fast amortization.

Moreover, in the present days, some cost-effective novel technologies related to advanced wastewater treatment for effective removal of the hazardous dyes and dye metabolites, and optimizing dyeing process are used in the textile industries worldwide especially, in the developed countries that follow specific regulatory guidelines; for example, EPA (1996) guidelines. These promising mitigation processes require continuous optimization that including better fixation of the less hazardous dyes to fabrics, use of minimal water, identification of the potential carcinogenic, and/or mutagenic dyes/derivatives, and genetic improvements for resistant culture mediums/treatments and reduction of sludge production, along with economic photo-electrochemical method and engineering advancement, probably all in the form of an integrated system. The knowledge of new aspects and strong legislation for implementation are highly essential for such kind of successful mitigation strategies. In contrast, the scenario of the developing and underdeveloped countries are more, or less similar due to lack of the regulatory guidelines, monitoring and implementation as well as loopholes of the legislations (Ferronato and Toretta, 2019; Babazadeh et al., 2018). In reality, many countries also misuse resources such as, raw materials, additives, byproducts/derivatives, water, energy etc. along with lacking of awareness among the people that aggravate environmental pollution and health hazards.

Apart from these, several enzymes such as, cytochrome-P450, prostglandin H synthase, N-acetyltransferase, glutathione transferase, glucouronyl transferase, sulfotransferase and *o*-acetylase play critical role in metabolic activation, or detoxification of xenobiotics including arylamines; thereby regulate the onset of carcinogenesis/mutagenesis and species specificity of the arylamine-inducing diseases (Murata and Kawanish, 2011), which needs to be investigated more in the context of dyes and bleaching agents used in the textile industry. Although oxidative DNA damage by arylamines contributes to chemical carcinogenesis (Murata and Kawanish, 2011), the in-depth molecular mechanism of these agents for tumorigenesis of the reproductive system is yet to be explored, which would certainly be beneficial for development of novel translational strategy of the complex textile waste. Further, the potential harmful effects and actual caution level of azo dyes on human reproduction are yet to be identified as till today most of the results are based on the experimental models. Therefore, in-depth molecular mechanism of genotoxic and epigenotoxic potentials of the textile dyes and dye intermediates, bleaching agents, additives etc. are the utmost global need for the development of next-generation mitigation strategy targeting human fertility potential in order to accelerate societal as well as individual benefits.

ABBREVIATIONS

2-EE, 2-Ethoxyethanol; **2-ME**, 2-methoxyethanol; **2-NA**, 2-Naphthylamine; **4-ABP**, 4-Aminobiphenyl; **8-oxo-dG**, 7,8-Dihydro-8-oxo-2′deoxyguanosine; **AOP**, Advanced oxidation process; **BOD**, Biological oxygen demand; **COD**, Chemical oxygen demand; **EAA**, Ethoxyacetic acid; **EU**, European Union; **$FeSO_4$**, Ferrous sulphate; **H_2O_2**, Hydrogen peroxide; **HR**, Heart rate; **HIS**, Heat stress index; **IARC**, International agency for research on cancer; **MAA**, Methoxyacetic acid; **–N=N–**, Azo bonds; **NADH**, Nicotinamide adenine dinucleotide hydrogen; **$-NH_2$**, Amino group; **NH_2^+**, Nitrenium ions; **NO_2**, Nitro group; **nZVI**, Nano zero valent iron; **$OH^{\cdot}$**, Hydroxyl free radicals; **PBTA**, 2-Benzotriazoles fenil benzotriazol; **ROS**, Reactive oxygen species; **–SH**, Sulphydryl group; **SO_3**, Sulphonate group; **$-SO_4$**, Sulphate; **SOD**, Superoxide dismutase; **TDS**, Total dissolved solids; **TiO_2**, *Titanium dioxide;* **TSS**, Total suspended solids; **UV**, Ultra violet; **WHO**, World Health Organization.

AUTHOR CONTRIBUTION

NN conceived, designed and wrote the book chapter along with literature survey. GM partially contributed to literature survey for section 2-3.

FUND DISCLAIMER

No grant received from any funding agency in public, commercial, or not-for-profit sector.

CORRESPONDING AUTHORS

Nibedita Naha

ACKNOWLEDGMENT

Ms. Vishakha Shrimali, Ph.D. student, Biochemistry Div., ICMR-NIOH for digitalization of the illustration; and Gujarat University for Ph.D. registration of Mr. Gokul Manickavachagam, DST-SERB (New Delhi) Jr. Research Fellow.

REFERENCES

Abo El Gheit, R. E., Oun, A. M., & Al Arshal, A. (2020). The effect of maternal exposure to textile industry-induced pollution on pregnancy and its outcome. *International Journal of Medical Arts*, *2*, 162–172.

Abou-Okeil, A., El-Shafie, A., & El Zawahry, M. M. (2010). Eco-friendly laccase–hydrogen peroxide/ultrasound-assisted bleaching of linen fabrics and its influence on dyeing efficiency. *Ultrasonics Sonochemistry*, *17*(2), 383–390. doi:10.1016/j.ultsonch.2009.08.007 PMID:19748814

Akarslan, E., & Demiralay, H. (2015). Effects of textile materials harmful to human health. *Acta Physica Polonica A*, *128*(2B), B-407–B-408. doi:10.12693/APhysPolA.128.B-407

Amin, T., Afrin, M., Haque, Z., & Islam, R. (2016). Effects of textile dye waste water on reproductive system of mice and their progeny. *International Journal of Scientific and Engineering Research*, *7*, 565–570.

Anjaneyulu, Y., Chary, N. S., & Raj, D. S. S. (2005). Decolorization of industrial effluents - available methods and emerging technologies: A review. *Reviews in Environmental Science and Biotechnology*, *4*(4), 245–273. doi:10.100711157-005-1246-z

Arslan, I., Balcioglu, A., & Tuhkanen, T. (1999). Oxidative treatment of simulated dye-house effluent by UV and near-UV light assisted fenton's reagent. *Chemosphere*, *39*(15), 2767–2783. doi:10.1016/S0045-6535(99)00211-8

Augugliaro, V., Prevol, A. B., Loddo, V., Marci, G., Palmisano, L., Pramauro, E., & ... , (2000). Photodegradation kinetics of aniline, 4-ethylaniline, and 4-chloroaniline in aqueous suspension of polycrystalline titanium dioxide. *Research on Chemical Intermediates*, *26*(5), 413–426. doi:10.1163/156856700X00426

Ayadi, I., Souissi, Y., Jlassi, I., Peixoto, F., & Mnif, W. (2016). Chemical Synonyms, molecular structure and toxicological risk assessment of synthetic textile dyes: A critical review. *Journal of Developing Drugs*, *5*, 151.

Baban, A., Yediler, A., & Ciliz, N. K. (2010). Integrated water management and CP implementation for wool and textile blend processes. *CLEAN Soil Air Water*, *38*(1), 84–90. doi:10.1002/clen.200900102

Babazadeh, T., Nadrian, H., Mosaferi, M., & Allahverdipour, H. (2018). Identifying challenges and barriers to participating in the source separation of waste program in Tabriz, northwest of Iran: A qualitative study from the citizens' perspective. *Resources*, *7*(3), 53. doi:10.3390/resources7030053

Babel, S., & Kurniawan, T. A. (2003). Low-cost adsorbents for heavy metals uptake from contaminated water: A review. *Journal of Hazardous Materials*, *97*(1-3), 219–243. doi:10.1016/S0304-3894(02)00263-7 PMID:12573840

Bafana, A., Devi, S. S., & Chakrabarti, T. (2011). Azo dyes: Past, present and the future. *Environmental Reviews*, *19*(NA), 350–370. doi:10.1139/a11-018

Barani, H., & Montazer, M. (2008). A review on applications of liposomes in textile processing. *Journal of Liposome Research*, *18*(3), 249–262. doi:10.1080/08982100802354665 PMID:18770074

Barathi, S., Karthik, C., Nadanasabapathi, S., & Padikasan, I. A. (2020). Biodegradation of textile dye Reactive Blue 160 by *Bacillus firmus* (Bacillaceae: Bacillales) and non-target toxicity screening of their degraded products. *Toxicology Reports*, *7*, 16–22. doi:10.1016/j.toxrep.2019.11.017 PMID:31871898

Bedford, J. M. (1983). Significance of the need for sperm capacitation before fertilization in Eutherian mammals. *Biology of Reproduction*, *28*(1), 108–120. doi:10.1095/biolreprod28.1.108 PMID:6338941

Benigni, R., & Passerini, L. (2002). Carcinogenicity of the aromatic amines from structure activity relationships to mechanisms of action and risk assessment. *Mutation Research*, *511*(3), 191–206. doi:10.1016/S1383-5742(02)00008-X PMID:12088717

Benoff, S., Jacob, A., & Hurley, I. R. (2000). Male infertility and environmental exposure to lead and cadmium. *Human Reproduction Update*, *6*(2), 107–121. doi:10.1093/humupd/6.2.107 PMID:10782569

Bisschops, I. A. E., & Spanjers, H. (2003). Literature review on textile wastewater characterization. *Environmental Technology*, *24*(11), 1399–1411. doi:10.1080/09593330309385684 PMID:14733393

Bornick, H., Eppinger, P., Grischek, T., & Worch, E. (2001). Simulation of biological degradation of aromatic amines in river bed sediments. *Water Research*, *35*(3), 619–624. doi:10.1016/S0043-1354(00)00314-6 PMID:11228957

Börnick, H., & Schmidt, T. C. (2006). Amines. In T. Reemtsma & M. Jekel (Eds.), *Organic pollutants in the water cycle - properties, occurrence, analysis and environmental relevance of polar compounds* (pp. 181–208). Wiley-VCH. doi:10.1002/352760877X.ch8

Canle, L. M., Santaballa, J. A., & Vulliet, E. (2005). On the mechanism of TiO_2-photocatalyzed degradation of aniline derivatives. *Journal of Photochemistry and Photobiology A Chemistry*, *175*(2-3), 192–200. doi:10.1016/j.jphotochem.2005.05.001

Carmen, Z., & Daniela, S. (2012). Textile organic dyes - characteristics, polluting effects and separation/elimination procedures from industrial effluents: A critical overview. In T. Puzyn (Ed.), *Organic pollutants ten years after the Stockholm convention - Environmental and analytical update*. doi:10.5772/32373

Carneiro, P. A., Osugi, M. E., Sene, J. J., Anderson, M. A., & Zanoni, M. V. B. (2004). Evaluation of color removal and degradation of a reactive textile azo dye on nanoporous TiO thin-film electrodes. *Electrochimica Acta*, *49*(22-23), 3807–3820. doi:10.1016/j.electacta.2003.12.057

Chang, W., Tran, H., Park, D., Zhang, R., & Ahn, D. (2009). Ammonium nitrogen removal characteristics of zeolite media in a Biological Aerated Filter (BAF) for the treatment of textile wastewater. *Journal of Industrial and Engineering Chemistry*, *15*(4), 524–528. doi:10.1016/j.jiec.2009.01.009

Chequer, F. M. D., Lizier, T. M., Felicio, R., Zanoni, M. V. B., Debonsi, H. M., Lopes, N. P., & ... , (2011). Analyses of the genotoxic and mutagenic potential of the products formed after the biotransformation of the azo dye Disperse Red 1. *Toxicology In Vitro*, *25*(8), 2054–2063. doi:10.1016/j.tiv.2011.05.033 PMID:21907275

Chia, M. A., & Musa, R. I. (2014). Effect of indigo dye effluent on the growth, biomass production and phenotypic plasticity of Scenedesmus quadricauda (Chlorococcales). *Journal of the Brazilian Academy of Science*, *36*(1), 419–428. doi:10.1590/0001-3765201420130225 PMID:24676177

Christie, R. (2001). *Colour chemistry*. The Royal Society of Chemistry.

Chu, W., Choy, W. K., & So, T. Y. (2007). The effect of solution pH and peroxide in the TiO_2-induced photocatalysis of chlorinated aniline. *Journal of Hazardous Materials*, *141*(1), 86–91. doi:10.1016/j.jhazmat.2006.06.093 PMID:16916576

Chung, K. T. (2015). Occurrence, uses and carcinogenicity of arylamines. *Frontiers in Bioscience*, *7*(2), 367–393. doi:10.2741/e737 PMID:25553383

Chung, K. T., & Cerniglia, C. E. (1992). Mutagenicity of azo dyes: Structure-activity relationships. *Mutation Research*, *277*(3), 201–220. doi:10.1016/0165-1110(92)90044-A PMID:1381050

Chung, K. T., Stevens, S. E. J., & Cerniglia, C. E. (1992). The reduction of azo dyes by the intestinal microflora. *Critical Reviews in Microbiology*, *18*(3), 175–197. doi:10.3109/10408419209114557 PMID:1554423

Claxton, L., Hughes, T., & Chung, K. (2001). Using base-specific Salmonella tester strains to characterize the types of mutation induced by benzidine and benzidine congeners after reductive metabolism. *Food and Chemical Toxicology*, *39*(12), 1253–1261. doi:10.1016/S0278-6915(01)00072-2 PMID:11696399

Couto, S. R. (2009). Dye removal by immobilised fungi. *Biotechnology Advances*, *27*(3), 227–235. doi:10.1016/j.biotechadv.2008.12.001 PMID:19211032

Das, S., & Dash, H. R. (2014). Microbial bioremediation: A potential tool for restoration of contaminated areas. In D. S. Elsevier (Ed.), *Microbial Biodegradation and Bioremediation* (pp. 1–21). doi:10.1016/B978-0-12-800021-2.00001-7

Das, S., & Dash, H. R. (2017). *Handbook of metal-microbe interactions and bioremediation*. CRC Press. doi:10.1201/9781315153353

De Oliveira, G. A. R., de Lapuente, J., Teixido, E., Porredon, C., Borras, M., & de Oliveira, D. P. (2015). Textile dyes induce toxicity on zebrafish early life stages. *Environmental Toxicology and Chemistry*, *35*(2), 429–434. doi:10.1002/etc.3202 PMID:26267709

Dwevedi, S. (2013). Effect of textile dyes on Spirulina platensis. *Journal of Chemical and Pharmaceutical Research*, *5*, 66–80.

EPA. (1996). *Textile Mills Effluent Guidelines*. https://www.epa.gov/eg/textile-mills-effluent-guidelines

Eswaramoorthi, S., Dhanapal, K., & Chauhan, D. (2008). *Advanced in textile waste water treatment: The case for UV-ozonation and membrane bioreactor for common effluent treatment plants in Tirupur, Tamil Nadu, India*. Environment with People's Involvement and Coordination in India.

EWA. (2005). *Efficient use of water in the textile finishing industry. Official Publication of European Water Association*. EWA.

Fernandes, F. H., Bustos-Obregon, E., & Salvadori, D. M. F. (2015). Disperse Red 1 (textile dye) induces cytotoxic and genotoxic effects in mouse germ cells. *Reproductive Toxicology (Elmsford, N.Y.)*, *53*, 75–81. doi:10.1016/j.reprotox.2015.04.002 PMID:25883024

Ferraz, E. R. A., Umbuzeiro, G. A., de-Almeida, G., Caloto-Oliveira, A., Chequer, F. M. D., Zanoni, M. V. B., Dorta, D. J., & Oliveira, D. P. (2011). Differential toxicity of Disperse Red 1 and Disperse Red 13 in the ames test, HepG2 cytotoxicity assay, and daphnia acute toxicity test. *Environmental Toxicology*, *26*(5), 489–497. doi:10.1002/tox.20576 PMID:20549607

Ferrero, F., & Periolatto, M. (2012). Ultrasound for low temperature dyeing of wool with acid dye. *Ultrasonics Sonochemistry*, *19*(3), 601–606. doi:10.1016/j.ultsonch.2011.10.006 PMID:22055328

Ferronato, N., & Torretta, V. (2019). Waste Mismanagement in Developing Countries: A Review of Global Issues. *International Journal of Environmental Research and Public Health*, *16*(6), 1060. doi:10.3390/ijerph16061060 PMID:30909625

Finklea, H. O. (1998). *Semiconductor electrodes*. Elsevier.USA, p-199. ISBN: 0-444-42926-3.

Firmino, P. I. M., Silva, M. E. R., Cervantes, F. J., & Santos, A. B. (2010). Colour removal of dyes from synthetic and real textile wastewaters in one- and two-stage anaerobic systems. *Bioresource Technology*, *101*(20), 7773–7779. doi:10.1016/j.biortech.2010.05.050 PMID:20542688

Flores, E. R., Perez, F., & Torre, M. (1997). Scale-up of Bacillus thuringiensis fermentation based on oxygen transfer. *Journal of Fermentation and Bioengineering*, *83*(6), 561–564. doi:10.1016/S0922-338X(97)81137-3

Foley, G. L., & Schlafer, D. H. (1994). Bacterial endotoxemia and reproductive effects in ruminants. *The Veterinary Clinics of North America. Food Animal Practice*, *10*(3), 491–501. doi:10.1016/S0749-0720(15)30534-X PMID:7728632

Foster, P. M. D., Blacburn, D. M., Moore, R. B., & Llovd, S. C. (1986). Testicular toxicity of 2-methyoxy-acetaldehyde, a possible metabolite of ethylene glycol monomethyl ether, in the rat. *Toxicology Letters*, *32*(1-2), 73–80. doi:10.1016/0378-4274(86)90051-2 PMID:3738932

Fukushima, M., Tatsumi, K., & Morimoto, K. (2010). The fate of aniline after a photo-fenton reaction in an aqueous system containing iron (III), humic acid, and hydrogen peroxide. *Environmental Science & Technology*, 34.

Glaze, W. H., Kang, J. W., & Chapin, D. H. (1987). The chemistry of water treatment processes involving ozone, hydrogen peroxide and ultraviolet radiation. *Ozone Science and Engineering*, *9*, 335–342.

Gray, L. E. Jr, & Ostby, J. S. (1993). The effect of prenatal administration of azo dyes on testicular development in the mouse: A structure activity profile of dyes derived from benzidine, dimethyl benzidine or dimethoxy benzidine. *Fundamental and Applied Toxicology*, *20*(2), 177–183. doi:10.1006/faat.1993.1024 PMID:8449389

Gray, L. E. Jr, Ostby, J. S., Kavlock, R. J., & Marshall, R. (1992). Gonadal effects of fetal exposure to the azo dye Congo red in mice: Infertility in female but not male offspring. *Fundamental and Applied Toxicology*, *19*(3), 411–422. doi:10.1016/0272-0590(92)90180-P PMID:1459372

Guaratini, C. C. I., & Zanoni, M. V. B. (2000). Textile dyes. *Quimica Nova*, *23*(1), 71–78. doi:10.1590/S0100-40422000000100013

Gupta, R. S., Sharma, R., Chaudhary, R., Yadav, R. K., & Khan, T. I. (2003). Effect of textile waste water on the spermatogenesis of male albino rats. *Journal of Applied Toxicology*, *23*(3), 171–175. doi:10.1002/jat.862 PMID:12794938

Hardin, B. D. (1983). Reproductive toxicity of the glycol ethers. *Toxicology*, *27*(2), 91–102. doi:10.1016/0300-483X(83)90014-8 PMID:6351353

Hassaan, M., El Nemr, A., & Madkour, F. F. (2017). Advanced oxidation processes of mordant violet 40 dye in freshwater and seawater. *Egyptian Journal of Aquatic Research*, *43*(1), 1–9. doi:10.1016/j.ejar.2016.09.004

Hassaan, M. A., & El Nemr, A. (2017). Advanced oxidation processes for textile wastewater treatment. *International Journal of Photochemistry and Photobiology*, *2*, 85–93.

Hassaan, M. A., El Nemr, A., & Madkour, F. F. (2016). Application of ozonation and UV-assisted ozonation for decolorization of direct yellow 50 in seawater. *Pharmaceutical Chemistry Journal*, *3*, 131–138.

Hemminki, K., Franssila, E., & Vainio, H. (1980). Spontaneous abortions among female chemical workers in Finland. *International Archives of Occupational and Environmental Health*, *45*(2), 123–126. doi:10.1007/BF01274131 PMID:7364488

HSE. (2016). *Dyes and chemicals in textile finishing: An introduction. Dyeing and Finishing Information Sheet No 1- Health and Safety Executive (HSE) information sheet*. HSE.

ILO. (2015). *Textiles, clothing, leather and footwear sector.* http://www.ilo.org/global/industriesand-sectors/textiles-clothing-leather-footwear/lang--ja/index.htm

Ishaya, K. S., Maracus, D. N. K., & Issac, J. (2011). The influence of waste water on soil chemical properties on irrigated fields in Kaduna South Township, North Central Nigeria. *Journal of Sustainable Development in Africa*, *13*, 91–101.

Josephy, P. D., & Novak, M. (2013). Reactive electrophilic metabolites of aromatic amine and amide carcinogens. *Frontiers in Bioscience*, *S5*(1), 341–359. doi:10.2741/S376 PMID:23277055

Joshi, V. J., & Santani, D. D. (2012). Physicochemical characterization and heavy metal concentration in effluent of textile industry. *Universal Journal of Environmental Research and Technology*, *2*, 93–96.

Karthikeyan, S., Jambulingam, M., Sivakumar, P., Shekhar, A. P., & Krithika, J. (2006). Impact of textile effluents on fresh water fish *Mastacembelus arimatus. Journal of Chemistry*, *3*, 303–306.

Kasai, Y. (2011). Molecular technologies for analysis of petroleum bioremediation. In A. Koukkou (Ed.), *Microbial Bioremediation of non-metals: Current Research* (pp. 233–252). Caister Acedemic Press.

Kdasi, A., Idris, A., Saed, K., & Guan, C. (2004). Treatment of textile wastewater by advanced oxidation processes: A review. *Global Nest International Journal*, *6*, 222–230.

Kestioglu, K., Yonar, T., & Azbar, N. (2005). Feasibility of physico-chemical treatment and advanced oxidation processes (AOPs) as a means of pretreatment of olive mill effluent (OME). *Process Biochemistry*, *40*(7), 2409–2416. doi:10.1016/j.procbio.2004.09.015

Khai, N. M., Tuan, P. T., Vinh, C. N., & Oborn, I. (2008). Effects of using wastewater as nutrient sources on soil chemical properties in periurban agricultural systems. *VNU Journal of Science: Earth and Sciences (Paris)*, *24*, 87–95.

Khan, A., & Malik, A. (2013). Environmental and health effects of textile industry wasterwater. In A. Malik, E. Grohman, & R. Akhtar (Eds.), *Environmental deterioration and human health: Natural and anthropogenic determinants* (pp. 55–71). Springer, Netherlands.

Kirk-Othmer. (2004). *Encyclopedia of chemical technology* (5th ed., Vol. 7). Wiley-Interscience.

Klaine, S. J., & Lewis, M. A. (1995). Algae and plant toxicity testing. In D. J. Hoffman, B. A. Rattner, G. A. Burton Jr, & J. Cairns Jr., (Eds.), *Handbook of ecotoxicology* (p. 371). CRC Press.

Klemola, K., Pearson, J., Wright, A. V., Liesivuori, J., & Lindstrom-Seppa, P. (2007). Evaluating the toxicity of reactive dyes and dyed fabrics with the hepa-1 cytotoxicity test. *AUTEX Research Journal*, *7*, 224–230.

Konstantinou, I. K., & Albanis, T. A. (2004). TiO2-assisted photocatalytic degradation of azo dyes in aqueous solution: Kinetic and mechanistic investigations - A review. *Applied Catalysis B: Environmental*, *49*(1), 1–14. doi:10.1016/j.apcatb.2003.11.010

Lellis, B., Fávaro-Polonio, C. Z., Pamphile, J. A., & Polonio, J. C. (2019). Effects of textile dyes on health and the environment and bioremediation potential of living organisms. *Biotechnology Research and Innovation*, *3*(2), 275–290. doi:10.1016/j.biori.2019.09.001

Lesoin, L., Crampon, C., Boutin, O., & Badens, E. (2011). Preparation of liposomes using the supercritical anti-solvent (SAS) process and comparison with a conventional method. *The Journal of Supercritical Fluids*, *57*(2), 162–174. doi:10.1016/j.supflu.2011.01.006

Levine, W. G. (1991). Metabolism of azo dyes: Implication for detoxication and activation. *Drug Metabolism Reviews*, *23*(3-4), 253–309. doi:10.3109/03602539109029761 PMID:1935573

Lin, S. H., & Chen, M. L. (1997). Purification of textile wastewater effluents by a combined Fenton process and the ion exchange. *Desalination*, *109*(2), 121–130. doi:10.1016/S0011-9164(97)00058-1

Liu, F., Gentles, A., & Theodorakis, C. W. (2007). Arsenate and perchlorate toxicity, growth effects, and thyroid histopathology in hypothyroid zebrafish Danio rerio. *Chemosphere*, *71*(7), 1369–1379. doi:10.1016/j.chemosphere.2007.11.036 PMID:18164748

Makena, P. S., & Chung, K. T. (2007). Evidence of oxidative genotoxicity induced by 4-aminobiphenyl, benzidine, and benzidine congeners. *Environmental and Molecular Mutagenesis*, *48*, 404–413. doi:10.1002/em.20288 PMID:17370336

Manimaran, D., Sulthana, A. S., & Elangovan, N. (2018). Reactive black 5 induced developmental defects via potentiating apoptotic cell death in Zebrafish (*Danio rerio*) embryos. *Pharmacy and Pharmacology International Journal*, *6*, 449–451.

Marimuthua, S., Antonisamya, A. J., Malayandia, S., Rajendrana, K., Tsaib, P. C., & Pugazhendhi, A. (2020). Silver nanoparticles in dye effluent treatment: A review on synthesis, treatment methods, mechanisms, photocatalytic degradation, toxic effects and mitigation of toxicity. *Journal of Photochemistry and Photobiology. B, Biology*, *205*, 111823. doi:10.1016/j.jphotobiol.2020.111823 PMID:32120184

Mathur, N., Bhatnagar, P., & Sharma, P. (2012). Review of the mutagenicity of textile dye products. *Universal Journal of Environmental Research and Technology*, *2*, 1–18.

Mehta, R., & Yadav, K. (2013). Soil contamination due to textile effluent-case study on the printing cluster of Jaipur. *Journal of the Textile Association*, *73*, 367–370.

Menashe, I., Figueroa, J. D., Garcia-Closas, M., Chatterjee, N., Malats, N., Picornell, A., & et al., (2012). Large-scale pathway-based analysis of bladder cancer genome-wide association data from five studies of European background. *PLoS One*, *7*(1), e29396. doi:10.1371/journal.pone.0029396 PMID:22238607

Mirghani, M. E. S., Kabbashi, N. A., Qudsieh, I. Y., & Elfaki, F. A. (2008). A new method for the determination of toxic dye using FTIR spectroscopy. *IIUM Engineering Journal*, *9*, 27–38. doi:10.31436/iiumej.v9i2.98

Mojiri, A. (2011). Effects of municipal waste water on physical and chemical properties of saline soil. *Journal of Biological and Environmental Sciences*, *5*, 71–76.

Monograph, I. A. R. C. (1998). Aromatic amines, in: Some aromatic amines, hydrazine and related substances, N-nitroso compounds and miscellaneous alkylating agents. *IARC Monographs on the evaluation of the carcinogenic risk of chemicals to humans*, *4*.

Monograph, I. A. R. C. (2010). Some aromatic amines, organic dyes and related exposures. *IARC Monographs on the Evaluation of Carcinogenic Risks to Humans*, *99*. PMID:21528837

Moore, S. B., & Ausley, L. W. (2004). Systems thinking and green chemistry in the textile industry: Concepts, technologies and benefits. *Journal of Cleaner Production*, *12*(6), 585–601. doi:10.1016/S0959-6526(03)00058-1

Moosvi, S., Kher, X., & Madamwar, D. (2007). Isolation, characterization and decolorization of textile dyes by a mixed bacterial consortium JW2. *Dyes and Pigments*, *74*(3), 723–729. doi:10.1016/j.dyepig.2006.05.005

Moss, E. J., Thomas, L. V., Cook, M. W., Walters, D. G., Foster, P. M. D., Creasv, D. M., & ... (1985). The role of metabolism in 2-methoxyethanol-induced testicular toxicity. *Toxicology and Applied Pharmacology*, *79*(3), 480–489. doi:10.1016/0041-008X(85)90145-0 PMID:4035689

Murata, M., & Kawanish, S. (2011). Mechanisms of oxidative DNA damage induced by carcinogenic arylamines. *Frontiers in Bioscience*, *16*(1), 1132–1143. doi:10.2741/3739 PMID:21196222

O'Neill, C., Hawkes, F. R., Hawkes, D. L., Lourenço, N. D., Pinheiro, H. M., & Delée, W. (1999). Colour in textile effluents - sources, measurement, discharge consents and simulation: A review. *Journal of Chemical Technology and Biotechnology (Oxford, Oxfordshire)*, *74*(11), 1009–1018. doi:10.1002/(SICI)1097-4660(199911)74:11<1009::AID-JCTB153>3.0.CO;2-N

OECD. (2017). The future of global value chains: Business as usual or a "new normal"? *OECD Publishing*, No. 41.

Ohnishi, S., Murata, M., Degawa, M., & Kawanishi, S. (2001). Oxidative DNA damage induced by an N-hydroxy metabolite of carcinogenic 4-aminoaminoazobenzene. *Japanese Journal of Cancer Research*, *92*(1), 23–29. doi:10.1111/j.1349-7006.2001.tb01043.x PMID:11173540

Oliveira, G. A. R., Ferraz, E. R. A., Chequer, F. M. D., Grando, M. D., Angeli, J. P. F., Tsuboy, M. S., Marcarini, J. C., Mantovani, M. S., Osugi, M. E., Lizier, T. M., Zanoni, M. V. B., & Oliveira, D. P. (2010). Chlorination treatment of aqueous samples reduces, but does not eliminate, the mutagenic effect of the azo dyes Disperse Red 1, Disperse Red 13 and Disperse Orange 1. *Mutation Research*, *703*(2), 200–208. doi:10.1016/j.mrgentox.2010.09.001 PMID:20851777

Oudiz, D., & Zenick, H. (1986). In vivo and in vitro evaluations of spermatoxicity induced by 2-ethoxyethanol treatment. *Toxicology and Applied Pharmacology*, *84*(3), 576–583. doi:10.1016/0041-008X(86)90263-2 PMID:3726877

Paras, M. K., Ekwall, D., Pal, R., Curteza, A., Chen, Y., & Wang, L. (2018). An exploratory study of Swedish charities to develop a model for the reuse-based clothing value chain. *Sustainibility*, *10*(4), 1176. doi 10.3390u10041176

Pereira, R. S. (2004). Identificação e caracterização das fontes de poluição em sistemas hídricos. *Revista Eletrônica de Recursos Hídricos*, *1*, 20–36.

Pereira, W. S., & Freire, R. S. (2005). Ferro zero: Uma nova abordagem para o tratamento de águas contaminadas com compostos orgânicos poluentes. *Quimica Nova*, *28*(1), 130–136. doi:10.1590/S0100-40422005000100022

Pramauro, E., Prevot, A. B., Augugliaro, V., & Palmisano, L. (1995). Photocatalytic treatment of laboratory wastes containing aromatic amines. *Analyst (London)*, *120*(2), 237–242. doi:10.1039/an9952000237

Pratt, M. (2000). Disperse blue dyes 106 and 124 are common causes of textile dermatitis and should serve as screening allergens for this condition. *American Journal of Clinical Dermatology*, *11*(1), 30–41. doi:10.1016/S1046-199X(00)90030-7 PMID:10684387

Püntener, A., & Pages, C. (2013). *European on certain azo dyes*. http://wwot.destcorp.com/documents/Azo_Dye_Information.pdf

Puvaneswari, N., Muthukrishnan, J., & Gunasekaran, P. (2006). Toxicity assessment and microbial degradation of azo dyes. *Indian Journal of Experimental Biology*, *44*, 618–626. PMID:16924831

Raman, C. D., & Kanmani, S. (2016). Textile dye degradation using nano zero valent iron: A review. *Journal of Environmental Management*, *177*, 341–355. doi:10.1016/j.jenvman.2016.04.034 PMID:27115482

Report, L. G. C. (1999). The risk of cancer caused by textiles and leather goods coloured with azo dyes. CSTEE plenary meeting, Brussels, Belgium.

Ronis, M. J., Badger, T. M., Shema, S. J., Roberson, P. K., & Shaikh, F. (1998). Effects on pubertal growth and reproduction in rats exposed to lead perinatally or continuously throughout development. *Journal of Toxicology and Environmental Health*, *53*(4), 327–341. doi:10.1080/009841098159312 PMID:9490329

Rusan, M., Hinnawi, S., & Rousan, L. (2007). Long term effect wastewater irrigation of forage crops on soil and plant quality parameters. *Desalination*, *215*(1-3), 143–152. doi:10.1016/j.desal.2006.10.032

Sakamoto, M., Ahmed, T., Begum, S., & Huq, H. (2019). Water Pollution and the Textile Industry in Bangladesh: Flawed Corporate Practices or Restrictive Opportunities? *Sustainability*, *11*(7), 1951–1964. doi:10.3390u11071951

Saletta, F., Matullo, G., Manuguerra, M., Bardelli, S. A., & Vineis, P. (2007). Exposure to the tobacco smoke constituent 4-aminobiphenyl induces chromosomal instability in human cancer cells. *Cancer Research*, *67*(15), 7088–7094. doi:10.1158/0008-5472.CAN-06-4420 PMID:17671175

Sandin, G., & Peters, G. M. (2018). Environmental impact of textile reuse and recycling: A review. *Journal of Cleaner Production*, *184*, 353–365. doi:10.1016/j.jclepro.2018.02.266

Sekar, P., Prasad, S. H., & Raman, M. D. (2009). Effect of textile dye industry effluent on the nutritive value of fresh water female crab *Spiralothelphusa hydrodroma*. *Journal of Applied Sciences Research*, *5*, 2041–2048.

Selvaraj, D., Leena, R., & Kamal, D. C. (2015). Toxicological and histopathological impacts of textile dyeing industry effluent on a selected teleost fish Poeciliareticulata. *Journal of Pharmacology and Toxicology*, *3*, 26–30.

Setyawan, H., Pratiwi, Q. C., Sjarifah, I., & Atmojo, T. B. (2017). Environmental heat stress enhances crystallization in urine. *IOP Conference Series: Earth and Environmental Science, 129*, 1-6.

Setyawan, H., Qodrijati, I., Widjanarti, M. P., Rinawati, S., Atmojo, T. B., & Fajariani, R. (2018). The impact of hot work climate on textile industry productivity. *IOP Conference Series: Earth and Environmental Science, 200*, 1-5. 10.1088/1755-1315/200/1/012053

Shaikh MA (2009). Environmental issues related with textile sector. *Pakistan Textile Journal*, 36-40.

Sheth, K. N., & Marvania, M. (2017). A critical study on heat stress in textile industry. *International Journal for Innovative Research in Science and Technology*, *4*, 47–49.

Shiozawa, T., Suyama, K., Nakano, K., Nukaya, H., Sawanishi, H., Oguri, A., Wakabayashi, K., & Terao, Y. (1999). Mutagenic activity of 2 phenylbenzotriazole derivatives related to a mutagen, PBTA-1, in river water. *Mutation Research*, *442*(2), 105–111. doi:10.1016/S1383-5718(99)00069-8 PMID:10393279

Soni, P., Sharma, S., Sharma, S., Kumar, S., & Sharma, K. P. (2006). A comparative study on the toxic effects of textile dye waste waters (untreated and treated) on mortality and RBC of fresh water fish *Gambusia affinis*. *Journal of Environmental Biology*, *27*, 623–628. PMID:17405321

Sousa, M. L. D., Moraes, P., Lopes, P. R. M., & Montagnolli, R. (2012). Contamination by Remazol Red Brilliant dye and its impact in aquatic photosynthetic microbiota. *Journal of Environmental Management and Sustainable Development*, *1*, 129–138.

Soyinka, O. O., Oritogun, K. S., Amballi, A. A., Olorundami, B. O., & Ogundahunsi, O. A. (2020). Effects of occupational exposure to VAT dye on fertility and pregnancy outcome of women dye users. *Arch Basic Applied Medicine*, *8*, 21–25.

Sripriya, L., Vijayalakshmi, M., Sumathy, R., & Sharmila, J. (2014). The impact of textile dyes on the biochemistry and histology of liver, a fresh water fish, tilapia, *Oreochromis mossambicus*. *Journal of Pharmaceutical and Biosciences*, *5*, 271–298.

Statista. (2019). https://www.statista.com/statistics/268184/leading-import-countries-worldwide/

Stiborova, M., Frei, E., & Schmeiser, H. H. (1992). Comparison of cytochrome P-450- and peroxidase-mediated activations of carcinogenic azo dyes and N-nitrosamines. *General Physiology and Biophysics*, *11*, 489–498. PMID:1291449

Suryavathia, S., Sharma, S., Sharma, S., Saxena, P., Pandey, S., Grover, R., & ... (2005). Acute toxicity of textile dye wastewaters (untreated and treated) of Sanganer on male reproductive systems of albino rats and mice. *Reproductive Toxicology (Elmsford, N.Y.)*, *19*(4), 547–556. doi:10.1016/j.reprotox.2004.09.011 PMID:15749270

Suteu, D., Zaharia, C., & Malutan, T. (2011). Removal of orange 16 reactive dye from aqueous solutions by waste sunflower seed shells. *Journal of the Serbian Chemical Society*, *76*(4), 607–624. doi:10.2298/JSC100721051S

Tufekci, N., Sivri, N., & Toroz, I. (2007). Pollutants of textile industry wastewater and assessment of its discharge limits by water quality standards. *Turkish Journal of Fisheries and Aquatic Sciences*, *7*, 97–103.

Umbuzeiro, G. A., Freeman, H. S., Warren, S. H., Oliveira, D. P., Terao, Y., Watanabe, T., & ... (2005). The contribution of azo dyes to the mutagenic activity of the Cristais river. *Chemosphere*, *60*(1), 55–64. doi:10.1016/j.chemosphere.2004.11.100 PMID:15910902

US Congress. (1985). Evidence for workplace hazards to reproductive function. In *Reproductive health hazards in the workplace* (p. 93). Office of Technological Assessment.

Vajnhandl, S., & Le Marechal, A. M. (2005). Ultrasound in textile dyeing and the decolouration/mineralization of textile dyes. *Dyes and Pigments*, *65*(2), 89–101. doi:10.1016/j.dyepig.2004.06.012

Veulemans, H., Steeno, O., Masschelein, R., & Groeseneken, D. (1993). Exposure to ethylene glycol ethers and spermatogenic disorders in man: A case-control study. *British Journal of Industrial Medicine*, *50*(1), 71–78. doi:10.1136/oem.50.1.71 PMID:8431394

Walsh, F. C. (2001). Electrochemical technology for environmental treatment and clean energy conversion. *Pure and Applied Chemistry*, *73*(12), 1819–1837. doi:10.1351/pac200173121819

Wang, L., Barrington, S., & Kim, J. W. (2007). Biodegradation of pentylamine and aniline from petrochemical wastewater. *Journal of Environmental Management*, *83*(2), 191–197. doi:10.1016/j.jenvman.2006.02.009 PMID:16678965

Wang, Y. Y., Tawfik, O., & Wood, G. W. (1998). Endotoxin-induced abortion in mice is mediated by activated fetal macrophages. *Journal of Leukocyte Biology*, *63*(1), 40–50. doi:10.1002/jlb.63.1.40 PMID:9469471

Wong, E. Y., Ray, R. M., Gao, D. L., Wernli, K. J., Li, W., Fitzgibbons, E. D., & et al., (2009). Dust and chemical exposures, and miscarriage risk among women textile workers in Shanghai, China. *Occupational and Environmental Medicine*, *66*(3), 161–168. doi:10.1136/oem.2008.039065 PMID:18805889

Wong, M. (2008). *Development and Trends in the Chinese Garment and Textile Industry. MFA+3 Labour rights in a changing garment industry*. https://en.archive.maquilasolidarity.org/ sites/maquilasolidarity.org/files/GM-MFA3-TrendsofChineseTextileIndustry-2008-09-28.pdf

Zaharia, C., Suteu, D., Muresan, A., Muresan, R., & Popescu, A. (2009). Textile wastewater treatment by homogeneous oxidation with hydrogen peroxide. *Environmental Engineering and Management Journal*, *8*(6), 1359–1369. doi:10.30638/eemj.2009.199

Zhou, S. Y., Liang, Y. X., Chen, Z. Q., & Wang, Y. L. (1988). Effects of occupational exposure to low-level carbon disulfide (CS2) on menstruation and pregnancy. *Industrial Health*, *26*(4), 203–214. doi:10.2486/indhealth.26.203 PMID:3230053

Chapter 4
Chronodisruption and Loss of Female Reproductive Potential Due to Shift Work

Chandana Haldar
Department of Zoology, Banaras Hindu University, India

Jayita Pal Chowdhury
Department of Zoology, Banaras Hindu University, India

ABSTRACT

Entrainment of 24 hrs light/dark cycle is nowadays getting altered in corporate sector and even at home due to over exposure to artificial light at night (ALAN) and is commonly denoted as chronodisruption. Chronodisruption interferes with the daily physiology that is regulated by SCN/pineal-melatonin/Hypothalamo-Pituitary axis. In females, reduction in sleep due to shift work desynchronizes HPG-HPA axis induces a stress like condition, accompanied by increased free radical generation in ovary leading to polycystic ovaries, depletion of ovarian reserve, luteal phase defects, endometriosis, implantation failure, etc. Evidence suggest that there is remedial rescue by management of circulatory melatonin due to its chronobiotic/antioxidant/anti-inflamatory/antiapoptotic potentials and its role in ovarian folliculogenesis and steroidogenesis. Advancement in melatonin research suggests that it could be beneficial for commercial use, that is, (1) as an endogenous synchronizer, (2) to improve the quality of oocyte, and (3) for promoting success in the frequency of IVF.

INTRODUCTION

Movement of the earth brings predictable daily and seasonal environmental challenges. The mammalian physiology was evolutionarily programmed to a 24 hour light/dark cycle and the environmental photoperiod (being seasonal) is the absolute factor that governs it (Kennaway et al., 2012). The extra ordinarily rampant use of artificial light at night (ALAN) or light at night (LAN) truncates/manipulates the duration of darkness (as an exposure to artificial light after sunset or before sunrise) and also by

DOI: 10.4018/978-1-7998-4480-8.ch004

interrupting the darkness at night. Thus, the natural entrainment of light/dark cycle is nowadays getting altered due to longer exposure to ALAN condition not only in corporate areas but also at home due to excessive usage of light emitting electronic devices. ALAN not only has widespread ecological effects like changing patterns of breeding seasons in wild, semi-wild and domestic animal species (Russart & Nelson, 2018)but also has increased the susceptibility of humans to a number of physiological and reproductive complications leading to disorders/diseases.

The 24-hr rhythms in mammalian physiology are orchestrated by a primary clock in the suprachiasmatic nucleus (SCN) of the hypothalamus. This central circadian system is extremely sensitive and perceives the briefest exposure to light within the 24 hour cycle. Perception of light through lateral eyes, initiates an impulse at the retinal receptors that is transmitted via retino-hyothalamic tract to SCN- the master clock. This impulse alters the expression of a self-sustaining feedback loop of genes within the SCN that stimulates rhythmic production of a small indole-amine, melatonin from the pineal gland. The mammalian clock basically comprises of the interacting positive and negative transcriptional-/translational-feedback loops. This core clock loop involves commonly known clock genes *viz.* Clock, Bmal1, Per1, Per2, Per3, Cry1 and Cry2. The two transcription factors CLOCK (circadian locomotor output cycles kaput)(paralog to neuronal Per-Arnt-Sim (PAS) domain containing protein; NPAS2) and BMAL1 (brain and muscle arylhydrocarbon receptor nuclear transporter-like protein 1) (Ko& Takahashi, 2006) constitute the positive arm of the core clock loop.Protein products of the *period* genes, *per1, per2* and *per 3* (Per1,Per2 and Per3 respectively) and *cryptochrome* genes (*cry1*-CRY1 and cry2- CRY2) form the negative elements of the molecular circadian clock (Saha et al., 2019). The initiation of cellular clock function takes place with the hetero-dimerization of CLOCK and BMAL1, translocation of this heterodimer from the cytosol to the nucleus where it binds to the E-box cis-regulatory enhancer sequences in the promoters of per and cry genes, leading to the transcription of the elements of the negative arm of the loop (Husse et al., 2015). This increases the cytoplasmic level of PER1, PER2 and PER3, each protein reaching its peak level at different circadian time points (Jung et al., 2003). The PER and CRY proteins undergo post translational modifications like phosphorylation by Casein kinases, get accumulated in adequate amount and heterodimerize with each other. They get translocated to the nucleus and inhibit the transcription of their respective genes by blocking transcriptional activity of CLOCK/ BMAL enhancer complex (Tosini et al., 2008). As a result, the levels of PER (PER1, PER2, and PER3) and CRY (CRY1 and CRY2) proteins in the cytoplasm get depleted which then leads to removal of the inhibition of transcription of clock and bmal1 genes by the heterodimers of PER and CRY that cyclically causes CLOCK and BMAL-1 mediated stimulation of *per* and *cry* genes (Barnes et al., 2003). The entire process of nuclear translocation of the repressor complex, brought about by concomitant phosphorylation and protein degradation spans over approximately 24 hours, thus defining the period of the oscillator or a single oscillation (Sellix & Menaker, 2011). Additionally, a secondary loop of interlocking transcriptional regulators comprising of a repressor REV-ERBα and an enhancer- retinoic acid-like orphan receptor α(RORα) regulates BMAL1 expression (Sellix, 2015). The core clock genes also control output genes. Also known as clock-controlled genes, the clock output genes are the effector molecules of the clock that mediate temporal control of the core clock over cell-type–specific gene expression and physiology (Hamada et al., 2004; Kennaway et al., 2003).

While constant exposure to bright light can disrupt or completely abolish circadian rhythms, bright or reduced light levels are equally disruptive (Fonken & Nelson, 2014). Studies have found that a brief pulse of light can transiently induce expression of Period 1 (Per1) and alter the phase of the molecular clock SCN. A single 30 minute pulse of light during the dark phase sufficiently activates the neurons of the SCN in Siberian hamsters (Bedrosian et al., 2013). The SCN coordinates other central circadian oscillators (e.g., the hypothalamus and pituitary gland) that drive rhythms in peripheral tissues like ovary and uterus, but also other endocrine organs such as the adrenal, thyroid, adipose tissue etc. that secondarily regulates female reproductive physiology (Sáenz de Miera et al., 2017) This primary or circadian clock (SCN) is thus an integral and regulatory part of the female/male reproductive system. The photic input pathway from the retina to the SCN drives pineal melatonin secretion at night. This melatonin signal is sculpted by photoperiod (or precisely, the scotoperiod/ dark phase of the light dark cycle) and provides an internal endocrine representation for external photoperiod.

Light at night was first introduced as an endocrine disruptor by Russart and Nelson (2018). LAN/ ALAN mediated chronodisruption affects the important neuroendocrine rhythms of the body, especially that of thehypothalamo-pituitary-gonadal (HPG) axis. In females, ovarian activity display a regular estrous/menstrual cycle driven mostly by HPG axis that gets modified by environmental photoperiod, that is evident from studies in seasonal breeders (Simonneaux, 2015). Occupational over-exposure to light or reduction in sleep duration due to the shift work is responsible for the desynchronization of normal hormonal rhythm especially in females, as the production and action of the reproductive hormones is cyclic. Reproductive areas of brain in females clearly suggest that there is a neural timing system that drives and orchestrates neuroendocrine events ultimately leading up to the pre-ovulatory luteinizing hormone (LH) surge and ovulation. The SCN has been shown to be necessary for normal functioning of the hypothalamic pituitary gonadal (HPG) axis, and rhythms of clock-gene expression have been recorded in brain region controlling both the HPG and hypothalamic-pituitary-adrenal (HPA) axes of the females(Gamble et al., 2013). The earliest and best described circadian oscillator in the HPG axis is the pituitary gland where the rhythmic gene expression of prolactin and gonadotropin releasing hormone (GnRH) receptors has been shown to be mediated via clock-gene regulatory elements (Sellix & Menaker, 2011). Pineal gland dysfunction/desynchronization was linked to the definite proof that melatonin inhibits pituitary GnRH(Carter et al., 2014). In each of these areas, presence of melatonin receptors (MT1/MT2) has been documented proposing a direct chronobiotic action melatonin in female reproduction. Mammalian melatonin synthesis peaks between midnight and 0400 hours (varying among species from tropical/temperate zone) and exposure to light during this time inhibits melatonin secretion for the entire night. Thus, exposure to light outside the normal light-dark cycle not only has direct effects on the normal body physiology but also disrupts the benefits of a normal melatonin rhythm. Very short term LAN is capable of disrupting the rhythmic expression of three essential clock genes, Per1, Per2, and Cry2 in SCN and also in peripheral clocks(Fonken et al., 2013).

Experimental evidences suggest that melatonin can be implemented as a possible remedy towards chronodisruption induced alterations in physiology, not only by its virtue as a chronobiotic/endogenous 'time keeper but also due to its popularity as an immunomodulatory, anti-inflammatory, anti-apoptotic and free radical scavenging agent (Loren et al., 2017; Tan et al., 2003, 2015a). ALAN/LAN exposure strongly interferes with the internal melatonin rhythm and level thus curbing its multifarious potentials. Reports also suggest that circulatory melatonin level and its receptor regulates folliculogenesis and steroidogenesis in ovary(Sellix & Menaker, 2011). Melatonin is proposed to regulate GnRH secretion in females via two complementary mechanisms: (a) change in the steroid negative feedback on GnRH release and (b) direct steroid-independent modulation of GnRH secretion(Yoshikawa et al., 2009). On the other hand, direct action of melatonin on ovarian function in vitro and in vivo strongly supported the rescuing potentials of melatonin from oxidative insult (oxidative stress) under different stressful conditions(Volt et al., 2016).

As hormones mediate individual responses to changing environment, the endocrine system might be one of the first system affected, as well as being the first line of defence to ameliorate any negative impact. In this regard, advancement in melatonin research has highlighted the fact that it could be a protector/rescuer of ovary from stress/pathological conditions and could be beneficial for clinical trials. In addition, reproductive hazards of females, disturbed pregnancy issues have been associated mostly with working nights or rotating shifts and have been documented to have an increased risk of irregular menstrual cycles, endometriosis, infertility, miscarriage, low birth weight or pre-term delivery, and reduced incidence of breastfeeding. As duration of light/dark regulates the presence and level of melatonin in circulation that further regulates HPG axis and many other physiological functions, therefore, the regulation of level of melatonin could also be a remedy of disruption of neuroendocrine function of females (Bhatti et al., 2013).

ALAN/LAN AS ENDOCRINE DISRUPTOR

There is always increasing evidences for a role of ALAN and LAN in human health, including disruption of circadian regulation and melatonin signalling, metabolic dysregulation, cancer risk, and disruption of other hormonally-driven systems (HPG axis). These effects are not limited to humans; domesticated

Table 1. Impact of chronodisruption on female reproductive physiology under different clinical and experimental conditions.

Clinical / Experimental condition	*Species*	*Physiological Manifestation*	*Source*
Shift work (permanent and rotating)	Human	Changes in rhythmic hormone secretion *e.g.* melatonin, cortisol	Weibel et al., 1997 Hennig et al., 1998 Boivin and James, 2002
Light exposure during sleep	Human	Higher body weight, waist circumference, BMI, obesity	McFadden et al., 2014
Night shift work in nurses (undifferentiated)	Human	Shortening or lengthening of menstrual cycles, irregular menstrual cyclicity, amenorrhea, dysmenorrhea, scanty or heavy menstrual flow *etc.*	Uehata and Sasakawa, 1982 Labyak et al., 2002 Chung et al., 2005 Wan and Chung, 2012
Shift work during pregnancy	Human	Increased risk of preterm birth, miscarriage	Bonde et al., 2013, Fernandez et al., 2016
Shift work/ rotating shift work	Human	Obese individuals having magnified risk ofendometrial cancer, fatal ovarian cancer	Vishwanathan et al., 2007 Carter et al., 2014
Exposure to ALAN	Mice Rat	Impaired glucose tolerance, increased body mass	Fonken et al., 2010 Borniger et al., 2014 Figueiro et al., 2017 Opperhuizen et al., 2017
Constant light exposure	Mice	Increased LH, FSH, E2, decreased P4 Decrease in ovarian follicle number by increasing apoptosis and DNA damage of ovarian cells	Li et al., 2019
Prolonged exposure to light during pregnancy	Mice	Higher rates of embryo resorption and poor pup survival	Endo and Watanabe, 1989
Chronic jet lag models	Rat	Pups develop metabolic dysfunctions like obesity, hyperinsulinemia,	Varcoe et al., 2011

animals as well as wildlife are also exposed to LAN, and at risk for disrupted circadian rhythms that support the role of LAN as an endocrine disruptor in female reproductive physiology (Table 1.) and to be considered in treatments and lifestyle suggestions. ALAN or LAN leads to a stress like condition and is responsible for increased circulatory level of corticosteroids (Gómez-Acebo et al., 2015) that are widely accepted as inhibitors of reproduction, especially in females.Stress is also accompanied by generation of free radicals (Reactive Oxygen Species-ROS) due to mitochondrial dysfunction. Together, all these factors in female lead to varieties of reproductive ailments/condition like polycystic ovaries, luteal phase defects, depletion of ovarian reserve, endometriosis, pre- and post-pregnancy related problem, ovary/uterine cancer etc. (Table 1).

CHRONODISRUPTION:- REFLECTIONS IN FEMALE REPRODUCTIVE SYSTEM

The ovulatory cycle comprising of folliculogenesis and steroidogenesis is critical factor in female reproductive success. Chronodisruption has a potential impact on the normal follicular development and has been found to lower the antral follicles count (Eskew et al., 2017). Sleep cycle disturbances in females demonstrated by variation in mid-cycle sleep point has been shown to disrupt circadian rhythms and subsequently melatonin secretion. Melatonin is a powerful antioxidant that has a protective effect at the level of the ovary. Chronodisruption may contribute to decreased nocturnal melatonin secretion, leading to increased oxidative stress within the ovary thus may reduce mature/antral follicle no. This is one probable mechanism contributing to the negative impact of chronodisruption on antral follicle count as a marker of ovarian reserve(Reiter et al., 2014).

Constant Light Exposure, Sleep Deprivation and PCOS

Polycystic ovary syndrome (PCOS) is a complex endocrine and metabolic disorder associated with ovulatory dysfunction, abdominal obesity, and infertility and it affects 5%–10% of women of reproducing age (Franks, 1995) but the pathogenesis is yet to be completely clarified. Earlier investigations have reported occurrence of thinner endometrium and relatively small uteri in women with PCOS and oligoamenorrhea (Leonhardt et al., 2012). Sleep deprivation/restriction has become widespread in modern society. Poor sleep habits or negligence of sleep hygiene/quality may be an important lifestyle factor in pathogenesis of PCOS. Many women with PCOS have psychological problems and sleep disorders that significantly reduce sleep quality (Shreeve et al., 2013). Constant light may lead to partial sleep deprivation and low melatonin level. Physiological stress responses and elevated cortisol response in the hypothalamic-pituitary-adrenal (HPA) axis occurs during sleep deprivation (Minkel et al., 2014). The risk for obstructive sleep apnea (OSA) is higher in obese women with PCOS(Mokhlesi et al., 2012) that further affects sleep quality.

Exposure to constant light leads to inhibition of the pineal gland functions while pinealocyte cell activity increases when exposed to constant darkness. Melatonin plays a static role in the maintenance of proper follicular function and is thus important for ovulation and progesterone production (Maganhin et al., 2013). Constant light induces continuous estrous to anovulatory (CEA) syndrome and blockage of pineal gland activity in rodents. Chronic treatment with melatonin is able to overcome the anovulatory state in rats in approximately 70% cases. The luteinizing effect of melatonin can significantly be counteracted by feeding the animals with a tryptophan-poor diet or by administering methiothepin (a blocker of central serotoninergic receptors (Turek et al., 2005) as melatonin elicits luteinization in constant light induced anovulatory rats through the brain serotoninergic system (Trentini et al., 1978). Further, exposure to constant light regimens accelerates age-related inhibition of the estrous function in females (Anisimov et al., 2013). Researchers have shown that the exposure of rats to continuous light can induce PCOS. The LH surges that trigger ovulation are controlled by cyclic light-dark photoperiods. A disturbance of the balanced light-dark cycles within a 24-hour period by constant exposure to light thus disrupts normal cycling and inhibits ovulation (Weber & Adler, 1979) leading to anovulation. In fact, exposure of female rats to a constant light environment was developed as an alternative approach for inducing PCOS (Singh, 1969;Baldissera et al., 1991). Rats under continuous light environment exhibit both ovarian and uterine augmentation; with disordered estrous cycle and polycystic ovary-like changes, accompanied with fur loss, lethargy and high serum testosterone suggesting that continuous light can lead to the occurrence of PCOS in female rats.

Ovarian Development/apoptosis

Constant light exposure induces constant estrous anovulatory (CEA) syndrome and blockage of pineal gland activity. Constant light at first alters the serum sex hormone levels by increasing FSH, LH and E2 level. This significantly decreases progesterone level and hence, increases the risk of apoptosis due to increased DNA damage in ovarian cells. It also decreases the peripheral clock gene expressions in ovarian tissue (Li et al., 2019). Experiments suggest that genotoxic stress is generated due to long light resulting in DNA damage and apoptosis in reproductive tissues (Li et al., 2019). A significant decrease of catalase activity was compensated by an increase in the amplitude of circadian rhythms of activity of this enzyme in animals by photoperiod. The presence of light rhythm of a smaller amplitude and shorter period does manifest a distinctive feature of daily fluctuations of activity of antioxidant enzymes in ovaries of rats (BehramKandemir et al., 2017; Wilking et al., 2013). ALAN has been considered as a potential cause for mitochondrial dysfunction and thereby production of ROS. It can be predicted that the daily rhythms of activity of antioxidant enzymes are altered in dependence from surrounding light (LAN/ALAN) conditions for regulation of reproductive function.

Figure 1. Schematic representation of effect of chronodisruption on female fertility.

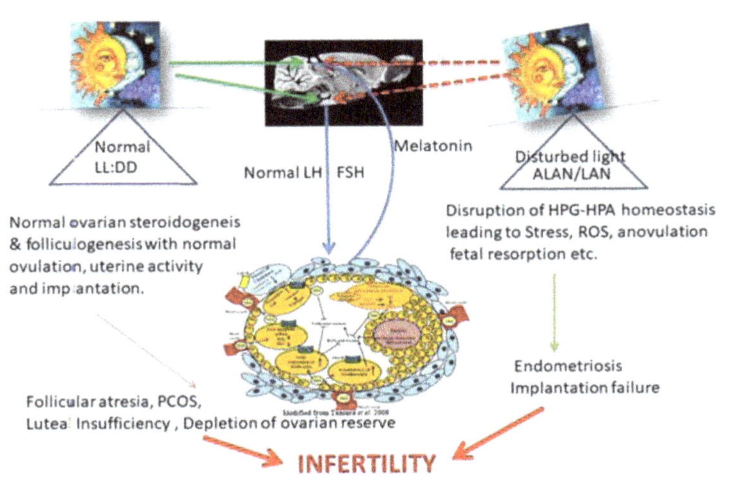

Ovarian ROS in Female Fertility

An increase in the number of atretic follicles after pinealectomy was demonstrated in mammalian ovaries(Soares et al., 2003). Although the exact mechanism was not understood, the cause was attributed to absence of the free radical scavenging attribute of melatonin. Ever since, multiple reports have emerged on the direct effect of ROS and melatonin on oocyte maturation (Tamura et al., 2013). The importance of ROS has been further established in a study conducted on cows with cystic ovarian disease (COD) showed that the follicular fluid content of ROS was significantly lower than healthy cows (Rizzo et al., 2009a). The anouvulatory conditions that lead to cyst formation in such animals may be attributed to the inhibition of follicular rupture due to insufficient ROS concentrations that failed induce the apoptosis of the cells of the follicular wall, thus inhibiting ovulation (Rizzo et al., 2009a). A sharp increase in concentrations of ROS was detected in oestrus phase when compared with the different phases of the oestrous cycle in female canids(Rizzo et al., 2009b).

Experiments suggest that not only animals but human being also become stressed and anxious as a result of major changes in their environment. Understanding of the role of ROS in female infertility is still incomplete. Women with idiopathic infertility had higher concentrations of malondialdehyde and lower antioxidant levels in the peritoneal fluid than the fertile women (Polak et al., 2001). The results indicate that the pro-oxidant/antioxidant balance is disturbed in infertile women, which may result in oxidative stress and damage to oocytes or the zygote. The contact of the ovary and the fallopian tubes with the peritoneal fluid maximize the risk of oxidative damage caused by the ROS present. The fact that oxidative and antioxidant systems are present in various female reproductive tissues suggests that infertility and certain reproductive disease such as endometriosis and hydrosalpinx may be caused, at least in part, by oxidative stress(Murphy. et al., 1998a,b; Bedaiwy, 2002a,b;Agarwal et al., 2003).Peritoneal fluid in women with endometriosis and unexplained infertility had been observed to have higher ROS concentrations that negatively affected sperm function(Oak et al., 1985; Wang et al., 1997). Here, ROS may be produced by red blood cells, macrophages, endometrial cells and debris from menstrual reflux which is also implicated in the pathogenesis of endometriosis. Activation of macrophages in patients with endometriosis is associated with an increase in scavenger receptor activity (SCRs) (Murphy et al., 1998a,b) and also responsible for increased production of ROS in patients with endometriosis(Zeller et al., 1987;Murphy et al., 1998a,b). Oxidative modification of substances in peritoneal fluid and ectopic endometrial tissue have been found to be higher which further supports the theory that ROS plays a role in pathology of endometriosis (Van Langendonckt et al., 2002).

Patients with endometriosis have also been reported to have increased auto-antibodies to modified proteins due to oxidative stress. The expression of defensive antioxidant enzymes such as SOD and glutathione peroxidase is altered in endometrial tissue in these women. Although in a few studies a difference in peritoneal ROS concentrations between healthy women and those with endometriosis was not detected (Wang et al., 1997; Bedaiwy et al., 2002a), it is proposed that that elevated ROS concentrations may be a local phenomenon that occurs at the site of disease and may not be detectable in peritoneal fluid (Van Langendonckt et al., 2002).

Light at Night Increases Risk of Oncogenesis

Ovarian Cancer

The biological mechanism underlying association of abnormal light via irregular circulatory melatonin and ovarian malignancies are likely complex. Population based studies revealed an intimate relationship between light at night and an increased risk for cancers in multiple organs including ovary of females of different age groups in women undergoing shift work (Carter et al., 2014). In today's social scenario, when the presence of LAN is indispensable and unavoidable, the risk is increasing alarmingly.

Breast Cancer

Observational studies report a higher risk of breast cancer(Megdal et al., 2005) among female night workers. Night shift work decreases serum melatonin levels, which in turn may enhance tumorigenesis has consistently been suggested by animal and in vitro studies(Schernhammer & Schulmeister, 2004). In addition to its potential anti- estrogenic activities(Stevens, 1987), melatonin has been shown to modulate aromatase activity in mammary tumors(Cos et al., 2005) women. Investigating breast cancer risk among females chronically exposed to LAN showed an average increase of about 50%(Jia et al., 2013; Megdal et al., 2005) A significant increase in risk of breast cancer was observed in nurses who had worked at least three nights per month over a period of 30 years, the risk increasing as much as 36% with the number of years spent working night shifts(Lie et al., 2007; Schernhammer et al., 2003) Increased in endometrial and ovarian or fatal ovarian cancers were noted in similar cases of long term shift work in various studies across the world (Carter et al., 2014). Women who worked rotating night shifts of at least three consecutive nights for 15 years had an increased risk of colorectal cancer, but a recent study indicates that short duration of sleep might be the contributing factor(Schernhammer et al., 2003; Thompson et al., 2011).

The relationship between melatonin and prevention of cancer was proposed as early as 1978, where melatonin was considered for treatment of breast cancer(Cohen et al., 1978). Later, the tumour enhancing effects of pinealectomy (Tamarkin et al., 1981), and its prevention by melatonin administration was extensively studied(Anisimov et al., 1997; Cini et al., 1998; Cos et al., 1998)as both MT1 and MT2 receptors were localized in human breast cancer xenografts. The LAN induced suppression of melatonin rhythm hence is the most likely cause for the cases discussed above. Exposure to diffused LAN amplifies inflammatory responses thereby challenging the anti-inflammatory action of melatonin. Inflammatory cytokines such as IL-1, IL-6 accelerate tumour progression (Luboshits et al., 1999; Tricot, 2000; Vidal-Vanaclocha et al., 2000)and thus the heightened inflammatory response due to LAN exposure provide a suitable environment for tumorigenesis and cancer. In this aspect, the anti-oestrogenic role of melatonin cannot be ignored especially in cases of oestrogen dependent cancers. Melatonin interacts with estrogen receptor α(Blask et al., 2005) decreases signal of estrogen. Decreased melatonin and an increased estradiol concentration were detected in women engaged in shift work(Schernhammer & Schulmeister, 2004) while a significant inverse relationship was found between plasma melatonin concentrations and estrogen-receptor positive breast cancer(Danforth et al., 1985). Hence, increased endogenous signalling of estrogens and a perpetually high inflammatory condition is the most possible mechanism in LAN induced melatonin suppression in hormone-dependent tumour progression.

Endometrial cancer

Shift work involving circadian disruption was accepted as carcinogenic(Lie et al., 2011). Melatonin has several oncostatic properties, including possible anti-estrogenic and anti-aromatase activity, and seems to be linked with fat metabolism/obesity. Night workers have lower levels of melatonin, which may predispose them to develop cancer. Endometrial cancer risk is influenced significantly by hormonal and metabolic factors; therefore, it has been hypothesized that female night workers may have an increased risk of endometrial cancer(Viswanathan et al., 2007). Study suggests that of the women provided with rotating night shift work developed invasive endometrial cancer when controlling for endometrial cancer risk factors. Women of younger age had a significantly increased risk of endometrial cancer. Cross-sectional studies have shown lower melatonin levels in women with endometrial cancer (Hussain et al 2007). An MT2 melatonin receptor subtype that may mediate the cancer-protective effect of melatonin has been described in a human endometrial cancer cell line (Kobayashi et al., 2003)suggesting the first evaluation of the relationship between night work and endometrial cancer risk in a large prospective cohort of pre- and postmenopausal females. Increased in endometrial and ovarian or fatal ovarian cancers were noted in cases of long term shift work in various studies across the world(Carter et al., 2014; Zhao et al., 2016). Women who worked rotating night shifts of at least three consecutive nights for 15 years had an increased risk of colorectal cancer, but a recent study indicates that short duration of sleep might be the contributing factor(Schernhammer et al., 2003; Thompson et al., 2011). Women working rotating night shifts for a long duration have a significantly increased risk of endometrial cancer, particularly if they are obese.

Influence of Light Shift on Pregnancy

Night work during pregnancy may confer an increased risk of miscarriage due to alteration in the normal 24 hour maternal rhythm. Epidemiological studies have found that night work during pregnancy is associated with an increased risk of preterm birth (Bonde et al., 2013; Jin et al., 2004; Whelan et al., 2007). Recent studies demonstrate that night work is a likely risk factor for poor pregnancy outcomes including miscarriage (Fernandez et al., 2016). During pregnancy the circadian rhythmicity of HPA axis is consistent with circadian regulation of the clock-controlled plasminogen activator inhibitor 1 (important for embryonic development), uterine circadian clock-gene regulation, and maternal placental circadian clock-gene regulation in rodent models (Gamble et al., 2013). Pregnant mice, when housed on a 22- or 26-h day conditions (prolonged exposure to light, outside the normal photoperiodic entrainment period) results in a higher incidence of embryo resorption and poor pup survival (Endo & Watanabe, 1989). Chronic jet lag models of mice (repeated 6-h phase advances or delays) exhibited a significant decrease in the number of full-term pregnancies after successful mating (Summa et al., 2012). Studies have also explained an important role for circadian regulation of prolactin in maintaining pregnancy in rodents (Gamble et al., 2013). Mice with mutations in the Clock gene have insufficient prolactin release, suggesting that disruption of the circadian clock could impact pregnancy outcome through dysregulation of prolactin (Williams &Kriegsfeld, 2012).

Adult mammals are in direct contact with the photic environment, and translate this signal via the hormone melatonin, to time their own changes in physiology and behaviour (Reiter, 1993). Fetal SCN and eyes are non-functional and their responses towards light are controlled by the maternal circadian system by a phenomenon called 'maternal photoperiodic programming' (MPP) (Sáenz de Miera et al., 2017; Varcoe et al., 2018). Destruction of maternal SCN has been found to abolish fetal hormonal rhythms (Reppert& Schwartz, 1986). Melatonin synthesized in maternal pineal is the key player that regulates circadian system in fetus and newborn offspring, in accordance to the light conditions affecting the mother (Endo & Watanabe, 1989; Reppert& Schwartz, 1983) These signals about light regimes are transferred to the fetus from the pregnant female with melatonin that passes through the placental barrier during embryonic development (Iwasaki et al., 2005; Tan, et al., 2014) and breast milk during the neonatal period (Engler et al., 2012).Thus, maternal photoperiodic programming is a useful paradigm for understanding how in utero programming of hypothalamic function leads to life-long effects on growth, reproduction, health and disease in mammals, including humans (Summa et al., 2012).

Circadian dysregulation during pregnancy may not only affect the health of the mother, but may also affect the health of the pup even after maturity. For example, rats that underwent chronic jet lag (12-h shifts) during pregnancy produced pups with metabolic problems such as obesity, hyperleptinemia, and modulation of glucose tolerance/insulin insensitivity (Gamble et al., 2013). Overall, the link between shift work and pregnancy outcome is consistent with findings in mice with genetic knockout of the Clock gene. Only 57% of these mutant mice deliver viable pups and the rest either reabsorb their pups or go into a long, extended labor that does not end in delivery(Miller et al., 2004)

Light Influences on Metabolism/Obesity and Female Fertility

Chronodisruption has been suggested to cause obesity by causing metabolic derangements. Clock-gene mutant mice exhibit reduction in insulin production, impaired glucose tolerance, and develop obesity(Marcheva et al., 2010; Turek et al., 2005; Zvonic et al., 2006), hypoinsulinemia and diabetes (Marcheva et al., 2010)leading to chronic ailments of the reproductive system that is more evident in females. The most probable pathway that contributes to metabolic dysregulation that resulting from circadian disruption is the neuropeptides that regulate hunger centre in the hypothalamus. Orexin, secreted by neurones in the lateral hypothalamus promotes wakefulness and food-seeking behaviour in animals in response to ghrelin signalling(Yamanaka et al., 2003), The SCN is also connected to the orexin neurones via indirect pathways, and ablation of the SCN abolishes orexin rhythmicity(Zhang et al., 2004). In humans, chronic partial sleep loss/sleep disorders have been linked with obesity and metabolic syndrome (Garaulet et al., 2010). Obese women working rotating night shifts doubled their baseline risk of endometrial cancer compared with obese women who did no night work (De Bacquer et al., 2009). Based on these studies, it has been proposed that the increased risk is attributable to the effects of melatonin on hormonal and metabolic factors. Melatonin further seems to play an important role in fat metabolism, and increased adiposity is associated with an increased risk of endometrial cancer.

UNDERLYING MOLECULAR BASIS OF LOSS OF PREPRODUCTIVE POTENTIAL IN FEMALES DUE TO CHRONDISRUPTION

Circadian or chronodisruption primarily alters the functioning of the self-sustaining molecular clock within the central pace maker/ SCN as well as the independent peripheral clocks. Broadly, chronodisruption is desynchrony of the central and the peripheral rhythm generators that are otherwise aligned with the environmental light/dark cycle maintaining a homeostasis in the physiology. Most of the information related to the physiological significance of the different components of the molecular clock has been obtained by studies on mutant rodents. Female *Bmal1* knockout mice not only exhibit loss of the circadian rhythm of behaviour and gene expression both in the SCN and periphery, but also have failure of embryo implantation (Ratajczaket al., 2009; Boden et al., 2010). They display decreased steroidogenesis due to reduced expression of steroidogenic acute regulatory protein (Alvarez et al., 2008), abnormally long estrous cycles, but produce viable and fertilizable ova with a higher incidence of implantation failure (compared to wild types) due to reduced progesterone secretion from the corpus luteum (Ratajczaket al., 2009)

Clock gene mutant animals having altered circadian period length and impaired circadian rhythmicity have increased fetalresorption rate and non-productive labor compared to wild type animals (Goldstein & Smith, 2016). Abnormal *Clock* expression has been noted in spontaneously aborted mice fetuses (Li et al., 2015; Li & Wang,2015) while *Per* mutant animals demonstrated frequent postimplantation pregnancy loss (Pilorz&Steinlechner, 2008).Dominant negative *clockΔ19* mutant mice have long erestrous cycles with prolonged estrus phase and a reduced amplitude or complete absence of a proestrous LH surge (Miller et al., 2004; Dolatshad et al., 2006).The ovaries of such mice are morphologically normal, they regularly mate and produce fertilizable ova, but a significant proportion of embryos and full-term fetuses are reabsorbed (Miller et al., 2004). Irregular estrous cycles in *clock* mutant mice have been attributed to abnormal rhythms of vasopressin expression in SCN neurons and reduced AVP1a receptor in SCN (Miller et al, 2006). Mice with mutations in the *Clock* gene also have insufficient prolactin release, which may impact pregnancy outcome through dysregulation of prolactin (Williams and Kriegsfield, 2012). Such mutant/knockout rodents also serve as models of several metabolic dysfunctions associated with chronodisruption. Changes in behavior rhythms along with hyperphagy, weight gain, hyperinsulinemia have been seen in *Clock* mutants (Turek et al., 2005), *Cry1/ 2*(van der Horst et al., 1999; Vitaterna et al., 1999; Barclay et al., 2013) and *Reverbα/β* double knockouts (Bugge et al., 2012; Cho et al., 2012; Solt et al., 2012) while Bmal1knockouts show hyperlipidemia (Shimba et al., 2011).*Per1/2* double knockouts show leptin resistance and *Per 2* have altered lipid metabolism, hypotriglyceridemia, weight gain (Grimaldi et al.,2010; Kettner 2015; Yang et al., 2009).Experimentation on wild types has also yielded promising findings that may help uncover the exact mechanism of clock mediated chronodisruption of peripheral clocks. Constant (24-h) light exposure decreased the expression of murine ovary circadian clock-related genes (Li et al., 2019). Expression of *Rorγ, Bmal1, Cry1* and *Per1* significantly decreased after 4 days of continuous light exposure while *Rorα, Rorβ, Rorγ, Per1, Per2* and *Ck1ε* was significantly decreased after 8 and 12 days of experimentation when compared to controls maintained in a normal photoperiodic cycle (12L:12D).

In case of humans observations related to misalignment of clocks in the reproductive organs are scarce. However, plenty of evidences are available in support of an apparent disruption of peripheral clocks. Real time quantitative polymerase chain reaction analysis of lymphocytes revealed a significantly higher gene expression of *BMAL1, CLOCK, NPAS2, PERI, PER2*, and *REVERBα*, lower expression of *PER3, CRY1* and *CRY2* in shift workers beside having significantly higher 17-ß-estradiol levels in comparison to day time working nurses (Bracci et al., 2014). In human peripheral blood mononuclear cells, circadian clock-gene expression is rhythmic (Takata et al., 2002; Boivin et al., 2003; Kusanagi et al., 2004; Teboul et al., 2005) with *hPER1/2* peaking at the sleep-to-wake transition and *hBMAL1* peaking at the end of the wake period(James et al., 2007). *hPER1/2* rhythms have been shown to shift under simulated conditions of shift work (Boivin et al.,2003).A more direct observation related to chronodisruption and loss of female reproductive potential lies in the peripheral clock function in the uterus. Both the non-pregnant and gravid human uterine myometrium were noted to express *BMAL1* and *PER2*(Beesley et al., 2015).

The expression of *PER2* is rhythmic in human endometrial stromal cells and is downregulated during decidualization (Muter et al., 2015). Endometrial biopsies during the mid luteal phase from patients with recurrent miscarriagesdemonstrated an inverse correlation between PER2 transcript levels and number of failed pregnancies (Muter et al., 2015).Premature downregulation of *PER2* disrupts the decidual response which suggests that intact circadian timing of reproductive events subsequent to ovulation may be required for successful conception (Goldstein & Smith, 2016).

Fertility disorders like PCOS are often characterised by disruption of neuroendocrine and endocrine axes leading to irregular patterns and timing of ovarian steroid hormone secretion (Franks and Hardy, 2010; Nisenblat and Norman, 2009; Padmanabhan and Veiga-Lopez, 2011).Hormonal imbalances associated with such conditions may cause or be caused by altered timing of clock gene expression in peripheral tissues such as the ovary(Amaral et al., 2014) Excessively high serum androgen level is a common clinical feature of PCOS (Shi and Vine, 2012) that causes dramatic change in the timing system, particularly the timing of clock gene expression in the ovarian follicle and liver (Sellix et al., 2013). Reproductive endocrinopathies arising due to developmental programming by abnormal steroid environments (e.g., PCOS), can alter the timing of the clock in tissues associated with both reproductive and metabolic function (Goldstein & Smith, 2016).Another aspect of the molecular clock that draws attention is the homeostatic control of the cell division. The circadian oscillators are involved in the major pathways of the cell cycle *viz. Per1* and *Per 2* repress G1-S transition while *Clock* and *Bmal1* regulate G2-M transition (Matsuo et al., 2003; Fu et al., 2005; Hua et al., 2006; Borgs et al., 2009). BMAL1 has been predicted to prevent cells from entering S-phase under conditions of oxidative stress (Geyfman et al., 2012). Reports on the increased risk/susceptibility towards several forms of cancers in shift/ night workers and due to prolonged ALAN exposure further highlight an underlying molecular basis of circadian disruption and the disorders associated with it.

MELATONIN: A REMEDIAL RESCUE

The earliest studies on melatonin showed that elevated levels of melatonin during short day conditions was associated with suppression of reproductive functions hence, the melatonin was inadvertently thought to have an anti-gonadotrophic effect(Reiter et al., 2009).Interestingly melatonin concentrations detected in fluid collected from the human ovarian follicle was found to exceed those in simultaneously collected blood samples (Brzezinski et al., 1987; Reiter et al., 2009), exhibiting a 24-hour rhythm in the follicle (Rönnberg et al., 1990). These obtained values were reported to increase with follicular growth, being highest in those ready to be ovulated (Nakamura et al., 2003)allowing studies with a fresh approach with respect to the reproductive effects. The female reproductive system has, since then being an area of interest for such thoughts. Additionally, melatonin receptors have been identified in both the granulosa and theca cells along with the corpus luteum, indicate that melatonin acts directly on ovarian physiology (Tamura et al., 2012). It has not been clearly established regarding the high concentration of melatonin in ovary is actually synthesised or a derivative of the circulation. Recent studies support that the ovary as a whole-the granulosa cells, including those making up the cumulus oophorus(Amireault & Dubé, 2005; El-Raey et al., 2011), and the oocyte(Sakaguchi et al., 2013) carry the active molecular machinery for melatonin synthesis.Melatonin in the ovary also may control progesterone production by the transforming granulosa cells after ovulation(Tamura et al., 1998). The melatonin produced by the ovary like many other extra-pineal organs is totally utilized locally as antioxidant and as an autocrine or paracrine agent(Tan et al., 2003).

Melatonin, Oocyte Quality and Maturation

An increase in the number of atretic follicles after pinealectomy was demonstrated in rat ovaries (Soares et al., 2003). Although the exact mechanism was not understood, the cause was attributed to absence of the free radical scavenging attribute of melatonin. Ever since, multiple reports have emerged on the direct effect of ROS and melatonin on oocyte maturation(Tamura et al., 2008). Dose dependent addition of H_2O_2 in oocyte culture medium as a source of ROS affected the oocyte maturation in the denuded oocytes from immature mice treated with PMSG were cultured in the incubation medium with various concentrations of H_2O_2. The percentage of the mature oocytes (MII stage oocytes with a first polar body) obtained was significantly decreased by the addition of H_2O_2. Melatonin dose dependently removed the inhibitory effect of H_2O_2 on oocyte maturation (Tamura et al., 2012). The study further analyzed the intra-cellular role of melatonin, by incubating oocytes with dichlorofluorescein (DCF-DA) in presence of both H_2O_2 and melatonin. High fluorescence intensities were observed in the presence of H_2O_2, otherwise absent in those grown without H_2O_2. This fluorescence intensity was significantly decreased by melatonin treatment.

Rodriguez-Osorio et al.(2007)demonstrated the effects of melatonin on in vitro porcine embryo development by melatonin supplementation (10^9 M) that had a positive effect on the fertilization rates of inseminated porcine embryos that were cultured. While melatonin did not increase rate of blastocyst, the number of blastocyst cells in the melatonin-supplemented group was significantly higher than in the control group. Oocytes recovered from porcine follicles incubated in the medium with melatonin (10^7 M), fertilization rate, blastocyst rate and the number of blastocyst cells were significantly higher than that of the control (Pang et al., 2013).

The beneficial effects of melatonin on bovine embryo development was observed not in a low oxygen environment but in a high oxygen environment where free radicals are easily produced which is in support of thought that the effect of melatonin on embryo development is an effect of its role as an antioxidant(Loren et al., 2017). Bovine oocytes cultured *in vitro* with 10^6-10^9 M melatonin for 24 h during maturation exhibited up-regulated MnSOD and Cu-ZnSOD in cumulus cells along with decreasesd fragmentation of the oocytes (Rodrigues-Cunha et al., 2016). In murine pronuclear embryos culture melatonin present at a concentration of 10^7 M during promoted embryo development, blastocyst rate, hatching rate and blastocyst cell number. It also upregulated SOD and BCL2 and downregulated CAS3 and p53(Su et al., 2015; Wei et al., 2013). It was also shown that the interaction between melatonin and FSH maintains ultrastructural integrity and stimulates further growth of cultured caprine preantral follicles(Rocha et al., 2013). Experimental evidences available regarding the use of melatonin for assisted reproductive techniques suggest that melatonin can be used as an antioxidant to prevent ROS production during gamete/embryo handling and preservation and optimize the gamete quality and viability strengthening the ground the use of melatonin in maintenance of normal outcome of ART in terms of pregnancy, beginning with the sperm and oocyte quality, continuing with embryo implantation, and finally to fetal development and parturition.

Oral supplementation of melatonin during the ovarian stimulation for the IVF cycle has been shown to reduce intra-follicular oxidative damage and increases fertilization rates and embryo quality (Taketani et al., 2011). In a study of women with sleep disturbances undergoing IVF, treatment with 3 mg of oral melatonin at night from day 3–5 of their cycle of metaphase II oocytes and percentage of Grade-I embryos(Tamura et al., 2008). Melatonin supplementation also increased progesterone levels in patients undergoing IVF-ET, from the time of their HCG trigger throughout the luteal phase(Tamura et al., 2008). Melatonin has been found to increase serum progesterone levels in patients with luteal phase defect(Wei et al., 2013). Low concentrations of melatonin in culture media improved nuclear maturation rate of immature human MI oocytes(Wang & Dey, 2006), implantation rate and although insignificant, an increase in clinical pregnancy rate(Gao et al., 2012).

Melatonin, an Assistant in Reproduction

Experimental evidences available regarding the use of melatonin for assisted reproductive techniques suggest that melatonin can be used as an antioxidant to prevent ROS production during oocyte and embryo handling and during preservation optimizing the gamete quality and viability. Such findings have strengthened the ground for the use of melatonin in maintenance of normal outcome of ART in terms of pregnancy, beginning with gamete quality, embryo implantation, and finally fetal development and parturition.

Oxidative stress during pre-implantation period within the uterus or in the embryo culture media impairs the embryo quality thereby minimizing the chances of success (Shang et al., 2003). The use of melatonin in embryo culture *in vitro* has been found to reduce the apoptosis and oxidative stress(Tan et al., 2015b). Here, melatonin not only counteracts oxidative stress in embryo, but also has been reported to increase the rate of blastocyst formation(Takahashi, 2012). This protective effect melatonin of on the embryos is probably exerted by its capacity to reduce mitochondrial damage and apoptosis and a simultaneous increase expression of antioxidant genes. Melatonin treated blastocysts possessed a higher number of inner cell mass (ICM) and trophoectoderm (TE) cells (Moshkdanian et al., 2017). Local estradiol levels and p53 expression in uterus has been observed to increase after melatonin treatment. p53 has been proposed to be a possible downstream element of melatonin receptor MT1/MT2 activation that might regulate leukemia inhibitory factor (LIF) expression creating a better uterine micro-environment for implantation and therefore pregnancy outcome(Zhang et al., 2017). The differential expression of MT1 and MT2 receptors in rat uterus has been found to affect the cyclic rhythm of my-endometrial contraction (Morgan, 1992). Melatonin treatment when from the first IVF-ET cycle when continued through pregnancy has shown improved pregnancy outcome(Rizzo et al., 2010). The cytotrophoblast has been reported to produce melatonin locally even after establishment of the placenta suggesting its significance in maintenance of pregnancy(Iwasaki et al., 2005).

CONCLUSION

The purpose of this review is to explore the hypothesis that misalignment of reproductive hormones with environmental light/dark cycle and or sleep-wake rhythms can disrupt puberty, menstrual cycle, pregnancy and parturition. Rotating shift work schedule thus leading to sleep-wake rhythm disruption in females/pregnant females may underscore pathology of female reproductive potentiality. The role of the circadian clock is indispensible for regulation of the female reproductive physiology and it's desynchrony is a major cause of several pathological conditions. The phenomenon of chronodisruption in literature is commonly studied in photoperiodic animals while exploring potential mechanisms behind it. Important gaps in the present knowledge explaining the major impact of light on physiology/pathophysiology of reproduction may serve as intentions for future investigation. Detailed studies are necessary in order to define the actual mechanism by which shift work/ALAN induced circadian misalignment (Fig. 2) impairs reproductive capacity of females, but it also needs to identify strategies for the future that can minimize such effects.

Figure 2. Actual mechanism by which shift work/ALAN induced circadian misalignment and impairs reproductive capacity of females.

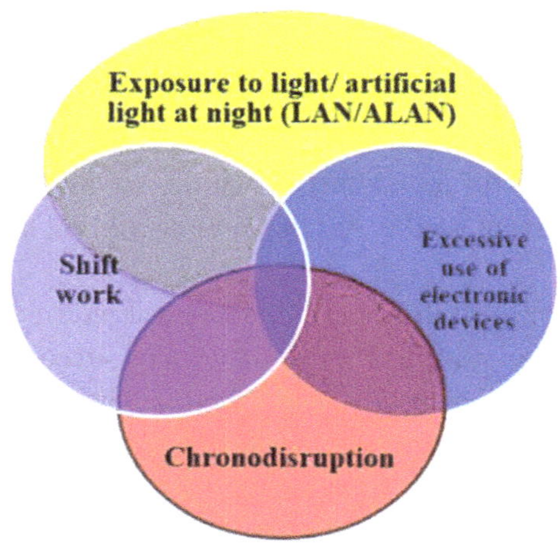

AUTHOR CONTRIBUTION

CH conceived and designed the book chapter. CH, JPC performed literature survey and wrote the book chapter.

FUND DISCLAIMER

UGC, New Delhi for BSR Faculty Fellowship to Professor Chandana Haldar; and ICMR (New Delhi) for Sr. Research Fellowship to Ms. Jayita Pal Chowdhury.

CORRESPONDING AUTHORS

Chandana Haldar

REFERENCES

Agarwal, A., Saleh, R. A., & Bedaiwy, M. A. (2003). Role of reactive oxygen species in the pathophysiology of human reproduction. *Fertility and Sterility*, *79*(4), 829–843. doi:10.1016/S0015-0282(02)04948-8 PMID:12749418

Alvarez, J. D., Hansen, A., Ord, T., Bebas, P., Chappell, P. E., Giebultowicz, J. M., Williams, C., Moss, S., & Sehgal, A. (2008). The circadian clock protein BMAL1 is necessary for fertility and proper testosterone production in mice. *Journal of Biological Rhythms*, *23*(1), 26–36. doi:10.1177/0748730407311254 PMID:18258755

Amaral, F. G., Castrucci, A. M., Cipolla-Neto, J., Poletini, M. O., Mendez, N., Richter, H. G., & Sellix, M. T. (2014). Environmental control of biological rhythms: Effects on development, fertility and metabolism. *Journal of Neuroendocrinology*, *26*(9), 603–612. doi:10.1111/jne.12144 PMID:24617798

Amireault, P., & Dubé, F. (2005). Serotonin and Its Antidepressant-Sensitive Transport in Mouse Cumulus-Oocyte Complexes and Early Embryos1. *Biology of Reproduction*, *73*(2), 358–365. doi:10.1095/biolreprod.104.039313 PMID:15858217

Anisimov, V. N., Popovich, I. G., & Zabezhinski, M. A. (1997). Melatonin and colon carcinogenesis: I. Inhibitory effect of melatonin on development of intestinal tumors induced by 1,2-dimethylhydrazine in rats. *Carcinogenesis*, *18*(8), 1549–1553. doi:10.1093/carcin/18.8.1549 PMID:9276629

Anisimov, V. N. A., Vinogradova, I. V., Panchenko, A. G., Popovich, I. A., & Zabezhinski, M. (2013). Light-at-Night-Induced Circadian Disruption, Cancer and Aging. *Current Aging Science*, *5*, 170–177. doi:10.2174/18746098112050300002 PMID:23237593

Barclay, J. L., Shostak, A., Leliavski, A., Tsang, A. H., Jöhren, O., Müller-Fielitz, H., Landgraf, D., Naujokat, N., van der Horst, G. T. J., & Oster, H. (2013). High-fat diet induced hyperinsulinemia and tissue-specific insulin resistance in Cry-deficient mice. *American Journal of Physiology. Endocrinology and Metabolism*, *304*(10), E1053–E1063. doi:10.1152/ajpendo.00512.2012 PMID:23531614

Barnes, J. W., Tischkau, S. A., Barnes, J. A., Mitchell, J. W., Burgoon, P. W., Hickok, J. R., & Gillette, M. U. (2003). Requirement of mammalian timeless for circadian rhythmicity. *Science*, *302*(5644), 439–442. doi:10.1126cience.1086593 PMID:14564007

Bedaiwy, M. A., Falcone, T., Goldberg, J., Attaran, M., Nelson, D., & Agarwal, A. (2002). Prediction of endometriosis with serum and peritoneal fluid markers: A prospective controlled trial. *Fertility and Sterility*, *17*, 426–431. PMID:11821289

Bedrosian, T. A., Vaughn, C. A., Galan, A., Daye, G., Weil, Z. M., & Nelson, R. J. (2013). Nocturnal light exposure impairs affective responses in a wavelength-dependent manner. *The Journal of Neuroscience: The Official Journal of the Society for Neuroscience*, *33*(32), 13081–13087. doi:10.1523/JNEUROSCI.5734-12.2013 PMID:23926261

Beesley, S., Lee, J., & Olcese, J. (2015). Circadian clock regulation of melatonin MTNR1B receptor expression in human myometrial smooth muscle cells. *Molecular Human Reproduction*, *21*(8), 662–671. doi:10.1093/molehr/gav023 PMID:25939854

Behram Kandemir, Y., Aydin, C., & Gorgisen, G. (2017). The effects of melatonin on oxidative stressand prevention of primordial follicle loss via activation of mTOR pathway in the rat ovary. *Cellular and Molecular Biology*, *63*(2), 100–106. doi:10.14715/cmb/2017.63.2.16 PMID:28364790

Bhatti, P., Cushing-Haugen, K. L., Wicklund, K. G., Doherty, J. A., & Rossing, M. A. (2013). Nightshift work and risk of ovarian cancer. *Occupational and Environmental Medicine*, *70*(4), 231–237. doi:10.1136/oemed-2012-101146 PMID:23343856

Blask, D. E., Brainard, G. C., Dauchy, R. T., Hanifin, J. P., Davidson, L. K., Krause, J. A., Sauer, L. A., Rivera-Bermudez, M. A., Dubocovich, M. L., Jasser, S. A., Lynch, D. T., Rollag, M. D., & Zalatan, F. (2005). Melatonin-depleted blood from premenopausal women exposed to light at night stimulates growth of human breast cancer xenografts in nude rats. *Cancer Research*, *65*(23), 11174–11184. doi:10.1158/0008-5472.CAN-05-1945 PMID:16322268

Boden, M. J., Varcoe, T. J., Voultsios, A., & Kennaway, D. J. (2010). Reproductive biology of female Bmal1 null mice. *Reproduction (Cambridge, England)*, *139*(6), 1077–1090. doi:10.1530/REP-09-0523 PMID:20200203

Boivin, D. B., James, F. O., Wu, A., Cho-Park, P. F., Xiong, H., & Sun, Z. S. (2003). Circadian clock genes oscillate in human peripheral blood mononuclear cells. *Blood*, *102*(12), 4143–4145. doi:10.1182/blood-2003-03-0779 PMID:12893774

Boivin, D. B., & James, F. O. (2002). Circadian adaptation to night-shift work by judicious light and darkness exposure. *Journal of Biological Rhythms*, *17*(6), 556–567. doi:10.1177/0748730402238238 PMID:12465889

Borniger, J. C., Maurya, S. K., Periasamy, M., & Nelson, R. J. (2014). Acute dim light at night increases body mass, alters metabolism, and shifts core body temperature circadian rhythms. *Chronobiology International*, *31*(8), 917–925. doi:10.3109/07420528.2014.926911 PMID:24933325

Brzezinski, A., Seibel, M. M., Lynch, H. J., Deng, M. H., Wurtman, R. J. (1987). Melatonin in human preovulatory follicular fluid. *Journal of Clinical Endocrinology and Metabolism, 64*(4), 865–867.

Bracci, M., Manzella, N., Copertaro, A., Staffolani, S., Strafella, E., Barbaresi, M., Copertaro, B., Rapisarda, V., Valentino, M., & Santarelli, L. (2014). Rotating-shift nurses after a day off: Peripheral clock gene expression, urinary melatonin, and serum 17-β-estradiol levels. *Scandinavian Journal of Work, Environment & Health*, *40*(3), 295–304. doi:10.5271jweh.3414 PMID:24402410

Bugge, A., Feng, D., Everett, L. J., Briggs, E. R., Mullican, S. E., Wang, F., Jager, J., & Lazar, M. A. (2012). Rev-erbα and Rev-erbβ coordinately protect the circadian clock and normal metabolic function. *Genes & Development*, *26*(7), 657–667. doi:10.1101/gad.186858.112 PMID:22474260

Carter, B. D., Diver, W. R., Hildebrand, J. S., Patel, A. V, Gapstur, S. M. (2014). Circadian Disruption and Fatal Ovarian Cancer. *American Journal of Preventive Medicine, 46*(3), S34–S41.

Cho, H., Zhao, X., Hatori, M., Yu, R. T., Barish, G. D., Lam, M. T., Chong, L. W., Ditacchio, L., Atkins, A. R., Glass, C. K., Liddle, C., Auwerx, J., Downes, M., Panda, S., & Evans, R. M. (2012). Regulation of Circadian Behavior and Metabolism by Rev-erbα and Rev-erbβ. *Nature*, *485*(7396), 123–127. doi:10.1038/nature11048 PMID:22460952

Chung, F. F., Yao, C. C., & Wan, G. H. (2005). The associations between menstrual function and lifestyle/working conditions among nurses in Taiwan. *Journal of Occupational Health*, *47*(2), 149–156. doi:10.1539/joh.47.149 PMID:15824480

Cini, G., Coronnello, M., Mini, E., & Neri, B. (1998). Melatonin's growth-inhibitory effect on hepatoma AH 130 in the rat. *Cancer Letters*, *125*(1-2), 51–59. Advance online publication. doi:10.1016/S0304-3835(97)00480-1 PMID:9566696

Cohen, M., Lippman, M., & Chabner, B. (1978). Role of pineal gland in aetiology and treatment of breast cancer. *Lancet*, *2*, 1001–1002. PMID:81980

Cos, S., Fernández, R., Güézmes, A., & Sánchez-Barcelo, E. J. (1998). Influence of melatonin on invasive and metastatic properties of MCF-7 human breast cancer cells. *Cancer Research*, *58*, 4383–4390. PMID:9766668

Cos, S., Martínez-Campa, C., Mediavilla, M. D., & Sánchez-Barceló, E. J. (2005). Melatonin modulates aromatase activity in MCF-7 human breast cancer cells. *Journal of Pineal Research*, *38*(2), 136–142. doi:10.1111/j.1600-079X.2004.00186.x PMID:15683469

Danforth, D. N. Jr, Tamarkin, L., Mulvihill, J. J., Bagley, C. S., & Lippman, M. E. (1985). Plasma melatonin and the hormone-dependency of human breast cancer. *Journal of Clinical Oncology*, *3*(7), 941–948. doi:10.1200/JCO.1985.3.7.941 PMID:4020407

De Bacquer, D., Van Risseghem, M., Clays, E., Kittel, F., De Backer, G., & Braeckman, L. (2009). Rotating shift work and the metabolic syndrome: A prospective study. *International Journal of Epidemiology*, *38*(3), 848–854. doi:10.1093/ije/dyn360 PMID:19129266

Dolatshad, H., Campbell, E. A., O'Hara, L., Maywood, E. S., Hastings, M. H., & Johnson, M. H. (2006). Developmental and reproductive performance in circadian mutant mice. *Human Reproduction (Oxford, England)*, *21*(1), 68–79. doi:10.1093/humrep/dei313 PMID:16210390

El-Raey, M., Geshi, M., Somfai, T., Kaneda, M., Hirako, M., Abdel-Ghaffar, A. E., Sosa, G. A., El-Roos, M. E. A. A., & Nagai, T. (2011). Evidence of melatonin synthesis in the cumulus oocyte complexes and its role in enhancing oocyte maturation in vitro in cattle. *Molecular Reproduction and Development*, *78*(4), 250–262. doi:10.1002/mrd.21295 PMID:21381146

Eskew, A., Reschke, L., Broughton, D. E., Schulte, M., & Jungheim, E. (2017). Chronodisruption and decreased antral follicle count. *Fertility and Sterility*, *108*(3), e323. doi:10.1016/j.fertnstert.2017.07.955

Figueiro, M. G., Radetsky, L., Plitnick, B., & Rea, M. S. (2017). Glucose tolerance in mice exposed to light-dark stimulus patterns mirroring dayshift and rotating shift schedules. *Scientific Reports*, *7*(1), 40661. doi:10.1038rep40661 PMID:28079162

Fonken, L. K., Workman, J. L., Walton, J. C., Weil, Z. M., Morris, J. S., Haim, A., & Nelson, R. J. (2010). Light at night increases body mass by shifting the time of food intake. *Proceedings of the National Academy of Sciences of the United States of America*, *107*(43), 18664–18669. doi:10.1073/pnas.1008734107 PMID:20937863

Fonken, L. K., Aubrecht, T. G., Meléndez-Fernández, O. H., Weil, Z. M., & Nelson, R. J. (2013). Dim light at night disrupts molecular circadian rhythms and increases body weight. *Journal of Biological Rhythms*, *28*(4), 262–271. doi:10.1177/0748730413493862 PMID:23929553

Fonken, L. K., & Nelson, R. J. (2014). The effects of light at night on circadian clocks and metabolism. *Endocrine Reviews*, *35*(4), 648–670. doi:10.1210/er.2013-1051 PMID:24673196

Franks, S. (1995). Polycystic ovary syndrome. *The New England Journal of Medicine*, *333*(13), 853–861. doi:10.1056/NEJM199509283331307 PMID:7651477

Franks, S., & Hardy, K. (2010). Aberrant follicle development and anovulation in polycystic ovary syndrome. *Annales d'Endocrinologie*, *71*(3), 228–230. doi:10.1016/j.ando.2010.02.007 PMID:20362969

Gao, C., Han, H. B., Tian, X. Z., Tan, D. X., Wang, L., Zhou, G., Zhu, S.-E., & Liu, G.-S. (2012). Melatonin promotes embryonic development and reduces reactive oxygen species in vitrified mouse 2-cell embryos. *Journal of Pineal Research*, *52*(3), 305–311. doi:10.1111/j.1600-079X.2011.00944.x PMID:22225541

Garaulet, M., Ordovás, J. M., & Madrid, J. A. (2010). The chronobiology, etiology and pathophysiology of obesity. *International Journal of Obesity*, *34*(12), 1667–1683. doi:10.1038/ijo.2010.118 PMID:20567242

Gómez-Acebo, I., Dierssen-Sotos, T., Papantoniou, K., García-Unzueta, M. T., Santos-Benito, M. F., & Llorca, J. (2015). Association between exposure to rotating night shift versus day shift using levels of 6-sulfatoxymelatonin and cortisol and other sex hormones in women. *Chronobiology International*, *32*(1), 128–135. doi:10.3109/07420528.2014.958494 PMID:25216206

Goldstein, C. A. & Smith, Y. R. (2016). Sleep, Circadian Rhythms, and Fertility. *Current Sleep Medicine Reports*.

Grimaldi, B., Bellet, M. M., Katada, S., Astarita, G., Hirayama, J., Amin, R. H., Granneman, J. G., Piomelli, D., Leff, T., & Sassone-Corsi, P. (2010). PER2 controls lipid metabolism by direct regulation of PPARγ. *Cell Metabolism*, *12*(5), 509–520. doi:10.1016/j.cmet.2010.10.005 PMID:21035761

Hamada, T., Antle, M. C., & Silver, R. (2004). Temporal and spatial expression patterns of canonical clock genes and clock-controlled genes in the suprachiasmatic nucleus. *The European Journal of Neuroscience*, *19*(7), 1741–1748. doi:10.1111/j.1460-9568.2004.03275.x PMID:15078548

Gamble, K. L., Resuehr, D., & Johnson, C. H. (2013). Shift work and circadian dysregulation of reproduction. *Frontiers in Endocrinology*, *4*, 1–9. doi:10.3389/fendo.2013.00092 PMID:23966978

Hennig, J., Kieferdorf, P., Moritz, C., Huwe, S., & Netter, P. (1998). Changes in cortisol secretion during shiftwork: Implications for tolerance to shift work? *Ergonomics*, *41*(5), 610–621. doi:10.1080/001401398186784 PMID:9613222

Husse, J., Eichele, G., & Oster, H. (2015). Synchronization of the mammalian circadian timing system: Light can control peripheral clocks independently of the SCN clock: alternate routes of entrainment optimize the alignment of the body's circadian clock network with external time. *BioEssays*, *37*(10), 1119–1128. doi:10.1002/bies.201500026 PMID:26252253

Iwasaki, S., Nakazawa, K., Sakai, J., Kometani, K., Iwashita, M., Yoshimura, Y., & Maruyama, T. (2005). Melatonin as a local regulator of human placental function. *Journal of Pineal Research*, *39*(3), 261–265. doi:10.1111/j.1600-079X.2005.00244.x PMID:16150106

James, F. O., Boivin, D. B., Charbonneau, S., Belanger, V., & Cermakian, N. (2007). Expression of clock genes in human peripheral blood mono nuclear cells throughout the sleep/wake and circadian cycles. *Chronobiology International*, *24*(6), 1009–1034. doi:10.1080/07420520701800736 PMID:18075796

James, F. O., Cermakian, N., & Boivin, D. B. (2007). Circadian rhythms of melatonin, cortisol and clock gene expression during simulated night shiftwork. *Sleep*, *30*(11), 1427–1436. doi:10.1093leep/30.11.1427 PMID:18041477

Jia, Y., Lu, Y., Wu, K., Lin, Q., Shen, W., Zhu, M., Huang, S., & Chen, J. (2013). Does night work increase the risk of breast cancer? A systematic review and meta-analysis of epidemiological studies. *Cancer Epidemiology*, *37*(3), 197–206. doi:10.1016/j.canep.2013.01.005 PMID:23403128

Kennaway, D. J., Varcoe, T. J., & Mau, V. J. (2003). Rhythmic expression of clock and clock-controlled genes in the rat oviduct. *Molecular Human Reproduction*, *9*(9), 503–507. doi:10.1093/molehr/gag067 PMID:12900508

Kennaway, D. J., Boden, M. J., & Varcoe, T. J. (2012). Circadian rhythms and fertility. *Molecular and Cellular Endocrinology*, *349*(1), 56–61. doi:10.1016/j.mce.2011.08.013 PMID:21872642

Kettner, N. M., Mayo, S. A., Hua, J., Lee, C., Moore, D. D., & Fu, L. (2015). Circadian Dysfunction Induces Leptin Resistance in Mice. *Cell Metabolism*, *22*(3), 448–459. doi:10.1016/j.cmet.2015.06.005 PMID:26166747

Ko, C. H., & Takahashi, J. S. (2006). Molecular components of the mammalian circadian clock. *Human Molecular Genetics*, *15*(suppl_2), R271–R277. doi:10.1093/hmg/ddl207 PMID:16987893

Kobayashi, Y., Itoh, M. T., Kondo, H., Okuma, Y., Sato, S., Kanishi, Y., Hamada, N., Kiguchi, K., & Ishizuka, B. (2003). Melatonin binding sites in estrogen receptor-positive cells derived from human endometrial cancer. *Journal of Pineal Research*, *35*(2), 71–74. doi:10.1046/j.1439-0434.2002.00691_2.x-i1 PMID:12887647

Kolbe, I., & Oster, H. (2019). Chronodisruption, metabolic homeostasis,and the regulation of inflammation in adipose tissues. *The Yale Journal of Biology and Medicine*, *92*, 317–325. PMID:31249492

Kusanagi, H., Mishima, K., Satoh, K., Echizenya, M., Katoh, T., & Shimizu, T. (2004). Similar profiles in human period1 gene expression in peripheral mononuclear and polymor-pho nuclear cells. *Neuroscience Letters*, *365*(2), 124–127. doi:10.1016/j.neulet.2004.04.065 PMID:15245792

Labyak, S., Lava, S., Turek, F., & Zee, P. (2002). Effects of shift work on sleep and menstrual function in nurses. *Health Care for Women International*, *23*(6-7), 703–714. doi:10.1080/07399330290107449 PMID:12418990

Leonhardt, H., Gull, B., Kishimoto, K., Kataoka, M., Nilsson, L., Janson, P. O., Stener-Victorin, E., & Hellström, M. (2012). Uterine morphology and peristalsis in women with polycystic ovary syndrome. *Acta Radiologica*, *53*(10), 1195–1201. doi:10.1258/ar.2012.120384 PMID:23081959

Li, D. X., Romans, S., De Souza, M. J., Murray, B., & Einstein, G. (2015). Actigraphic and self-reported sleep quality in women: Associations with ovarian hormones and mood. *Sleep Medicine*, *16*(10), 1217–1224. Advance online publication. doi:10.1016/j.sleep.2015.06.009 PMID:26429749

Li, R., Cheng, S., & Wang, Z. (2015). Circadian clock gene plays a key role on ovarian cycle and spontaneous abortion. *Cellular Physiology and Biochemistry*, *37*(3), 911–920. doi:10.1159/000430218 PMID:26390085

Li, Y., Cheng, S., Li, L., Zhao, Y., Shen, W., & Sun, X. (2019). Light-exposure at night impairs mouse ovary development via cell apoptosis and DNA damage. *Bioscience Reports*, *39*(5), BSR20181464. doi:10.1042/BSR20181464 PMID:30962269

Lie, J. A. S., Andersen, A., & Kjærheim, K. (2007). Cancer risk among 43 000 Norwegian nurses. *Scandinavian Journal of Work, Environment & Health*, *33*(1), 66–73. doi:10.5271jweh.1066 PMID:17353967

Lie, J. A. S., Kjuus, H., Zienolddiny, S., Haugen, A., Stevens, R. G., & Kjærheim, K. (2011). Night work and breast cancer risk among norwegian nurses: Assessment by different exposure metrics. *American Journal of Epidemiology*, *173*(11), 1272–1279. doi:10.1093/aje/kwr014 PMID:21454824

Loren, P., Sánchez, R., Arias, M. E., Felmer, R., Risopatrón, J., & Cheuquemán, C. (2017). Melatonin scavenger properties against oxidative and nitrosative stress: Impact on gamete handling and in vitro embryo production in humans and other mammals. *International Journal of Molecular Sciences*, *18*(6), 1119. doi:10.3390/ijms18061119 PMID:28613231

Luboshits, G., Shina, S., Kaplan, O., Engelberg, S., Nass, D., Lifshitz-Mercer, B., Chaitchik, S., Keydar, I., & Ben-Baruch, A. (1999). Elevated expression of the CC chemokine regulated on activation, normal T cell expressed and secreted (RANTES) in advanced breast carcinoma. *Cancer Research*, *59*, 4681–4687. PMID:10493525

Maganhin, C. C., Fuchs, L. F. P., Simões, R. S., Oliveira-Filho, R. M., De Jesus Simões, M., Baracat, E. C., & Soares, J. M. Jr. (2013). Effects of melatonin on ovarian follicles. *European Journal of Obstetrics, Gynecology, and Reproductive Biology*, *166*(2), 178–184. doi:10.1016/j.ejogrb.2012.10.006 PMID:23102587

Marcheva, B., Ramsey, K. M., Buhr, E. D., Kobayashi, Y., Su, H., Ko, C. H., Ivanova, G., Omura, C., Mo, S., Vitaterna, M. H., Lopez, J. P., Philipson, L. H., Bradfield, C. A., Crosby, S. D., Jebailey, L., Wang, X., Takahashi, J. S., & Bass, J. (2010). Disruption of the clock components CLOCK and BMAL1 leads to hypoinsulinaemia and diabetes. *Nature*, *466*(7306), 627–631. doi:10.1038/nature09253 PMID:20562852

McFadden, E., Jones, M. E., Schoemaker, M. J., Ashworth, A., & Swerdlow, A. J. (2014). The relationship between obesity and exposure to light at night: Cross-sectional analyses of over 100,000 women in the breakthrough generations study. *American Journal of Epidemiology*, *180*(3), 245–250. doi:10.1093/aje/kwu117 PMID:24875371

Megdal, S. P., Kroenke, C. H., Laden, F., Pukkala, E., & Schernhammer, E. S. (2005). Night work and breast cancer risk: A systematic review and meta-analysis. *European Journal of Cancer*, *41*(13), 2023–2032. doi:10.1016/j.ejca.2005.05.010 PMID:16084719

Miller, B. H., Olson, S. L., Levine, J. E., Turek, F. W., Horton, T. H., & Takahashi, J. S. (2006). Vasopressin regulation of the proestrous luteinizing hormone surge in wild-type and Clock mutant mice. *Biology of Reproduction*, *75*(5), 778–784. doi:10.1095/biolreprod.106.052845 PMID:16870944

Miller, B. H., Olson, S. L., Turek, F. W., Levine, J. E., Horton, T. H., & Takahashi, J. S. (2004). Circadian clock mutation disrupts estrous cyclicity and maintenance of pregnancy. *Current Biology*, *14*(15), 1367–1373. doi:10.1016/j.cub.2004.07.055 PMID:15296754

Minkel, J., Moreta, M., Muto, J., Htaik, O., Jones, C., Basner, M., & Dinges, D. (2014). Sleep deprivation potentiates HPA axis stress reactivity in healthy adults. *Health Psychology*, *33*(11), 1430–1434. doi:10.1037/a0034219 PMID:24818608

Mokhlesi, B., Scoccia, B., Mazzone, T., & Sam, S. (2012). Risk of obstructive sleep apnea in obese and nonobese women with polycystic ovary syndrome and healthy reproductively normal women. *Fertility and Sterility*, *97*(3), 786–791. doi:10.1016/j.fertnstert.2011.12.024 PMID:22264851

Morgan, M. A. (1992). Different patterns of myometrial activity and 24-h rhythms in myometrial contractility in the gravid baboon during the second half of pregnancy. *Biology of Reproduction*, *46*(6), 1158–1164. doi:10.1095/biolreprod46.6.1158 PMID:1391314

Moshkdanian, G., Moghani-Ghoroghi, F., Pasbakhsh, P., Nematollahi-Mahani, S. N., Najafi, A., & Kashani, I. R. (2017). Melatonin upregulates ErbB1 and ErbB4, two primary implantation receptors, in pre-implantation mouse embryos. *Iranian Journal of Basic Medical Sciences.*, *20*, 655–661. PMID:28868121

Murphy, A. A., Palinski, W., Rankin, S., Morales, A. J., & Parthasarathy, S. (1998). Macrophage scavenger receptor(s) and oxidatively modified proteins in endometriosis. *Fertility and Sterility*, *69*(6), 1085–1091. doi:10.1016/S0015-0282(98)00088-0 PMID:9627297

Muter, J., Lucas, E. S., Chan, Y. W., Brighton, P. J., Moore, J. D., Lacey, L., Quenby, S., Lam, E. W. F., & Brosens, J. J. (2015). The clock protein period 2 synchronizes mitotic expansion and decidual transformation of human endometrial stromal cells. *The FASEB Journal*, *29*(4), 1603–1614. doi:10.1096/fj.14-267195 PMID:25573754

Nakamura, Y., Tamura, H., Takayama, H., & Kato, H. (2003). Increased endogenous level of melatonin in preovulatory human follicles does not directly influence progesterone production. *Fertility and Sterility*, *80*(4), 1012–1016. doi:10.1016/S0015-0282(03)01008-2 PMID:14556825

Nisenblat, V., & Norman, R. J. (2009). Androgens and polycystic ovary syndrome. *Current Opinion in Endocrinology, Diabetes, and Obesity*, *16*(3), 224–231. doi:10.1097/MED.0b013e32832afd4d PMID:19390322

Oak, M. K., Chantler, E. N., Vaughan Williams, C. A., & Elstein, M. (1985). Sperm survival studies in peritoneal fluid from infertile women with endometriosis and unexplained infertility. *Clinical Reproduction and Fertility*, *3*, 297–303. PMID:3841649

Opperhuizen, A. L., Stenvers, D. J., Jansen, R. D., Foppen, E., Fliers, E., & Kalsbeek, A. (2017). Light at night acutely impairs glucose tolerance in a time-, intensity- and wavelength-dependent manner in rats. *Diabetologia*, *60*(7), 1333–1343. doi:10.100700125-017-4262-y PMID:28374068

Padmanabhan, V., & Veiga-Lopez, A. (2011). Developmental origin of reproductive and metabolic dysfunctions: Androgenic versus estrogenic reprogramming. *Seminars in Reproductive Medicine*, *29*(03), 173–186. doi:10.1055-0031-1275519 PMID:21710394

Pilorz, V., & Steinlechner, S. (2008). Low reproductive success in Per1 and Per2 mutant mouse females due to accelerated ageing? *Reproduction (Cambridge, England)*, *135*(4), 559–568. doi:10.1530/REP-07-0434 PMID:18367514

Pang, Y. W., An, L., Wang, P., Yu, Y., Yin, Q. D., Wang, X. H., Xin-Zhang, Qian-Zhang, Yang, M.-L., Min-Guo, Wu, Z.-H., & Tian, J.-H. (2013). Treatment of porcine donor cells and reconstructed embryos with the antioxidant melatonin enhances cloning efficiency. *Journal of Pineal Research*, *54*(4), 389–397. doi:10.1111/jpi.12024 PMID:24325731

Polak, G., Koziol-Montewka, M., Gogacz, M., Blaszkowska, I., & Kotarski, J. (2001). Total antioxidant status of peritoneal fluid in infertile women. *European Journal of Obstetrics, Gynecology, and Reproductive Biology*, *94*(2), 261–263. doi:10.1016/S0301-2115(00)00352-3 PMID:11165736

Ratajczak, C. K., Boehle, K. L., & Muglia, L. J. (2009). Impaired steroidogenesis and implantation failure in Bmal1-/- mice. *Endocrinology*, *150*(4), 1879–1885. doi:10.1210/en.2008-1021 PMID:19056819

Reiter, R. J., Tamura, H., Tan, D. X., & Xu, X. Y. (2014). Melatonin and the circadian system: Contributions to successful female reproduction. In *Fertility and Sterility* (Vol. 102, Issue 2, pp. 321–328).

Reiter, R. J., Tan, D.-X., Manchester, L. C., Paredes, S. D., Mayo, J. C., & Sainz, R. M. (2009). Melatonin and Reproduction Revisited. *Biology of Reproduction, 81*(3), 445–456.

Reppert, S. M., & Schwartz, W. J. (1986). Maternal suprachiasmatic nuclei are necessary for maternal coordination of the developing circadian system. *The Journal of Neuroscience: The Official Journal of the Society for Neuroscience*, *6*(9), 2724–2729. doi:10.1523/JNEUROSCI.06-09-02724.1986 PMID:3746430

Reppert, S. M., & Schwartz, W. J. (1983). Maternal coordination of the fetal biological clock in utero. *Science*, *220*(4600), 969–971. doi:10.1126cience.6844923 PMID:6844923

Rizzo, A., Roscino, M. T., Minoia, G., Trisolini, C., Spedicato, M., Mutinati, M., Pantaleo, M., Jirillo, F., & Sciorsci, R. L. (2009). Serum levels of reactive oxygen species (ROS) in the bitch. *Immunopharmacology and Immunotoxicology*, *31*(2), 310–313. doi:10.1080/08923970802683954 PMID:19235534

Rizzo, P., Raffone, E., & Benedetto, V. (2010). Effect of the treatment with myo-inositol plus folic acid plus melatonin in comparison with a treatment with myo-inositol plus folic acid on oocyte quality and pregnancy outcome in IVF cycles. A prospective, clinical trial. *European Review for Medical and Pharmacological Sciences*, *14*, 555–561. PMID:20712264

Rocha, R. M. P., Lima, L. F., Alves, A. M. C. V., Celestino, J. J. H., Matos, M. H. T., Lima-Verde, I. B., Bernuci, M. P., Lopes, C. A. P., Báo, S. N., Campello, C. C., Rodrigues, A. P. R., & Figueiredo, J. R. (2013). Interaction between melatonin and follicle-stimulating hormone promotes in vitro development of caprine preantral follicles. *Domestic Animal Endocrinology*, *44*(1), 1–9. doi:10.1016/j.domaniend.2012.07.001 PMID:22920266

Rodrigues-Cunha, M. C., Mesquita, L. G., Bressan, F., del Collado, M., Balieiro, J. C. C., Schwarz, K. R. L., de Castro, F. C., Watanabe, O. Y., Watanabe, Y. F., de Alencar Coelho, L., & Leal, C. L. V. (2016). Effects of melatonin during IVM in defined medium on oocyte meiosis, oxidative stress, and subsequent embryo development. *Theriogenology*, *86*(7), 1685–1694. doi:10.1016/j.theriogenology.2016.05.026 PMID:27471183

Rodriguez-Osorio, N., Kim, I. J., Wang, H., Kaya, A., & Memili, E. (2007). Melatonin increases cleavage rate of porcine preimplantation embryos in vitro. *Journal of Pineal Research*, *43*(3), 283–288. doi:10.1111/j.1600-079X.2007.00475.x PMID:17803526

Rönnberg, L., Kauppila, A., Leppäluoto, J., Martikainen, H., & Vakkuri, O. (1990). Circadian and seasonal variation in human preovulatory follicular fluid melatonin concentration. *The Journal of Clinical Endocrinology and Metabolism*, *71*(2), 492–496. doi:10.1210/jcem-71-2-493 PMID:2380343

Russart, K. L. G., & Nelson, R. J. (2018). Light at night as an environmental endocrine disruptor. *Physiology & Behavior*, *190*, 82–89. doi:10.1016/j.physbeh.2017.08.029 PMID:28870443

Saha, S., Singh, K. M., & Gupta, B. B. P. (2019). Melatonin synthesis and clock gene regulation in the pineal organ of teleost fish compared to mammals: Similarities and differences. *General and Comparative Endocrinology*, *279*, 27–34. doi:10.1016/j.ygcen.2018.07.010 PMID:30026020

Sáenz de Miera, C., Bothorel, B., Jaeger, C., Simonneaux, V., & Hazlerigg, D. (2017). Maternal photoperiod programs hypothalamic thyroid status via the fetal pituitary gland. *Proceedings of the National Academy of Sciences of the United States of America*, *114*(31), 8408–8413. doi:10.1073/pnas.1702943114 PMID:28716942

Sakaguchi, K., Itoh, M. T., Takahashi, N., Tarumi, W., & Ishizuka, B. (2013). The rat oocyte synthesises melatonin. *Reproduction, Fertility, and Development*, *25*(4), 674–682. doi:10.1071/RD12091 PMID:22951050

Schernhammer, E. S., & Schulmeister, K. (2004). Melatonin and cancer risk: Does light at night compromise physiologic cancer protection by lowering serum melatonin levels? *British Journal of Cancer*, *90*(5), 941–943. doi:10.1038j.bjc.6601626 PMID:14997186

Schernhammer, E. S., Laden, F., Speizer, F. E., Willett, W. C., Hunter, D. J., Kawachi, I., Fuchs, C. S., & Colditz, G. A. (2003). Night-shift work and risk of colorectal cancer in the Nurses' Health Study. *Journal of the National Cancer Institute*, *95*(11), 825–828. doi:10.1093/jnci/95.11.825 PMID:12783938

Sellix, M. T., & Menaker, M. (2011). Circadian clocks in mammalian reproductive physiology: Effects of the "other" biological clock on fertility. *Discovery Medicine*, *11*, 273–281. PMID:21524381

Sellix, M. T., Murphy, Z. C., & Menaker, M. (2013). Excess androgen during puberty disrupts circadian organization in female rats. *Endocrinology*, *154*(4), 1636–1647. doi:10.1210/en.2012-2066 PMID:23417420

Sellix, M. T. (2015). Circadian clock function in the mammalian ovary. *Journal of Biological Rhythms*, *30*(1), 7–19. doi:10.1177/0748730414554222 PMID:25367899

Shi, D., & Vine, D. F. (2012). Animal models of polycystic ovary syndrome: A focused review of rodent models in relationship to clinical phenotypes and cardiometabolic risk. *Fertility and Sterility*, *98*(1), 185–193. doi:10.1016/j.fertnstert.2012.04.006 PMID:22607890

Shimba, S., Ogawa, T., Hitosugi, S., Ichihashi, Y., Nakadaira, Y., Kobayashi, M., Tezuka, M., Kosuge, Y., Ishige, K., Ito, Y., Komiyama, K., Okamatsu-Ogura, Y., Kimura, K., & Saito, M. (2011). Deficient of a clock gene, brain and muscle Arnt-like protein-1 (BMAL1), induces dyslipidemia and ectopic fat formation. *PLoS One*, *6*(9), e25231. doi:10.1371/journal.pone.0025231 PMID:21966465

Shreeve, N., Cagampang, F., Sadek, K., Tolhurst, M., Houldey, A., Hill, C. M., Brook, N., MacKlon, N., & Cheong, Y. (2013). Poor sleep in PCOS; Is melatonin the culprit? *Human Reproduction (Oxford, England)*, *28*(5), 1348–1353. doi:10.1093/humrep/det013 PMID:23438443

Simonneaux, V., & Bahougne, T. (2015). A multi-oscillatory circadian system times female reproduction. *Frontiers in Endocrinology (Lausanne)*, *6*, 157. doi:10.3389/fendo.2015.00157 PMID:26539161

Singh, K. B. (1969). Induction of polycystic ovarian disease in rats by continuous light. II. Observations on mating, pregnancy, and the postpartum period. *American Journal of Obstetrics and Gynecology*, *104*(7), 1004–1007. doi:10.1016/0002-9378(69)90692-9 PMID:5794828

Soares, J. M., Simões, M. J., Oshima, C. T. F., Mora, O. A., De Lima, G. R., & Baracat, E. C. (2003). Pinealectomy changes rat ovarian interstitial cell morphology and decreases progesterone receptor expression. *Gynecological Endocrinology*, *17*(2), 115–123. doi:10.1080/gye.17.2.115.123 PMID:12737672

Soares, J. M. Jr, Masana, M. I., Erşahin, C., & Dubocovich, M. L. (2003). Functional melatonin receptors in rat ovaries at various stages of the estrous cycle. *The Journal of Pharmacology and Experimental Therapeutics*, *306*(2), 694–702. doi:10.1124/jpet.103.049916 PMID:12721330

Solt, L. A., Wang, Y., Banerjee, S., Hughes, T., Kojetin, D. J., Lundasen, T., Shin, Y., Liu, J., Cameron, M. D., Noel, R., Yoo, S. H., Takahashi, J. S., Butler, A. A., Kamenecka, T. M., & Burris, T. P. (2012). Regulation of circadian behaviour and metabolism by synthetic REV-ERB agonists. *Nature*, *485*(7396), 62–68. doi:10.1038/nature11030 PMID:22460951

Stevens, R. G. (1987). Electric power use and breast cancer: A hypothesis. *American Journal of Epidemiology*, *125*(4), 556–561. doi:10.1093/oxfordjournals.aje.a114569 PMID:3548332

Su, J., Wang, Y., Xing, X., Zhang, L., Sun, H., & Zhang, Y. (2015). Melatonin significantly improves the developmental competence of bovine somatic cell nuclear transfer embryos. *Journal of Pineal Research*, *59*(4), 455–468. doi:10.1111/jpi.12275 PMID:26331949

Takahashi, M. (2012). Oxidative stress and redox regulation on in vitro development of mammalian embryos. *The Journal of Reproduction and Development*, *58*(1), 1–9. doi:10.1262/jrd.11-138N PMID:22450278

Takata, M., Burioka, N., Ohdo, S., Takane, H., Terazono, H., Miyata, M., Takanori, S., Hisashi, S., Yasushi, F., Katsuyuki, T., & Eiji, S. (2002). Daily expression of mRNAs for the mammalian Clock genes Per2 and clock in mouse suprachiasmatic nuclei and liver and human peripheral blood mononuclear cells. *Japanese Journal of Pharmacology*, *90*(3), 263–269. doi:10.1254/jjp.90.263 PMID:12499581

Taketani, T., Tamura, H., Takasaki, A., Lee, L., Kizuka, F., Tamura, I., Taniguchi, K., Maekawa, R., Asada, H., Shimamura, K., Reiter, R. J., & Sugino, N. (2011). Protective role of melatonin in progesterone production by human luteal cells. *Journal of Pineal Research*, *51*(2), 207–213. doi:10.1111/j.1600-079X.2011.00878.x PMID:21585519

Tamarkin, L., Cohen, M., Reichert, C., Reichert, C., Lippman, M., & Chabner, B. (1981). Melatonin Inhibition and Pinealectomy Enhancement of 7,12-Dimethylbenz(a)anthracene-induced Mammary Tumors in the Rat. *Cancer Research*, *41*, 4432–4436. PMID:6796259

Tamura, H., Nakamura, Y., Takiguchi, S., Kashida, S., Yamagata, Y., Sugino, N., & Kato, H. (1998). Melatonin directly suppresses steroid production by preovulatory follicles in the cyclic hamster. *Journal of Pineal Research*, *25*(3), 135–141. doi:10.1111/j.1600-079X.1998.tb00551.x PMID:9745981

Tamura, H., Takasaki, A., Miwa, I., Taniguchi, K., Maekawa, R., Asada, H., Taketani, T., Matsuoka, A., Yamagata, Y., Shimamura, K., Morioka, H., Ishikawa, H., Reiter, R. J., & Sugino, N. (2008). Oxidative stress impairs oocyte quality and melatonin protects oocytes from free radical damage and improves fertilization rate. *Journal of Pineal Research*, *44*(3), 280–287. doi:10.1111/j.1600-079X.2007.00524.x PMID:18339123

Tamura, H., Takasaki, A., Taketani, T., Tanabe, M., Kizuka, F., Lee, L., Tamura, I., Maekawa, R., Asada, H., Yamagata, Y., & Sugino, N. (2013). Melatonin as a free radical scavenger in the ovarian follicle. *Endocrine Journal*, *60*(1), 1–13. doi:10.1507/endocrj.EJ12-0263 PMID:23171705

Tan, D. X., Manchester, L. C., Esteban-Zubero, E., Zhou, Z., & Reiter, R. J. (2015). Melatonin as a potent and inducible endogenous antioxidant: Synthesis and metabolism. *Molecules (Basel, Switzerland)*, *20*(10), 18886–18906. doi:10.3390/molecules201018886 PMID:26501252

Tan, D. X., Manchester, L. C., Hardeland, R., Lopez-Burillo, S., Mayo, J. C., Sainz, R. M., & Reiter, R. J. (2003). Melatonin: A hormone, a tissue factor, an autocoid, a paracoid, and an antioxidant vitamin. *Journal of Pineal Research*, *34*(1), 75–78. doi:10.1034/j.1600-079X.2003.02111.x PMID:12485375

Teboul, M., Barrat-Petit, M. A., Li, X. M., Claustrat, B., Formento, J. L., Delaunay, F., Lévi, F., & Milano, G. (2005). Atypical patterns of circadian clock gene expression in human peripheral blood mono nuclear cells. *Journal of Molecular Medicine*, *83*(9), 693–699. doi:10.100700109-005-0697-6 PMID:16096739

Thompson, C. L., Larkin, E. K., Patel, S., Berger, N. A., Redline, S., & Li, L. (2011). Short duration of sleep increases risk of colorectal adenoma. *Cancer*, *117*(4), 841–847. doi:10.1002/cncr.25507 PMID:20936662

Trentini, G. P., Mess, B., De Gaetani, C. F., & Ruzsás, C. (1978). Effect of melatonin on induction of ovulation in the light- induced constant estrous-anovulatory syndrome and possible role of the brain serotoninergic system. *Journal of Endocrinological Investigation: Official Journal of the Italian Society of Endocrinology*, *1*(4), 305–310. doi:10.1007/BF03350974 PMID:574524

Tricot, G. (2000). New insights into role of microenvironment in multiple myeloma. *TheLancet*, *355*(9200), 248–250. doi:10.1016/S0140-6736(00)00019-2 PMID:10675068

Tosini, G., Pozdeyev, N., Sakamoto, K., & Iuvone, P. M. (2008). The circadian clock system inthe mammalian retina. *BioEssays*, *30*(7), 624–633. doi:10.1002/bies.20777 PMID:18536031

Turek, F. W., Joshu, C., Kohsaka, A., Lin, E., Ivanova, G., McDearmon, E., Laposky, A., Losee-Olson, S., Easton, A., Jensen, D. R., Eckel, R. H., Takahashi, J. S., & Bass, J. (2005). Obesity and metabolic syndrome in circadian Clock mutant mice. *Science*, *308*(5724), 1043–1045. doi:10.1126cience.1108750 PMID:15845877

Uehata, T., & Sasakawa, N. (1982). The fatigue and maternity disturbances of night workwomen. *Journal of Human Ergology*, *11*, 465–474. PMID:7188479

van der Horst, G. T., Muijtjens, M., Kobayashi, K., Takano, R., Kanno, S., Takao, M., De Wit, J., Verkerk, A., Eker, A. P. M., Van, L. D., Buijs, R., Bootsma, D., Hoeijmakers, J. H. J., & Yasui, A. (1999). Mammalian Cry1 and Cry2 are essential for maintenance of circadian rhythms. *Nature*, *398*(6728), 627–630. doi:10.1038/19323 PMID:10217146

Van Langendonckt, A., Casanas-Roux, F., & Donnez, J. (2002). Oxidative stress and peritoneal endometriosis. *Fertility and Sterility*, *77*(5), 861–870. doi:10.1016/S0015-0282(02)02959-X PMID:12009336

Varcoe, T. J., Wight, N., Voultsios, A., Salkeld, M. D., & Kennaway, D. J. (2011). Chronic phase shifts of the photoperiod throughout pregnancy programs glucose intolerance and insulin resistance in the rat. *PLoS One*, *6*(4), e18504. doi:10.1371/journal.pone.0018504 PMID:21494686

Varcoe, T. J., Gatford, K. L., & Kennaway, D. J. (2018). Maternal circadian rhythms and the programming of adult health and disease. *American Journal of Physiology. Regulatory, Integrative and Comparative Physiology*, *314*(2), R231–R241. doi:10.1152/ajpregu.00248.2017 PMID:29141950

Vidal-Vanaclocha, F., Fantuzzi, G., Mendoza, L., Fuentes, A. M., Anasagasti, M. J., Martin, J., Carrascal, T., Walsh, P., Reznikov, L. L., Kim, S. H., Novick, D., Rubinstein, M., & Dinarello, C. A. (2000). IL-18 regulates IL-1β-dependent hepatic melanoma metastasis via vascular cell adhesion molecule-1. *Proceedings of the National Academy of Sciences of the United States of America*, *97*(2), 734–739. doi:10.1073/pnas.97.2.734 PMID:10639148

Viswanathan, A. N., Hankinson, S. E., & Schernhammer, E. S. (2007). *Night Shift Work and the Risk of Endometrial Cancer*. Academic Press.

Vitaterna, M. H., Selby, C. P., Todo, T., Niwa, H., Thompson, C., & Fruechte, E. M. (1999). Differential regulation of mammalian period genes and circadian rhythmicity by cryptochromes 1 and 2. *Proceedings of the National Academy of Sciences of the United States of America*, *96*, 12114-12119. 10.1073/pnas.96.21.12114

Volt, H., García, J. A., Doerrier, C., Díaz-Casado, M. E., Guerra-Librero, A., Lõpez, L. C., Escames, G., Tresguerres, J. A., & Acuña-Castroviejo, D. (2016). Same molecule but different expression: Aging and sepsis trigger NLRP3 inflammasome activation, a target of melatonin. *Journal of Pineal Research*, *60*(2), 193–205. doi:10.1111/jpi.12303 PMID:26681113

Wan, G. H., & Chung, F. F. (2012). Working conditions associated with ovarian cycle in a medical center nurses: A Taiwan study. *Japan Journal of Nursing Science*, *9*(1), 112–118. doi:10.1111/j.1742-7924.2011.00191.x PMID:22583945

Wang, H., & Dey, S. K. (2006). Roadmap to embryo implantation: Clues from mouse models. *Nature Reviews. Genetics*, *7*(3), 185–199. doi:10.1038/nrg1808 PMID:16485018

Wang, Y., Goldberg, J., Sharma, R. K., Agarwal, A., & Falcone, T. (1997). Importance of reactive oxygen species in the peritoneal fluid of women with endometriosis or idiopathic infertility. *Fertility and Sterility*, *68*(5), 826–830. doi:10.1016/S0015-0282(97)00343-9 PMID:9389810

Weber, A. L., & Adler, N. T. (1979). Delay of constant light-induced persistent vaginal estrus by 24-hour time cues in rats. *Science*, *204*(4390), 323–325. doi:10.1126cience.571146 PMID:571146

Wei, D., Zhang, C., Xie, J., Song, X., Yin, B., Liu, Q., Hu, L., Hao, H., Geng, J., & Wang, P. (2013). Supplementation with low concentrations of melatonin improves nuclear maturation of human oocytes in vitro. *Journal of Assisted Reproduction and Genetics*, *30*(7), 933–938. doi:10.100710815-013-0021-2 PMID:23737216

Weibel, L., Spiegel, K., Gronfier, C., Follenius, M., & Brandenberger, G. (1997). Twenty-four-hour melatonin and core body temperature rhythms: Their adaptation in night workers. *The American Journal of Physiology*, *272*, R948–R954. PMID:9087659

Wilking, M., Ndiaye, M., Mukhtar, H., & Ahmad, N. (2013). Circadian rhythm connections to oxidative stress: Implications for human health. *Antioxidants and Redox Signaling*, *19*, 192-208.

Yamanaka, A., Beuckmann, C. T., Willie, J. T., Hara, J., Tsujino, N., Mieda, M., Tominaga, M., Yagami, K. I., Sugiyama, F., Goto, K., Yanagisawa, M., & Sakurai, T. (2003). Hypothalamic orexin neurons regulate arousal according to energy balance in mice. *Neuron*, *38*(5), 701–713. doi:10.1016/S0896-6273(03)00331-3 PMID:12797956

Yang, S., Liu, A., Weidenhammer, A., Cooksey, R. C., McClain, D., Kim, M. K., Aguilera, G., Abel, E. D., & Chung, J. H. (2009). The role of mPer2 clock gene in glucocorticoid and feeding rhythms. *Endocrinology*, *50*(5), 2153–2160. doi:10.1210/en.2008-0705 PMID:19179447

Yoshikawa, T., Sellix, M., Pezuk, P., & Menaker, M. (2009). Timing of the ovarian circadian clock is regulated by gonadotropins. *Endocrinology*, *150*(9), 4338–4347. doi:10.1210/en.2008-1280 PMID:19520783

Zeller, J. M., Henig, I., Radwanska, E., & Dmowski, W. P. (1987). Enhancement of Human Monocyte and Peritoneal Macrophage Chemiluminescence Activities in Women With Endometriosis. *American Journal of Reproductive Immunology and Microbiology*, *13*(3), 78–82. doi:10.1111/j.1600-0897.1987.tb00097.x PMID:3605484

Zhang, L., Zhang, Z., Wang, F., Tian, X., Ji, P., & Liu, G. (2017). Effects of melatonin administration on embryo implantation and offspring growth in mice under different schedules of photoperiodic exposure. *Reproductive Biology and Endocrinology*, *15*(1), 78. doi:10.118612958-017-0297-7 PMID:28969693

Zhang, S., Zeitzer, J. M., Yoshida, Y., Wisor, J. P., Nishino, S., Edgar, D. M., & Mignot, E. (2004). Lesions of the suprachiasmatic nucleus eliminate the daily rhythm of hypocretin-1 release. *Sleep*, *27*(4), 619–627. doi:10.1093leep/27.4.619 PMID:15282996

Zhao, M., Wan, J., Zeng, K., Tong, M., Lee, A. C., Ding, J., & Chen, Q. (2016). The reduction in circulating melatonin level may contribute to the pathogenesis of ovarian cancer: A retrospective study. *Journal of Cancer*, *7*(7), 831–836. doi:10.7150/jca.14573 PMID:27162542

Zvonic, S., Ptitsyn, A. A., Conrad, S. A., Scott, L. K., Floyd, Z. E., Kilroy, G., Wu, X., Goh, B. C., Mynatt, R. L., & Gimble, J. M. (2006). Characterization of peripheral circadian clocks in adipose tissues. *Diabetes*, *55*(4), 962–970. doi:10.2337/diabetes.55.04.06.db05-0873 PMID:16567517

Chapter 5
Impact of Heat Stress on Embryonic Implantation

Khursheed Ahmad Wani
https://orcid.org/0000-0001-5577-3455
Government Degree College, Bijbehara, India

Jamila Irfan
Government Womens College, Srinagar, India

Junaid Ahmad Malik
https://orcid.org/0000-0003-4411-2015
Government Degree College, Bijbehara, India

ABSTRACT

One of the expected effects of mammalian developmental defects is a rise in air temperature. Heat stress adversely affects embryo oogenesis, oocyte maturation, fertilization, and implantation. The number of defects caused by heat stress in all mammals is almost identical, but each species has its own particular sensitivity to specific defects. It suggests that genotype may have a significant effect on the type of defect, its occurrence, and its extent. By heat output and loss, the body temperature is usually preserved in a restricted range, but the equilibrium can be disrupted by illness, inadequate nutrition, and severe environmental temperature. Elevated maternal temperature during pregnancy, either by fever or any other means of heating, may result in embryo death, retarded development, abortion, and many embryonic defects such as cell proliferation, migration, differentiation and apoptosis or programmed cell death, structural and functional defects, and changes in maternal physiology. Maternal heat stress also reduces the levels of placental hormones. This chapter deals with the heat stress effect on reduction of reproductive function, implantation defects, etc. of different animals and humans.

DOI: 10.4018/978-1-7998-4480-8.ch005

INTRODUCTION

A rise in mammalian body temperature is caused by exposure to hot environment. Prosser and Heath (1991) studied that most mammals have 35-39^{0}C body temperature and these temperatures are sustained above ambient temperatures by metabolic heat production. According to Webster and Edwards (1984), different genotypes have distinctly different sensitivities to some forms of defects in a population, and there may be variations between strains in the gestational times at which the defect may be caused. Maternal heat exposure after fertilization reduces the risk of pregnancy, embryonic death before implantation and disturbs the intrauterine environment both for embryos and uterine tissue. During summer, heat stress induced by high ambient temperature may lead to increased internal heat rates, and can decrease animal growth, milk production, and fertility. The impacts of ambient temperature and humidity on the physiology of domesticated animals have been studied in many studies. Berman et al. (1985) suggested that the furthest reaches of ambient temperatures, at which high milk producing dairy cows should sustain a steady internal heat level (38.5^{0}C) of 25-26^{0}C, and that protocol should be developed to restrict the rise in internal heat levels at natural temperatures above 25°C.

High ecological temperature raises the amount of internal heat in bovine lactating to around 40°C (Sartori et al., 2002). It may be due to the additional heat created in relation to the milk produced and less heat loss due to the lower temperature contrast between body and environment. Milk production declines substantially as the internal heat level increases (Bohmanova et al., 2002; Ravagnolo et al., 2006). However, in pigs, heat demand reduces food consumption and body weight (Huynh et al., 2005). In addition to productivity, heat stress significantly affects the regenerative ability of well-developed creatures with a decrease in pregnancy levels in dairy cattle (Garcia et al., 2006) and pigs (Wettemann and Bazer, 1985). It is therefore indicative that climate change, especially temperature, can have a significant impact on embryonic implantation. Hence, this chapter is an approach to study the impact of heat stress on embryonic implantation in different animals and humans.

EFFECTS OF HIGH TEMPERATURE

The closeness between a portion of the impacts of heat and the impacts of ionizing radiation, especially in their targeting of multiplying cells, have been accounted by Wanner and Edwards (1983). Also, harm to veins and the placenta will prompt further disturbance of developmental procedures. These impacts can be changed somewhat by the cell 'heat shock' reaction, which gives some security to uncovered undeveloped organisms. Various components can prompt premature birth and fetal development retardation. Killing of cells warmed during division is a significant element of the heat damaged embryonic brain and face and can be found immediately after short portion of heat. Close to a moderate rise of temperature (2-3^{0}C) given to guinea pig embryos during early stages of brain development (day 21 post-origination), mitotic cells show clustering of chromosomes and die. At higher heights (3-4^{0}C), cells in S-stage (DNA replication stage) die by apoptosis from 6 hours after exposure, significantly higher temperature may cause prominent degrees of this type of cell death. The imperfections coming about because of limited regions of cell death are normally very distorting and self-evident, for instance neural tube defects (NTD). Heat at later stages, for instance in day 21 guinea pig undeveloped organisms, causes a more extensive, progressively diffuse circulation of cell death and results in a proportional miniature of the normal brain (Edwards et al., 2003).

In spite of the fact that the cell division cycle is deferred by heat, this doesn't seem to cause clear imperfections in brain or facial improvement except if accompanied by higher degrees of cell death. Sometimes a little portion of heat (42^{0}C for 10 min) was given to 9.5 day rodent embryo in culture, a defensive heat shock reaction was started and this introduction was followed by a higher portion of heat (43^{0}C for 7.5 min), that would ordinarily cause extreme imperfections in an unprotected embryo, only couple of cell deaths were found and no deformities of the head followed these exposures (Walsh et al., 1993; 1997). Both the underlying, defensive introduction and the resulting progressively extreme exposure were followed by suspension of ordinary cell protein synthesis and suspension of cell expansion yet didn't cause defective development of the neural tube and face. Effective neural tube closure and development of the head face and different structures from the neural plate may depend, on a threshold number of cells being accessible to shape these structures. Disturbance of gene expression without noteworthy cell death may be the cause for vertebral and rib defects in mice. Cell death has been noted in the region of the developing vertebral segment of rodents after a high portion of heat (Breen et al., 1999).

Cell membranes are upset by heat during neural tube closure and this may be one of the components prompting the development of rosettes (blister like structures) which structure encephaloceles in the brain (Hicks and D'Amato, 1966; Smith et al., 1992) and syringomyelia (cysts) influencing the spinal cord (Edward, 1971). Interruption of the ventricular covering of the cerebral sides of the equator and breakdown of junctional edifices were likewise found in the cerebrum of multi day warmed guinea pig embryos (Wanner et al., 1976). Investigations of cells in culture have additionally exhibited harm to membranes (Lepock, 1982), microtubules and different components of the cytoskeleton (Coss et al., 1982; van Bergenen et al., 1985). Aggravations of cell relocation are accepted to bring about abnormalities of the heart and face (neural peak cell movement) and to cause Hirschsprung ailment, a deformity of the innervation of the intestine, and to bring about ectopic homes of neurones in the cerebrum.

Neural peak cells start relocation at about the phase of neural cylinder conclusion with numerous goals, including the face and head, heart, digestive tract, skin, adrenals and somewhere else. Neuronal cells start to relocate from their causes around the cerebral ventricles at the hour of arrangement of the cortical plate at days 12-13 in rodents, 21-22 in guinea pigs and weeks 7-8 in humans, not long after neuronal expansion begins. The multiplication of neuronal antecedent cells was stifled and their movement to the cortical plate was decelerated altogether in the cerebrums of fetal mice warmed on day 13 or 14 (Hinoue et al., 2001). The ensuing procedures of neuronal multiplication, separation and relocation are vital in the further improvement of the cerebrum, particularly the cortex, and have been demonstrated to be intensely delicate to ionizing radiation in enormous quantities of investigations of test creatures and uncovered people, prompting conduct shortages and decreases in IQ (Kimler, 1998). The impact of heat on procedures like corticogenesis has been less completely investigated. Spillage and interstitial oedema (Nilsen, 1985) may cause vascular disturbance to the endothelium of vessels. It has been proposed as the reason for various imperfections, following introduction during late early stage or ahead of schedule to-mid-fetal stages including hypoplasia of appendages and toes, exomphalos (umbilical hernia), Moebius disorder (harm to neurones of various brainstem cores) and arthrogryposis (harm to motor neurons of the spinal cord with loss of appendages or body trunk).

Abortions could be due to severe embryonic damage that is incompatible with continuing development or to maternal factors including increased uterine activity, placental damage and changes in the endocrine status of the mother. The temperature of pregnant monkeys that were near term was raised to 41-42^{0}C, causing a two-fold increment in uterine movement during exposure, which may start premature births not long after exposure (Morishima, 1975). Kline et al. (1985) interviewed group of 1057 ladies to test a potential relationship between maternal fever (37.8⁰C or more) and unconstrained fetus removal in ladies who delivered at or after 28 weeks incubation and a group of 1835 ladies who had an unconstrained premature birth before 28 weeks. Some of the babies are karyotyped and, among the euploid hatchlings (ordinary chromosome supplement), fevers happened in 18% of the premature births and in 7.1% of the controls, suggesting fever during pregnancy is a hazard factor for unconstrained premature birth (Kline et al, 1985).

IMPACT ON HUMAN

Elevated maternal temperature during pregnancy may result in embryo death, delayed development, abortion, and various embryonic defects such as cell proliferation, migration, differentiation, and apoptosis, or programmed cell death. Increased maternal temperature during pregnancy can result in embryonic or fetal death, structural and functional defects, either through fever or some other forms of heating and changes in maternal physiology. The elevated temperature of 38^0C for more than 10 minutes can lead to neural tube defects in foetus and miscarriages (Smith et al, 1992). According to Edward et al. (2005) an elevation of maternal body temperature by 2^0C for at least 24 hours during fever can cause a range of developmental defects. Walsh et al. (1994) studied that if the embryo persists at 42^0C serious defects occur. It is probable that the protective response may be triggered at lower temperature elevations without a time of recovery at normal temperature. Prolonged or prolonged heat doses usually result in decreased placenta and fetal weights due to reduced maternal blood flow to the uterus and suspension of the standard protein synthesis in both the placenta and foetus.

The prenatal development in humans is divided into three stages;

1. *Pre–implantation period,* which extends from the time of fertilization to settling of embryo with the wall of uterus.
2. *Organogenesis period,* in which organs are formed from embryonic cells
3. *Foetal period,* which are characterized by growth of already formed organs takes place.

When embryos are exposed to heat before implantation in the wall of the uterus, they are either lost or fail to implant. Hyperthermia causes death at any stage, but during organogenesis, various processes such as proliferation, differentiation, migration are particularly susceptible, resulting in developmental deficiencies in any organ or structure and damage to cell membrane. In particular early rapid phase hyperthermia may cause defective developmental cell proliferation which forms the organs during sensitive periods of organogenesis. Similar structures are susceptible to damage at various stages of embryonic development. In pre-implantation stages, elevated heat can cause embryonic death before organs begin to form. The embryo is vulnerable to damage to the brain, eyes, ears, neck, vertebrae and ribs during the early stages of organogenesis, when neural plate and neural tubes begin to develop. Throughout the mid-embryonic stage, neural cells forming the brain start to proliferate leading to microencephaly that affects all parts of the brain. Also defective spinal cord development, small or missing toes, tooth defects, umbilical hernia, cataract, split iris, blindness, leading problems and seizures are very common.

Hypothermia also causes loss of motor neurons in the spinal cord during the fetal phase, leading to the development of arthrogryposis showing stiffness in limb joints, spine, muscle atropy and paralysis of several cranial nerves (Edward et al., 2003). A number of studies are focused on the effects of fever, while others provide evidence on other causes of maternal hyperthermia such as hot tub use, saunas or extreme hot work. During the first 12 weeks of pregnancy, McDonald (1958; 1961) found substantial rises during pregnancies and malformations in children of mothers with febrile disease. Of a group of 27 women working in hot laundry conditions, four had a child with significant abnormalities. The defects included anencephaly, hydrocephalus, heart defects, cleft lip and palate. Maternal hyperthermia can also increase the occurrence of teratologies (Graham et al., 1998). Even in the absence of these severe developmental defects, hyperthermia can lead to changes in physiological function in adulthood, suggesting maternal heat stress also influence the physiology of the offspring during adulthood (Barker, 2007).

EFFECT ON COW

Al Katanni et al. (2002) observed a high incidence of early embryonic mortality during hot seasons. Elevated temperature has a direct effect on follicular development and oocyte competence according to authors. Cows activate adipose stores during heat stress, resulting in high rates of non-esterified fatty acids (Wheelock et al., 2010), which have been shown to decrease oocyte and embryo quality in vitro studies (Wathes et al., 2007). During heat stress when in vitro hydrogen peroxide is applied to cow embryos, super oxide anion and hydroxyl radicals are elevated (Sakatani et al., 2004), which impairs embryonic pre-implantation growth (Guerin et al., 2001). Apoptosis is also caused by heat stress (Roth and Hansen, 2004). Maternal heat stress in cows decreases the placental hormone levels, disrupting placental activity and causing it embryonic death before implantation (Erdman and Iwaniuk, 2017). These factors could disrupt the intrauterine environment for both embryos and uterine tissue. According to Soto and Smith (2009), summer heat reduces luteal activity, resulting in decreased levels of progesterone in luteinized granulose cells, theca cells and plasma (Pleat et al., 1981).

The pre-implantation embryo is prone to maternal heat stress but as development progresses, the sensitivity decreases. Ealy et al. (1993) observed in cattle that day 1 after oestrus, when embryos were one to two cells, exposure of lactating cows to heat stress decreased the proportion of embryos that were blastocyst stage 8 after oestrus. Nevertheless, heat stress on days 3 (8-16 cells), 5 (morula) and 7 (blastocyst) did not influence the proportion of embryos that were blastocysts on day 8 (Ealy et al., 1993). According to Wolfensen et al. (2000) any effects of elevated temperature on embryonic survival *in utero*, might have been the result of changes in maternal physiology rather than a direct effect on the embryo. Reduced section of placental hormones due to heat stress may result in reduced milk yields in cows (Collier et al., 1982). Thus inadequate nutrition for the neonate during gestation could possibly be one consequence of maternal heat stress. Higher body temperatures in the summer can decrease development, milk production and fertility in cows. According to Berman et al. (1985), the upper limit of ambient temperature at which high yielding dairy cows can sustain a healthy body temperature is 25-26°C and practices should be implemented at ambient temperature above 25°C to reduce the increase in body temperature. Despite elevated body temperature, milk production declines dramatically (Bohmanova et al., 2007). In dairy cows, high temperature influences ovaries by inhibiting follicular growth and oocyte production. This also decreases inhibin rates by increasing the size decrease of the first wave dominant follicle and the development of the second dominant follicle (Roth et al., 2000).

IMPACT ON SHEEP

During gestation, heat stress causes decreased fetal development. Wallace et al. (2005) exposure of pregnant sheep to heat stress causes decreased fetal and placental weights and concentrations of placental hormones in the blood; growth effects are greater in mid gestation than in later gestation. Those heat stress effects on placental function reflect redistribution of blood to the periphery and reduced perfusion of the placental vascular bed (Alexander et al., 1987). Heat stress has more impact during mid gestation than late gestation, according to Thureen et al. (1992), since angiogenesis in the former phase is more severe. Maternal heat stress also decreases the transport ability of glucose through the placenta and this effect includes decreased expression of GLUT8 genes in cotyledonary placenta (Limesand et al., 2004).

IMPACT ON GUINEA PIG

Effects of hyperthermia on the pre-natal development of the guinea pigs caused abortions between 11-18 days. High incidence of neural tube defects and skeletal defects in embryos on days 23 or 24 were also observed (Cawdell et al., 1992). In embryos of 21 days, neuroepithelial cells in mitosis died following elevation of maternal body temperature by at least 2^{0}C. Pregnant pigs exposed to ambient temperature of 35-36.7^{0}C on days 1-5 showed severe embryonic loss. Increase of body temperature by heat stress also decreases food intake and body weight (Renaudeau and Noblet, 2001). Rising body temperature by heat stress often decreases food intake and body weight (Renaudeau and Noblet, 2001). In guinea pigs, heat stress *in utero* reduces learning activity in adulthood (Jonson et al., 1976). During cell division, a threshold dose of heat causes cell death, damages embryonic brain and face. Within minutes of a moderate elevation of temperature 2-3^{0}C given to guinea pig embryos during early stages of brain development, cells in the brain in mitosis show clumping of chromosomes and die (Upfold et al., 1989).

IMPACT ON RAT AND MOUSE

Heat stress affects the reproductive functions significantly with decreased pregnancy in rats. Heat stress increases gonadotropin receptor levels and granulose cell aromatase production and oestradiol concentrations of follicular fluid (Shimizu et al., 2005). Poor trophoectoderm (TE) development, particularly TE cell number and TE lineage-specific gene expression, is more likely to be caused during pregnancy and decreased developmental competency in heat stressed embryos. It is well known that exposure to high temperature of early embryos has an adverse effect on cleavage, blastocyst yield or early fetal growth in mice (Rivera and Hansen, 2001) and the heat stress effects varied depending upon experimental conditions, including species, strain, sex, duration of heat stress, timing of heat stress and culture medium composition (Sakatani et al., 2012). The mammalian testicular temperature changes under the central body temperature somewhere in the range between 2-8^{0}C (Durairajanayagam et al., 2015), and its intensity is important to ensure the production between ordinary sperm and semen.

Scrotal heat stress (SHS) can occur in males and may affect sperm quality. Occupational heat production, for example, that experienced by proficient drivers and cooks, can essentially increment scrotal temperature (Bujan et al., 2000) and likely lead to increasingly second rate sperm quality (Figa-Talamanca et al., 1996). The SHS effects on the testis, epididymis, and spermatozoa have been studied in diverse classes of mammals, as well as man (Alves et al., 2016; Banks et al., 2005; Cai et al., 2011; Jung and Schuppe, 2007; Kanter et al., 2013; Lucio et al., 2016; Lue et al., 2002; Rao et al., 2015; Rockett et al., 2001). Typical heat treatment in these examinations comprised of the utilization of transient heat at 42⁰C to 43⁰C for 30 minutes (Banks et al., 2005; Cai et al., 2011; Ghasemi et al., 2009; Perez-Crespo et al., 2008; Rockett et al., 2001). However, the heat to which they are exposed is typically constant rather than intermittent for sterile males subjected to SHS effects. A regular and mild scrotal heat stress (fmSHS) at a suitable temperature is required for efficient emulation of human SHS conditions. As the mouse 's entire spermatogenesis cycle is around five weeks (Nagy et al., 2003), SHS should be encouraged to provide warmth tension for at least five weeks to ensure that all sperm extracted is provided.

The results obtained from Wan et al. (2020), envisioned four complementary experiments;

1. SHS at 39⁰C for 30 minutes once every week for five consecutive weeks is a proper mouse model that can reflect the human fmSHS situation.
2. fmSHSis known to decline the fertilization capability of spermatozoa, spoil the subsequent embryo progress, and decreases the two-cell embryo implantation rate during IVF.
3. fmSHS is also a root cause for the aberration in the sex ratio of the progenies by creating an uneven proportion of X/Y chromosome–bearing spermatozoa.
4. fmSHS not only affects sperm conventional qualities but likewise extends to its X/Y chromosome–bearing ratio and its utility in deep levels after fertilization.

Figure 1. Schematic diagram of heat stress on reproductive output of male and female mammals (Adopted from Takahashi, 2011).

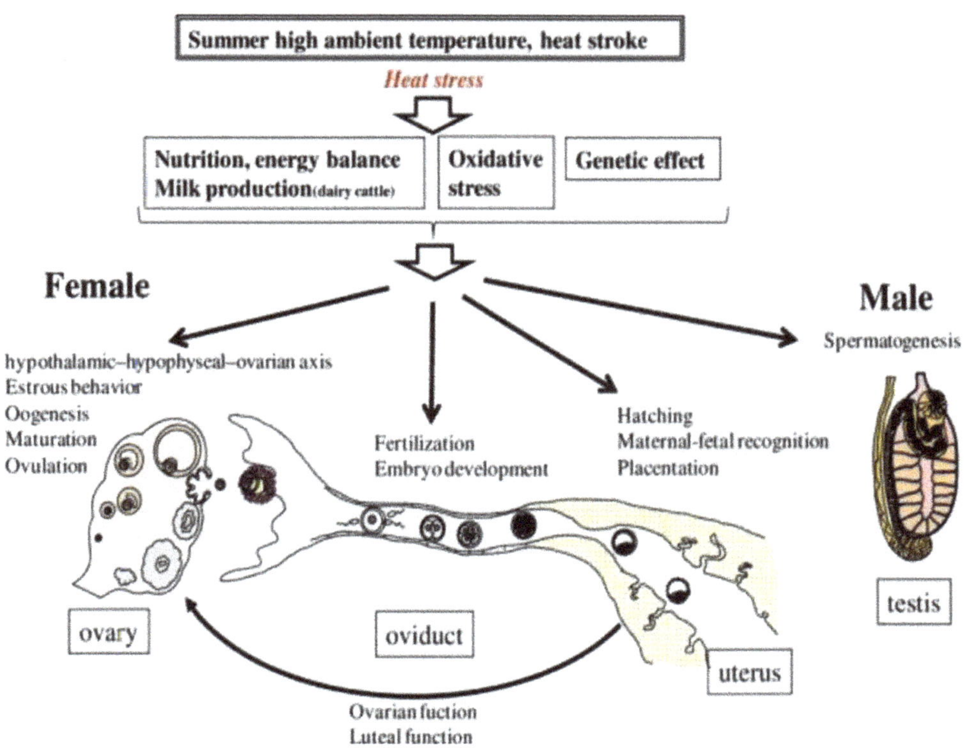

Wan et al. (2020), inferred that SHS at 39^{0}C for 30 minutes once a week for five back to back weeks is a frequent, and mild SHS (fmSHS) that imitates the most basic human scrotal warmth condition. fm-SHS not just lessens the fertilization capacity of spermatozoa yet in addition debilitates typical embryo development in the IVF process. Furthermore, fmSHS initiates progeny sex proportion distortion by evolving the X/Y-bearing spermatozoa proportion. These outcomes may assist with revealing new insight into the infertility treatment of males with scrotal heat risk and the wellbeing doubts of the posterity spread from their parents.

The overall scenario of heat stress on male and female reproductive outcome is shown in Fig. 1.

CONCLUSION

Climate change main has negative impacts on physiological and mental capacities of most mammalian organisms including humans. Animals can be exposed to high temperature levels which can affect their biological activities. Furthermore, high ambient temperatures cause a decrease in the duration and strength of the estrus by disrupting the ovarian ability just as the pregnancy rate after art decreases. Subsequently, an awareness of the impacts of climate change embryo implantation is crucial in order to increase domestic animal production. In dairy cows, high temperatures decrease appetite, weight gain and milk yield. This even unfavorably affects all genders' intellectual execution. It decreases spermatogenic action in males while in females it unfavorably impacts oogenesis, development of oocytes, development of fertilization, and rate of implantation. Recognition and assessment of the deteriorating effects of climate change on conceptional organs and cells will help improve animal regenerative capacity.

AUTHOR CONTRIBUTION

JI, KAW, JAM equally contributed to conceive and design the book chapter along with literature survey and wrote the book chapter.

FUND DISCLAIMER

No grant received from any funding agency in public, commercial, or not-for-profit sector.

CORRESPONDING AUTHOR

Khursheed Ahmad Wani

REFERENCES

Al. Katanani, Y. M., Paula Lopes, F. F., & Hansen, P. J. (2002). Effect of season and exposure to heat stress on oocyte competence in Holstein cows. *Journal of Dairy Science*, *85*(2), 390–396. doi:10.3168/jds.S0022-0302(02)74086-1 PMID:11913699

Alexander, G., Hales, J. R., Stevens, D., & Donnelly, J. B. (1987). Effects of acute and prolonged exposure to heat on regional blood flows in pregnant sheep. *Journal of Developmental Physiology*, *9*, 1–15. PMID:3559062

Alves, M. B., Andrade, A. F., Arruda, R. P., Batissaco, L., Florez-Rodriguez, S. A., Oliveira, B. M., Torres, M. A., Lanconi, R., Ravagnani, G. M., Prado Filho, R. R., Vellone, V. S., Losano, J. D., Franci, C. R., Nichi, M., & Celeghini, E. C. (2016). Recovery of normal testicular temperature after scrotal heat stress in ramsassessed by infrared thermography and its effects on seminal characteristics and testosterone bloodserum concentration. *Theriogenology*, *86*(3), 795–805.e792. doi:10.1016/j.theriogenology.2016.02.034 PMID:27045627

Banks, S., King, S. A., Irvine, D. S., & Saunders, P. T. (2005). Impact of a mild scrotal heat stress on DNAintegrity in murine spermatozoa. *Reproduction (Cambridge, England)*, *129*(4), 505–514. doi:10.1530/rep.1.00531 PMID:15798026

Barker, D. J. (2007). The origins of the developmental origins theory. *Journal of Internal Medicine*, *261*(5), 412–417. doi:10.1111/j.1365-2796.2007.01809.x PMID:17444880

Berman, A., Folman, Y., Kaim, M., Mamen, M., Herz, Z., Wolfenson, D., Arieli, A., & Graber, Y. (1985). Upper critical temperatures and forced ventilation effects for high yielding dairy cows in a subtropical climate. *Journal of Dairy Science*, *68*(6), 1488–1495. doi:10.3168/jds.S0022-0302(85)80987-5 PMID:4019887

Bohmanova, J., Misztal, I., & Cole, J. B. (2007). Temperature humidity indices as indicators of milk production losses due to heat stress. *Journal of Dairy Science*, *83*(4), 2120–2125. doi:10.3168/jds.2006-513 PMID:17369235

Breen, J. G., Claggett, T. W., Kimmel, G. L., & Kimmel, C. A. (1999). Heat shock during rat embryo development in vitro results in decreased mitosis and abundant cell death. *Reproductive Toxicology (Elmsford, N.Y.)*, *13*(1), 31–39. doi:10.1016/S0890-6238(98)00056-2 PMID:10080297

Bujan, L., Daudin, M., Charlet, J. P., Thonneau, P., & Mieusset, R. (2000). Increase in scrotal temperaturein car drivers. *Human Reproduction (Oxford, England)*, *15*(6), 1355–1357. doi:10.1093/humrep/15.6.1355 PMID:10831568

Cai, H., Ren, Y., Li, X. X., Yang, J. L., Zhang, C. P., Chen, M., Fan, C. H., Hu, X. Q., Hu, Z. Y., Gao, F., & Liu, Y. X. (2011). Scrotal heat stress causes a transient alteration in tight junctions and induction of TGF-betaexpression. *International Journal of Andrology*, *34*(4pt1), 352–362. doi:10.1111/j.1365-2605.2010.01089.x PMID:20633196

Cawdell, J., Upfold, J., Edwards, M., & Smith, M. (1992). Neural tube and other developmental anomalies in the guinea pig following maternal hyperthermia during early neural tube development. *Teratogen Carcinogen Mutagen, 12*, 1-9.

Collier, R. J., Doelger, S. G., Head, H. H., Thatcher, W. W., & Wilcox, C. J. (1982). Effects of heat stress during pregnancy on maternal hormone concentrations, calf birth weight and postpartum milk yield of Holstein cow. *Journal of Animal Science*, *54*(2), 309–319. doi:10.2527/jas1982.542309x PMID:7076593

Coss, R. A., Dewey, W. C., & Bamburg, J. R. (1982). Effects of hyperthermia on dividing Chinese hamster ovary cells and microtubules in vitro. *Cancer Research*, *42*, 1059–1071. PMID:7199378

Durairajanayagam, D., Agarwal, A., & Ong, C. (2015). Causes, effects and molecular mechanisms oft-esticular heat stress. *Reproductive Biomedicine Online*, *30*(1), 14–27. doi:10.1016/j.rbmo.2014.09.018 PMID:25456164

Ealy, A. D., Drost, M., & Hansen, P. J. (1993). Developmental changes in embryonic resistance to adverse effects of maternal heat stress in cows. *Journal of Dairy Science*, *76*(10), 2899–2905. doi:10.3168/jds.S0022-0302(93)77629-8 PMID:8227617

Edwards, J. L., Saxton, A. M., Lawrence, J. L., Payton, R. R., & Dunlap, J. R. (2005). Exposure to a physiologically relevant elevated temperature hastens in vitro maturation in bovine oocytes. *Journal of Dairy Science*, *88*(12), 4326–4333. doi:10.3168/jds.S0022-0302(05)73119-2 PMID:16291624

Edwards, M. J. (1971). The experimental production of arthrogryposis multiplex congenita in guinea pigs by maternal hyperthermia during gestation. *The Journal of Pathology*, *104*(4), 221–229. doi:10.1002/path.1711040403 PMID:5135658

Edwards, M. J., Saunders, R. D., & Shiota, K. (2003). Effects of heat on embryos and foetuses. *International Journal of Hyperthermia*, *19*(3), 295–324. doi:10.1080/0265673021000039628 PMID:12745973

Edwards, M. J., Walsh, D. A., & Li, Z. (1999). Hyperthermia, teratogenesis and the heat shock response in mammalian embryos in culture. *The International Journal of Developmental Biology*, *41*, 345–358. PMID:9184344

Figa-Talamanca, I., Cini, C., Varricchio, G. C., Dondero, F., Gandini, L., Lenzi, A., Lombardo, F., Angelucci, L., DiGrezia, R., & Patacchioli, F. R. (1996). Effects of prolonged autovehicle driving on malereproduction function: A study among taxi drivers. *American Journal of Industrial Medicine*, *30*, 750–758. doi:10.1002/(SICI)1097-0274(199612)30:6<750::AID-AJIM12>3.0.CO;2-1 PMID:8914722

Garcia-Ispierto, I., Lopez-Gatius, F., Santolaria, P., Yaniz, J. L., Nogareda, C., Lopez-Bejar, M., & Rensis, F. (2005). Relationship between heat stress during the peri-implantation period and early fetal loss in dairy cattle. *Theriogenology*, *2006*(65), 799–807. PMID:16085298

Ghasemi, N., Babaei, H., Azizallahi, S., & Kheradmand, A. (2009). Effect of long-term administration ofzinc after scrotal heating on mice spermatozoa and subsequent offspring quality. *Andrologia*, *41*(4), 222–228. doi:10.1111/j.1439-0272.2009.00920.x PMID:19601933

Graham, J. M. Jr, Edwards, M. J., & Edwards, M. J. (1998). Teratogen update: Gestational effects of maternal hyperthermia due to febrile illness and result patterns of defects in human. *Teratology*, *58*(5), 209–221. doi:10.1002/(SICI)1096-9926(199811)58:5<209::AID-TERA8>3.0.CO;2-Q PMID:9839360

Guerine, P., Mouatassim, S., & Menezo, Y. (2001). Oxidative stress and protection against reactive oxygen species in the pre-implantation embryo and its surroundings. *Human Reproduction Update*, *7*(d), 175–189. doi:10.1093/humupd/7.2.175 PMID:11284661

Hicks, S. P., & D'Amato, C. P. (1966). Effects of ionizing radiation on mammalian development. In D. H. M. Woollam (Ed.), *Advances in Teratology* (pp. 196–243). Logos Press.

Hinoue, A., Fushiki, S., Nishimura, Y., & Shiota, K. (2001). In utero exposure to brief hyperthermia interferes with the production and migration of neocortical neurons and induces apoptotic cell death in the fetal mouse brain. *Dev Brain Res*, *132*, 59–67.

Huynh, T. T., Aarnink, A. J., Verstegen, M. W., Gerrits, W. J., Heetkamp, M. J., Kemp, B., & Canh, T. T. (2005). Effects of increasing temperatures on physiological changes in pigs at different relative humidities. *Journal of Animal Science*, *83*(6), 1385–1396. doi:10.2527/2005.8361385x PMID:15890816

Inchul, C., & Amarnath, D. (2015). Effects of prolonged exposure of mouse embryos to elevated temperature on embryonic developmental competence. *Reproductive Biomedicine*, *31*(2), 171–179. doi:10.1016/j.rbmo.2015.04.017

Jockenhovel, F., Grawe, A., & Nieschlag, E. (1990). A portable digital data recorder for long-termmonitoring of scrotal temperatures. *Fertility and Sterility*, *54*, 694–700. doi:10.1016/S0015-0282(16)53832-1 PMID:2209892

Jonson, K. M., Lyle, J. G., Edwards, M. J., & Penny, R. H. C. (1976). Effect of prenatal heat stress on brain growth and serial discrimination reversal learning in the guinea pig. *Brain Research Bulletin*, *1*(1), 135–150. doi:10.1016/0361-9230(76)90056-3 PMID:974792

Jung, A., & Schuppe, H. C. (2007). Influence of genital heat stress on semen quality in humans. *Andrologia*, *39*(6), 203–215. doi:10.1111/j.1439-0272.2007.00794.x PMID:18076419

Kanter, M., Aktas, C., & Erboga, M. (2013). Heat stress decreases testicular germ cell proliferation andincreases apoptosis in short term: An immune histochemical and ultrastructural study. *Toxicology and Industrial Health*, *29*(2), 99–113. doi:10.1177/0748233711425082 PMID:22082826

Kimler, B. F. (1998). Prenatal irradiation: A major concern for the developing brain. *International Journal of Radiation Biology*, *73*(4), 423–434. doi:10.1080/095530098142266 PMID:9587081

Kline, J., Stein, Z., Susser, M., & Warburton, D. (1985). Fever during pregnancy and spontaneous abortion. *American Journal of Epidemiology*, *121*(6), 832–842. doi:10.1093/oxfordjournals.aje.a114054 PMID:4014176

Lepock, J. R. (1982). Involvement of membranes in cellular responses to hyperthermia. *Radiation Research*, *92*(3), 433–438. doi:10.2307/3575914 PMID:7178412

Limesand, S., Regnault, T.R., & Hay, W.W. (2004). Characterization of of glucose transporter 8 (GLUT8) in the ovine placenta of normal and growth restricted fetuses. *Placenta, 25*,70 – 77.

Lucio, A. C., Alves, B. G., Alves, K. A., Martins, M. C., Braga, L. S., Miglio, L., Alves, B. G., Silva, T. H., Jacomini, J. O., & Beletti, M. E. (2016). Selected sperm traits are simultaneously altered after scrotal heatstress and play specific roles in in vitro fertilization and embryonic development. *Theriogenology*, *86*(4), 924–933. doi:10.1016/j.theriogenology.2016.03.015 PMID:27087533

Lue, Y. H., Lasley, B. L., Laughlin, L. S., Swerdloff, R. S., Hikim, A. P., Leung, A., Overstreet, J. W., & Wang, C. (2002). Mild testicular hyperthermia induces profound transitional spermatogenic suppression throughincreased germ cell apoptosis in adult cynomolgus monkeys (Macacafascicularis). *Journal of Andrology*, *23*, 799–805. PMID:12399525

McDonald, A. D. (1958). Maternal health and congenital defect. *The New England Journal of Medicine*, *258*(16), 767–773. doi:10.1056/NEJM195804172581601 PMID:13541657

McDonald, A. D. (1961). Maternal health in early pregnancy and cogeital defect. Final report on a prospective inquiry. *British Journal of Preventive & Social Medicine*, *15*, 154–166.

Morishima, H. O., Glaser, B., Biermann, W. H., & James, L. S. (1975). Increased uterine activity and fetal deterioration during maternal hyperthermia. *American Journal of Obstetrics and Gynecology*, *121*(4), 531–538. doi:10.1016/0002-9378(75)90087-3 PMID:807107

Nagy, A., Gertsenstein, M., Vintersten, K., & Behringer, R. (2003). *Manipulating the mouse embryo: a laboratory manual* (Vol. 10). Cold Spring Harbor Laboratory Press.

Nilsen, N. O. (1985). Vascular abnormalities due to hyperthermia in chick embryos. *Teratology*, *30*(2), 237–251. doi:10.1002/tera.1420300211 PMID:6495224

Otak, E. M., & Schull, W. J. (1998). Review: radiation-related brain damage and growth retardation among the prenatally exposed bomb survivors. *International Journal of Radiation Biology.*, *74*, 159–71.

Ozawa, M., Tabayashi, D., Latief, T. A., Shimizu, T., & Kanal, Y. (2005). Alterations in follicular dynamics and Steroidogenic abilities induced by heat stress during follicular recruitment in goats. *Reproduction (Cambridge, England)*, *2005*(129), 621–630. doi:10.1530/rep.1.00456 PMID:15855625

Perez-Crespo, M., Pintado, B., & Gutierrez-Adan, A. (2008). Scrotal heat stress effects on sperm viability,sperm DNA integrity, and the offspring sex ratio in mice. *Molecular Reproduction and Development*, *75*(1), 40–47. doi:10.1002/mrd.20759 PMID:17474098

Pleet, H., Graham, J. M., & Smith, D. W. (1981). Central nervous system and facial defects associated with maternal hyperthermia at 4 to 14 weeks gestation. *Pediatrics*, *67*, 785–789. PMID:7232042

Prosser, C. L., Heath, J. E., & Prosser, C. L. (1991) Temperature. *Comparative animal physiology environmental and metabolic animal physiology*. New York: Wiley.

Rao, M., Zhao, X. L., Yang, J., Hu, S. F., Lei, H., Xia, W., & Zhu, C. H. (2015). Effect of transient scrotalhyperthermia on sperm parameters, seminal plasma biochemical markers, and oxidative stress inmen. *Asian Journal of Andrology*, *17*(4), 668–675. doi:10.4103/1008-682X.146967 PMID:25652627

Renaudeau, D., & Noblet, J. (2001). Effects of exposure to high ambient temperature and dietary protein level on sow milk production and performance of piglets. *Journal of Animal Science*, *79*(6), 1540–1548. doi:10.2527/2001.7961540x PMID:11424692

Rivera, R. M., & Hansen, P. J. (2001). Development of cultural bovine embryos after exposure to high temperature in the physiological range. *Reproduction (Cambridge, England)*, *121*, 107–115. doi:10.1530/rep.0.1210107 PMID:11226033

Rockett, J. C., Mapp, F. L., Garges, J. B., Luft, J. C., Mori, C., & Dix, D. J. (2001). Effects of hyperthermia onspermatogenesis, apoptosis, gene expression, and fertility in adult male mice. *Biology of Reproduction*, *65*(1), 229–239. doi:10.1095/biolreprod65.1.229 PMID:11420244

Roth, Z., & Hansen, P. J. (2009). Involvement of apoptosis in disruption of developmental competence of bovine oocytes by heat shock during maturation. *Biology ofReproduction*, *71*(6), 1898–1906. doi:10.1095/biolreprod.104.031690 PMID:15306551

Sakatani, M., Alvarez, N. V., Takahashi, M., & Hansen, P. J. (2012). Consequences of physiological heat shock beginning at the zygote stage on embryonic development and expression of stress response genes in cattle. *Journal of Dairy Science*, *95*(6), 3080–3091. doi:10.3168/jds.2011-4986 PMID:22612944

Sakatani, M., Kobayashi, S. I., & Takahashi, M. (2004). Effects of heat shock on invitro development and intracellular oxidative state of bovine pre-implantation embryos. *Molecular Reproduction and Development*, *67*(1), 77–82. doi:10.1002/mrd.20014 PMID:14648877

Sartori, R., Sartor-Bergfelt, R., Mertens, S. A., Guenther, J. N., Parrish, J. J., & Wiltbank, M. C. (2002). Fertilization and early embryonic development in heifers and lactating cows in summer and lactating and dry cows in winter. *Journal ofDairy Science*, *85*(11), 2803–2812. doi:10.3168/jds.S0022-0302(02)74367-1 PMID:12487447

Shimizu, T., Ohshima, I., Ozawa, M., Takahashi, S., Tajima, A., Shiota, M., Miyazaki, H., & Kanai, Y. (2005). Heat stress diminishes gonadotropin receptor expression and enhances susceptibility to apoptosis of rat granulose cells. *Reproduction*, *129*, 46-472.

Smith, M. S. R., Upfold, J. B., Edwards, M. J., Shiota, K., & Cawdell-Smith, J. (1992). The induction of neural tube defects by maternal hyperthermia; a comparison of the guinea-pig and human. *Neuropathology and Applied Neurobiology*, *18*(1), 71–80. doi:10.1111/j.1365-2990.1992.tb00765.x PMID:1579201

Soto, P., & Smith, L. C. (2009). BH4 peptide derived from Bel –Xl and Bax inhibitor peptide suppresses apoptotic mitochondrial changes in heat stressed bovine oocytes. *Molecular Reproduction and Development*, *76*(7), 637–646. doi:10.1002/mrd.20986 PMID:19062170

Takahashi, M. (2011). Heat stress on reproductive function and fertility in mammals. *Reproductive Medicine and Biology*, *11*(1), 37–47. doi:10.100712522-011-0105-6 PMID:29699104

Thureen, P.J., Trembler, K.A., Meschia, G., Makowski, E.L., & Wilkening, R.B. (1992). Placental glucose transport in heat induced fetal growth retardation. *American Journal of Physiology, 263*, R578 – R585.

Upfold, J., Smith, M. S. R., & Edwards, M. J. (1989). Quantitative study of the effects of maternal hyperthermia on cell death and proliferation in the guinea pig brain on day 21 of pregnancy. *Teratology*, *39*(2), 173–179. doi:10.1002/tera.1420390209 PMID:2928964

Van Bergenen, H. (1985). Studies on a possible relationship between alterations in the cytoskeleton and induction of heat shock protein synthesis in mammalian cells. *International Journal of Hyperthermia*, *1*, 59–83.

Wallace, J. M., Regnault, T. R., Limesand, S. W., Hay, W. W. Jr, & Anthony, R. V. (2005). Investigating the causes of low birth weight in contrasting ovine paradigms. *The Journal of Physiology*, *565*(1), 19–26. doi:10.1113/jphysiol.2004.082032 PMID:15774527

Walsh, D., Li, K., Wass, J., Dolnikov, A., Zeng, F., & Edwards, M. (1993). Heat-shock gene expression and cell cycle changes during mammalian embryonic development. *Development & Genetics, 14*, 127–36.

Walsh, D., Zhe, L., Zeng, F., Yan, W., & Edwards, M. (1994). Heat shocks genes and cell cycle regulation in early mammalian development. *Environmental Medicine*, *38*, 1–6.

Wan, X., He, X., Liu, Q., Duan, P., & Li, H. (2020). Frequent and mild scrotal heat stress impairs embryo development, implantation, and offspring sex ratio in mice. *Reproductive Biomedicine Online*, *40*(5), 617–626. doi:10.1016/j.rbmo.2020.01.020 PMID:32276888

Wanner, R. A., Edwards, M. J., & Wright, R. G. (1976). The effect of hyperthermia on the neuroepithelium of the 21-day guinea-pig foetus: Histologic and ultrastructural study. *The Journal of Pathology*, *118*(4), 235–244. doi:10.1002/path.1711180406 PMID:1271135

Wathes, D. C., Fenwick, M., Cheng, Z., Bourne, N., Liewellyn, S., & Morris, D. G. (2007). Influence of negative energy balance on cyclicity and fertility in the high producing dairy cow. *Theriogenology*, *68*(Suppl 1), S232–S241. doi:10.1016/j.theriogenology.2007.04.006 PMID:17475319

Webster, W. S., & Edwards, M. J. (1984). Hyperthermia and the induction of neural tube defects in mice. *Teratology*, *29*(3), 417–425. doi:10.1002/tera.1420290313 PMID:6463904

Wettemann, R. P., & Bazer, F. W. (1985). Influence of environmental temperature on prolificacy of pigs. *Journal of Reproduction and Fertility. Supplement*, *1985*(33), 199–208. PMID:3910825

Wheelock, J. B., Rhoads, R. P., Vanbaala, M. J., Sanders, S. R., & Baumgard, I. H. (2010). Effects of heat stress on energetic metabolism in lactating Holstein cows. *Journal of Dairy Science*, *93*(2), 644–655. doi:10.3168/jds.2009-2295 PMID:20105536

Wolfensen, D., Roth, Z., & Meidan, R. (2000). Impaired reproduction in heat stressed cattle: basic and applied aspects. *Animal Reproduction Science*, *60-61*, 535-547.

Chapter 6
Impact of Heat Stress and Nutritional Stress on Early Embryo Development

Golden Gokhale
https://orcid.org/0000-0001-7804-9355
Devi Ahilya Vishwavidyalaya, India

Pir Mohammad Ishfaq
https://orcid.org/0000-0002-3534-8248
Dr. Harisingh Gour Central University, Sagar, India

Gurudutt Sharma
Devi Ahilya Vishwavidyalaya, India

ABSTRACT

The nutritional, physiological, and reproductive function has detrimental effects on heat stress, which is found in many species of mammals. High ambient temperature in mammals cause a decrease in the length and intensity of estrus by disturbing ovarian function as well as decreasing pregnancy rate after artificial insemination. The effects of nutritional stress on developing oocytes in the ovarian follicle and in the reproductive tract on early embryos are because of the environment where a breeding female lives before conception and at the early stages of pregnancy. Maturity of oocyte, blastocyst yield, prenatal survival, and the number of offspring born alive are affected by change in consumption and quantity of the food taken during the pre-mating period. To improve reproductive efficiency and offspring quality, it is necessary to detect and evaluate the deteriorating effects of heat stress on reproductive organs and cells and to plan nutrition related strategies.

DOI: 10.4018/978-1-7998-4480-8.ch006

INTRODUCTION

Heat stress is a climatic element which disturbs equilibrium of body temperature. Heat stress is described as an environment change that acts to drive the temperature above the set factor temperature and its effect on the reproduction. Heat stress all through the summer disrupts numerous reproductive tactics ensuing in a pronounced depression of conception rate in dairy cows worldwide. The rise of internal body temperature for the duration of the summer season is considered accountable for the impaired reproduction. On the other hand, because of the nutritional desires of the growing population, there may be a vital call for extra green livestock production mainly in the developing countries. Nutrition plays a critical role during the process of pregnancy in ruminates. Among numerous internal and external factors, nutrients have the efficiency to modify the micro-surroundings of the oocyte and the embryo, making it more adverse to ideal fertilization and pre-implantation of the embryonic growth. Understanding the effect of nutritional stress on the oocyte function, embryo development and reciprocal signaling networks among the embryo and uterus will result in the reduction of early embryonic mortality (Wale and Gardner, 2016). Maternal adaptations also keep the nutritional microenvironment favorable for oocyte development and for the formation of the embryo (Velazquez, 2015). This maternal-embryo relation takes place via several dietary mediators. Although adaptive responses to malnutrition by way of both the mother and the embryo may make way for the blastocyst formation, the resultant of the embryo may be compromised, leading to early pregnancy (Berg et al., 2010). This chapter will mainly determine the influence of heat stress and nutritional stress on early embryo development.

IMPACT OF CLIMATIC SITUATIONS

Climate is an aggregate of factors that consist of temperature, humidity, rainfall, air movement, radiation, barometric pressure and ionization. Climatic zones fluctuate and depend on latitude, winning winds, evaporative conditions, availability of water, elevation, proximity to mountains and other factors (Berner and Berner, 2012). The variety and duration of the ambient temperature is essentially dependent on latitude, with latitudes towards the equator experiencing maximum heat stress (Chandler and Sohl, 2000). A range of changes occur inside the animal as a result of heat stress, especially with elevated body temperature of >39.16°C (regular is 38.61°C) and increased respiration rates of >70-80/ minute.

Increased Maintenance of Energy Requirement

Dairy cows spark off mechanisms in striving conditions to dissipate the excess heat and holding the body temperature. The extended respiration rate is one example. The maintenance of energy requirement may additionally increase by 20-30% in animals underneath heat stress. This decreases the consumption of power available for productive functions which include milk production. Blood drift to the skin will increase during starving conditions to dissipate heat. At the same time, blood flow to the center of the body will get decreased.

Feed Nutrient Utilization

An accelerated lack of sodium and potassium is usually related to heat stress. This is due to the loss related to the increased respiratory rate. This can shift the acid-base balance and result in a metabolic alkalosis that can lead to the decrease in the efficiency of nutrient utilization.

Dry Matter Intake

Dry matter consumption decreases in dairy cows subjected to heat stress. This melancholy in dry matter intake may be both for a brief time period and long term relying on the duration of heat stress. A 10 to 20% decrease in the intake of dry matter of is not unusual in industrial dairy herds.

Milk Production

There is generally low milk production in cows under heat stress. This low milk production can be either transitory or longer time period depending on the duration and severity of heat stress. These decreases in milk production can range from 10 to 25%. If heat stress lowers milk production during early lactation in dairy cows, ability of the milk production for the lactation could be decreased. Dairy cows in later lactation stage may additionally recover slowly from the adverse effects of heat stress.

Reproduction

Heat stress has additionally been suggested to lower reproductive performance in dairy cows. There are some of modifications in the overall reproductive performance in dairy cattle. The outcome of the reproduction may be extended and impact the animal for months after the heat stress exposure.

IMPACT OF HEAT STRESS

Adverse impacts are observed in the case of dairy cows because of heat stress such as extended rate of respiration, sweating and peripheral blood flow that has a vital impact on natural and inorganic composition of milk (Nardone et al., 2010). During heat stress conditions, the consumption of dry matter is decreased as a result, it limits the delivery of vitamins to mammary glands due to which milk production is also decreased. Elevated rate of respiration decreases the extent of circulating carbon dioxide (CO_2), which transforms the blood carbonic acid to bicarbonate equilibrium (Perry and Gilmour, 1993). As an end result, urinary bicarbonate excretion increase and blood pH becomes unstable which can lead to a number of metabolic changes in the dairy cows. At temperature below the required temperature, heat requirement will increase and less energy is stored. Therefore, in the animals that are deficient in feed, body fat is catabolized to provide energy (Council, 1981). Below vital temperature an animal loses ample amounts of heat per unit surface regardless of its dietary level (Silanikove, 2000). Protein metabolism is not improved under these stressed conditions even though heat manufacturing is accelerated to about two times (Alonso et al., 2012).

Male Reproduction

The outcome of heat stress on male mammalian reproduction primarily focus on delaying the sexual maturity of younger animals, letting down the libido of animals, lowering the ejaculate quantity and sperm counts, declining the sperm density, leading to sperm motility and integrity rate of acrosome (Zvi, 2017). Spermatogenesis is controlled with the aid of hypothalamic-pituitary-gonadal (HPG) axis and has to be hatched by using epididymis (Wirbisky and Freeman, 2015). The suitable temperature is essential for spermatogenesis. Studies have revealed that the impact of the environment on male reproduction requires more attention both in its own right and majorly because of the alteration in male physiology may additionally contribute to the reduction in female fertility observed in harsh environments.

- ***Ejaculatory quantity and sperm count***

High environmental temperatures tend to have a detrimental effect on ejaculatory quantity and sperm count. Studies have revealed that heat stress lowers the animal ejaculatory content and sperm count and as the temperature goes higher the extra apparent the decline is (Betteridge, 1986). During the extreme high temperature (40ºC), there might be no addition and recovery, which indicates that the outcome of heat stress to mammal reproduction are extremely obvious. Although animals exhibit some regulations, still they can't repair to the original degree in a quick time. However, studies have revealed that there's little effect of hyperthermia on the ejaculatory amount in Holstein bulls (Wolfenson et al, 1988). By studying the effects of the excessive heat on the semen series in Ashkenazi rabbit, French rabbit and hybrid rabbits, it has been determined that there is no significant effect of hyperthermia on French rabbit and hybrid rabbit possibly because of the insufficient time of exposure to heat or of the weak point of intensity; even as the semen amount decreased substantially in Ashkenazi rabbit, which indicates that exclusive types have varying resistance to heat stress (El-Tarabany et al, 2015). Similar studies confirmed that Jerseys are slightly more tolerant to heat stress than Brown Swiss that, in comparison, is more tolerant than Holstein-Friesians (Vermunt and Tranter, 2011).

- ***Sperm density***

The scrotum and testicles have low ability for heat regulation when exposed to heat stress. This heat stress affects the function of spermatogenesis, inhibit the sperm production and lead to a critical decline in sperm density. It has been revealed by previous studies that Holstein bulls and Brown Swiss farm animals were monitored in a heat chamber with a circulatory temperature from 27.8ºC to 40ºC and after 7 days their sperm density and sperm amount declined (Wahl and Reif, 2009). It's revealed that if the water buffalos were given shower, their sperm density could boom from 1.1 billion/ml to 1.5 billion/ml which indicates that, the effects of heat stress on sperm density might decline with temperature drop (Nandi et al., 1998). Studies suggest that it was dangerous for spermatogenesis underneath heat stress and might lead to reduction in sperm amount (Foote, 1978).

- ***Sperm motility and acrosome integrity rate***

Sperm motility decrease significantly at an excessive temperature. Studies showed that heat stress decreased testicular germ proliferation and expanded apoptosis (Setchell, 2018). Breeding boars were placed in an environment of 33.4–37.7°C and of 80% humidity. An increase of their sperm count and a sizeable decrease of their sperm motility are observed after 2-5 weeks of the experiment with the purpose that hyperthermia may have multiplied the respiration of sperms in epididymis and has fed on a huge number of metabolic substrates, thereby having substantially decreased the sperm power when discharged. As the acrosome is positioned at the head of the sperm and incorporates protease, hyaluronidase, penetrating enzyme and different other enzymes related to fertilization, it is easy to be degenerated and dropped from the top due to the heat stress (Gwatkin, 2012). It has been revealed that the frozen semen has high motility with a low chance of pregnancy after exposure to excess heat, reason is probably that the acrosome is damaged (Holt, 2000).

Female Reproduction

Female mammalian reproduction may be disturbed by way of oocyte differentiation, development, implantation, delivery, sexual functions and secondary sexual characteristics due to the heat stress which can also shorten the estrus of female or anestrus, thereby affecting the mating time, thus reducing the pregnancy rate (Conaway, 1971). Previous studies confirmed that first conception (CRI1) and average conception rates (CR) were lowest within summer and maximum in winter (Takahashi, 2012). In addition, heat stress can cause abortion inside the implantation of females after their smooth breeding. The ovarian pool of oocytes is also sensitive to expanded temperature (Fig. 1). A stage-dependent pattern of

Figure 1. Stage-dependent pattern of resistance/sensitivity of the ovarian pool of follicles and enclosed oocytes to heat stress.

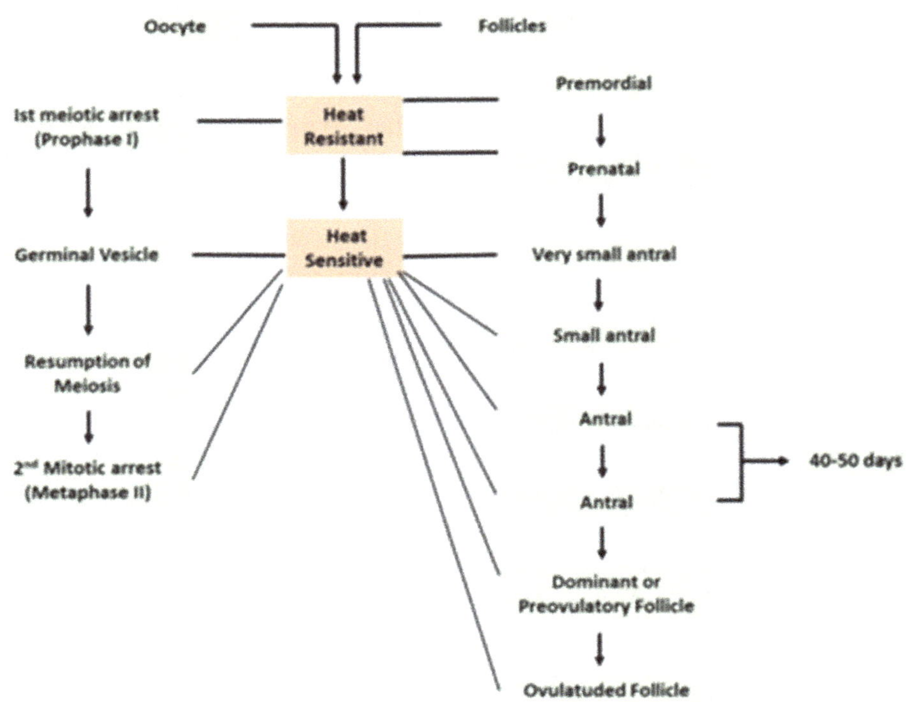

resistance and sensitivity to heat stress of the follicles and their enclosed oocytes have been studied (Roth, 2008). The oocyte acquires its developmental capability in a stepwise way at some stage in follicular improvement and therefore, heat stress cause perturbations in follicular functioning, that can result in reduced competence of its enclosed oocyte.

- ***Oocyte quality***

Alterations in oocyte quality, because of heat stress in all circumstances involve deviations in styles of folliculogenesis. Follicular dominance is decreased in cows exposed to heat stress, so that there may be an increase in variety of large follicles on the ovary. Extended heat stress can result in the ovulation of an aged, poor exceptional oocyte, which is associated with low fertilization and embryonic mortality (Santos et al., 2004). Additionally high temperature negatively influences pre-attachment embryos (Putney et al., 1989). Marked differences had been cited between Bostaurus and Bosindicus on the developmental ability, competence of oocytes and embryos (Zi et al., 2009). Higher resistance of embryos derived from Bosindicus cows to heat stress has been observed in previous studies (Silva et al., 2013). There are direct consequences of improved temperature (heat stress) on the competence of the oocytes undergoing maturation to develop into a blastocyst following fertilization (Ashraf et al., 2014).

- ***Follicular choice and development***

The first reproductive project dealing with the heat stressed cows is about the altered follicular development. Heat stressed cows lower feed intake inflicting less common pulses of the luteinizing hormone (LH), ensuing in longer follicular waves. This lengthening of the follicular wave leads to the selection and ovulation of multiple, smaller dominant follicles. Follicles are useful for producing estrogen, a hormone that causes cows to reveal signs and symptoms of heat. Smaller follicles produce less estrogen than larger ones; therefore, resulting in much less oestrus activity (Mehrotra and Redford, 2005). Ovarian follicles contain oocytes as well as somatic cells that synthesize estradiol. Estradiol has plenty of actions that encompass inflicting oestrus and the LH surge. Heat stress impairs follicle choice and increases the period of follicular waves, which reduces the power of oocytes (Bethea et al, 2008). It additionally allows for a couple of dominant follicles to develop, explaining the improved conceiving in summer season. The somatic cells within the follicles (theca and granulosa cells) are also broken due to the heat stress (Chen et al., 2009).

- ***Uterine environment***

Heat stress compromises uterine environment with decreased blood flow to the uterus and expanded uterine temperature, which can lead to implantation failure and embryonic mortality (Rivera and Hansen, 2001). These consequences are associated with the production of heat-shock proteins via the endometrium in the course of heat stress and reduced production of interferon-tau (IFN-Ꚍ) at the conceptus. Moreover, heat stress can affect endometrial prostaglandin secretion leading to untimely luteolysis and embryo loss (Fitzgerald, 2011). Earlier studies have found huge variations among Brahman and Holstein cows in endometrial responses at extended temperature (Malayer and Hansen, 1990).

Reproductive Hormones

- ***Luteinizing hormone (LH)***

Since the main elements regulating ovaries are gonadotropin liberating hormone from the hypothalamus and the gonadotropins, LH and FSH from the anterior pituitary gland, a few reports are available that suggest the impact of heat stress at the secretion of these hormones. The impact of heat stress on the pre-ovulatory surge of LH is similarly controversial; a reduction of the endogenous LH surge via heat stress changed in heifers (Khan, 2016). It has been revealed that these differences are related to pre-ovulatory estradiol levels because of the amplitude of LH pulses and gonadotropin-releasing hormone (GnRH). Induced pre-ovulatory plasma LH surges are decreased in cows with low plasma concentrations of estradiol but not in cows with excessive plasma concentrations of estradiol (Courtney, 1984). Most studies report that LH secretion is reduced due to heat stress, therefore a conclusion is derived that in summer time, the dominant follicle develops in a low LH surroundings which results in decreased estradiol secretion from the dominant follicle, leading to poor expression of estrus hence, decreased fertility. Plasma inhibin concentrations in summer time are decreased in heat stressed cows (De and Scaramuzzi, 2003) and in cyclic buffaloes in India reflecting decreased folliculogenesis and it has been observed that an enormous percentage of plasma inhibin comes from small and medium length follicles.

- ***Follicle stimulating hormone (FSH)***

FSH secretion increases due to heat stress and is related to a bigger range of follicles growing inside the ovaries. In agreement with this statement other studies confirmed a pronounced decrease in plasma inhibin in heat stressed cows, which in turn brought an increase in plasma FSH level and acknowledged to stimulate follicle increase within the ovaries (Spicer and Echternkamp, 1986; Wolfenson et al., 1995). These alterations might give an explanation for the significant upward thrust of double ovulation and the marked rise in calving of twins following summer insemination.

- ***Progesterone***

The corpus luteum (CL) secretes progesterone, which is vital for embryonic development. Luteal insufficiency refers to the reputation of CL that does not secrete adequate quantities of progesterone to help pregnancy and it has therefore long been associated with low fertility in cows and other female animals (Geisert et al., 1992). Under the situations of summer time heat stress, provision of exogenous progesterone to speed up sub-optimal endogenous progesterone concentrations may increase (Kerbler et al., 1997).

- ***Estradiol***

Heat stress inhibited the secretion of gonadotropins to a greater degree in cows with low plasma concentrations of estradiol compared to those with excessive concentrations (De and Scaramuzzi, 2003). These observations suggest that excessive concentrations of estradiol can counteract the effect of heat stress. Alternatively the neuroendocrine mechanism controlling gonadotropin secretion is greater sensitive to heat stress while concentrations of plasma estradiol are low. It has been found that heat stress may also act at once on the ovary to decrease its sensitivity to gonadotropin stimulation (Bridges et al., 2005). Also the somatic cells inside the follicles (theca and granulosa cells) may be broken because of the heat stress. In the course of steroid production, the thecal cells and granulosa cells are determined to be at risk of heat stress (Roth and Wolfenson, 2016). Regardless of an appropriate mechanism, any alteration inside the secretory activity of the follicle and perhaps the corpus luteum, because of heat stress might be important.

- ***Prostaglandin***

It is proven, that heat stress can increase secretion from endometrium and in some instances early activation of luteolysis mechanism, leading to failure in implantation (Khodaei et al., 2011; Spitz, Croxatto and Robbins, 1996).

Reproductive Efficiency

The deleterious consequences on reproductive efficiency induced by means of heat stress compromise the fertility of females and males (Fig. 2). High body temperature leads to increased uterine temperature.

Figure 2. Long-term effects of seasonal heat stress on the hypothalamic-pituitary-ovarian axis during reduce fertility of lactating cows.

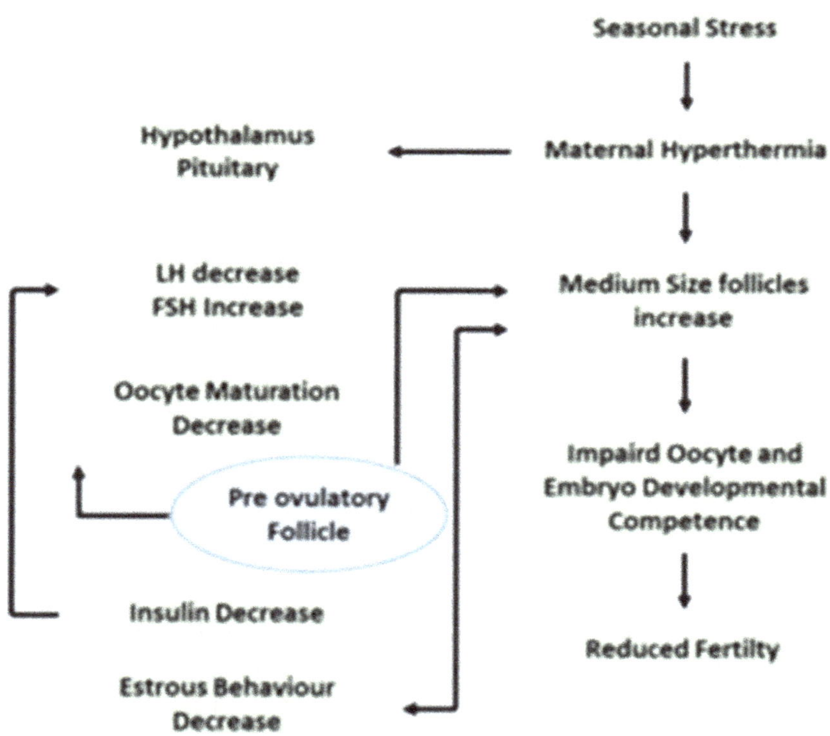

The blood supply is directed to the body periphery to combat heat loss; therefore, the blood flow in the uterus is decreased (Allen et al., 2013). Consequently, there is a decrease in gestation rate, embryonic loss, and decreased availability of vitamins and hormones due to decreased blood supply to uterus. This makes the reproductive environment unsuitable for the sperm to fertilize the ovule and the embryo development (Bridges et al.. 2013). With regard to heat stress results on bulls, the subsequent are discovered: semen decreased, ejaculate attention reduced, motility, vigor and counting of live spermatozoa are reduced. During the period of high temperature, males lower in reproductive performance and their libido (Alves et al., 2017). The defects found during the external examination are testicular diameter changes in texture and mass. Given that mammalian oocytes acquire their developmental capability in a stepwise manner at some point of follicular development, perturbations within the physiology of the follicle-enclosed oocyte can potentially cause an oocyte with reduced competence for fertilization and next development (Moussa et al., 2015).

Nuclear and Cytoplasmic Maturation

In vitro research suggests that heat stress (40°C to 41.5°C) induces changes in each nuclear and cytoplasmic event associated with oocyte maturation (Fig. 3). Exposing bovine oocytes to heat stress at some stage in maturation impairs both microtubulin and microfilaments, which might be a concerning factor in nuclear and organullar transport (Leibfried et al., 1997). As a result most heat stressed oocytes fail to undergo maturation and are arrested at the metaphase (M) I to (M) II stages. It is found, that an extra percentage of heat-stressed oocytes advanced to Metaphase I by means of 8 hour and arrested at Metaphase II at sixteen and 18 hour of maturation, indicating that heat stress multiplied the in vitro maturation of bovine oocytes (Hallberg et al., 2019). In vitro fertilization performed at 18–20 hrs vs. 24 hrs of maturation restored to some extent heat stressed oocytes that had developed to the blastocyst stage (Edwards, Saxton, Lawrence, Payton and Dunlap, 2005). Heat stress is suggested to induce perturbations

Figure 3. Direct and indirect effects of heat stress on oocyte developmental competence.

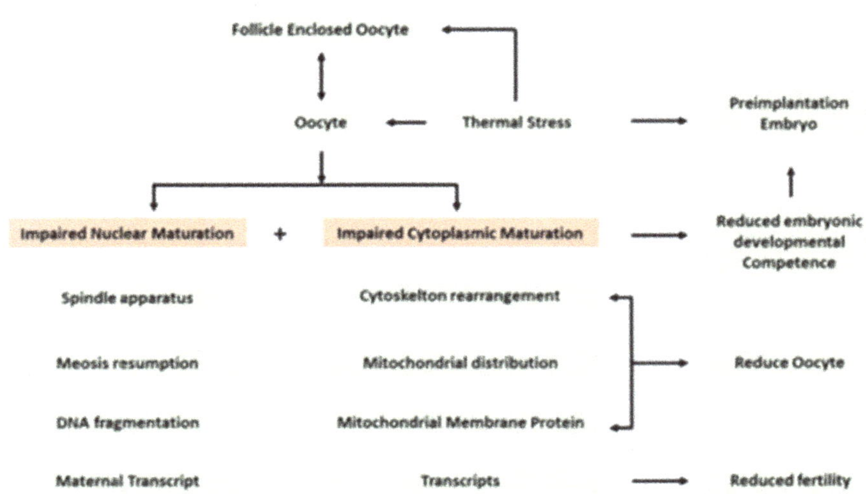

inside the spindle in porcine and bovine oocytes (Rispoli et al., 2011). Cytoplasmic maturation consists of multiple events along with reorganization of the cytoplasmic organelles (i.e., cortical granules, mitochondria and endoplasmic reticulum) through the actions of cytoskeletal microfilament and microtubules. Further it's revealed that heat shock impairs cortical granule translocation (Yamada and Isaji, 2011). Heat prompted alterations in both nuclear and cytoplasmic events related to oocyte maturation most possibly underline the reduced share of oocytes that undergo heat fertilization and further cleave and develop to blastocysts (Edwards et al., 2005).

Apoptosis in Oocytes

Heat stress is found to bring changes in mitochondrial characteristics associated with the activation of apoptotic cascades within the oocyte through intrinsic pathways (Fig. 4). Exposing oocytes to 41°C at some point of maturation has been observed to increase the percentage of oocytes expressing high interest of caspases 2, 3, and 7 thereby, showing nuclear fragmentation (Yang and Rajamahendran, 2002). Similarly, heat stress during maturation has increased the percentage of oocytes with low mitochondrial membrane potential and increased the proportion of these oocytes with terminal deoxynucleotidyl transferase (dUTP), nick cease labeling (TUNEL) high quality chromatin (Vandaele and Van, 2011). Specific inhibitors, fumonisin B1 to block dihydroceramide synthase and desipramine hydrochloride to sphingomyelinase, are used to inhibit ceramide formation via de novo and hydrolytic pathways, respectively (Kalo and Roth, 2015). Use of inhibitors alleviated the impact of heat stress, supporting the fact of heat stress triggered apoptosis through ceramide formation. Furthermore, administering the anti-apoptotic molecule sphingosine-1-phosphate (S1P) to modulate the ceramide-S1P balance in oocyte maturation is found to reduce the share of TUNEL-oocytes and increase the proportion of oocytes that have fertilized and developed to blastocysts. It's observed by the previous workers that manipulating sphingomyelin metabolism may offer several opportunities for modifying oocyte responses to heat stress. In particular, ceramide can be produced *de novo* from condensation of serine with palmitoyl-CoA (Henry et al., 2013)

Figure 4. Impact of heat shock on oocyte, zygote, blastocyst and morula.

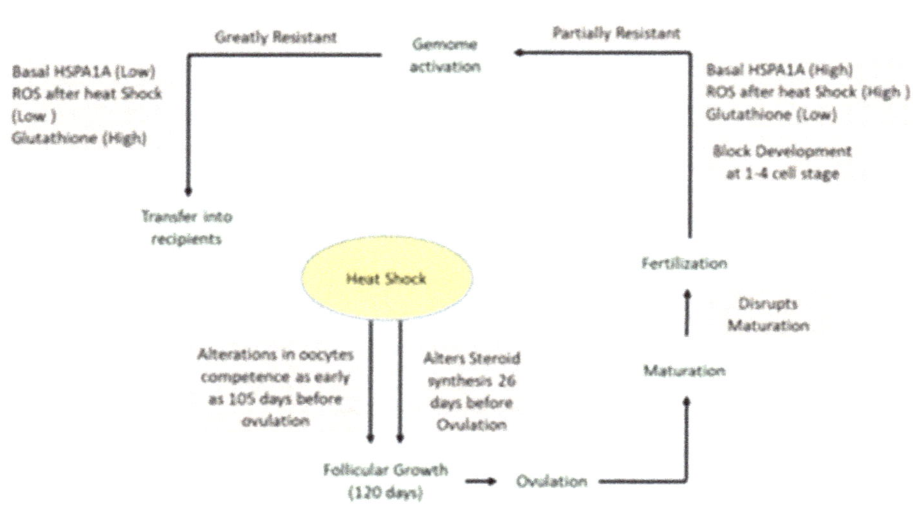

and therefore, dietary manipulations that have an effect on the membrane fatty acid composition or reduce ceramide formation, would possibly enhance oocyte quality. In vitro supplementation of palm oil with calcium salt of long chain fatty acids has increased the percentage of oocytes developing to blastocysts (Santos et al., 2008). However, management of nutritional fatty acids to lactating cows in summer has not been found to improve developmental competence of oocytes that have been aspirated and subjected to *in-vitro* maturation and fertilization (Leroy et al., 2005). Other molecules had been proven to minimize the impact of heat stress *in-vitro*. Inhibition of apoptosis with the caspase inhibitor, z-DEVD-fmk is found to decrease the deleterious impact of heat stress on oocyte developmental competence (Roth and Hansen, 2004). Anti-apoptotic peptides, the BH4 domain of Bcl-xL and Bax inhibitor peptide has reduced the share of DNA fragmented oocytes (Soto and Smith, 2009).

Fertilization and Early Embryo Development

It is possible that exposure of spermatozoa to increased temperatures, even as in uterus or oviduct of a hyper thermic female may lead to compromise sperm survival. While survival in uterus has not been examined, the frozen spermatozoa aren't greatly affected by heat stress. There are many influences of heat stress on embryo in general. Because of the overheating of reproductive tract and the increasing body temperature, the uterus is not conductive to spermatovum improvement. Increased temperature boost up heat dissipation and increase the blood volume of the skin so that the blood flowing through uterus will reduce, therefore causing a relative shortage of oxygen, moisture and nutrients that are important for embryonic development and ensuing in a retardation of embryonic increase. Heat stimulation also cause immunity decline, appetite decrease and result in malnutrition as a way to inhibit the embryo improvement. Furthermore, heat stress affects the secretion of prostaglandins (PG) within the uterus, lead to the dissolution of the corpus luteum and bring about reduced embryo development. Heat stress can have indirect negative consequences on oocyte and subsequent embryonic improvement. In addition, elevated non esterified fatty acid (NEFA) concentrations have an effect on the expression of genes related to lipid metabolism in various tissues, which include the mammary gland, oocytes and the embryo. While the effect of heat stress entails alterations in the follicle and its enclosed oocyte, pre-implanted embryos are also unstable to increased temperature (Roth, 2008). Two-cellular stage embryos are extra sensitive to heat stress than the ones at four and eight-cellular stages. Embryos at later developmental stages (i.e., morula, blastocyst) are extra immune to heat stress (Ealy et al., 1995). One of the capabilities of embryonic development is that the embryo becomes greater resistant to heat stress as pregnancy advances.

IMPACT OF NUTRITIONAL STRESS

Vitamin Stress

Nutritional deficiencies make contributions to extended postpartum anestrus. The interplay with genetic, environmental or management elements additionally impact the duration of anestrus (Patterson et al., 1992). There is an inverse relationship among energy stability and the time at which ovarian interest resumes postpartum, such as insufficient nutrient consumption, leads to loss of weight and altered body condition and finally cessation of estrous cycles. Prolonged postpartum anestrus is the main component proscribing reproductive efficiency (Short et al., 1990). During anestrus, ovulation does not arise, thereby not-withstanding ovarian follicular development, because developing follicles do no longer mature largely due to inadequate nutrient availability. On the other hand, supplementation of vitamin E, recognized to have antioxidant capability have not improved bovine embryos tolerance to heat stress. Similarly, supplementation of vitamins A and C did not have any useful impact in heat stressed cows. Effects of heat stress on the oocyte maturation can be reduced by provision of antioxidants (Hansen, 2018).

Earlier studies revealed that melatonin implants starting at 220 days of gestation to cows in summer time shortened the postpartum period and reduced the prevalence of cows, experiencing greater than three breedings consistent with conception (Creighton, 2004). The peak stage of melatonin within the blood of cows receiving implants became low (i.e, 260-300 pM) and it is probably because, melatonin became performing hormone as opposed to as an antioxidant. Interactions exist among the factors affecting the fertility in order, so that the combined effects are additive. Energy is a crucial nutrient for dairy cows both before and after calving and there is no substitute for energy inside the diet of ruminants. A stability of strength and protein is required, even before calving and inside the dry duration period (Vandaele and van, 2011). The negative impact of a relatively immoderate intake of protein is enhanced by strength deficiency at the same time. Under these conditions, a large amount of ammonia is flowing through the liver that results in liver damage. In such cases there may be static ovaries, anaphrodisia, purulent endometritis and embryonic mortality. This may be as a result from the alterations of uterine secretions (Lotthammer, 1991). Uterine remedies with antibiotics or other preparations in those cases aren't very successful until nutrients are balanced and therefore they are no longer economic.

Excessive energy intake for the duration of overdue lactation and the dry period can cause "fats cow" problems which decreases reproductive efficiency inside the subsequent lactation. When heifers are fed insufficient amounts of energy they reach sexual maturity very late (Bindari et al., 2013). If energy poor feed is fed to heifers which have begun to have ordinary estrous cycles, they prevent cycling. An example is heifers fed diets composed specially of poor excellent hay. They regularly do not show signs of estrus in winter time. If grains are provided in the diet, or they're placed on true pasture, normal estrous cycle resumes as they start to consume adequate amounts of energy (Diskin and Sreenan, 2000). Proper management is required in feeding of nutrients before or after calving. Animals with excess body temperature, Body condition score (BCS >7), have overall lower reproductive performance and more calving difficulties than animals in normal body temperature (BCS 5-6) (Burkholder, 2000) Excessive energy intake all through the past due lactation and dry periods can lead to "fat cow" problems. Cows that are over stressed once they calve have a higher incidence of retained placenta, major uterine infections and cystic ovaries. They also have a higher prevalence of metabolic disorders and feature a possible tendency to move off feed.

Mineral Stress

For many years, phosphorus (P) deficiency is considered one of the prominent causes of infertility. It has been tested that phosphorus deficiency reduced total intake food, which prompted a decreased energy supply and decreased weight gain in heifers (Pradhan and Nakagoshi, 2008). In the case of stress, the negative effect of P deficiency is enhanced. The Ca:P ratio is also considered to be vital. The frequency of non-infectious endometritis is extended when Ca:P ratio decreases. Furthermore a decreased content of manganese is found in uterine tissue. This additionally demonstrates the significance of calcium, which has functions for the uterine performance, especially after calving for involution of the uterus .It's advised to keep Ca:P ratio intake over 2:1 with marginal phosphorous delivery which should be higher in stressed conditions. A deficiency of sodium and correlated extra potassium can lower fertility via abnormal oestrus cycles, endometritis and follicular cysts (Melendez, 2006). The Na:K ratio needs to be kept in range of 10:1. Sodium supplementation via salt is very reasonably priced and must be given ad libitum. Of the minerals, manganese and selenium might also influence reproduction. Manganese supply is correlated with pH of the soil because high pH values inhibit uptake from the soil. A deficiency of manganese results in anaphrodisia, endometritis and abortions (Lotthammer, 1991). Daily supply of mangnese according to cow should be 1000 mg.

Copper is a necessary component of a number of enzymes including superoxide dismutase, lysyl oxidase and thiol oxidase. These enzymes feature to eliminate free radicals that increase tissue susceptibility to bacterial infections, increase energy and elasticity of connective tissues and blood vessels and increase strength of horns such as inside the claw (Hoof), minimizing lameness (Sykes and Grace, 1984). Symptoms of copper deficiency include early embryonic death, resorption of embryo, increased number of the retained placentas and necrosis of the placenta (Soetan et al., 2010). Dairy cows with better serum copper level had significantly much less days to first service and fewer services per conception (Ingraham, Kappel, Morgan and Srikandakumar, 1987). Proper copper supplementation is needed for production of quality semen (Talukdar et al., 2016).

Manganese is an activator of enzyme system in the metabolism of carbohydrate, fats, protein and nucleic acids. Manganese is considered to have vital functions as it is necessary for cholesterol synthesis (Hansen et al., 2006) which in turn is needed for synthesis of the steroids, estrogen, progesterone and testosterone. Insufficient steroid manufacturing results in reduced circulating concentrations of reproductive hormones, resulting in low sperm count in adult males and abnormal estrus cycles in females. The corpus luteum has excessive manganese content material and can be tormented by manganese supplementation. A deficiency in manganese can be related to suppression of estrus, cyclic ovaries and reduced conception (Kumar et al., 2011).

Fat and Fasting Glucose Stress

The impact of fats in farm animals is a focus of great research as fatty acids and cholesterol are substrates for hormone synthesis (Behrman and Armstrong, 1969; Nafikov and Beitz, 2007). Increasing fat level within the weight reduction plan may additionally increase the potential of reproductive hormones (progesterone, prostaglandins) or fats may also act immediately at the reproductive axis. Therefore, the effects of fats can be independent or additive to energy availability. Cattle diets generally comprise less than 2 or 3% fats. High fat diets for livestock include 5% to 8%. Exceeding the dietary fat impairs rumen function. Lactating cows are the primary animals to be supplemented fats due to their extended energy requirements, and the problem that is worrying, is getting these cows rebred. Previous research indicates that feeding excessive fats diets to breeding heifers and postpartum cows, elevated progesterone manufacturing and the lifespan of the corpus luteum (CL) (Staples et al., 1998). Higher progesterone levels throughout the luteal phase generally bring about advanced fertility. Increasing dietary fats also results in an increased follicular increase. Moreover, small and medium sized follicles are present in cows and heifers that are fed excessive fat diets (Beam and Butler, 1997). In addition, the increased follicular growth is regularly observed by accelerated estrogen and progesterone production. Further, postnatal adult glucose homeostasis, determined by fasting has been observed to get affected in stressed compared to unstressed offsprings. The effect of gender has been observed to be significant, fasting glucose being higher in males. While, the effect of heat stress on fasting glucose has been found to be significant in case of females than in males.

Estrous Cycle

In the estrous cycle, energy is the basic necessity to regulate the reproductive hormones which is acquired by the nutrition but sometimes the imbalancing of hormones is observed due to the continuous nutritional stress. Nutritional stress has many side effects on the functioning of the estrous cycle. Nutritional stressed animals are reported to have the higher concentration of the FSH, as compared to the unaffected animals (Meng et al., 2016). The concentration of the LH secretion is found decreased due to which the dominant follicles do not undergo maturation phase as well as ovulation phase. Nutritional stress also affects the function of corpus luteum as a result it is compromised and also reduces the weight and diameter of the corpus luteum (Reed and Carr, 2018). The impact on progesterone level is variable and increased or often no difference in progesterone concentration is experienced (Ibtisham et al., 2018).

Follicular Development

Maternal mal-nutrition is reported to increase the incidence of DNA damage in fetal-oogonia, while from the early to mid-gestation it leads to the decreased fetal ovarian vascular development, possibly affecting the follicular quality (Smith et al., 2019). The nutritional stress is reported to disturb the hormone functions and corpus luteum cycle as well as decreases diameter and size of the follicles. The reduced size of the follicle is observed due to less secretion of the LH and estradiol hormone (Souza et al., 2018).

Oocycte Development

The over nutrition in the super ovulated animals is reported to degrade the oocyte quality and also has significant effects in blastocyst stage (Kruse et al., 2017). Studies reveal that the condition of undernutrition during the folliculogenesis, a phase characterized by the active angiogenesis and the protein synthesis results in poor quality of the oocyte (Abdelatty et al., 2018). Therefore, both in the preconception period as well as in the earliest stages of the embryonic development, the nutritional deficiencies may be responsible for the changes or complications of the reproductive process.

Early Embryonic Development

As the placenta develops, the blood flow between the mother and the embryo is established, the organs start developing, there is a major requirement of the nutrition during this phase (Knofler et al., 2019). The embryo development is sensitive towards nutritional alteration in the maternal uterine environment. During the period of early gestation, the placenta compensates for the nutritional restriction by increasing the number of cotyledons, while towards the end of the gestation period the size of the cotyledons is increased (Rainforth, 2019). This depicts the negative energy balance, due to the competition of energy with their offsprings. The increase of NEFA (Non esterified fatty acid) concentration takes place, that leads to the negative energy balance. This tends to decrease the oocyte capability to reach up to the blastocyst stage (Sutton et al., 2016). The nutritional stress also alters the function of secretion in uterus, as a result insulin growth factor and the glucose concentration are decreased but on the other hand NEFA

(Non esterified fatty acid) and growth hormones are increased. More recently, it has been evaluated that the prenatal malnutrition mostly occurs in the regulatory regions of the genes showing differential expression during the stage of early development (Navarro et al., 2017).

FUTURE PROSPECT OF RESEARCH

Proper identification of the genes linked with heat tolerance and by using these genes as the markers predominantly in the breeding programmes or marker assisted selection may be used to identify animals that may be adapted to heat stress considering the environment-genotype interactions in addition to the higher productivity. Moreover there is a need to carry systematic research on climate resilient animal agriculture for achieving sustainability in dairy farming system, especially in warm and humid climatic conditions.

CONCLUSION

Environment, in which breeding females live before conception and during the early stages of pregnancy has striking effects on the oocytes developing in the ovarian follicle and on the early embryos in the reproductive tract. Among the various environmental factors that are known to affect oocyte and embryo development, the altered nutrition during this critical phase are particularly well documented. Changes in the quantity of food being consumed or the composition of diet imposed mainly during the pre-mating period affect blastocyst yield, oocyte maturity, prenatal survival and the number of offsprings born alive. Most importantly, nutrition at the same time also affects the quality of embryos and the resultant off-springs, with strong evidence from variety of species depicting that peri-conception nutrition can change behavior, cardiovascular function and reproductive function throughout post-natal life. In livestock, it is very important to devise nutritional strategies that will improve reproductive potential and the quality of offsprings. Furthermore, the gross efficacy with which dietary nutrients are being given to the live-stock for performance could also be enhanced. Increased pregnancy rate of livestock could be attained by enhancing various managemental conditions. With increasing attention being paid worldwide on the climate change and food security it is essential that nutritional strategies must be employed to improve the reproductive outcome considering the environmental footprint of such development and the expected alterations in feedstuff availability.

AUTHOR CONTRIBUTION

GG, PMI conceived and designed the book chapter, performed literature survey and wrote the book chapter. GG, PMI, GDS drawn the illustrations.

FUND DISCLAIMER

No grant received from any funding agency in public, commercial, or not-for-profit sector.

CORRESPONDING AUTHORS

Golden Gokhale, Pir Mohammad Ishfaq,

REFERENCES

Abdelatty, A. M., Iwaniuk, M. E., Potts, S. B., & Gad, A. (2018). Influence of maternal nutrition and heat stress on bovine oocyte and embryo development. *International Journal of Veterinary Science and Medicine*, *6*(sup1), S1–S5. doi:10.1016/j.ijvsm.2018.01.005

Allen, J., Anderson, S., Collier, R., & Smith, J. (2013). *Managing heat stress and its impact on cow behavior*. Paper presented at the 28th Annual Southwest Nutrition and Management Conference.

Alonso-Fauste, I., Andres, M., Iturralde, M., Lampreave, F., Gallart, J., & Alava, M. A. (2012). Proteomic characterization by 2-DE in bovine serum and whey from healthy and mastitis affected farm animals. *Journal of Proteomics*, *75*(10), 3015–3030. doi:10.1016/j.jprot.2011.11.035

Alves, J. R. A., Andrade, T. A. A., Assis, D. M., Gurjao, T. A., Melo, L. R. B., & Souza, B. B. (2017). Productive and reproductive performance, behavior and physiology of cattle under heat stress conditions. *Journal of Animal Behaviour and Biometeorology*, *5*(3), 91–96. doi:10.31893/2318-1265jabb.v5n3p91-96

Ashraf, S., Shah, S. M., Saini, N., Dhanda, S., Kumar, A., Goud, T. S., & Upadhyay, R. (2014). Developmental competence and expression pattern of bubaline (Bubalusbubalis) oocytes subjected to elevated temperatures during meiotic maturation in vitro. *Journal of Assisted Reproduction and Genetics*, *31*(10), 1349–1360. doi:10.100710815-014-0275-3

Beam, S. W., & Butler, W. (1997). Energy balance and ovarian follicle development prior to the first ovulation postpartum in dairy cows receiving three levels of dietary fat. *Biology of Reproduction*, *56*(1), 133–142. doi:10.1095/biolreprod56.1.133

Behrman, H. R., & Armstrong, D. T. (1969). Cholesterol esterase stimulation by luteinizing hormone in luteinized rat ovaries. *Endocrinology*, *85*(3), 474–480. doi:10.1210/endo-85-3-474

Berg, D., Van Leeuwen, J., Beaumont, S., Berg, M., & Pfeffer, P. (2010). Embryo loss in cattle between Days 7 and 16 of pregnancy. *Theriogenology*, *73*(2), 250–260. doi:10.1016/j.theriogenology.2009.09.005

Berner, E. K. & Berner, R. A. (2012). *Global environment: water, air, and geochemical cycles*. Princeton University Press.

Bethea, C. L., Centeno, M. L., & Cameron, J. L. (2008). Neurobiology of stress-induced reproductive dysfunction in female macaques. *Molecular Neurobiology*, *38*(3), 199–230. doi:10.100712035-008-8042-z

Betteridge, K. (1986). Increasing productivity in farm animals. *Reproduction in Mammals, 5*, 1-47.

Bindari, Y. R., Shrestha, S., Shrestha, N., & Gaire, T. N. (2013). Effects of nutrition on reproduction-A review. *Advances in Applied Science Research* *4*(1), 421–429.

Bridges, G., Day, M., & Geary, T. (2013). Triennial Reproduction Symposium: deficiencies in the uterine environment and failure to support embryonic development. *Journal of Animal Science, 91*(7), 3002-3013.

Bridges, P., Brusie, M., & Fortune, J. (2005). Elevated temperature (heat stress) in vitro reduces androstenedione and estradiol and increases progesterone secretion by follicular cells from bovine dominant follicles. *Domestic Animal Endocrinology*, *29*(3), 508–522. doi:10.1016/j.domaniend.2005.02.017

Burkholder, W. J. (2000). Use of body condition scores in clinical assessment of the provision of optimal nutrition. *Journal of the American Veterinary Medical Association*, *217*(5), 650–654. doi:10.2460/javma.2000.217.650

Chandler, M. A., & Sohl, L. E. (2000). Climate forcings and the initiation of low latitude ice sheets during the Neoproterozoic Varanger glacial interval. *Journal of Geophysical Research, D, Atmospheres*, *105*(D16), 20737–20756. doi:10.1029/2000JD900221

Chen, Y., Breen, K., & Pepling, M. E. (2009). Estrogen can signal through multiple pathways to regulate oocyte cyst breakdown and primordial follicle assembly in the neonatal mouse ovary. *The Journal of Endocrinology*, *202*(3), 407–417. doi:10.1677/JOE-09-0109

Conaway, C. (1971). Ecological adaptation and mammalian reproduction. *Biology of Reproduction*, *4*(3), 239–247. doi:10.1093/biolreprod/4.3.239

Council, N. R. (1981). *Effect of environment on nutrient requirements of domestic animals*. National Academies Press.

Courtney, L. P. (1984). *The temporal relationship of the ovary and ovarian morphology for the onset and duratin of behavioralestrus in prostaglandin Fa, , -treated ewes*. Montana State University-Bozeman, College of Agriculture.

Creighton, K. W. (2004). *Heifer development systems for March-born heifers and improving pregnancy rates in June-calving cows*. The University of Nebraska- Lincoln.

De Rensis, F., & Scaramuzzi, R. J. (2003). Heat stress and seasonal effects on reproduction in the dairy cow-a review. *Theriogenology*, *60*(6), 1139–1151. doi:10.1016/S0093-691X(03)00126-2

Ealy, A. D., Howell, J. L., Monterroso, V. H., Arechiga, C. F., & Hansen, P. J. (1995). Developmental changes in sensitivity of bovine embryos to heat shock and use of antioxidants as thermoprotectants. *Journal of Animal Science*, *73*(5), 1401–1407. doi:10.2527/1995.7351401x

Edwards, J., Saxton, A., Lawrence, J., Payton, R., & Dunlap, J. (2005). Exposure to a physiologically relevant elevated temperature hastens in vitro maturation in bovine oocytes. *Journal of Dairy Science*, *88*(12), 4326–4333. doi:10.3168/jds.S0022-0302(05)73119-2

El-Tarabany, M. S., El-Bayomi, K., & Abdelhamid, T. (2015). Semen characteristics of purebred and crossbred male rabbits. *PLoS One, 10*(5), e0128435-e0128435. doi:10.1371/journal.pone.0128435

Foote, R. (1978). Factors influencing the quantity and quality of semen harvested from bulls, rams, boars and stallions. *Journal of Animal Science, 47*(suppl_II), 1-11.

Geisert, R., Morgan, G., Short, E., & Zavy, M. (1992). Endocrine events associated with endometrial function and conceptus development in cattle. *Reproduction, Fertility, and Development, 4*(3), 301–305. doi:10.1071/RD9920301

Gwatkin, R. (2012). *Fertilization mechanisms in man and mammals*. Springer Science and Business Media.

Hallberg, I., Kjellgren, J., Persson, S., Örn, S., & Sjunnesson, Y. (2019). Perfluorononanoic acid (PFNA) alters lipid accumulation in bovine blastocysts after oocyte exposure during in vitro maturation. *Reproductive Toxicology (Elmsford, N.Y.), 84*, 1–8. doi:10.1016/j.reprotox.2018.11.005

Hansen, P. (2018). Cellular and molecular basis of therapies to ameliorate effects of heat stress on embryonic development in cattle. *Animal Reproduction (AR), 10*(3), 322–333.

Hansen, S., Spears, J., Lloyd, K., & Whisnant, C. (2006). Growth, reproductive performance, and manganese status of heifers fed varying concentrations of manganese. *Journal of Animal Science, 84*(12), 3375–3380. doi:10.2527/jas.2005-667

Holt, W. (2000). Basic aspects of frozen storage of semen. *Animal Reproduction Science, 62*(1-3), 3–22. doi:10.1016/S0378-4320(00)00152-4

Ibtisham, F., Nawab, A., Li, G., Xiao, M., An, L., & Naseer, G. (2018). Effect of nutrition on reproductive efficiency of dairy animals. *Medycyna Weterynaryjna, 74*(1), 356–361. doi:10.21521/mw.6025

Ingraham, R., Kappel, L., Morgan, E., & Srikandakumar, A. (1987). Correction of subnormal fertility with copper and magnesium supplementation. *Journal of Dairy Science, 70*(1), 167–180. doi:10.3168/jds.S0022-0302(87)79991-3

Khan, I. (2016). *Evaluation of different stress indicators associated with repproductivecyclicity of crossbred dairy cows*. The University of Agriculture.

Khodaei-Motlagh, M., Shahneh, A. Z., Masoumi, R., & Derensis, F. (2011). Alterations in reproductive hormones during heat stress in dairy cattle. *African Journal of Biotechnology, 10*(29), 5552–5558.

Knöfler, M., Haider, S., Saleh, L., Pollheimer, J., Gamage, T. K., & James, J. (2019). Human placenta and trophoblast development: Key molecular mechanisms and model systems. *Cellular and Molecular Life Sciences, 76*(18), 1–18. doi:10.100700018-019-03104-6

Kruse, S., Bridges, G., Funnell, B., Bird, S., Lake, S., Arias, R., Amundson, O., Larimore, E., Keisler, D., & Perry, G. (2017). Influence of post-insemination nutrition on embryonic development in beef heifers. *Theriogenology, 90*, 185–190. doi:10.1016/j.theriogenology.2016.11.021

Kumar, S., Pandey, A. K., Razzaque, W., & Dwivedi, D. (2011). Importance of micro minerals in reproductive performance of livestock. *Veterinary World, 4*(5), 230. doi:10.5455/vetworld.2011.230-233

Leibfried-Rutledge, M. L., Dominko, T., & Critser, E. S. (1997). Tissue maturation in vivo and in vitro: gamete and early embryo ontogeny. In *Reproductive tissue banking* (pp. 23-138). Elsevier.

Leroy, J., Vanholder, T., Mateusen, B., Christophe, A., Opsomer, G., de Kruif, A., & Van Soom, A. (2005). Non-esterified fatty acids in follicular fluid of dairy cows and their effect on developmental capacity of bovine oocytes in vitro. *Reproduction (Cambridge, England), 130*(4), 485–495. doi:10.1530/rep.1.00735

Lotthammer, K. (1991). Influence of nutrition on reproductive performance of the milking/gestating cow in the tropics. *Feeding dairy cows in the tropics*, 36-47.

Malayer, J., & Hansen, P. (1990). Differences between Brahman and Holstein cows in heat-shock induced alterations of protein synthesis and secretion by oviducts and uterine endometrium. *Journal of Animal Science*, *68*(1), 266–280. doi 10.2527/1990.681266x

Mehrotra, V. P., & Redford, C. R. (2005). *Apparatus and method for detection of estrus and non-pregnancy*. Google Patents.

Melendez, P. (2006). Nutritional management of the transition period to optimize fertility in dairy cattle. *Proceedings 3rd Florida and Georgia Dairy Road Show Conference*.

Meng, L., Rijntjes, E., Swarts, H., Bunschoten, A., van der Stelt, I., Keijer, J., & Teerds, K. (2016). Dietary-induced chronic hypothyroidism negatively affects rat follicular development and ovulation rate and is associated with oxidative stress. *Biology of Reproduction*, *94*(4), 90, 91–11. doi:10.1095/biolreprod.115.136515

Moussa, M., Shu, J., Zhang, X., & Zeng, F. (2015). Maternal control of oocyte quality in cattle "a review". *Animal Reproduction Science*, *155*, 11–27. doi:10.1016/j.anireprosci.2015.01.011

Nandi, S., Chauhan, M. S., & Palta, P. (1998). Influence of cumulus cells and sperm concentration on cleavage rate and subsequent embryonic development of buffalo (Bubalusbubalis) oocytes matured and fertilized in vitro. *Theriogenology*, *50*(8), 1251–1262. doi:10.1016/S0093-691X(98)00224-6

Nardone, A., Ronchi, B., Lacetera, N., Ranieri, M. S., & Bernabucci, U. (2010). Effects of climate changes on animal production and sustainability of livestock systems. *Livestock Science*, *130*(1-3), 57–69. doi:10.1016/j.livsci.2010.02.011

Navarro, E., Funtikova, A. N., Fíto, M., & Schröder, H. (2017). Prenatal nutrition and the risk of adult obesity: Long-term effects of nutrition on epigenetic mechanisms regulating gene expression. *The Journal of Nutritional Biochemistry*, *39*, 1–14. doi:10.1016/j.jnutbio.2016.03.012

Patterson, D., Perry, R., Kiracofe, G., Bellows, R., Staigmiller, R., & Corah, L. (1992). Management considerations in heifer development and puberty. *Journal of Animal Science*, *70*(12), 4018–4035. doi:10.2527/1992.70124018x

Perry, S. F., & Gilmour, K. (1993). An evaluation of factors limiting carbon dioxide excretion by trout red blood cells in vitro. *The Journal of Experimental Biology*, *180*(1), 39–54.

Pradhan, R., & Nakagoshi, N. (2008). Reproductive disorders in cattle due to nutritional status. *Journal of International Development and Cooperation, 14*(1), 45-66.

Putney, D., Mullins, S., Thatcher, W., Drost, M., & Gross, T. (1989). Embryonic development in superovulated dairy cattle exposed to elevated ambient temperatures between the onset of estrus and insemination. *Animal Reproduction Science*, *19*(1-2), 37–51. doi:10.1016/0378-4320(89)90045-6

Rainforth, M. (2019). The effect of maternal nutritional restriction on fetal development and performance of offspring in beef cattle. *Alberta Academic Review*, *1*(2), 1–11. doi:10.29173/aar12

Reed, B. G., & Carr, B. R. (2018). The normal menstrual cycle and the control of ovulation. In *Endotext*. MDText. com, Inc.

Rispoli, L. A., Lawrence, J. L., Payton, R. R., Saxton, A. M., Schrock, G. E., Schrick, F. N., & Edwards, J. L. (2011). Disparate consequences of heat stress exposure during meiotic maturation: Embryo development after chemical activation vs fertilization of bovine oocytes. *Reproduction (Cambridge, England)*, *142*(6), 831–843. doi:10.1530/REP-11-0032

Rivera, R., & Hansen, P. (2001). Development of cultured bovine embryos after exposure to high temperatures in the physiological range. *Reproduction-Cambridge*, *121*(1), 107–115. doi:10.1530/rep.0.1210107

Roth, Z. (2008). Heat stress, the follicle, and its enclosed oocyte: Mechanisms and potential strategies to improve fertility in dairy cows. *Reproduction in Domestic Animals*, *43*, 238–244. doi:10.1111/j.1439-0531.2008.01168.x

Roth, Z. (2017). Effect of heat stress on reproduction in dairy cows: Insights into the cellular and molecular responses of the oocyte. *Annual Review of Animal Biosciences*, *5*(1), 151–170. doi:10.1146/annurev-animal-022516-022849

Roth, Z., & Hansen, P. (2004). Involvement of apoptosis in disruption of developmental competence of bovine oocytes by heat shock during maturation. *Biology of Reproduction*, *71*(6), 1898–1906. doi:10.1095/biolreprod.104.031690

Roth, Z., & Wolfenson, D. (2016). Comparing the effects of heat stress and mastitis on ovarian function in lactating cows: Basic and applied aspects. *Domestic Animal Endocrinology*, *56*, S218–S227. doi:10.1016/j.domaniend.2016.02.013

Santos, J., Bilby, T., Thatcher, W., Staples, C., & Silvestre, F. (2008). Long chain fatty acids of diet as factors influencing reproduction in cattle. *Reproduction in Domestic Animals*, *43*, 23–30. doi:10.1111/j.1439-0531.2008.01139.x

Santos, J., Thatcher, W., Chebel, R., Cerri, R., & Galvao, K. (2004). The effect of embryonic death rates in cattle on the efficacy of estrus synchronization programs. *Animal Reproduction Science*, *82*, 513–535. doi:10.1016/j.anireprosci.2004.04.015

Setchell, B. (2018). The effects of heat on the testes of mammals. *Animal Reproduction (AR)*, *3*(2), 81–91.

Short, R. E., Bellows, R. A., Staigmiller, R. B., Berardinelli, J. G., & Custer, E. E. (1990). Physiological mechanisms controlling anestrus and infertility in postpartum beef cattle. *Journal of Animal Science*, *68*(3), 799–816. doi:10.2527/1990.683799x

Silanikove, N. (2000). Effects of heat stress on the welfare of extensively managed domestic ruminants. *Livestock Production Science*, *67*(1-2), 1–18. doi:10.1016/S0301-6226(00)00162-7

Silva, C., Sartorelli, E., Castilho, A., Satrapa, R., Puelker, R., Razza, E., & Barros, C. M. (2013). Effects of heat stress on development, quality and survival of Bosindicus and Bostaurus embryos produced in vitro. *Theriogenology*, *79*(2), 351–357. doi:10.1016/j.theriogenology.2012.10.003

Smith, P., Juengel, J., Maclean, P., Rand, C., & Stanton, J.-A. L. (2019). Gestational nutrition 2: Gene expression in sheep fetal ovaries exposed to gestational under nutrition. *Reproduction (Cambridge, England)*, *157*, 13–25. doi:10.1530/REP-18-0358

Soetan, K., Olaiya, C., & Oyewole, O. (2010). The importance of mineral elements for humans, domestic animals and plants: A review. *African Journal of Food Science*, *4*(5), 200–222.

Soto, P., & Smith, L. C. (2009). BH4 peptide derived from Bcl-xL and Bax-inhibitor peptide suppresses apoptotic mitochondrial changes in heat stressed bovine oocytes. *Molecular Reproduction and Development*, *76*(7), 637–646. doi:10.1002/mrd.20986

Souza, A., Narciso, C., Batista, E., Carvalho, P., & Wiltbank, M. (2018). Effect of uterine environment on embryo production and fertility in cows. *Animal Reproduction (AR)*, *11*, 159–167.

Spicer, L., & Echternkamp, S. (1986). Ovarian follicular growth, function and turnover in cattle: A review. *Journal of Animal Science*, *62*(2), 428–451. doi:10.2527/jas1986.622428x

Spitz, I. M., Croxatto, H. B., & Robbins, A. (1996). Antiprogestins: Mechanism of Action and Contraceptive. *Annual Review of Pharmacology and Toxicology*, *36*(1), 47–81. doi:10.1146/annurev.pa.36.040196.000403

Staples, C., Burke, J., & Thatcher, W. (1998). Influence of supplemental fats on reproductive tissues and performance of lactating cows. *Journal of Dairy Science*, *81*(3), 856–871. doi:10.3168/jds.S0022-0302(98)75644-9

Sutton-McDowall, M. L., Wu, L. L., Purdey, M., Abell, A. D., Goldys, E. M., MacMillan, K. L., Thompson, J. G., & Robker, R. L. (2016). Nonesterified fatty acid-induced endoplasmic reticulum stress in cattle cumulus oocyte complexes alters cell metabolism and developmental competence. *Biology of Reproduction*, *94*(23), 21–29.

Talukdar, D., Talukdar, P., & Ahmed, K. (2016). Minerals and its impact on fertility of livestock: A review. *Agricultural Reviews (Karnal)*, *37*(4), 333–337. doi:10.18805/ag.v37i4.6464

Vandaele, L., & Van Soom, A. (2011). Intrinsic factors affecting apoptosis in bovine in vitro produced embryos. *Verhandelingen - Koninklijke Academie voor Geneeskunde van Belgie*, *73*(1-2), 79–104.

Velazquez, M. (2015). Impact of maternal malnutrition during the peri-conceptional period on mammalian preimplantation embryo development. *Domestic Animal Endocrinology*, *51*, 27–45. doi:10.1016/j.domaniend.2014.10.003

Vermunt, J. J., & Tranter, B. P. (2011). *Heat stress in dairy cattle–a review, and some of the potential risks associated with the nutritional management of this condition*. Paper presented at the review of AVA QLD Division Conference.

Wahl, R. L., & Reif, J. S. (2009). Temporal trends in bull semen quality: A comparative model for human health. *Environmental Research*, *109*(3), 273–280. doi:10.1016/j.envres.2008.10.012

Wale, P. L., & Gardner, D. K. (2016). The effects of chemical and physical factors on mammalian embryo culture and their importance for the practice of assisted human reproduction. *Human Reproduction Update*, *22*(1), 2–22. doi:10.1093/humupd/dmv034

Wirbisky, S. E., & Freeman, J. L. (2015). Atrazine exposure and reproductive dysfunction through the hypothalamus-pituitary-gonadal (HPG) axis. *Toxics*, *3*(4), 414–450. doi:10.3390/toxics3040414

Wolfenson, D., Flamenbaum, I., & Berman, A. (1988). Hyperthermia and body energy store effects on estrous behavior, conception rate, and corpus luteum function in dairy cows. *Journal of Dairy Science*, *71*(12), 3497–3504. doi:10.3168/jds.S0022-0302(88)79956-7

Wolfenson, D., Thatcher, W., Badinga, L., Savi, J., Meidan, R., Lew, B., & Berman, A. (1995). Effect of heat stress on follicular development during the estrous cycle in lactating dairy cattle. *Biology of Reproduction*, *52*(5), 1106–1113. doi:10.1095/biolreprod52.5.1106

Yamada, M., & Isaji, Y. (2011). Structural and functional changes linked to, and factors promoting, cytoplasmic maturation in mammalian oocytes. *Reproductive Medicine and Biology*, *10*(2), 69–79. doi:10.100712522-011-0079-4

Yang, M. Y., & Rajamahendran, R. (2002). Expression of Bcl-2 and Bax proteins in relation to quality of bovine oocytes and embryos produced in vitro. *Animal Reproduction Science*, *70*(3-4), 159–169. doi:10.1016/S0378-4320(01)00186-5

Zi, X.-D., Yin, R.-H., Chen, S.-W., Liang, G.-N., Zhang, D.-W., & Guo, C.-H. (2009). Developmental competence of embryos derived from reciprocal in vitro fertilization between yak (Bosgrunniens) and cattle (Bostaurus). *The Journal of Reproduction and Development*, 0906090175–090609017.

Chapter 7
Thermotolerance for Physiological and Endocrine Regulation of Embryo–Uterine Development

Sameena Khanday
Devi Ahilya Vishwavidyalaya, Indore, India

Rayees Ahmad
Jamia Millia Islamia, New Delhi, India

Guru Dutt Sharma
Devi Ahilya Vishwavidyalaya, Indore, India

ABSTRACT

Hyperthermia affects most aspects of reproductive performance in mammals by compromising the physiology of reproductive tract, through hormonal imbalance, disrupting the development and maturation of oocyte, causing embryonic mortality, abortion, growth retardation, and major developmental defects. Heat stress reduces the steroidogenic capacity of its theca and granulosa cells by altering the efficiency of follicular selection resulting in drop of luteinizing hormone and estradiol secretions from the dominant follicle in the plasma, reduced intensity, and duration of estrus expression. The mechanism for the developmental stage-dependent change in heat tolerance is considered to be the accumulation of antioxidants in embryos in response to heat-inducible production of reactive oxygen species. Morula or blastocysts can repair heat-induced misfolded or unfolded proteins or facilitate DNA damage induced apoptosis. Therefore, embryo transfer (ET) that can bypass the heat-sensitive stage could be a good solution to improve the conception rate under heat stress. However, further research is required to improve the reduction in pregnancy rates due to summer heat stress.

DOI: 10.4018/978-1-7998-4480-8.ch007

INTRODUCTION

Heat stress, also known as hypothermia is a condition in which the animal body temperature is elevated beyond normal temperature and is not able to sufficiently dissipate excess heat.It is a major significant factor contributing to reduced fertility in mammalians, which is a global issue and resulting in heavy economic losses. Researchers have secured major success and achievements mainly in reproductive technology; hormone levels, blood biochemical and immunological parameters and so on. This chapter reviews the effects and possible mechanisms of elevated maternal body temperature on reproductive efficiency of animals and humans. The early embryo–maternal dialogue that begins very early in the life of the embryo is criticalfor its own implantation. Successful embryo implantation involves the appropriate interaction between the receptive endometrium and an implantation-competent blastocyst, which occurs in a limited time period known as "the window of implantation" (Ma et al., 2003). A bidirectional communication is necessary for normal implantation thus the success of pregnancy, since disruptions will generate adverse development outcomes including decidualization and placentation, with potential loss of the pregnancy (Lim and Wang, 2010). Failure of implantation is an unsolved problem in reproductive medicine and is considered to be an important cause of infertility. Implantation is a complex biological process.

A disturbance in this complex embryo–maternal dialogue is the basic reasons for which majority of pregnancies are terminated at the end of the preimplantation period. This two way cumminication is precisely maintained by maternal hormones in particular, ovarian estrogens and progesterone (Conneely et al., 2002; Curtis Hewitt et al., 2002). The highly ordered events of cell migration, proliferation, specification or differentiation and programmed cell death (apoptosis) that illustrate the prenatal growth and development of the mammalian conceptus through to the potential effects of neo-natal and post-natal heat stress susceptibility. Maternal temperature elevations negatively affectseveral aspects related to pregnancy such as successful fertilization, establishment and maintenance of pregnancy such as production of reproductive hormones, uterine health, oocyte and embryo quality, and embryonic development. Earlier research has revealed that elevated maternal body temperatures induces embryonic and foetal mortality, growth retardation and developmental defects depends on the duration of the elevation in maternal body temperature and the specific phase of development during which the thermal dose is delivered. Direct heating of the embryo or foetus has shown a similar spectrum of effects signifying a direct effect of heat on development; disruption in maternal physiology may introduce additional factors complicating the overall interpretation. Studies have reported that raised maternal body temperatures in humans during pregnancy, either through fever or some other form of heating led to developmental defects. Two potential mechanisms associated with mitochondrial function-apoptosis and oxidative stress have been documented (for review, see Roth, 2017). Thermal stress not only affects cell function directly, but also leads to disruptions in the DNA or function of organelles by inducing oxidative stress (Roth, 2015; Lord and Averill, 2002). Oxidative stress agents like reactive oxygen species (ROS) is a major causative agent of DNA damage (leading to cell death), disrupt the mitochondrial function, lipid peroxidation, causing abnormal expression of genes and protein synthesis and ultimately result in apoptosis (Loven, 1998; Averill, 2002). Therefore, one factor that contrary influence the oocytes and embryos has been considered to be the increase of oxidative stress status originating from heat stress (Edwards et al., 2001; Hansen, 2007). Moreover, the change from maternal gene expression to zygotic/embryonic gene expression could be related with thermo tolerance in oocytes and embryos.

Exposure to the heat and tetratogens affects the foetus and the embryo in a variety of ways during development, and differ significantly depending on the sensitivity of the processes taking place. There is various time 'windows' in pre-natal development during which the sensitivity of the embryo or fetus to a teratogen is relatively high. Periods of cellular proliferation and migration are particularly susceptible. The process of pre-natal development of mammals is generally divided into three distinct periods: the pre-implantation period, extending from fertilization to settling of the embryo into the endometruim; the embryonic period or period of major organogenesis, marked by the formation of the main body structures; and the foetal period, during which growth of structures already formed takes place. Generally, exposure of embryos to thermal stress prior to implantation in the endometrial lining of the uterus either recovers entirely or is lost, either through the failure of the embryo to implantor embryonic death. This can leadto higher rates of pregnancy losses which may pass unseen in humans

Clearly, excessive maternal ambient temperature elevationsmay result in death at any stage, but multistep processes of organogenesis which involve cellular proliferation, differentiation, migration and apoptosis particularly contribute to the decrease in development and developmental defects in speciðc structures or organs. The developing brain, for example, seems to be extremely susceptible to a variety of extrinsic insults, including heat stress and ionizing radiation. However, cardiac, skeletal and neuromuscular defects have also been induced by heat. At the end of the embryonic period of, for example, humans (8 weeks), the embryo possesses 90% adult human anatomical structures The developing fetus is largely characterized by the growth of these structures; thermal stress mainly results in retarded growth and development, although heat induced effects are most common in central nervous system.

Thresholds thermal sensitivity therefore, varies during different developmental stages, in different organs or tissues. Therefore, such values are tissue specific and developmental stage-speciðc. It follows that there is no particular threshold dose for the development of particular abnormality, each malformation of a tissue or an organ has its own threshold dose of heat, and generally minimum value can only be identiðed by investigating the sensitivity of the most critical periods (Germain et al., 1985). In addition, there is a huge variability in the relative duration of developmental periods between different animal species as well as in the total duration of intra- uterine life. Furthermore, at any specific phase of development, the active status of maturation of any one structure, with respect to all others, show a considerable variability between species. Within a species, the different sensitivities differ significantly between closely related strains. This variation precludes the quantitative extrapolation across species and particularly from animals to humans (UNSCEAR, 1986).

HEAT STRESS ON REPRODUCTIVE EFFICIENCY OF FEMALE MAMMALIANS

Reproductive function of female mammalians will be distressed by oocyte differentiation, development, conception, delivery, reproductive functions and secondary sex characteristics because of the heat stress reduces the reproductive activity and percentage of conception rate by decreasing the oestrus occurrence and duration in females and frequency and behavior of mating and pregnancy rates. Study showed that first conception (CRI1) and overall conception rates (CR) were lowest in the summer and highest in winter (Slaem and Bouraoui, 2011). In addition; heat stress can cause miscarriages and late pregnancy complications in female mammalians after their easy breeding.

Reproductive Endocrinology

Studies on the effects of heat exposure-induced hypothermia on the activity of endocrine system and hormones involved in reproductive functions reveal significant difference and contradictory outcome. Sources of environmental and genetic variation may largely affect these results. Besides main difference organisms of different species, gender and age, different responses of heat stress may be found among more or less heat tolerant breeds and genotypes. Heat stress alters the efficiency of follicular selection and reduces the efficiency of the selected dominant follicle and this can be observed as reduced steroidogenesis of both theca and granulosa cells and result in drop in blood estradiol levels. Plasma progesterone concentrations can be increased or decreased might depend on the severity of the thermal stress and also on the metabolic aspects of the animal; these endocrine changes result in the reduction of follicular activity and alter the ovulatory pattern, leading to a decline in oocytes and abnormalities in pre-implantation embryos (Khodaei et al., 2011). Thermal stress can result in the secretion of adrenal hormones, and trigger the stress hormone system, central stress response system- the hypothalamic-pituitary-adrenal axis (HPA), thus enhance the release of a neuropeptide corticotrophin-releasing factor (CRF) in hypothalamic, and releasing it into the pituitary portal system so as to stimulate the secretion of pituitary adrenocorticotropic hormone (ACTH) which can increase the secretion of adrenal hormones in order to regulate the adaptive response to stress. However, the augment of adrenocorticotropic hormone (ACTH) secretion will in turn increase the secretion threshold of hypothalamic gonadotrophin releasing hormone (GnRH) and inhibit anterior pituitary gonadotrophs; Luteinizing hormone (LH) and follicle-stimulating hormone (FSH). The decline in LH secretion lead to the inhibition of follicular development and thereby block ovulation; meanwhile, estrogen level will also drop and inhibit the formation of corpus luteum and production of progesterone, and further lead to a variety of reproductive disorders. While endorphin secreted in the nervous system can inhibit hypothalamic-secreted GnRH directly, the progesterone produced by the adrenal cortex can LH surge and lead to estrus barrier and ovulation delay.

Besides, the decrease in the secretion of pancreatic insulin and thyroid hormones under heat stress adversely disrupts the fertility and reproductive efficiency. Elevation in temperature can decrease the production of major serum thyroid hormone-binding proteins, and lead to a significant drop of total protein, albumin and globulin in blood. Meanwhile, the decrease of thyroid hormone secretion may promote synthesis of glycogen, decline glucose level, and disturb the basal metabolic rate; while the blood glucose reduction will inhibit the insulin release. Furthermore, maternal heat stress during pregnancy also influence cortisol concentration in female guinea pigs and entail long-term consequences on the stress response of their offspring (Michel et al., 2011). Since evidences suggests that insulin play a universal role in the systemic maintaince of metabolism of three major nutrients in vivo, it also will, as is speculated, affect the synthesis of various hormones in vivo, esp. the reproductive hormones, thus regulate the reproductive functions during development, and these premature alterations could lead to peripheral and central malformations in both systems during puberty and adulthood.

The pituitary gland represents another target organ that is sensitive to heat stress and induce detrimental effects on reproductive processes. In ovariectomized ewes, hyperthermia induction (adjusted to result in rectal temperature 1.2°C higher than control) caused a reduction in critical pituitary hormone level; LH and 4-fold increase in plasma PRL on day 2 and day 3 after heat exposure, respectively (Schillo and Malven, 1978). In dairy cows, elevated temperatures during the hot months of the year in breeding locations caused reduction in the pulsatile secretion of LH (Wise et al., 1988) and dampen the preovulatory surge of LH (Gilad et al., 1993), which in turn, delay or disrupt the cascade of events leading to ovulation (Roth and Wolfenson, 2016). In goats, induced heat stress (38°C for 48 hr) reduced the follicular response to pulsatile pattern of LH activation, evaluated in vivo through estradiol secretion after administration of GnRH (Kanai et al., 2001).

Oocyte Maturation

It has been reported that exposing cows to elevated temperatures alters the oocyte's intrinsic quality, successs of fertilization and subsequent development of the embryo (Gendelman and Roth, 2012; (Rocha et al., 1998). Many in vitro studies were performed to understand the underlying mechanism of thermal stress on germinal vesicle (GV)-stage cumulus-oocyte complexes (COCs). Exposure to germinal vesicle stage cumulus-oocyte complexes to elevated temperatures from 40.0°C to 41.0°C is associated with oocyte impairments, quality, both nuclear and cytoplasmic maturation, increased abnormal spindle formation, and decreased developmental competence after IVF (Ju et al., 1999; Lopes et al., 2013). Thus, heat-stressed oocytes fail to mature and are arrested at the metaphase I stage (Roth, 2015; Payton et al., 2004). Some studies indicated that that oocytes are at high risk to the elevated temperature during the first 12 hrs of maturation because the exposure of oocytes to high thermal temperature would accelerate the progress of cytoplasmic and nuclear development (Edwards et al., 2001; Lannett and Hansen, 1996; Roth, 2005).

Heat stress in germinal vesicle (GV)-stage compromises the functions of cumulus-oocyte complexes (COCs) and oocytes through induction of apoptotic cascade (Nabenishi et al., 2012; Balboula et al., 2013) causes disruption of cytoskeleton elements (Roth and Hansen, 2005) and alters maternal transcription levels (Gendelman and Roth, 2012; Rispoli et al., 2013), ATP concentrations (Hooper et al., 2015) and mitochondrial functions. Heat stress affects not only oocyte function, but also have negative effects in cumulus cell functions such as progesterone secretion and matrix metallopeptidase 9 (MMP9) (Rispoli et al., 2013). These changes interfere in the synthesis of oocyte's proteins and disrupt the oocyte maturation and further preimplantation embryonic development (Roth, 2012; Rispoli et al., 2013; Hooper et al., 2015).

Heat stress also induces a state of oxidative stress exhibiting significant elevations in ROS levels in oocytes (Roth, 2012; Edwards et al., 2001; Nabenishi et al., 2012). ROS elicits damage to the mitochondrial DNA and initiate apoptosis or cellular organelle dysregulation such as the mitochondria (Roth, 2012). Thus, synthesis of ROS in heat-stressed oocytes would result in the production of poor quality of oocytes and disrupt viability. However, studies revealed that effects of antioxidant supplementation during in vitro maturation at elevated temperature enhanced oocyte maturation and developmental competence of embryos (Roth, 2012; Edwards et al., 2001). Glutathione (GSH), a crucial thiol-based tripeptide major antioxidant in the oocyte stage that has a pivotal role in scavenging ROS (Edwards et al., 2001; Hansen, 1999). Various antioxidants increased the glutathione levels and depletes the levels of mitochondrial reactive oxygen species of oocytes exposed to thermal stress. This indicates that the direct administration of antioxidants in heat stress conditions could decrease thermal-oxidative tolerance in oocytes and improve reproductive performance.

Fertilization

Molecular mechanisms of fertilization may be affected by elevated temperatures through direct and indirect sources. For the latter, it is clear whether the sperm or egg is the major source of the fertilization failure. Most of the adverse effects of thermal stress on reproductive traits can be explained by the increase in body temperature in heat-stressed cows. Small rise in maternal body temperature result in reduced preimplantation survival in cattle (Ulberg and Burfening, 1967). The elevation in maternal or fetal core body temperatures affects the reproductive tract and the premature development of embryo. One possible mechanism involves the direct impact of the elevated temperature on the developing embryo (Hansen, 2009). Second mechanism involves the effect of elevated thermal stress on the gut (leaky gut syndrome) mainly responsible for the disruption of the intestinal barrier function and the release of endotoxin into the circulation that can affect many aspects of animal production including embryonic development and pregnancy outcome (Baumgard and Rhoads, 2013; Abuajamieh et al., 2018).

Heat stress severely damage spermatozoa and cause sperm dysfunction, which, in turn, affects the fertilization potential, because elevated temperatures disrupt sperm function, cause reduction in sperm viability and integrity (Hendricks and Hansen, 2009; Perez-Crespo et al., 2008). On the other hand, exposure of in vitro matured bovine oocytes to a 42°C heat shock for 4 hr significantly reduced blastocyst development (Ju et al., 2005). Sperm-ovum fusion and rates of post-fusion events in hamsters were recorded within a temperature range from 4°C to 37°C. At 4°C to 10°C, acrosome-reacted spermatozoa bound to egg plasma membranes, but could not fuse with them. At 25°C or above, both binding and fusion took place. The post fusion events could occur over a broad temperature range but these events progressed rapidly with increasing temperature. However, eggs inseminated at 37°C and cultured at 25°C showed abnormal development of egg pronucleus (Hirao and Yanagimachi, 1978). Rabbits, inseminated with semen exposed to 38°C or 40°C did not impact the fertilization and cleavage rates, but embryonic survival rate was higher with semen incubated at 38°C (Burfening and Ulberg, 1968). Incubation of bovine sperm at 30°C and 38°C for 22 hr did not alter the sperm viability. However, temperatures less than 35°C did not reverse the impaired the acrosomal reaction when the temperature was above 40°C, and 38°C was recorded as optimal. Synthisis of hyaluronic acid by granulose and cumulus cells during expansion of in vitro matured oocytes, evaluated by incorporation of labeled glucosamine into glycosaminoglycans, is significantly decreased at 41°C but did not vary between temperatures of 35°C, 37°C, or 39°C (Lenz and First, 1983). Later experiments (Rivera and Hansen, 2001; Sakatani et al., 2015) showed that rate of fertilization was not impaired at 40°C; however, embryonic compaction and subsequent blastocyst formation were significantly decreases at 41°C. Although no significant difference was observed in rate of sperm penetration, polyspermy tended to be increased by heat stress during fertilization. The zona pellucidae of zygotes resulting from fertilization at 40°C for 6hr were more susceptible to digestive enzyme pronase solution. Moreover, these zygotes exhibited higher hydrogen peroxide levels(H_2O_2), higher transcription of heat-shock protein genes HSP70, but lower transcription of polyspermy preventing protein gene UCHL1 (Sakatani et al., 2015). (Rivera and Hansen, 2001) also suggested that heat stress-induced modifications in the cytoskeleton, as observed in fibroblasts, played a key role in failure of fertilization (Welch and Suhan, 1985), due to the disruption of sperm-oocyte fusion and the evolution of early embryo development.

Embryo Development

Negative effects of heat stress to embryonic development can occur either directly on the embryo or as a result of gametic mutations at different stages during gametogenesis. The initial stages of the embryonic development are more sensitive to heat stress and decreases as pregnancy progresses. Ealy et al. (1993) found one to two celled embryos from the lactating cows exposed to heat stress a day 1 after estrus, when embryos were one or two cells; the proportion of embryos reduced that developed to the blastocyst stage at day 8 after estrus. However, thermal stress at days 3(8-16 cells), 5(morula) and 7(blastocysts) had no effect on the proportion of embryos that were blastocysts at day 8. A similar pattern of developmental acquisition of thermal resistance is followed bypigs (Tompkins et al., 1967) and sheep (Dutt, 1964). The negative effects of heat shock on cultured embryos also are reduced at advanced stages of development, atleast in the cow (Krininger et al., 2002; Sakatani et al., 2004). In contrast, there is no large variation in sensitivity to elevated culture temperatures between various developmental stages mouse embryos; at the two-cell, four-cell and morula stages of development (Areehiga and Hansen1998).

Exposures of pre-implantation embryos to elevated temperature probably involve increased production of ROS. The best evidence for this idea comes from the studies of rodents mostly mice. Maternal hyperthermia resulted in increased ROS activity in oviducts and embryos and reduced glutathione concentration in recovered embryos (Ozawa et al., 2002; Matsuzuka et al., 2005). Furthermore, adverse effects of heat stress on embryonic development can be reduced by treatment of female mice with either melatonin (Matsuzuka et al., 2005) or vitamin E (Sakamoto et al., 2008). Female embryos survive are better able to survive from detrimental effects of elevated temperatures than males and this gender difference has been observed in response to reduced ROS production in females (Pe´rez-Crespo et al., 2005). Overproduction of ROS in response to elevated culture temperature has also been reported for cattle (Sakatani et al., 2004; 2008) and treatment with the antioxidant 2-meracaptoethanol has been reported to improve the negative effects of heat shock on embryo (Sakatani et al., 2008).

There are several reasons why embryos become more tolerant to negative effects of maternal heat stock as pregnancy progresses. Production of ROS during period of heat stress declines as bovine embryonic development advance (Sakatani et al., 2004) whereas an intracellular cytoplasmic antioxidant glutathione concentration increases (Lim et al., 1996). In addition, there is developmental regulation in the ability of the embryo to undergo the induced thermotolerance response, whereby exposure to a slight thermal elevation lead to enhance resistance to a subsequent more severe elevation in temperature. This mechanism does not develop until day 4 in cattle (Paula-Lopes and Hansen,2002) and the eight-cell stage inmice(Areechiga et al,1995). The induced thermotolerance involves synthesis of heat shock protein 70 (HSP70) exposure as molecular chaperone. This anti-apoptoticprotein, which inhibits apoptosis by stabilizing intracellular proteins and organelles (Brodsky and Chiosis 2006), can be induced by elevated temperature as early as the two-cell stage in mice (Christians et al.,1997) and cattle (Edwards and Hansen, 1996). Therefore, other molecular systems must be involved. Glutathione is required for induction of thermotolerance (Areechigaet al,1995) and changes in redox status may be an important determinant of resistance to heat stress.

The addition of astaxanthin, a keto-carotenoid belonging to a larger class of chemical compounds known as terpenes with higher antioxidant properties, to culture medium ameliorated the heat stress induced impairment of blastocyst development and recovered mitochondrial membrane potential and permeability in embryos subjected to heat shock treatment at various temperatures (Kuroki et al., 2013). On the other hand, advanced stage embryos, such as morula or blastocyst, develop thermotolerance, for which exposure to slight heat shock enhances the capacity to resist more severe heat stress. The attainment of such competence is associated with (a) the accumulation of antioxidants in embryos in response to heat-inducible generation of ROS, (b) the capacity to produce an antiapoptotic protein, HSP70, that occurs as early as 2-cell stage in mouse (Christians et al., 1997) and bovine (Edwards and Hansen, 1996) embryos. The successive refractory nature of heat stress with the advancement of embryo development may be due to the progressive reduction of ROS production and the increase in production of antioxidant glutathione during embryonic development (Arechiga and Hansen, 1995). This pattern exhibits distinct preference for energy metabolic pathways, as the embryo faces an increasingly intrauterine hypoxic environment.

In addition, there is a change in redox potential within the embryo, affecting not only the energy generation necessary for development, but also the activity of redox-sensitive transcription factors, which may change the pattern of gene expression through regulatory transcription factors (Harvey and Thompson, 2002). This shift in metabolic pathways would play a major role in regulating ROS concentrations within the cytoplasm with a reduction in intracellular glutathione in the embryo under heat stress (Ozawa et al., 2002). Embryos at morulae and blastocysts stage can repair heat-induced unfolded proteins or prevent DNA damage occurring in processes such as apoptosis. Therefore inhibition of apoptosis in bovine embryos with a caspase inhibitor increased the magnitude of the deleterious effects of elevated temperature on development (Paula-Lopes and Hansen, 2002b). Thus, if apoptotic process is restricted to the most damaged cells of the embryo, may allow the embryo to develop further after an environmental insult. In cattle, elevated temperature did not induce apoptosis until the 8–16 cell stage at day 4 after insemination (Paula-Lopes and Hansen, 2002). Some adverse effects of elevated temperature on embryonic survival in uterus during pregnancy could be the result of changes in maternal physiology rather than a direct effect on the embryo. In particular, there are reports that heat stress can reduce the concentrations of circulating reproductive hormones such as progesterone (Wolfenson et al., 2000).

Fetal Development

The mammalian fetus has a limited body's ability to thermo regulate its own body temperature; hence, he/she is entirely dependent upon the maternal thermoregulatory capacity. Under high thermal conditions, a significant increase in embryonic/fetal alterations leads to resorption of the embryo, abortion of the fetus or teratogenesis. In contrast to prolonged or excessive hyperthermia, short duration of exposure to maternal body temperature created with brief duration of radiofrequency exposures have not produced major fetal abnormalities (Chernovetz, et al., 1977). A clear dose-response association has been found in laboratory animals between temperature elevation and abnormalities (Lary, et al., 1983). These events lead to more severe endocrine and vascular abnormalities affecting the placenta, which lower the weight of placenta that further results in malnutrition, fetal hypoxia and eventually cause intrauterine fetal growth restriction (IUGR) (Wiltbank et al., 2016; Tao and Dahl, 2013; Wallace et al., 2005). Heat stress in sheep (40°C for 9 hr/day for 7–20 days) decreases blood flow to the placenta and restricts development of fetus, especially in the central trimester stage of pregnancy, rather than in the final third. Reduced placental perfusion might be attributable to an increase in vascular resistance caused by reduced angiogenesis (Galan et al., 2005). These conditions would also be seen in the cow, associated with decreased levels of placental hormones (Collier et al., 1982).

The negative effects of heat stress on production of prostaglandin E2 and PGF2α are studied in *in vitro* cultured bovine intrauterine epithelial and stromal cells. To evaluate the deleterious effects of heat shock endocrine functions, the cells were cultured at 38.5°C (control) or 40.5°C (heat stress). The disruptive effects of heat stress resulted in the significant increase in the production of both prostaglandins PGE2 and PGF2α in the stromal cells, but did not influence on production of prostaglandin PGE2 or PGF2α in the epithelial cells (Sakai et al., 2018). In dairy cattle, dry period (late gestation) is a critical period for fetal growth as well as for maternal welfare because of the physiological transition into the next period of lactation. Prepartum cooling heat stressed dairy cows in the dry period improves the immune status of transition cows and evidence suggests that altered peptide hormone PRL signaling in immune cells mediates the effects of heat stress on functioning of immune system (Tao and Dahl, 2013). Effects of maternal heat stress during pregnancy also affect offspring postnatal growth. Exposure of heat stress during late gestation, result in birth of calves with lower body weight at birth, shorter stature at weaning, and fail to gain the same height or weight at 12 months of age observed in calves from dams that have been cooled during the same gestational period. A portion of the retarded growth may result from the lower immune status observed in calves heat stressed in uterus, which begins with poorer apparent efficiency of absorption of immunoglobulin and extends to lower survival rates through puberty. A comparative study on reproductive performance of heat stressed calves versus utero cooled calves reveals that the cooled heifers require fewer services for conceiving and achieve pregnancy at an earlier age. Milk production in cows exposed to heat stress vs. those exposed to cold temperature during their fetal life also revealed a significant lower milk production in the first lactation (Dahl et al., 2016).

In humans, evidences suggest that even slight maternal thermal stress result in increased shunting of blood volume to the periphery as a heat dissipation mechanism. This could result in lowering heat exchange with the fetus by altered placental and umbilical blood perfusion and (Schroder and Power, 1997). In addition to environmental conditions that contribute to heat stress, many studies have highlighted potential effects on induction of fetal hyperthermia that might be derived from various diagnostic aids. Temperature elevations caused by routine basis ultrasound imaging are considered as negligible; however, some high-power Doppler ultrasound devices result in raising the fetal temperature by several degrees, with potential impact on fetal welfare (Ziskin and Morrissey, 2011).

Maternal-foetal Heat Exchange

Foetal body temperature is maternally dependent until birth because the foetus has almost no ability to regulate its own body temperature. A temperature variation of 0.58C is required to balance the rate at which heat is produced in the foetus with the rate at which it is degenerated to the mother. This close association between mother and foetus regarded a heat clamp; effectively prevents the foetus from independently regulating its body temperature before birth. Heat transfer pathways are the umbilical blood flow and through the foetal skin, amniotic/allantoic ñuid and uterine wall. Of these, bulk (85%) of foetal heat is generally thought to eliminate through the umbilical circulation (Power, 1989). Because the foetal and maternal blood ñows are relatively large and are separated by a thin placental barrier causes the thermal equilibrium to be reached more rapidly. If the umbilical cord is occluded, foetal temperatures rise quickly in relation to the extent of the occlusion, highlighting the importance of this site of heat exchange. However, efficiency of heat transfer is lower than predicted, probably due to the counter-current heat exchange mechanisms in the umbilical cord vessels and in the smaller vessels in the placenta. Factors influencing the body temperature of the foetus not only include maternal body temperature but also placental blood ñow. Heat stress from vigorous exercise result in the reduction of placental blood ñow (Erkkola et al., 1992). Short-term related exercises induce hyperthermia in hypertensive patients (Pirhonen et al., 1994). Long term hyperthermia contributes to reduction in uterine and placental blood flow blood ñow is further thought to reduce placental and foetal body weights (Murray and Katz, 1990). The importance of these affects in teratogenesis is not apparent, but they might disrupt with maternal/fetal heat exchange mechanism retarding the rate of heat transfer and embryonic development. The overall situation of heat stress in female mammalians is shown in Fig. 1.

Figure 1. Impact of heat stress on female reproductive outcome.

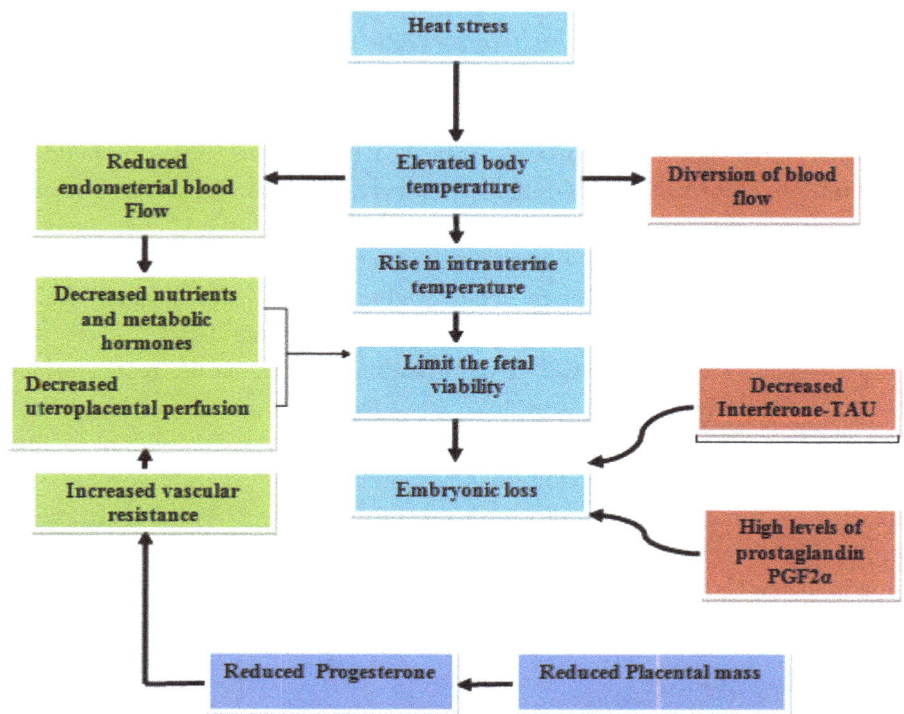

Heat Shock Response

Exposure to heat and a number of other toxic agents trigger the 'heat shock' response in the early embryonic and foetal life. The response is a survival mechanism and draws preference over other cellular activities. Cellular processes like cell proliferation and protein synthesis are suspended as the response is initiated. The constitutive (chaperone) heat shock proteins have pivotal role in cell survival and developmental processes, and exist in larger quantities in areas of the embryo and at times when growth and development of new organs are in progress. Experimental studies suggest that chaperone proteins might provide some degree of protection to embryos against the negative affects of low doses of heat, up to the threshold of 2–2.5°C. The induced response is initiated after a potentially damaging dose of heat. However, it cannot be induced in rabbit and mouse embryos between the 2-cell and blastocyst stages and prior to implantation and early embryonic loss usually follows heat exposure at this stage. The heat shock response has been demonstrated in cultured rat embryos (Walsh et al., 1989; Mirkes et al., 1997). At the earliest stages of organogenesis (day 9.5), a mild dose of heat (42°C for 10 min) induces a protective response against a subsequent, heavier dose of heat (43°C for 7.5 min) which would further result in severe abnormalities (Walsh et al., 1989). The protective response, synchronize with the synthesis of a number of induced HSP (heat shock proteins), is evident within 15 min after the induction of dose if embryos are allowed to make progress for at least 15 min at the normal culture temperature (38.5°C). If the embryos remain at 42°C, severe abnormalities occur. It is possible that the protective response might be induced at lower elevations of temperature, without a recovery period at normal temperature.

For this reason, the degree as well as the rate of temperature elevation might be significant reason in thermal damage. Synthesis of heat shock protein; HSP72 in 10.5 day cultured rat embryos was detectable within 2.5 hour after exposure to 41°C, 42°C and 43°C for 15 min, and for 2.5 min respectively (Mirkes et al., 1997). It persisted in measurable amounts for 24–48 hours, particularly noticeable during foetal development, long term or repetitive doses of heat commonly result in weight reduction of the placenta and foetuses that have been endorsed to various factors, but particularly reduced uterine blood ñow (Bell, 1987). It is also possible that foetal growth restriction following long term periods of heat stress might also be due to repeated induction of the heat shock response with protein synthesis inhibition in both the placenta and foetus.

Heat Shock Proteins

Few studies have reported that heat stress may impact reproduction by affecting the expression of heat shock proteins (Schroder and Power, 1997). Heat stress protein family (HSP) is a group of anti- stress proteins that are produced by cells in response to stress stimulation, and have a variety of physiological functions. Since heat stress can stimulate gene transcription and subsequent protein expression of HSP70, HSP70.2 and HSP105/110 and promote depletion of HSP70 in ovarian cells, Heat stress proteins can be indicators and mediators of heat stress. Experimental studies reported that rate of oocyte development and blastocyst formation would increaseat temperature of 43°C for 15 mins (Power, 1989). The reason might be that heat stress had induced an expression of a large number of HSPs. Over expression of heat stress, circulatory shockand cerebralischemia duringheat stroke which signifies the central role of HSP in cytoprotection. proteins provide protection against hypothermia (Gaughan et al., 2010). Using two-dimensional differential gel Electrophoresis analysis made of gametes at 32°C revealed the expression of Hsp70, Grp75, Grp78, Hsp70-2 and Hsc70t, while at 42–43°C, Hsp70-1 and Hsp70-3 could not be induced, therefore there is an association between HSP and co breeding activity of cow, might be due to reduction of HSP expression under thermal stress that can potentially affect the normal development of gametes (sperms, oocytes) and embryos, although needs further research.

Heat Stress on Appetite and Feed Intake

High environmental temperature during summer months' result in the inhibition of animals' feed central system result in remarkable depression of appetite, feed consumption and change metabolismthus influencing the hypothalamic-pituitary system, resulting in the sharp decline in secretion of GnRH and LH, and indirectly affect the reproduction. Metabolic modifications also destroy energy balance, decrease the plasma insulin, IGF-I and glucose levels (Hendrickx et al., 1979). Increasing the concentration of growth hormone (GH) and non-esterified fatty acid (NEFA), insulin acts as an important mediatorof ovarian follicular development, for IGF-I and glucose can stimulate ovarian follicle growth and embryo implantation; while glucose can also regulate LH secretion directly, so severe glycopenia will limit secretion of LH and ovulation. Moreover, prolactin is also a heat- sensitive hormone, and the secretion of prolactins at high ambient temperature inhibits follicular growth. Therefore, the reduction of plasma insulin, insulin like growth factor (IGF-I) and PRL, glucose concentrations are certainly leading causes of decline in reproductive performances in animals.

Reducing Degree of Heat Stress

An evident way for reducing the environmental-induced hyperthermia is to change the environment and management factors driving cows into hyperthermia. Extensive cooling management system with adequate shade, sprinklers, fans, and various evaporative cooling systems are often installed for dairy cattle during hot and dry climatic condition. Additionally, cooling cows during the summer increased can improve the conception rate. Although, large seasonal variations in reproductive performance can persist even in farm animals that have made broad use of such cooling system (Murray and Katz, 1990). Earlier stages of embryonic development are most sensitive to heat exposure providing cooling for a limited number of days during maximum sensitivity of the embryo to heat stress can moderately enhance the chances for establishing pregnancy (Otake and Schull, 1998). For example, pregnancy rate per AI for cows and heifers cooled for 8 days beginning after administration of prostaglandin (PGF2α) is 16% vs. 6% for controls (Hinoue et al., 2001). Failure of limited cooling to improve overall pregnancy rate more dramatically is probably a reflection of heat stress on maintenance and functioning of follicles and oocytes before the induction of cooling or soon after pregnancy or after the termination of cooling.

HEAT STRESS ON REPRODUCTIVE EFFICIENCY OF HUMANS

The human body maintains its core body temperature through homeostasis at around 37°C (±1°C) (Ziskin and Morrissey, 2011). Human body can tolerate internal temperatures below 35°C or above 41°C only for short duration. It is not clear exactly how pregnancy impacts on the women's heat tolerance, but altered pregnancy hormone levels and the increased circulatory demands of the mother and growing fetus may increase her susceptibility to thermal disorders. Severe maternal hyperthermia (over-heating) due to illness appears to increase the incidence of spontaneous abortions, severe fetal malformations or death, However, the human embryo and foetus can be vulnerable to elevated maternal temperatures during certain stages of cellular development such as cell proliferation, cell migration, cell differentiation, and apoptosis (programmed cell death) at different times for different organ structures. With limited capacity of the developing embryo and fetus, regulate temperature on its own that relies entirely on the maternal metabolic and thermoregulatory processes. Maternal elevated temperatures above the set-point of thermoregulation in animal models have been found to trigger developmental abnormalities. However, due to the difference in thermo neutral temperatures and the thermoregulatory mechanisms for humans, these effects and the thresholds cannot be directly extrapolated to humans. Though, earlier experimental studies have suggested that lower magnitude of maternal thermal stress, even if may be asymptomatic, could apparently improve the redirection of blood to the periphery (as heat loss mechanism) at the expense of placental and umbilical blood perfusion, which could ultimately lead to the reduction of heat transfer to the fetus (Ziskin and Morrissey, 2011).

Earlier review regarding the impacts of climate change on adverse pregnancy outcome, taking a wide range of environmental effects into consideration, presented significant relation between the pregnancy-related outcomes and the climate change variables such as eclampsia, pre- eclampsia, preterm birth, low birth weight, congenital cataract, sex ratio, and duration of gestation (Poursafa et al., 2015). It is estimated that approximately 10–15% of the congenital structural malformations are as a result of environmental influences on prenatal development (Brent, 2001; Gilbert-Barness, 2010). Environmental factors result in permanent structural and functional abnormality, growth restriction that could ultimately lead to pregnancy loss or embryonic failure (Gilbert-Barness, 2010). Maternal hyperthermia in early pregnancy; (e.g., fever because of an infectious disease) at 7 to 16 weeks of gestation can induce severe fetal congenital malformations in humans. Various studies have reported that teratogenic maternal hyperthermia during the early stage of the first trimester is most intense (Dreier et al., 2014; Fraser and Skelton, 1978; Peietal., 2015). Dadvand et al. (2011) found that exposure to hot weather or thermal stress during pregnancy was associated with 5 days' reduction in average gestational age. Earlier reports have suggested that preterm delivery is related to is the secretion heat-shock proteins and oxytocin and prolactin that result from the environmental heat load (Dreiling et al., 1991; Poursafa et al., 2015). Dehydration caused by thermal heat exposure has also been thought as another possible reason for preterm birth (Shoghag et al., 1997).

In general, hyperthermia is thought to be a potent teratogen in many experimental animals and maternal hyperthermia in pregnancy has been associated with congenital malformations in human offsprings (Edwards et al., 2003; Edwards, 2006; VanZutphen et al., 2012; Moretti et al., 2005). The central nervous system is known to be most thermo sensitive site in the brain (Edwards et al., 2003; Graham, 2005; Moretti et al., 2005). According to clinical studies, normal human body temperature is considered to be 37°C and 39°C is regarded as the threshold for potential damage (VanZutphen et al., 2012). However, the teratogenicity of heat is reliant on the association between the susceptible stage of development and ambient elevation of heat; the type of defect is entirely developmental stage specific (Graham, 2005; VanZutphen et al., 2012). Major developmental defects during any stage of pregnancy result in spontaneous abortion or fetal death (Parker et al., 2010), as it is reported earlier that major percentage of embryos with neural tube defects expires in the uterus (womb) prior to expulsion or delivery. Ambient heat exposure levels in hot regions of the world could therefore induce congenital malformations and miscarriage or stillbirths in humans (especially for outdoor workers, e.g., in agriculturalists), that would be expected from other causes acting alone, but few practical investigations in real-world settings are reported in the existing scientific literature.

HORMONAL TREATMENT AND ASSISTED REPRODUCTIVE (AI) TECHNOLOGIES

Hormonal treatments have the potential to control and decrease the heat stress effects in animals. Administering GnRH in the early stages of estrus coincides with the secretion of the low endogenous LH surge and improves ovulation or pregnancy rate. Administration of either GnRH agonist or human chorionic gonadotropin on day 5 of the estrous cycle results in ovulation or luteinization of the first wave dominant follicle and forms an accessory CL that increases the plasma progesterone levels to compensate its decrease in chronic heat stress (Wolfenson et al., 2000; Samal, 2013). Efficiency of timed AI during the summer months also improves summer fertility when associated with the administration of GnRH to induce anovulatory follicular recruitment. This protocol should be followed by prostaglandin (PGF2α) injection 7 days later for the regression of the CL which induces the final follicular maturation. Moreover, a second dose of GnRH 48 hours after injecting prostaglandin PGF2α may induce ovulation and AI of cows at 16 hours to ensure successful pregnancy (Pursley et al., 1998). The Ovsynch protocol in combination with AI showed successful estrus, ovulation and conception rate in buffaloes (Hoque et al., 2014). El-Tarabany and El-Tarabany, 2015 reported improvement in conception and pregnancy rates in Holstein cows using CIDRsynch and Presynch protocols under subtropical environmental conditions. The technique of embryo transfer (ET) can eliminate such heat stress and improves pregnancy rates during summer because embryos are transferred after the time at which they are more sensitive to heat stress. Compared to AI, heat stressed cows has been shown to increase the pregnancy rates by transfer of either frozen or unfrozen embryos collected from superovulated donors (Stewart et al., 2011).

CONCLUSION

In conclusion, heat stress shows its negative effects from different aspects of oocyte maturation to its subsequent embryonic development. Most of the detrimental effects of heat shock could be due to the oxidative stress originating from the elevation of temperature over an optimal level. Environmental stress and production of oxidative stress cause cellular and organelle damage directly or induce DNA damage can lead to cell death by processes such as apoptosis, alter expression of genes, and reduce developmental competence in preimplantation embryos. Oocytes, zygotes, and early stage embryos have low potential for scavenging ROS emissions and cannot mediate heat stress. On the other hand, it is likely that the advanced stage embryos such as such as morula or blastocysts after embryonic genome activation (EGA) would have higher ROS scavenging capacity and higher ability ability to repair heat-induced DNA damage or unfolded proteins. Although a number of evidences provide proof of the direct effects of heat stress on the preimplantation embryo, further studies would be required to reveal the underlying thermo tolerance mechanism and development of scientific techniques to manage heat stress and to improve the conception and pregnancy rates.

AUTHOR CONTRIBUTION

SK, GDS designed the book chapter. SK, RA performed literature survey and wrote the book chapter in consultation with GDS.

FUND DISCLAIMER

No grant received from any funding agency in public, commercial, or not-for-profit sector.

CORRESPONDING AUTHOR

Sameena Khanday

REFERENCES

Abuajamieh, M., Kvidera, S. K., Mayorga, E. J., Kaiser, A., Lei, S., Seibert, J. T., Horst, E. A., Sanz Fernandez, M. V., Ross, J. W., Selsby, J. T., Keating, A. F., Rhoads, R. P., & Baumgard, L. H. (2018). The effect of recovery from heat stress on circulating bioenergetics and inflammatory biomarkers. *Journal of Animal Science*, *96*, 4599–4610. doi:10.1093/jasky345 PMID:30476152

Arechiga, C. F., & Hansen, P. J. (1998). Response of preimplantation murine embryos to heat shock as modified by developmental stage and glutathione status. *In Vitro Cellular & Developmental Biology. Animal*, *34*(8), 655–659. doi:10.100711626-996-0016-8 PMID:9769152

Balboula, A. Z., Yamanaka, K., Sakatani, M., Kawahara, M., Hegab, A. O., Zaabel, S. M., & Takahashi, M. (2013). Cathepsin B activity has a crucial role in the developmental competence of bovine COCs exposed to heat-shock during in vitro maturation. *Reproduction (Cambridge, England)*, *146*, 407–417. doi:10.1530/REP-13-0179 PMID:23898216

Baumgard, L. H., & Rhoads, R. P. Jr. (2013). Effects of heat stress on postabsorptive metabolism and energetics. *Annual Review of Animal Biosciences*, *1*(1), 311–337. doi:10.1146/annurev-animal-031412-103644 PMID:25387022

Bell, A. W. (1987). Consequences of severe heat stress for fetal development. In J. R. S. Hales & D. A. B. Richards (Eds.), *Heat Stress, Physical Exertion and Environment* (pp. 313–333). Excerpta Medica.

Brent, R. L. (2001). The cause and prevention of human birth defects: What have we learned in the past 50 years? *Congenital Anomalies*, *41*(1), 3–21. doi:10.1111/j.1741-4520.2001.tb00869.x

Brodsky, J. L., & Chiosis, G. (2006). Hsp70 molecular chaperones: Emerging roles in human disease and identification of small molecule modulators. *Current Topics in Medicinal Chemistry*, *6*(11), 1215–1225. doi:10.2174/156802606777811997 PMID:16842158

Burfening, P. J., & Ulberg, L. C. (1968). Embryonic survival subsequent to culture of rabbit spermatozoa at 38 and 40 C. *Journal of Reproduction and Fertility*, *15*(1), 87–92. doi:10.1530/jrf.0.0150087 PMID:5643479

Chernovetz, M. E., Justesen, D. R., & Oke, A. F. (1977). A teratological study of the rat: Microwave and infrared radiations compared. *Radio Science*, *12*(6S), 191–197. doi:10.1029/RS012i06Sp00191

Christians, E., Michel, E., Adenot, P., Mezger, V., Rallu, M., Morange, M., & Renard, J. P. (1997). Evidence for the involvement of mouse heat shock factor 1 in the atypical expression of the HSP70.1 heat shock gene during mouse zygotic genome activation. *Molecular and Cellular Biology*, *17*(2), 778–788. doi:10.1128/MCB.17.2.778 PMID:9001232

Collier, R. J., Doelger, S. G., Head, H. H., Thatcher, W. W., & Wilcox, C. J. (1982). Effects of heat stress during pregnancy on maternal hormone concentrations, calf birth weight and postpartum milk yield of holstein cows. *Journal of Animal Science*, *54*(2), 309–319. doi:10.2527/jas1982.542309x PMID:7076593

Conneely, O. M., Mulac-Jericevic, B., DeMayo, F., Lydon, J. P., & O'Malley, B. W. (2002). Reproductive functions of progesterone receptors. *Recent Progress in Hormone Research*, *57*(1), 339–355. doi:10.1210/rp.57.1.339 PMID:12017551

Curtis Hewitt, S., Goulding, E. H., Eddy, E. M., & Korach, K. S. (2002). Studies using the estrogen receptor alpha knockout uterus demonstrate that implantation but not decidualization-associated signaling is estrogen dependent. *Biology of Reproduction*, *67*(4), 1268–1277. doi:10.1095/biolreprod67.4.1268 PMID:12297545

Dadvand, P., Nazelle, A., Triguero-Mas, M., Schembari, A., Cirach, M., Amoly, E., Figueras, F., Basagana, X., & Ostro, B. (2012). Nieuwenhuijsen, M. Surrounding greenness and exposure to air pollution during pregnancy: An analysis of personal monitoring data. *Environmental Health Perspectives*, *120*(9), 1286–1290. doi:10.1289/ehp.1104609 PMID:22647671

Dahl, G. E., Tao, S., & Monteiro, A. P. A. (2016). Effects of late-gestation heat stress on immunity and performance of calves. *Journal of Dairy Science*, *99*(4), 3193–3198. doi:10.3168/jds.2015-9990 PMID:26805989

Dreier, J. W., Andersen, A.-M. N., & Berg-Beckhoff, G. (2014). Systematic review and meta-analyses: Fever in pregnancy and health impacts in the offspring. *Pediatrics*, *133*(3), e674–e688. doi:10.1542/peds.2013-3205 PMID:24567014

Dreiling, C. E., Carman, F. S. III, & Brown, D. E. (1991). Maternal endocrine and fetal metabolic responses to heat stress. *Journal of Dairy Science*, *74*(1), 312–327. doi:10.3168/jds.S0022-0302(91)78175-7 PMID:2030175

Dutt, R. H. (1964). Detrimental effects of high ambient temperature on fertility and early embryo survival in sheep. *International Journal of Biometeorology*, *8*(1), 47–56. doi:10.1007/BF02186927 PMID:5888798

Ealy, A. D., Drost, M., & Hansen, P. J. (1993). Developmental changes in embryonic resistance to adverse effects of maternal heat stress in cows. *Journal of Dairy Science*, *76*(10), 2899–2905. doi:10.3168/jds.S0022-0302(93)77629-8 PMID:8227617

Edwards, J. L., & Hansen, P. J. (1996). Elevated temperature increases heat shock protein 70 synthesis in bovine two-cell embryos and compromises function of maturing oocytes. *Biology of Reproduction*, *55*(2), 340–346. doi:10.1095/biolreprod55.2.341 PMID:8828838

Edwards, J. L., King, W. A., Kawarsky, S. J., & Ealy, A. D. (2001). Responsiveness of early embryos to environmental insults: Potential protective roles of HSP70 and glutathione. *Theriogenology*, *55*(1), 209–223. doi:10.1016/S0093-691X(00)00455-6 PMID:11198083

Edwards, J. L., Saxton, A. M., Lawrence, J. L., Payton, R. R., & Dunlap, J. R. (2005). Exposure to a physiologically relevant elevated temperature hastens in vitro maturation in bovine oocytes. *Journal of Dairy Science*, *88*(12), 4326–4333. doi:10.3168/jds.S0022-0302(05)73119-2 PMID:16291624

Edwards, M. J. (2006). Review: Hyperthermia and fever during pregnancy. *Birth Defects Research. Part A, Clinical and Molecular Teratology*, *76*(7), 507–516. doi:10.1002/bdra.20277 PMID:16933304

Edwards, M. J., Saunders, R. D., & Shiota, K. (2003). Effects of heat on embryos and foetuses. *International Journal of Hyperthermia*, *19*(3), 295–324. doi:10.1080/0265673021000039628 PMID:12745973

El-Tarabany, M. S., & El-Tarabany, A. A. (2015). Impact of thermal stress on the efficiency of ovulation synchronization protocols in Holstein cows. *Animal Reproduction Science*, *160*, 138–145. doi:10.1016/j.anireprosci.2015.08.002 PMID:26297623

Erkkola, R. U., Pirhonen, J. P., & Kivijarvi, A. K. (1992). Flow velocity waveforms in uterine and umbilical arteries during submaximal bicycle exercise in normal pregnancy. *Obstetrician & Gynaecologist*, *79*, 611–615. PMID:1553187

Fraser, F. C., & Skelton, J. (1978). Possible teratogenicity of maternal fever. *Lancet*, *2*(8090), 634–634. doi:10.1016/S0140-6736(78)92859-3 PMID:80560

Galan, H. L., Anthony, R. V., Rigano, S., Parker, T. A., de Vrijer, B., Ferrazzi, E., & Regnault, T. R. H. (2005). Fetal hypertension and abnormal Doppler velocimetry in an ovine model of intrauterine growth restriction. *American Journal of Obstetrics and Gynecology*, *192*(1), 272–279. doi:10.1016/j.ajog.2004.05.088 PMID:15672036

Gaughan, J. B., Mader, T. L., Holt, S. M., Sullivan, M. L., & Hahn, G. L. (2010). Assessing the heat tolerance of 17 beef cattle genotypes. *International Journal of Biometeorology*, *54*(6), 617–627. doi:10.100700484-009-0233-4 PMID:19458966

Gendelman, M., & Roth, Z. (2012). In vivo vs. in vitro models for studying the effects of elevated temperature on the GV-stage oocyte, subsequent developmental competence and gene expression. *Animal Reproduction Science*, *134*(3-4), 125–134. doi:10.1016/j.anireprosci.2012.07.009 PMID:22898494

Germain, M. A., Webster, W. S., & Edwards, M. J. (1985). Hyperthermia as a teratogen: Parameters determining hyperthermia-induced defects in the rat. *Teratology*, *31*(2), 265–272. doi:10.1002/tera.1420310212 PMID:3992495

Gilad, E., Meidan, R., Berman, A., Graber, Y., & Wolfenson, D. (1993). Effect of heat stress on tonic and GnRH-induced gonadotrophin secretion in relation to concentration of oestradiol in plasma of cyclic cows. *Journal of Reproduction and Fertility*, *99*(2), 315–321. doi:10.1530/jrf.0.0990315 PMID:8107012

Gilbert-Barness, E. (2010). Review: Teratogenic causes of malformations. *Annals of Clinical and Laboratory Science*, *40*, 99–114. PMID:20421621

Hale, B. J., Hager, C. L., Seibert, J. T., Selsby, J. T., Baumgard, L. H., Keating, A. F., & Ross, J. W. (2017). Heat stress induces autophagy in pig ovaries during follicular development. *Biology of Reproduction*, *97*(3), 426–437. doi:10.1093/biolre/iox097 PMID:29025092

Hansen, P. J. (1999). Possible roles for heat shock protein 70 and glutathione in protection of the mammalian preimplantation embryo from heat shock. *Annual Review of Biomedical Science*, *1*, 5–29.

Hansen, P. J. (2007). To be or not to bedeterminants of embryonic survival following heat shock. *Theriogenology*, *68*, 40–S48. doi:10.1016/j.theriogenology.2007.03.013

Hansen, P. J. (2009). Effects of heat stress on mammalian reproduction. Phil Trans Royal Soc London. Series B. *Biological Science*, *364*(1534), 3341–3350. doi:10.1098/rstb.2009.0131

Hendricks, K. E. M., Martins, L., & Hansen, P. J. (2009). Consequences for the bovine embryo of being derived from a spermatozoon subjected to post-ejaculatory aging and heat shock: Development to the blastocyst stage and sex ratio. *The Journal of Reproduction and Development*, *55*(1), 69–74. doi:10.1262/jrd.20097 PMID:18957823

Hendrickx, A. G., Stone, G. W., Henrickson, R. V., & Matayoshi, K. (1979). Teratogenic effects of hyperthermia in the bonnet monkey (macaca radiata). *Teratology*, *19*(2), 177–182. doi:10.1002/tera.1420190208 PMID:112705

Hinoue, A., Fushiki, S., Nishimura, Y., & Shiota, K. (2001). In utero exposure to brief hyperthermia interferes with the production and migration of neocortical neurons and induces apoptotic cell death in the fetal mouse brain. *Brain Research. Developmental Brain Research*, *132*(1), 59–67. doi:10.1016/S0165-3806(01)00295-4 PMID:11744107

Hirao, Y., & Yanagimachi, R. (1978). Temperature dependence of sperm-egg fusion and post-fusion events in hamster fertilization. *The Journal of Experimental Zoology*, *205*(3), 433–437. doi:10.1002/jez.1402050312 PMID:702084

Hooper, L. M., Payton, R. R., Rispoli, L. A., Saxton, A. M., & Edwards, J. L. (2015). Impact of heat stress on germinal vesicle breakdown and lipolytic changes during in vitro maturation of bovine oocytes. *The Journal of Reproduction and Development*, *61*(5), 459–464. doi:10.1262/jrd.2014-168 PMID:26120041

Hoque, M. N., Talukder, A. K., Akter, M., & Shamsuddin, M. (2014). Evaluation of ovsynch protocols for timed artificial insemination in water buffaloes in Bangladesh. *Turkish Journal of Veterinary and Animal Sciences*, *38*, 418–424. doi:10.3906/vet-1302-35

Ju, J. C., Jiang, S., Tseng, J. K., Parks, J. E., & Yang, X. (2005). Heat shock reduces developmental competence and alters spindle configuration of bovine oocytes. *Theriogenology*, *64*(8), 1677–1689. doi:10.1016/j.theriogenology.2005.03.025 PMID:15951010

Ju, J. C., Parks, J. E., & Yang, X. (1999). Thermotolerance of IVM-derived bovine oocytes and embryos after short-term heat shock. *Molecular Reproduction and Development*, *53*(3), 336–340. doi:10.1002/(SICI)1098-2795(199907)53:3<336::AID-MRD9>3.0.CO;2-M PMID:10369394

Kanai, Y., Yagyu, N., & Shimizu, T. (2001). Hypogonadism in heat stressed goats: Poor responsiveness of the ovary to the pulsatile LH stimulation induced by hourly injections of a small dose of GnRH. *The Journal of Reproduction and Development*, *41*(2), 133–139. doi:10.1262/jrd.41.133

Khodaei, Motlagh, M., Shahneh, A. Z., Masoumi, R. and Derensis, F. (2011). Alterations in reproductive hormones during heat stress in dairy cattle. *African Journal of Biotechnology*, *29*, 5552–5558.

Krininger, C. E. III, Stephens, S. H., & Hansen, P. J. (2002). Developmental changes in inhibitory effects of arsenic and heat shock on growth of preimplantation bovine embryos. *Molecular Reproduction and Development*, *63*(3), 335–340. doi:10.1002/mrd.90017 PMID:12237949

Kuroki, T., Ikeda, S., Okada, T., Maoka, T., Kitamura, A., Sugimoto, M., & Kume, S. (2013). Astaxanthin ameliorates heat stress-induced impairment of blastocyst development in vitro:– astaxanthin colocalization with and action on mitochondria. *Journal of Assisted Reproduction and Genetics*, *30*(5), 623–631. doi:10.100710815-013-9987-z PMID:23536152

Lannett Edwards, J., & Hansen, P. J.Lannet- Edwards. (1996). Elevated temperature increases heat shock protein 70 synthesis in bovine 2-cell embryos and compromises function of maturing oocytes. *Biology of Reproduction*, *55*(2), 340–346. doi:10.1095/biolreprod55.2.341

Lary, J. M., Conover, D. L., Johnson, P. H., & Burg, J. R. (1983). Teratogenicity of 27.12-MHz radiation in rats is related to duration of hyperthermic exposure. *Bioelectromagnetics*, *4*(3), 249–255. doi:10.1002/bem.2250040305 PMID:6626274

Lenz, R. W., Ball, G. D., Leibfried, M. L., Ax, R. L., & First, N. L. (1983). In vitro maturation and fertilization of bovine oocytes are temperature-dependent processes. *Biology of Reproduction, 29*(1), 173–179. doi:10.1095/biolreprod29.1.173 PMID:6615963

Lim, H. J., & Wang, H. (2010). Uterine disorders and pregnancy complications: Insights from mouse models. *The Journal of Clinical Investigation, 120*(4), 1004–1015. doi:10.1172/JCI41210 PMID:20364098

Lim, J. M., Liou, S. S., & Hansel, W. (1996). Intracytoplasmic glutathione concentration and the role of β-mercaptoethanol in preimplantation development of bovine embryos. *Theriogenology, 46*(3), 429–439. doi:10.1016/0093-691X(96)00165-3 PMID:16727911

Lord-Fontaine, S., & Averill-Bates, D. A. (2002). Heat shock inactivates cellular antioxidant defenses against hydrogen peroxide: Protection by glucose. *Free Radical Biology & Medicine, 32*(8), 752–765. doi:10.1016/S0891-5849(02)00769-4 PMID:11937301

Loven, D. P. (1988). A role for reduced oxygen species in heat induced cell killing and the induction of thermotolerance. *Medical Hypotheses, 26*(1), 39–50. doi:10.1016/0306-9877(88)90111-9 PMID:2840559

Ma, W. G., Song, H., Das, S. K., Paria, B. C., & Dey, S. K. (2003). Estrogen is a critical determinant that specifies the duration of the window of uterine receptivity for implantation. *Proceedings of the National Academy of Sciences of the United States of America, 100*(5), 2963–2968. doi:10.1073/pnas.0530162100 PMID:12601161

Matsuzuka, T., Sakamoto, N., Ozawa, M., Ushitani, A., Hirabayashi, M., & Kanai, Y. (2005b). Alleviation of maternal hyperthermia-induced early embryonic death by administration of melatonin to mice. *Journal of Pineal Research, 39*(3), 217–222. doi:10.1111/j.1600-079X.2005.00260.x PMID:16150100

McMurray, R. G., & Katz, V. L. (1990). Thermoregulation in pregnancy: Implications for exercise. *Sports Medicine (Auckland, N.Z.), 10*(3), 146–158. doi:10.2165/00007256-199010030-00002 PMID:2237032

Michel, C. L, Chastel, O. & Bonnet, X. (2011). Ambient temperature and pregnancy influence cortisol levels in female guinea pigs and entail long-term effects on the stress response of their offspring. *General and Comparative Endocrinology, 171*, 275-282.

Mirkes, P. E., Cornel, L. M., Park, H. W., & Cunningham, L. M. (1997). Induction of thermotolerance in early postimplantation rat embryos is associated with increased resistance to hyperther- mia-induced apoptosis. *Teratology, 56*(3), 210–219. doi:10.1002/(SICI)1096-9926(199709)56:3<210::AID-TERA4>3.0.CO;2-4 PMID:9358608

Mirkes, P. E., & Dogget, B. (1992). Accumulation of heat shock protein 72 (hsp72) in postimplantation rat embryos after exposure to various periods of hyperthermia (40–438C) in vitro: Evidence that heat shock protein 72 is a biomarker of heat-induced embryotoxicity. *Teratology, 46*(3), 301–309. doi:10.1002/tera.1420460315 PMID:1523587

Moretti, M. E., Bar-Oz, B., Fried, S., & Koren, G. (2005). Maternal hyperthermia and the risk for neural tube defects in offspring systematic review and meta-analysis. *Epidemiology (Cambridge, Mass.), 16*(2), 216–219. doi:10.1097/01.ede.0000152903.55579.15 PMID:15703536

Nabenishi, H., Ohta, H., Nishimoto, T., Morita, T., Ashizawa, K., & Tsuzuki, Y. (2012). The effects of cysteine addition during in vitro maturation on the developmental competence, ROS, GSH and apoptosis level of bovine oocytes exposed to heat stress. *Zygote (Cambridge, England)*, *20*(3), 249–259. doi:10.1017/S0967199411000220 PMID:21729376

Otake, M., & Schull, W. J. (1998). Review: Radiation-related brain damage and growth retardation among the prenatally exposed bomb survivors. *International Journal of Radiation Biology*, *74*(2), 159–171. doi:10.1080/095530098141555 PMID:9712546

Ozawa, M., Hirabayashi, M., & Kanai, Y. (2002). Developmental competence and oxidative state of mouse zygotes heat-stressed maternally or in vitro. *Reproduction (Cambridge, England)*, *124*, 683–689. doi:10.1530/rep.0.1240683 PMID:12417007

Paula-Lopes, F. F., & Hansen, P. J. (2002). Heat-shock induced apoptosis in preimplantation bovine embryos is a developmentally-regulated phenomenon. *Biology of Reproduction*, *66*(4), 1169–1177. doi:10.1093/biolreprod/66.4.1169 PMID:11906938

Paula-Lopes, F. F., Lima, R. S., Satrapa, R. A., & Barros, C. M. (2013). Physiology and Endocrinology Symposium: Influence of cattle genotype (Bos indicus vs. Bos taurus) on oocyte and preimplantation embryo resistance to increased temperature. *Journal of Animal Science*, *91*(3), 1143–1153. doi:10.2527/jas.2012-5802 PMID:23296831

Payton, R. R., Romar, R., Coy, P., Saxton, A. M., Lawrence, J. L., & Edwards, J. L. (2004). Susceptibility of bovine germinal vesicle-stage oocytes from antral follicles to direct effects of heat stress in vitro. *Biology of Reproduction*, *71*(4), 1303–1308. doi:10.1095/biolreprod.104.029892 PMID:15201201

Pei, Zhu, H., Ye, R., Wu, J., Liu, J., Ren, A., Li, Z., & Zheng, X. (2015). Interaction between the SLC19A1 gene and maternal first trimester fever on offspring neural tube defects. *Birth Defects Research. Part A, Clinical and Molecular Teratology*, *103*(1), 3–11. doi:10.1002/bdra.23257 PMID:24917213

Perez-Crespo, M., Pintado, B., & Gutierrez-Adan, A. (2008). Scrotal heat stress effects on sperm viability, sperm DNA integrity, and the offspring sex ratio in mice. *Molecular Reproduction and Development*, *75*(1), 40–47. doi:10.1002/mrd.20759 PMID:17474098

Perez-Crespo, M., Ramirez, M. A., Fernandez-Gonzalez, R., & Gutierrez-Adan, A. (2005). Differential sensitivity of male and female mouse embryos to oxidative induced heat-stress is mediated by glucose-6-phosphate dehydrogenase gene expression. *Molecular Reproduction and Development*, *72*(4), 502–510. doi:10.1002/mrd.20366 PMID:16149081

Pirhonen, J. P., Vähä-Eskeli, K., Seppänen, A., Vuorinen, J., & Erkkola, R. (1994). Does thermal stress decrease uterine blood ñow in hypertensive pregnancies. *American Journal of Perinatology*, *11*(05), 313–316. doi:10.1055-2007-994542 PMID:7993506

Poursafa, P., Keikha, M., & Kelishadi, R. (2015). Systematic review on adverse birth outcomes of climate change. *Journal of Research in Medical Sciences*, *20*, 397–402. PMID:26109998

Power, G. G. (1989). Biology of temperature: The mammalian fetus. *Journal of Developmental Physiology*, *12*, 295–304. PMID:2701105

Pursley, J., Silcox, R. W., & Wiltbank, M. C. (1998). Effect of time of artificial insemination on pregnancy rates, calving rates, pregnancy loss and gender ratio after synchronization of ovulation in lactating dairy cows. *Journal of Dairy Science*, *81*(8), 2139–2144. doi:10.3168/jds.S0022-0302(98)75790-X PMID:9749378

Rispoli, L. A., Payton, R. R., Gondro, C., Saxton, A. M., Nagle, K. A., Jenkins, B. W., Schrick, F. N., & Edwards, J. L. (2013). Heat stress effects on the cumulus cells surrounding the bovine oocyte during maturation: Altered matrix metallopeptidase 9 and progesterone production. *Reproduction (Cambridge, England)*, *146*(2), 193–207. doi:10.1530/REP-12-0487 PMID:23744615

Rivera, R., & Hansen, P. (2001). Development of cultured bovine embryos after exposure to high temperatures in the physiological range. *Reproduction (Cambridge, England)*, *121*(1), 107–115. doi:10.1530/rep.0.1210107 PMID:11226033

Rocha, A., Randel, R. D., Broussard, J. R., Lim, J. M., Blair, R. M., Roussel, J. D., Godke, R. A., & Hansel, W. (1998). High environmental temperature and humidity decrease oocyte quality in Bos taurus but not in Bos indicus cows. *Theriogenology*, *49*, 657–665. doi:10.1016/S0093-691X(98)00016-8 PMID:10732044

Roth, Z. (2015). Physiology and endocrinology symposium: Cellular and molecular mechanisms of heat stress related to bovine ovarian function. *Journal of Animal Science*, *93*(5), 2034–2044. doi:10.2527/jas.2014-8625 PMID:26020299

Roth, Z. (2017). Effect of heat stress on reproduction in dairy cows: Insights into the cellular and molecular responses of the oocyte. *Animal Bioscience*, *5*(1), 151–170. doi:10.1146/annurev-animal-022516-022849 PMID:27732786

Roth, Z., & Hansen, P. J. (2005). Disruption of nuclear maturation and rearrangement of cytoskeletal elements in bovine oocytes exposed to heat shock during maturation. *Reproduction (Cambridge, England)*, *129*(2), 235–244. doi:10.1530/rep.1.00394 PMID:15695618

Roth, Z., & Wolfenson, D. (2016). Comparing the effects of heat stress and mastitis on ovarian function in lactating cows: Basic and applied aspects. *Domestic Animal Endocrinology*, *56*, 218–S227. doi:10.1016/j.domaniend.2016.02.013 PMID:27345320

Sakai, S., Hagihara, N., Kuse, M., Kimura, K., & Okuda, K. (2018). Heat stress affects prostaglandin synthesis in bovine endometrial cells. *The Journal of Reproduction and Development*, *64*(4), 64–2051. doi:10.1262/jrd.2018-051 PMID:29710018

Sakamoto, N., Ozawa, M., Yokotani-Tomita, K., Morimoto, A., Matsuzuka, T., Ijiri, D., Hirabayashi, M., Ushitani, A., & Kanai, Y. (2008). dl-α-Tocopherol acetate mitigates maternal hyperthermia-induced pre-implantation embryonic death accompanied by a reduction of physiological oxidative stress in mice. *Reproduction (Cambridge, England)*, *135*(4), 489–496. doi:10.1530/REP-07-0379 PMID:18367509

Sakatani, M., Kobayashi, S., & Takahashi, M. (2004). Effects of heat shock on in vitro development and intracellular oxidative state of bovine preimplantation embryos. *Molecular Reproduction and Development*, *67*(1), 77–82. doi:10.1002/mrd.20014 PMID:14648877

Sakatani, M., Yamanaka, K., Balboula, A. Z., Takenouchi, N., & Takahashi, M. (2015). Heat stress during in vitro fertilization decreases fertilization success by disrupting anti-polyspermy systems of the oocytes. *Molecular Reproduction and Development*, *82*(1), 36–47. doi:10.1002/mrd.22441 PMID:25462627

Samal, L. (2013). Heat stress in dairy cows - reproductive problems and control measures. *International Journal of Livestock Research*, *3*(3), 14–23.

Schillo, K. K., Alliston, C. W., & Malven, P. V. (1978). Plasma concentrations of luteinizing hormone and prolactin in the ovariectomized ewe during induced hyperthermia. *Biology of Reproduction*, *19*(2), 306–313. doi:10.1095/biolreprod19.2.306 PMID:719089

Schroder, H. J., & Power, G. G. (1997). Engine and radiator: Fetal and placental interactions for heat dissipation. *Experimental Physiology*, *82*(2), 403–414. doi:10.1113/expphysiol.1997.sp004035 PMID:9129954

Shoghag, L., Hudson, S., Applewhite, L., Feldman, J., & Minkoff, H. L. (1997). An association between the heat-humidity index and preterm labor and delivery: A preliminary analysis. *American Journal of Public Health*, *87*(7), 1205–1207. doi:10.2105/AJPH.87.7.1205 PMID:9240114

Slaem, M. B., & Bouraoui, R. (2011). Heat stress in Tunisia: Effects on dairy cows and potential means of alleviating. SA. *Journal of Animal Science*, *39*, 256–259.

Stewart, B. M., Block, J., Morelli, P., Navarette, A. E., Amstalden, M., Bonilla, L., Hansen, P. J., & Bilby, T. R. (2011). Efficacy of embryo transfer in lactating dairy cows during summer using fresh or vitrified embryos produced in vitro with sexsorted semen. *Journal of Dairy Science*, *94*(7), 3437–3445. doi:10.3168/jds.2010-4008 PMID:21700029

Tao, S., & Dahl, G. E. (2013). Invited review: Heat stress effects during late gestation on dry cows and their calves. *Journal of Dairy Science*, *96*(7), 4079–4093. doi:10.3168/jds.2012-6278 PMID:23664343

Tompkins, E. C., Heidenreich, C. J., & Stob, M. (1967). Effect of post-breeding thermal stress on embryonic mortality in swine. *Journal of Animal Science*, *26*(2), 377–380. doi:10.2527/jas1967.262377x

Ulberg, L. C., & Burfening, P. J. (1967). Embryo death resulting from adverse environment on spermatozoa or ova. *Journal of Animal Science*, *26*(3), 571–577. doi:10.2527/jas1967.263571x PMID:6068502

UNSCEAR. (1986). Biological effects of pre-natal irradiation. Annex C. In *United Nations Scientiðc Committee on the Effects of Atomic Radiation, Genetic and Somatic Effects of Ionising Radiation. UNSCEAR Report to the General Assembly with Scientific Annexes* (pp. 263–366). United Nations.

Van Zutphen, A. R., Fletcher, B. A., & Hwang, S. A. (2012). A population-based case–control study of extreme summer temperature and birth defects. National Institute of Environmental Health Sciences. *Environmental Health Perspectives*, *120*(10), 1443–1449. doi:10.1289/ehp.1104671 PMID:23031822

Wallace, J. M., Regnault, T. R. H., Limesand, S. W., Hay, W. W. Jr, & Anthony, R. V. (2005). Investigating the causes of low birth weight in contrasting ovine paradigms. *The Journal of Physiology*, *565*(1), 19–26. doi:10.1113/jphysiol.2004.082032 PMID:15774527

Walsh, D., Zhe, L., Zeng, F., Yan, W., & Edwards, M. (1994). Heat shock genes and cell cycle regulation in early mammalian development. *Environmental Medicine*, *38*, 1–6.

Welch, W. J., & Suhan, J. P. (1985). Morphological study of the mammalian stress response: Characterization of changes in cytoplasmic organelles, cytoskeleton, and nucleoli, and appearance of intranuclear actin filaments in rat fibroblasts after heat-shock treatment. *The Journal of Cell Biology*, *101*(4), 1198–1211. doi:10.1083/jcb.101.4.1198 PMID:3900086

Wiltbank, M. C., Baez, G. M., Garcia-Guerra, A., Toledo, M. Z., Monteiro, P. L. J., Melo, L. F., & Sartori, R. (2016). Pivotal periods for pregnancy loss during the first trimester of gestation in lactating dairy cows. *Theriogenology*, *86*(1), 239–253. doi:10.1016/j.theriogenology.2016.04.037 PMID:27238438

Wise, M. E., Armstrong, D. V., Huber, J. T., Hunter, R., & Wiersma, F. (1988). Hormonal alterations in the lactating dairy cow in response to thermal stress. *Journal of Dairy Science*, *71*(9), 2480–2485. doi:10.3168/jds.S0022-0302(88)79834-3 PMID:3183143

Wolfenson, D., Roth, Z., & Meidan, R. (2000). Impaired reproduction in heat-stressed cattle: Basic and applied aspects. *Animal Reproduction Science*, *60*, 535–547. doi:10.1016/S0378-4320(00)00102-0 PMID:10844222

Ziskin, M. C., & Morrissey, J. (2011). Thermal thresholds for teratogeni- city, reproduction, and development. *International Journal of Hyperthermia*, *27*(4), 374–387. doi:10.3109/02656736.2011.553769 PMID:21591900

KEY TERMS AND DEFINITIONS

Blastocyst: A thin–walled hollow structure in early embryonic development that contains a cluster of cells called the inner cell mass from which the embryo arises.

Estradiol: A female sex hormone produced by the ovaries, adrenal gland, and also the placenta during pregnancy.

Hyperthermia: Abnormally high body temperature caused by a failure of the heat-regulating mechanisms of the body to deal with the heat coming from the environment.

Oxidative Stress: A condition where the levels of ROS significantly overwhelm the capacity of antioxidant defenses, leading to potential damage in a biological system.

Prostaglandin F2-Alpha: A stable prostaglandin that stimulates the contraction of uterine and bronchial smooth muscle and produces vasoconstriction (tightening) in some blood vessels. It is used for the induction of abortion, for evacuation of the uterus after a missed abortion.

Reactive Oxygen Species (ROS): A phrase used to describe a number of reactive molecules and free radicals derived from molecular oxygen. ROS play a key role as a messenger in normal cell signal transduction and cell cycling.

Chapter 8
Neuroendocrine Cross-Talk and Axial Regulation of Embryonic Implantation:
Impact of Climate Change

Younis Ahmad Hajam
Career Point University, Hamirpur, India

Rajesh Kumar
Department of Biosciences, Division Zoology, Career Point University, India

Javid Ahmad Malik
Department of Zoology, Guru Ghasidas Vishwavidyalaya, India

Mohd Rafi Wani
Department of Forestry, Wild Life, and Environmental Sciences, Guru Ghasidas Vishwavidyalay, India

Ankush Sharma
https://orcid.org/0000-0001-7208-2998
Department of Zoology, Sri Sai University, India

ABSTRACT

Early embryonic mortality has become a major issue of reproductive wastages and key constraint. The survivability of embryo during early embryonic life mostly depends on secretion of hormones. Development of embryo involves the coordinated function of different hormones synthesized in different endocrine glands. Exploring the complex intricate cellular and molecular mechanism interlaying between the neuroendocrine axial system environments shall facilitate to develop new strategies to augment survivability of embryo and success rate. Therefore, the chapter focused to explore the hidden cross talk and regulatory mechanisms to unfold the coordinating role of axial system in the embryonic implantation and also to explore the impact of climate change on embryonic implantation.

DOI: 10.4018/978-1-7998-4480-8.ch008

INTRODUCTION

Reproduction is important process for the perpetuation and propagation for all the species to continue their generations. In most mammalian species, a zygote is formed in vivo following fertilization of an egg by a sperm. After the successful fertilization of egg by sperm in the fallopian tube, it moves through the fallopian tube and reaches in the uterus. During this time the fertilized egg divides and produces multicellular structure termed as blastocyst. It then attaches itself to the endometrium, this adherence of blastocyst considered as initial step of implantation process (Wilcox et al., 1999). This whole implantation process is controlled by endocrine hormones (estrogen and progestrone), which prepares the endometrium for the attachment of blastocyst. Estrogen and progesterone increase the receptivity of the endometrium for the successful implantation of blastocyst (Lessey, 2011). During the process of embryonic implantation different types of changes occurs in the endometrium.

ENDOMETRIUM AND MENSTRUAL CYCLE

The endometrium is made up of two layers viz, functionalis and basalis layers, both of these layers undergoes a sequence of changes during each ovulatory cycle that makes it momentarily capable to implantation. The functionalis layer makes the upper two thirds of the endometrium in which proliferation, secretion, and degradation occurs, however the basalis layer encompasses the lower one third and acts as site for the regeneration and renewal of tissues. In the proliferative phase during which the ovarian follicles grow and produces estrogen in higher quantity, which helps in the regeneration of the functionalis layer and hence the new growth of glands, stroma, and endothelial cells occurs. Next process is ciliogenesis during which the cells becomes ciliated present around the opening of glands, estradiol also helps in this process and starts on Day 7 or 8 of an ideal 28-day menstrual cycle (Ludwig and Spornitz et al., 1991). During the preovulatory17b-estradiol increases which leads to more proliferation and differentiation of epithelial cells of uterus (Norwitz et al., 2001). When ovulation occurs, the corpus luteum gets formed and secretion of progesterone starts. The progesterone acts on the endometrium which in turn increases the secretion of glycoproteins and peptides into the endometrial cavity in actively promotive fashion. Endometrial epithelial tissue proliferation ceases during this phase, because progesterone-helps in the blockade of receptor expression of estrogen and stimulation of 17b-hydroxysteroid dehydrogenase and sulfotransferase actions, which helps in the conversion of potent estradiol into estrone and then excreted out (Gurpide et al. 1976, Falany and Falany, 1996). About 7 days following the secretion of luteinizing hormone (LH) surge occurs, the secretory activity reaches upto the peak, the stroma of endometrial becomes tremendously edematous, and proliferation of blood vessels occurs in response to the sex steroids and secretion of prostaglandins (local factors).

Decidualization phase starts during the late in the luteal phase under the influence of progesterone, during this phase rate of mitosis increases and stromal cells undergoes differentiation. Along with decidualization progesterone-dependent infiltration of specific leukocyte subsets takes places into the stroma of endometrial, including natural killer cells, T cells, and macrophages (Bulmer et al., 1991). Different cytokines are used during the process of implantation, including Colony-stimulating factor (CSF)-1, has been shown to be expressed in human endometrium in the mid-proliferative and mid-secretory phases

(Bartocci et al., 1986). Studies reported that CSF-1 might be facilitating the attachment of blastocyst (Giudice, 1999). The infiltration of leukocytes into the endometrium is facilitated by Interleukins during the progesterone mediated decidualization (Bulmer et al., 1991). Likewise, reports also suggesting that the interleukins might have ability facilitate the cross-talk between the embryo and endometrium. Since steroid receptors are absent in the leukocytes, these chemoattractant cytokines (chemokines), including interleukin (IL)-8 and Monocyte Chemoattractant Protein-1 (MCP-1), facilitates steroid-dependent infiltration of leukocytes into the endometrium (Wood et al., 1997). These primary chemokines lead to the inductive signal for the result for the release of other cytokines, such as leukemia inhibiting factor (LIF) and IL-1 and the growth factor heparin-binding epidermal growth factor (HB-EGF) (Cross et al., 1994). IL-8 and MCP-1 are expressed by endometrial glandular and luminal epithelial cells (Arici et al., 1993, Arici et al., 1995), where they are up-regulated by progesterone during the window of implantation (Cabellero-Campo et al., 2002). This steroid-mediated infiltration of leukocytes occurs indirectly, becausein these cells receptors for estrogen or progesterone are seemed to absent (King et al., 1996).

STEROID HORMONE MECHANISM

The steroid hormones carry out their functions by finding with the intracellular receptors to regulate their gene expression of downstream effectors, such as peptide hormones, cytokines, and growth factors (Norwitz et al., 2001). The receptors for estrogen and progesterone are mainly localized in the nucleus of cell, even though some nucleocytoplasmic shuttling does occur (Speroff et al., 1999). Binding of ligand to these steroid receptors causes dimerization and consequently hormone receptor complex binds with the hormone responsive elements on DNA and finally causes activation or repression of transcription target genes (Speroff et al., 1999). There are two subtypes (a and b; and A and B) of receptors for estrogen and progesterone respectively. Estrogen receptor (ER)-a is expressed by endometrial epithelial and stromal cells during the proliferative phase, but its expression decreases in the secretory phase (Lessey et al., 1988). The stromal expression of ER-a regulates the cellular proliferation of the endometrial epithelium in response to estrogen (Kurita et al., 1998). However, ER-b is expressed only in the glandular epithelial cells at the lower rate (Matsuzaki et al., 1999) and ER-a modulates expression of genes in the uterus (Wiehua et al., 2000).

Both ER-a and -b sometimes forms homo- or heterodimers. The response of a cell towards the estrogen stimulation relies on the relative presence of the ER subtype, the type of estrogen, and the targeted response element (Speroff et al., 1999). Likewise, the relative number of progesterone receptor (PR)-A and –B within a target cell decides if activation of gene on the stimulation by hormone, because PR-A are predominantly suppressing the transcriptional activation by PR-B (Graham et al., 1996). During the proliferative and secretory phases of the menstrual cycle PR-A is expressed in the stroma and epithelium; nevertheless, epithelial levels of PR-A steadily declines during the secretary phase (Stavreus-Evers et al., 2001). During proliferative phase PR-B is present in glandular and stromal nuclei (Wang et al.,1998). Estrogen and growth factors increase the level of PR estrogens and growth factors, while as progesterone exposure decrease the PR-B level (Kraus and Katzenellenbogen, 1993). In the luminal epithelium expression of ER-b also down-regulates PRs level (Wiehua et al., 2000). Decreased regulation of PR during the window of implantation is a primary requirement for endometrial receptivity.

CELLULAR AND MOLECULAR MARKERS OF IMPLANTATION

Implantation is multifaceted, hormonally-regulated processes that require synchronization between the developing embryos and distinguish endometrium. This whole process is mediated by molecular cross-talk between the embryo and endometrium (Glasser et al., 1991). Different studies have reported potential receptive markers present in the endometrial acts as predictors of successful implantation and, in doing so, have facilitates to define the cellular and molecular mechanisms by which implantation occurs. These markers include pinopodes, cell adhesion molecules, cytokines, homeobox (HOX) genes, growth factors, matrix metalloproteinases, and their inhibitors. Different clinical conditions in which implantation is impaired (eg, hydrosalpinx) are connected with normal estrogen and progesterone levels; this indicates that the downstream effectors of these hormones are deregulated. Previous studies reported that there are 39 Hox (mouse)/HOX (human) genes, having similar 183-base pair DNA sequence. The homeobox, is extremely conserved 61–amino acid domain that encodes homeodomain (McGinnis and Krumlauf, 1992).

Different transcription factors are responsible for embryonic development for determining the regional body pattern along with formation of anterior–posterior body axis and also the reproductive tract as well (Taylor, 2000). In developing oviduct particularly Hoxa-9 is expressed Hoxa-10 in the uterus, Hoxa-11 in the lower uterine segment and cervix, and Hoxa-13 in the upper vagina (Taylor et al.,1997). Both HOXA-10 is expressed in the endometrial glands where it is up-regulated by 17b-estradiol, whereas HOXA-11 is expressed in the stroma upregulated by progesterone (Taylor et al., 1998) throughout the menstrual cycle (Taylor et al., 1998; Taylor et al., 1999); their levels increase maximally during the mid-secretory phase at the time of implantation (Taylor et al., 1997). The effects of both the steroid hormones (estrogen and progesterone) are direct action of their receptors (ER or PR) by binding to the regulatory sites of the Hoxa-10 or Hoxa-11 genes (Taylor et al., 1999; Ma et al., 1998). The continued expression of Hox/HOX genes in the female reproductive tract make possible the growth and differentiation of the endometrium, and thereby allows for the maintenance of developmental plasticity, which is essential for successful implantation. Maximum level of HOXA-10 is observed during the mid-luteal phase and EMX2 expression decreases; this down-regulation occurs as a result of HOXA-10 binding to the regulatory region of theEMX2 gene (Troy et al., 2003).

Growth Factors

Growth factors are the proteins that bind to specific receptors, after binding they helps in the differentiation of cells or their proliferation. These growth factors include implantationrelated factors HB-EGF (Yoo et al., 1997; Birdsall et al.,1996) and amphiregulin (Das et al., 1995). In some species (mouse) expression of HB-EGF has been observed spatially and temporally at the site of blastocyst implantation (Das et al., 1994), and hence, it might be helping in the attachment of blastocyst attachment. Whereas in some other species (women), HB-EGF has also been found to be expressed during the window of implantation (Yoo et al., 1997, Birdsall et al.,1996). Therefore, HB-EGF (growth factor) might be playing its role stimulation of growth and development of human (Martin et al., 1998) and mouse (Das et al., 1994) blastocyst in vitro. Earlier studies reported that HB-EGF regulates the expression of avb3 in the endometrium (Lessey et al., 2000). Similarly, several other growth factors, expression of endometrial HB-EGF is controlled by steroid hormones. This can be explained implantation can be delayed in theabsence of estrogen, on the other hand, when estrogen is supplemented, expression of HB-EGF increases at the site of blastocyst apposition which in turn helps in the activation of blastocyst (Das et al., 1994). Amphiregulin also belongs with EGF family reported to be essential for the endometrial receptivity initially throughout the uterine epithelium in mouse (Das et al., 1995).

Transforming growth factor (TGF)-b, limits the invasion of trophoblast during implantationhence act as ''maternal restraints'' (Graham et al., 1993). Expression of TGF-b1 in the endometrial glands and stroma enhances during the secretory phase, it suppresses the rate of cytotrophoblasts proliferation, therefore forces them to differentiate into a noninvasive phenotype, and stimulate protease inhibitors (e.g., plasminogen activator inhibitor [PAI] and tissue inhibitors of matrix metalloproteases [TIMP]-1]) (Graham et al., 1992). Like insulin which helps in the promotion of growth, differentiation of cells and also regulates the metabolism in cells, insulin-like growth factors (IGF)-I and -II are single-chain polypeptides also mediates the same functions locally. Insulin-like growth factor binding protein (IGFBP)-1 secreted by the secretory endometrium and decidua (Zhou and Bondy, 1992; Han et al., 1996) acts one more restraint for the invasion of trophoblast through its binding IGF-I and IGF-II, thereby blocking their functions. The role of IGFBP-1 stimulatesinvasion of trophoblast in vitro conditions (Irving and Lala, 1995; Hamilton et al., 1992). In addition, IGFBP-1also helps in the recognition of embryo and the processes related with initial stage of implantation as it interacts directly with integrins (e.g., a5b1) expressed in the cytotrophoblasts (Irving and Lala, 1995; Irwin and Giudice 1998).

These changes include increased glandular sensitivity and more irregularity in shape including papillary or saw-tooth appearance, vacuoles move towards the supra-nuclear site of epithelial cells in order to visualize the content within the lumen of uterus. The stroma of uterus becomes looser and more edematous, and also uterine stroma gets decidualized (Diaz-Gimeno et al. 2013). On the surface of epithelial are present hair-like microvilli which are known as pinopods or uterodomes. These pinopods or uterodomes fuses transiently to form a single flowerlike projection on the membrane and appears on the luminal surface of endometrial epithelial cells during the formation of window of implantation (WOI) (Nikas 1999; Nikas and Aghajanova 2002). In humans pinopods are present on the endometrial are considered the markers of uterine receptivity for implantation (Gordon 1975; Psychoyos 1986). These pinapods are divided into different classes developing pinopods, fully developed pinopods, and regressing pinopods according to their morphology (Nikas and Aghajanova 2002). The involvement of pinopods with endometrial receptivity has been associated with success rate of implantation has been observed all patients with plentiful pinopods (>50% of endometrial surface) became pregnant, however rates of pregnancy rate are less 50% in the patients having lesscount of pinopods group (20–50%), and none of the patients with few pinopods (<20%) were pregnant (Nikas and Aghajanova 2002). Other studies revealed that number of pinopods and uterine receptivity are associated with the expression of potential molecular markers of receptivity on the pinopods implantation is the (Aplin et al. 1996; Nardo et al. 2003).

Formations of epithelial plaques are the initial endometrial responses for the occurrence of successful implantation of the blastocyst and the initiation of pregnancy in primates (Su and Fazleabas, 2015). The formation of epithelial plaque is first reaction which leads to the morphological changes including conversion of epithelia lining the uterine lumen and changes in the necks of the uterine glands, which are distinguished by the hypertrophy, hyperplasia, and rounded acinar multicellular pad (Enders 1991). This epithelial plaque reaction in theendometrium has been notedin the early days (1 day) following implantation in primary implantation sites (Enders et al., 1983; Tarara et al., 1987). Chronic gonadotropin induces changes in the epithelial plaque which influences the whole luminal surface of the endometrium (Fazleabas et al., 1999). In some cases, if epithelial plaques do not occur it may cause infertility (Rosario et al., 2005).

Embryo-Mother Cross-Talk

Cross-talk between the embryo and mother is essential for the organization and establishment of successful pregnancy. During mating due to the stimulation of cervix, induction of prolactin secretion begins from the pituitary gland as well as development of the corpus luteum (CL) of pregnancy. Lactogen secreted from the placenta helps in the maintenance of pregnancy after implantation of embryo. In addition to this embryo secretes chorionic gonadotropin (hCG), which functions like luteinizing hormone (LH) and it also helps in the maintenance of pregnancy till the formation of luteal-placental occurs. However, in some other animals (cattle and sheep) embryo secretes interferon (IFN-τ) linked to the type I IFNs. IFN-τ stops the secretion of prostaglandin F2α (PGF2α) from the uterine endometrium by represses the expression of estrogen and oxytocin receptor. However, in some animals (pig) estrogen secreted from the conceptus is a factor for maternal recognition of pregnancy. Estrogen leads to transfer in the direction of PGF2α secretion from endocrine to exocrine and restrain the regression of corpus leutum (CL). Hence, development and establishment of pregnancy is a species specific (Kimura, 2005).

Implantation

Embryo implantation is the first step for embryo-maternal interaction and is important for further embryo development (Chen et al., 2009; Dey, 2010; Wang and Dey, 2006). Implantation and the establishment of pregnancy are critical for the propagation of the species, but yet remain the limiting steps in human and primate reproduction. Successful implantation requires a competent blastocyst and a receptive endometrium during a specific window of time during the menstrual cycle to initiate the bilateral communication required for the establishment of a successful pregnancy. Embryo implantation has been generally considered as the first step towards embryo-maternal interaction and is vital for future embryo development (Chen et al., 2009; Dey, 2010; Wang and Dey, 2006). The success of mammalian pregnancy mostly depends on the easily implantation of embryo into the maternal endometrium/deciduas. This process is highly regulated and coordinated process and requires the adaptation of maternal function to the needs of developing fetus (Liu and Li, 2019). Implantation involves synchronization among the developing embryo and endometrium. The dialog between embryo and endometrium and the receptivity of the final is under the control of the sex steroids, estrogen and progesterone, as well as other hormones, such as prolactin, calcitonin, and human chronic gonadotropin (hCG).

In human's embryo implantation is a complex process that involves various steps such as attachment of blastocyst, adhesion and attachment of embryo into the endometrial tissue. Under normal conditions to keep uniformity in different steps, the blastocyst-stage embryo and the uterine endometrium must work in collaborative way during the window of implantation (León-Olea et al., 2014). Now it is a well-developed and concrete concept that neuroendocrine system is crucial for successful reproduction in mammals, the cells of neuroendocrine system has ability to synthesize and release different hormones into the gonads (testis and ovaries), when they perceive any signal from neurons (nervous system) (Margalioth et al., 2006). Remarkably Hypothalamus is considered as metronome of reproduction, because primary function of hypothalamus is to receive neural signals from the brain. These neural signals are converted into the endocrine output that causes the discharge of GnRH and other modulatory factors (Liu and Li, 2019). Pituitary gland is a connecting link between the brain and neuroendocrine organs, pituitary glands comprises two lobes viz anterior lobe (adenohypophysis) and posterior lobe (neurohypophysis). Anterior lobe is associates with the regulation of the hypothalamic-pituitary-gonadal axis

(HPG axis), hypothalamic-pituitary-thyroid axis (HPT), and hypothalamic-pituitary-adrenal axis (HPA axis), whereas posterior lobe mostly regulates the hypothamic-neurohypophysal system (Musumeci et al., 2015). Therefore, role of neuroendocrine and axial system is very important for the implantation.

AXIAL REGULATION OF IMPLANTATION

Metronome of Reproduction: Hypothalamic-Pituitary-Gonadal (HPG) Axis

The hypothalamus constitutes about 10,000 gonadotropin releasing hormone (GnRH) neurons that drive the menstrual cycle by secreting GnRH in a pulsatile pattern (Rance et al., 1994). The key functions of the hypothermic GnRH neurons obtain neural signals from the brain and convert them into endocrine output (pulsatile release of GnRH). The arcuate nucleus is a neuroendocrine transducer that transforms different electrical signals into an endocrine signal (Barbieri, 2014). The rate and amplitude of GnRH pulses are calculated by many neuroendocrine modulators such as kisspeptin, leptin, ovarian steroids and peptides. These neuroendocrine signals develop an association between the environment and reproductive status (Barbieri, 2014). Hypothalamic-pituitary gonadal axis with different major members such as gonadotropin releasing hormone (GnRH), the gonadotropin (luteinizing hormone (LH), follicular-stimulating hormone (FSH), and gonadal hormones are recognized as a central controlling system in human reproduction (Wu et al., 2015). Hypothalamic gonadotropin-releasing hormone is the main regulator of the HPG axis, because of their hormonal secretion in pulsatile manner, which plays predominant role in the endocrine control of reproduction and probable effects of hormonal regulation on implantation (Yu, et al., 2011). Mainly two GnRH molecules (GnRH I and GnRH isoform II) are present in humans, both of these GnRH molecules are very important during the implantation (Wu, et al., 2015; Barbieri, 2014). However, studies also reported that in humans type II GnRH receptors shows specific expression of the receptors in the anterior pituitary and also these receptors might function together to regulate the biosynthesis and secretion of both LH and FSH (Millar, 2003; Millar et al., 2001).

Pituitary-Gonadal Axis and Functions of GnRH in Implantation

When GnRH is secreted from the hypothalamus in pulsatile pattern via HPG axis, GnRH then binds to the GnRH receptors (GnRHs) on the pituitary gland, which in response synthesizes and secretes LH and FSH in pulse to give positive response to the GnRH signal (Liu and Li, 2019). Consequently, steroid hormones including progesterone and estrogen are released from the gonads and creating a loop by generating positive or a negative feedback effect on GnRH releasing and gonadotropin (Barbieri, 2014). For successful implantation estrogen and progesterone regulates the proliferation of uterine cells and are very essential for the changes in both blastocyst and epithelial cells of uterus. Therefore, imbalance in estrogen and progesterone during the luteal phase might cause to implantation defects (Lessey and Young, 2014).

Earlier studies reported that estrogen is not necessary for embryo implantation, because its receptors disappear during the time of implantation (Ghosh and Sengupta, 1994; Donaghay and Lessey, 2007). Nevertheless, recent studies revealed that exposure of estrogen can drive the progesterone to make the endometrium receptive for the embryo implantation in prescribed time period (Groll et al., 2009; Young, 2013). Furthermore, it has been confirmed that estrogen expression regulates the early growth response

1 (Egr1) as the downstream target through the leukemia inhibitory factor-signal transducer and activator of transcription 3 (LIF-STAT3) signaling pathway in the uterus of mouse. Moreover, it also regulates decasualization of stromal cells through the regulation of Wnt4 (Liang et al., 2014). However, some studies suggest that estrogen acts as activator to close the implantation window through insulin-like growth factor 1 pathway and make the endometrium receptive for the implantation of blastocyst (Yoshinaga, 2014; Kobayashi, 2017).

During mid-luteal phase of the reproductive cycle progesterone production reaches upto its peak value, and at this stage is secretory endometrium is fully prepared for the embryo implantation (Barbieri, 2014). Progesterone is produced during luteal phase, which in turn stimulates proliferation of the lining of the uterus to prepare for implantation by blocking the production of matrix metalloproteinase (MMP) and stimulates the formation of tissue factor (TF) and plasminogen activator inhibitor 1 (PAI-1) (Liu and Guo, 2011, Nordengren et al., 2004). It has been reported that improved implantation can be achieved by using the support of progesterone and improved pregnancy rate confirmed that lower the progesterone level lower can be implantation (van der et al., 2015, Tan et al., 2011). Further, research reports suggested that progesterone is very important for the successful occurrence of implantation and maintenance of pregnancy (Yoshinaga, 2014).

Role of GnRH in Implantation via Peripheral Reproductive Tissue

The role of GnRH/GnRH receptor expression system and GnRH II in the female reproductive tissues very clear. Thus, besides its distinguished functions on the HPG axis, GnRH and GnRH II might be also regulating extra-pituitary reproductive tissues through the local GnRHRs (Tan and Bukulmez, 2011). GnRH and GnRH II have a dynamic manner during the process of implantation; they might be regulating process of trophoblast local invasion and implantation of embryo through the regulation of the degradation of the extracellular matrix by proteolytic enzymes of the endometrial stroma and the motility of decidual endometrial stromal cells (Wu et al., 2015). During the luteal phase of the menstrual cycle, high levels of both GnRH and GnRHRare expressed in the endometrium. Therefore, the blastocyst will prefer to attach the endometrial epithelial surface and go through implantation.

During the luteal phase of the menstrual cycle, high levels of both GnRH and GnRHR are expressed in the endometrium; thus, the blastocyst will prefer to attach the endometrial epithelial surface and go through implantation (Maggi, et al., 2016, Yu et al., 2011). In mammals, the GnRH II is widely distributed more in peripheral tissues than that of GnRH I, it is also reported that GnRH II carries some extra functions and possible regulator in endometrial environment (Millar et al., 2001). Further, studies reported that GnRH directly regulates the behavior of cells of endometrium. Failure of embryo implantation might be due to the dysfunction of human decidual cells of endometrial stromal cell mortality through different pathways (Huang et al., 2013). Tissue remodeling and angiogenesis are two characteristics of implantation; both are regulated through the action of MMPs and tissue inhibitors of MMPs (TIMs) (Naruse et al., 2009). Further, studies suggested that GnRH II might be a contributory factor for the regulation of cellular motility of endometrial decidual stromal cells through the binding of GnRH-IRs. This in turn stimulates the expression of MAPK-mediated MMP-2 and MMP-9 proteases. This protease particularly degrades the basement membrane and then facilitates the invasion and migration of decidual endometrial cells of stroma (Wu et al., 2015).

Hypothalamus-Pituitary-Adrenal (HPA) Axis

The role of HPG axis in implantation and infertility is known clearly and a well-established. Nevertheless, the role of HPA axis cannot be ignored, because this axis also plays very important role in the reproductive tract. After, receiving neuronal signals, the corticotrophin-releasing hormone (CRH) secreted from the hypothalamus (ACTH) and consequently acts on the adrenal cortex to produce glucocorticoids (GC) hormones (mostly cortisol in case of humans). Ultimately, these glucocorticoid hormones will act back on the hypothalamus and pituitary (to decrease the production of CRH and ACTH production) in a negative feedback cycle (Wu et al., 2015). The hormones within the HPA axis can act at differentlevels of the HPG axis and uterus as well as other peripheral tissues.

Role of CRH in Implantation

CRH is a leading hormone of the HPA axis, as its expression is mediated by its membrane receptors including CRH-R1 and CRH-R2 (Hillhouse and Grammatopoulos, 2006). Previous studies reported that by upregulated expression of CRHR1 and macrophages indicates that CRH plays very important role in implantation (Vergetaki et al., 2011). In mammals many GnRH containing neurons in the hypothalamus have been found to express in both CRH receptors (CRH-R1, CRH-R2) which have ability to exert an estradiol-dependent effect of directly influencing GnRH release in mice and rats. These all lead to equivalent changes in serum LH levels and finally contribute to the variation of response implantation process (Phumsatitpong and Moenter, 2018). In addition to autocrine and paracrine modulatory role of CRH expressed in the peripheral tissues of the female reproductive system, including the endometrium (Kalantaridouet al., 2010). Elevated level of CRH in the endometrium during early stages of pregnancy and killing activated T-lymphocytes helps to support and maintain pregnancy (Makrigiannakis et al., 2011). Moreover, CRH possesses the ability to down-regulate the expression of carcinoembryonic antigen-related cell adhesion molecule 1 (CEACAM1) by extra villous trophoblast cells, thereby controlling proper trophoblast invasion during implantation (Bamberger et al., 2006).

ACTH Functions in Implantation

In addition to downstream effects from CRH, ACTH also has been some impact on implantation. ACTH has stimulatory impact on the HPG axis by suppressing LH secretion that results in impaired implantation. In contrast to long term treatment results in elevation of progesterone, which may cause a positive feedback and stimulate the increase of LH and FSH within the HPG axis (Brann and Mahesh, 1991). Exogenous ACTH produced deffers implantation effects, which result in more chances of miscarriage and embryonic resorption (Chatterjee and Harper, 1970).

Glucocorticoid Functions in Implantation

Glucocorticoids (GC) quantification is a specific marker of stress and leads to well-characterized profound reproduction functions by reducing the HPG axis and affecting the peripheral tissue (Whirledge et al., 2015). Administration of glucocorticoids could develop pregnancy rates and outcomes both in animals and human beings (Whirledge et al., 2015; Boomsma et al., 2007). Essential hormone secreted from HPG axis, GC functions of declining the release of GnRH from the hypothalamus, directly and indirectly suppressing the GnRH pulses or suppressing upstream regulators of GnRH release, these hormones in turn producing its effects on the anterior pituitary (Geraghty and Kaufer, 2015). GnRH-comprising neurons also express glucocorticoids receptors (GR) and have an effect of directly suppressing GnRH release. This causes decrease in the level of LH in serum (Mohsen, 2017). Further, studies reported that long term treatment of corticosterone (CORT) can directly suppress the GnRH release in the hypothalamus and effect on implantation and reproduction (Phumsatitpong and Moenter, 2018; Gore et al., 2006). Kisspeptin (KISS1) and RFamide relatedpeptide-3 (RFRP-3) modulates hormones upstream of GnRH (Ciechanowska et al., 2018). Kisspeptin (KISS1) and RFamide related peptide-3 (RFRP-3) express GRs, confirming that they are potentially able to respond to GC and suppress the release of GnRH, which can cause rise and fall of LH and FSH from the pituitary (Takumi et al., 2012).

Along to the secondary signaling influence on the pituitary, glucocorticoid possesses potential to suppress its secretion directly through various mechanisms in the pituitary gland, such as to adopt the sensitivity of the pituitary to alter secretion of GnRH and moreover reducing the synthesis and secretion of LH and FSH from the pituitary (Whirledge and Cidlowski, 2013). Nevertheless, some studies reported that glucocorticoid directly on the pituitary in secretion and synthesis varies at high level (Du Ruisseau et al., 1979). GC demonstrated both simulative and suppressive affects on HPG axis it (Michael et al., 1993). GC can amend the expression of the LH receptor (LHR) on the gonads, and also it suppresses the synthesis of testosterone (T), estrogen (E2), and progesterone (P). Therefore, it can easily dysregulate these signals and cause reflective fertility problems in both ovarian function during ovulation and uterine function during fertilization, implantation, and pregnancy (Whirledge and Cidlowski, 2013). Mifepristone, a strong high-affinity GR antagonist, it effects the signaling of GC that decreases the regulatory effect on embryo attachment and invasion (Couzinet et al., 1986). Although, recent reports revealed that GR signaling has direct on uterine physiology. Further, it has been suggested that reduced blastocyst implantation rate and subsequent defects in decasualization of cell of stroma were found out in a uterine-specific GR in knockout (uterine GR KO) mouse. GC being as a chief regulator suppresses the expression of various important genes and finally influencing the uterine environment (Gore et al., 2006).

Neuroendocrine Integration

It is integration between the nervous system and endocrine system and plays an important role in the reproductive system. There is different axial regulatory pathway through which reproduction can be regulated which includes hypothalamo-pituitary-thyroid axis and hypothamo-neurohypophyseal system. Hypothamo-pituitary thyroid axis includes the three major hormones such as thyrotropin-releasing hormone (TRH), thyrotropin stimulating hormone (TSH), and thyroid hormone (TH: T3 and T4) (Musumeci et al., 2015). During the pregnancy attack of thyroid disease (hypothyroidism or hyperthyroidism) can lead to the subfertility and recurrent miscarriage (Twig et al., 2012). In *in vivo* reports, revealed that hypothyroidism have twin effects on pregnancy, by influencing implantation at an extremely early stage and regulating placental development at the afterward stage (Piccirilli et al., 2018). However, in vitro studies reported that TH is associated in the bidirectional cross talk between the blastocyst and endometrium during implantation (Liu and Li, 2019). In addition to these thyroid receptors both TH receptors (TR) and TSH receptors (TSHR) are reported to be expressed in thefeto-maternal crossing point, endometrium and trophoblast during the implantation, which can affect the feto-maternal crossing point (Colicchia et al. 2014; van den et al., 2011). Even though expression level of TSHR, TRa1, TRa2 and TRb1 increases in the endometrial cells during implantation (Sayem et al., 2017). TH synthesis is associated with different transcript, such as prolyl 4-hydrosylase beta (P4HB), it is a molecular chaperone involved in endocytosis of immature thyroglobulin (Tg) molecule thyroid peroxidase (TPO), and Tg, which has been observed in the endometrium (Catalano et al., 2007). It has been reported that hypothyroidic females have low level of two key regulatory molecules (ISP1 and ISP2) essential for embryonic implantation both are present in the uterus (Piccirilli et al., 2018).

Moreover, endometrium and the trophoblast are affected by TH through directly or indirectly because TH affects the synthesis and the activity of implantation-mediated molecules (Colicchia et al. 2014). TH has prominent role in the angiogenesis and remodeling of endometrial tissues, which is important for the successful implantation, decidualization, and placentation. Although role of TH has been also in the angiogenesis through genomic as well as non-genomic mechanism in different animals (Yalcin et al., 2010). In addition to proliferative and endometrial tissue maturative role of TH. TH also could have significant effect on the regulation of the HPG axis and consequently on implantation (Redmond, 2004). Further, different cell adhesion molecules (CAMs), such as mucins (Carson et al., 1994; Feinberg et al., 1997) and trophinin (Fukuda et al., 1995), are used during the attachment phase of implantation, these molecules help in the attachment of blastocyst to the wall of the endometrium which is mediated by receptor-mediated model of implantation (Yoshinaga, 1989). Most reported molecule among CAMs is integrins, it is a heterodimeric glycoprotein that comprises noncovalently connected a and b subunits (Lessey et al., 2002). 20 different types ofintegrin heterodimers have been identified, formed from 4a and 9b subunits (Lindhard et al., 2002). Integrins are remarkablereceptors present on the surface of cells receptor, they bind with low affinity and are present in huge numbers; this allows for ligand motility without loss of attachment (Kodaman and Taylor, 2004).

TH regulates integrin $\alpha\nu\beta3$ (a class of cell adhesion molecules (CAM) that communicate with extracellular matrix (ECM) ligands, matrix metalloproteinases (MMPs), and CAMs, playing an important role in the cascade of processes for the successful implantation (Tei et al., 2003). Therefore, it might be concluded that TSH as well as TH are necessary agents' in the mechanism of regulating implantation (Liu and Li, 2019). Nevertheless, least research reports revealed effect of TRH on the implantation process (Liu and Li, 2019). Hormones synthesized from the hypothamus and stored in the posterior pituitary gland plays very important role in the neuroendocrine modulation and implantation. Studies reported that uterine contraction are important for the successful implantation, receptors of oxytocin and vasopressin are reported to be expressed in the uterus, which are essential for the contractile activity of uterus (van den et al., 2011). The receptor for oxytocin and vasopressin belongs to the class I family of G-protein coupled receptors, which helps in the contraction through the central or peripheral pathways (Chini and Fanelli, 2011). Studies suggested that oxytocin and vasopressin antagonist treatment is found to be successful for the priming of the uterus for implantation and subsequently increases the rate of implantation after IVF (Pierzynski, 2011; Chou et al., 2011; Khairy et al., 2016, Kim, et al., 2016).

Embryo implantation is the first step for the embryo-maternal interaction and is important for further embryo development (Chen et al., 2009a; Dey, 2010; Wang and Dey, 2006). Embryo implantation has been commonly considered as the first step towards embryo-maternal interaction and is critical for further embryo development (Chen et al., 2009; Dey, 2010; Wang and Dey, 2006). In the past decades, there have been tremendous advances towards understanding the physiological and molecular events during embryo implantation, including the establishment of uterine receptivity, blastocyst activation, and embryo-uterine interactions (Dey et al., 2004; Lim and Wang, 2010; Wang and Dey, 2006). However, much less progresses have been made towards understanding the mechanisms that control intrauterine embryo distribution, which are equally important for successful embryo implantation (Chen et al., 2011b).

The pre-implantation embryo is about 0.1 mm in diameter and much smaller compared to the size of the uterine cavity, and thus, the precise spatiotemporal transport and orientation of such a small object within the uterine cavity apparently involves fine-tuned mechanisms that coordinate the floating embryo near the uterine wall. The optimal intrauterine embryo distribution should provide the blastocyst(s) with appropriate localization and orientation within the intra-uterine environment in order to establish and maintain a successful pregnancy.

CLIMATE CHANGE ON AXIAL REGULATION OF IMPLANTATION

In 2007, the Intergovernmental Panel on Climate Change (IPCC) produced different evidences about the global warming and the effect of human activities on global climate change. The chapter deals with the impact of climate change on implantation, infant and maternal mortality, poorer reproductive health, especially in tropical, developing countries. The temperature of earth and oceans has increased in the whole world over last century, particularly following the progression of industrial revolution (IPCC, 2007). The Antarctic ice cap is becoming thinner and the Artic sea ice extent is decreasing day by day. In the 20th century, global sea levels heightened. The incidence of adverse weather conditions, including droughts and heavy precipitation, has increased from last 30 years. In the atmosphere level of 'greenhouse gases' such as carbon dioxide (CO_2), methane (methane), had attained the peaks for 650,000 years. The utilization of fossil fuels, use of land for agricultural purposes is accountable for the increment of greenhouse gases during the last 250 years.

The use of fossil fuels, land use, and agricultural activities are accountable for the increased concentrations of greenhouse gases over the last 250 years. It has been estimated that since 1910 the average change in global temperature is 0.748^{0}C. The highest increase has been recorded in the 1970s (Costello et al., 2009). Moreover, hypothesized models and interpretations that current scenario indicates that the average temperature in the whole world is increasing between 0.15^{0}C and 0.38^{0}C per decade (IPCC, 2007). The impact of global warming on the local climate has different in different regions, countries, and particular areas (Hyatt et al., 2010). It has been projected that highest increase in temperature has been observed in the Arctic areas. Nevertheless, normally temperatures in these regions are originally low, but this increased temperature has adverse impacton human health. Precipitationwill differ in frequency, amount, intensity, and kind with some regionsare drier and other areas are wetter. Rising humidity in hot areas adds to the heat stressand health risks (Parsons, 2002). All species on the Earth will be influenced by climatechange. It has been reported that 'Climate change is the major global healththreat of the 21st century' (Costello et al., 2009). In developing countries, the populations in tropical regions are mostly affecteddue to climate change because of poverty, poor sanitation, poorpopulation health status, high population density, poorhealth care systems, and political instability, that is,limited government ability to cope with external crisis.The potential of health care systems to counter to anincrease in the load of climate-sensitive outcomes play a significant role to examine the impact of climate change on health (Rylander et al., 2013). Number of factors have identified in which climate change can affect human health: deficiency of food and safe drinking water, poor sanitation, population migration, changing disease pattern and morbidity, more recurrentgreat weather events, andlack of shelter (Costello et al., 2009). Extreme heat exposure has also beenidentified as apromising risk for human health by others (Kjellstrom, 2008). Pregnant women, the developing fetus, and youngchildren are regard as the more sensitive members ofour species and are, in addition, already marginalized inmany countries. They will therefore be most vulnerable tothe environmental effects of climate change (Rylander et al., 2013).

MATERNAL HEALTH AND MILLENNIUM DEVELOPMENT GOALS

The World Health Organization (WHO) reported maternal health as the health of women during pregnancy, childbirth, and the postpartum period. Every year, about 536,000 women diesbecauseof problems during pregnancy, childbirth, or the 6 weeks following delivery (Homer et al., 2009). Most women die of blood loss or hypertension, which account for half of all maternal deaths (the United Nations Millennium Development Goals; UN MDGs). Other indirect causes, such as malaria, HIV/ AIDS, and heart diseases, account for 18% of all maternal deaths, while as complications during childbirth comprises about 11%. Many maternal deaths are preventable, and 99% of all maternal deaths are associated to childbearing or giving birth occurs in developing countries (Costello et al., 2009). Thus, climate change will have a substantial impact on the health and survival of the next generation among already challenged populations. There is limited knowledge regarding which regions will be most heavily affected (Rylander et al., 2013).

CLIMATE CHANGE EFFECTS ON DEVELOPMENT OF FOETUS

The environment affects the life on this earth directly or indirectly. The environmental pollution, global warming, climate change, rapid industrialization etc. have direct consequences on the living beings. The most affected age group due to environmental factors is the children below the age of 10 years. Children at any stages of growth and development need to be protected from environmental health hazards. They need safe and healthy environment to attain optimum growth and development. However physical, chemical, biological and social environments have changed throughout decades and children of today are living in a very different environment than from their grandparents and parents. Today children are at most risk of being exposed to new chemicals, gases and other pollutants that are mostly not tested in children. The majority of chemicals being getting used in industries nowadays have not been studied adequately for their impacts on human health specifically fetus. Many of these chemicals are persistent and bioaccumulative, remaining in the human body long after the exposure. The traces of toxic pollutants used to enter the body of pregnant mothers through food chain and thus reached the fetus through placenta.

The couples working in radiation laboratories, hazardous chemical industries etc. may get exposed to harmful environment. Parental exposures occurred before the conception of fetus threatens the fetus both due to two reasons: - (i) the maternal or paternal reproductive organs may get affected due to chemical hazards, and (ii) chemicals that can be accumulated in the mother's body before pregnancy may cross over placental barrier during pregnancy. This may lead to the health effects prenatally and at birth such as miscarriage, stillbirth, low birth weight, birth defects, diseases such as metabolic syndrome and cardiovascular dysfunction. This strongly establishes a causal link between environmental exposures and multifactorial health outcomes. Therefore, different developmental periods have shown varying sensitivity to environmental cues. The information on the impact of environmental toxins on human health, or in other words for establishing a relationship between possible health effects and environmental factors comes from studies examining the long-term effects of chemical exposures due to accidents, or natural disasters, occupational hazards, animal and epidemiologic studies.

REASONS FOR PRE-IMPLANTATION EMBRYO SENSITIVENESS TO ENVIRONMENT

There are many possible reasons for sensitiveness of embryo to its environment. Some of them are as follow:

(i) The pre-implantation embryo consists of a small number of cells, which later give rise to all the tissues of the conceptus; this pool of cells is, therefore, vulnerable by virtue of its limited supply and essential future contribution.
(ii) Second, the preimplantation embryo's main task is to generate two pools of cells capable of forming the embryo (the future fetus) and placental components of the conceptus.

In 2004, a study was carried out by the Environmental Working Group (EWG) in the US, examined the presence of 413 different environmental toxins in the cord blood from a total of 10 randomly selected newborns from different hospitals. 287 of these toxins tested positive with an average of approximately 200 toxic chemicals detected in each cord blood sample (http://www.ewg.org/reports/bodyburden2). Among these, are several substances defined as "carcinogens" by the Environment Protection Agency (EPA) such as eight different type of perflourinated chemicals used in fast food industry (due to their water and grease resistance) and textile/fabric industry (including teflon chemicals); dozens of brominated chemicals used for flame retardation and their by-products; and pesticides. One-hundred eighty of these 287 chemicals have previously been reported to cause cancer in human and animals, 217 have been associated with birth anomalies, and 217 are neurotoxic (Rhind, 2010). This piece of information directly calls into the question the effects of this mixture of toxic substances on the developing fetus, with possibly dreadful answers due to the fact that prenatal exposure might be associated with more serious, permanent, and/or irreversible damage as compared to exposure in later stages of life. This complex mixture incorporating hundreds of different carcinogenic, neurotoxic, and developmental toxic chemicals in the cord blood has not been very well studied in terms of its impact on the programmed fetal growth and on the later stages of development.

On a daily basis, the background exposure to environmental pollutants is mostly due to phthalates, plastic products such as bisphenol A, polychlorinated biphenyls (PCBs), lead, mercury, cadmium, asbestos, dioxins, polyhalogenated aromatic hydrocarbons (PHAHs), solvents, pesticides, inhaled toxins (tobacco smoke, ozone, particulate matter), and chlorinated disinfection byproducts (DBPs). Most of these chemicals are "endocrine disruptors"; in other words, they are chemical substances that disrupt the production, release, transport, binding, and/or breakdown of natural hormones that are essential for the growth and development at the time of conception and thereafter. Chronic low dose exposure to certain endocrine disruptors may lead to developmental functional disorders that are not observed with high dose exposure (Franceschini, 2008). The multifactorial nature of the emergent health effects is a major obstacle to establish a cause and effect relationship for an individual environmental toxin. Also the term "environmental" is usually regarded as a vague, arbitrary, and subjective attribute when it comes to define a pathological condition.

DEVELOPMENTAL PLASTICITY AND ENVIRONMENTAL CHALLENGES

All the developmental processes starting from embryonic stage to upto death of living beings are dynamic events that allow a high degree of developmental plasticity rather than following a predetermined static 'programmed'. It has long been observed that embryos taken out of their normal in vivo environment within the mother and placed into in vitro culture often delay their developmental progress. Some changes make take place in their gene expression and metabolic profiles according to the in vitro conditions. However, most of embryos still retain their capacity to implant and continue to develop into viable offspring when transferred back into the mother. In vitro culture media usually consist of a buffered salt solution with a limited diversity of nutrient and growth factor supplements. Thus, the in vitro culture conditions are very drastic changes, which the embryo is provided when compared with the 'normal' conditions in vivo. It is therefore not physiologically relevant. Majority of such children have not yet reached adulthood and follow-up studies are still going on, it is difficult to reach a firm conclusion. However, a number of reports on increased birth defects and altered metabolic and cardiovascular physiology associated with in vitro fertilization/intracytoplasmic sperm injection treatments have been published worldwide (Ceelen et al., 2007, 2008, 2009; Hvidtjørn et al., 2009; Knoester et al., 2007; Manipalviratn et al., 2009, Miles et al., 2007; Sakka et al, 2009). More recently, animal models show that more subtle environmental changes such as, slightly altering single media supplements in vitro or specific maternal diets or body conditions in vivo also influence development and adverse adult outcomes (Kwong et al., 2000; Watkins et al., 2008; Ecker et al., 2004; Sinclair et al., 2007; Samuelsson et al., 2008; Mitchell et al., 2009). Such developmental plasticity underlies the biological principle of adaptive responses to predicted conditions. Measurable changes can become evident before implantation and may be transient, or long lasting.

CONSEQUENCES OF ENVIRONMENTAL POLLUTION ON EMBRYO PRE-IMPLANTATION

Anthropogenic pollutants comprise a wide range of synthetic organic compounds and heavy metals, which are dispersed throughout the environment, usually at low concentrations. Exposure to any pollutants can act additively or synergistically to upset multiple physiological systems at all ages particularly in the developing fetus. In some of the animals like sheep, organs affected by pollutant exposure include the ovary, testis, hypothalamus and pituitary gland and bone. It has been reported that the effects of exposure include changes in organ weight and gross structure, histology, and gene and protein expression but these changes are not reflected in changes in reproductive performance under the conditions tested. Effects of pollutants on the thyroid gland, immune, cardiovascular and obesogenic systems have not been shown explicitly, in ruminants, but research work on other species suggests that these systems can also be disturbed. Exposure to a mixture of anthropogenic pollutants has significant effects on a wide variety of physiological systems, including the reproductive system.

Environmental pollutants include a wide range of chemicals, mostly man-made, which are present in products that are used on a daily basis, such as detergents, pesticides and adhesives or derived from the combustion of hydrocarbon fuels. Although concentrations of these chemicals in the environment are generally low and may not have adverse effects when acting in isolation. However, combinations of these chemicals can interfere with physiological systems and the capacity of animals to reproduce. Thus, both the viability of populations and the health and welfare of individuals could be compromised by exposure to environmental levels of pollutants. A better understanding of the potential effects of environmental exposures on fetal and childhood growth is of marked public health significance because of:- (i) increasing exposure levels of a variety of pollutants resulting from increased industrialization; (ii) the known capability of environmental pollutants to readily transport across the placenta; (iii) plausible mechanisms linking environmental exposures to impaired fetal and childhood growth; and (iv) and the influence of early life exposures on the risk of both childhood and adult diseases. Therefore, an environmental pollutant like, perfluorinated compounds (PFCs), heavy metals, polyhalogenated aromatic hydrocarbons (PHAHs), and air pollutants have direct or indirect influence on fetus and childhood growth and development.

The endogenous activity of the gonads is driven by the actions of the hypothalamus and pituitary gland, both of which are steroid-sensitive. The studies carried out in sheep and goats have reported significant effects of exposure to chemicals with known endocrine disrupting effects, including octylphenol (Sweeney *et al.*, 2000; Wright *et al.*, 2002), bisphenol A (Evans *et al.*, 2004; Savabieasfahani *et al.*, 2006), methoxyclor (Savabieasfahani *et al.*, 2006), PCB153 (Lyche *et al.*, 2004; Oskam *et al.*, 2005) and valporate (Krogenaes *et al.*, 2008), on the hypothalmic-pituitary (HP) axis. The majority of these studies have investigated the effects of EDCs on the HP axis during development, that is, during gestation and/or lactation, the most sensitive 'windows', when the endocrine system can be permanently altered (IPCS, 2002).

Fetal development is a critical window of exposure-related susceptibility because the etiology of diseases in adulthood may have a fetal origin and may be attributed to adverse effects of in utero environmental exposures. Effect of black carbon (BC) particles as part of combustion-derived particulate matter in human placentae has been reported by Bove (2019) using white-light generation under femtosecond pulsed illumination. BC is identified in all screened placentae, within average (SD) particle count of 0.95×10^4 (0.66×10^4) and 2.09×104 (0.9×10^4) particles/$m^3$for low and high exposed mothers, respectively. Furthermore, the placental BC load is positively associated with mothers' residential BC exposure during pregnancy (0.63–2.42 μg/m^3). It was concluded that BC particles accumulate on the fetal side of the placenta suggests that ambient particulates could be transported towards the fetus and represents a potential mechanism explaining the detrimental health effects of pollution from early life onwards (Bove, 2019).

Developing brain of fetus is a target organ for environmental pollutants and other associated toxic agents as described before. Like other organs, brain also goes through a series of programmed and scheduled developmental processes during embryonic life such as cell proliferation, migration, cell differentiation, axon formation, gliogenesis, synaptogenesis, apoptosis, and myelination etc. Under normal conditions, all these developmental changes occur in a perfect manner at pre-determined timeframes within the brain along with the involvement of neurotransmitters (acetylcholine, norepinephrine, dopamine, serotonin, gamma-amino butyric acid/GABA, glutamate and aspartate), hormones (mainly thyroid hormones, steroid hormones, growth hormones, sex hormones), neurotropins, growth factors, and cellular receptors in the process (Rice 2000). Prenatal exposure to methyl mercury interferes with neurogenesis (Rodier, 1984) with an eventual disruption of neural migration as well. Ethanol (Miller, 1993) and methyl mercury (Choi, 1989), two developmental neurotoxic agents, have been shown to adversely affect both of the two above mentioned processes, resulting in disrupted cellular differentiation.

CONCLUSION

There is a strong relationship between fetal environmental, toxic chemical exposure, adverse pregnancy and childhood health outcome. All these possible connection between fetal toxic environmental exposures and increase in some childhood diseases and disorders such as, some childhood malignancies, neuro-developmental disorders and learning difficulties i.e., Attention-Deficit/Hyperactivity Disorder (ADHD), reduced memory and attention, decreased verbal ability, impaired information processing, reduced psychomotor development, adverse behavioral and emotional effects, decreased sustained activity, increased depressed behavior and other endocrine and sexual developmental disorders is an emerging area of concern.

Environmental pollutants and toxicants can also have adverse effects on animal health especially on reproductive function, by either direct or indirect effects on numerous organs and systems. However, evidences of the relationships between exposure and physiological effects are meager, particularly for ruminants, revealing the fact that levels of exposure to each individual chemical are generally very low and they do not act individually. Nevertheless, concerns remain that there may be subtle perturbations of reproductive function and since some of the observed changes in physiological function may be expressed in subsequent generations, even without further exposure to pollutants, there may be even greater cause for concern indicating that some trends in health/performance may only be observed on a population basis rather than an individual basis.

AUTHOR CONTRIBUTION

YAH conceived the book chapter. YAH, RK designed and wrote the book chapter. JAM, AS, MRW performed literature survey. YAH, MRW, AS arranged the references. All authors shared equal contribution.

FUND DISCLAIMER

No grant received from any funding agency in public, commercial, or not-for-profit sector.

CORRESPONDING AUTHOR

Younis Ahmad Hajam

REFERENCES

Aplin, J. D., Spansewick, C., Behzad, F., Kimber, S. J., & Vićovac, L. (1996). Molecular aspects of implantation: Integrins β5, β3 and αv are apically distributed in endometrial epithelium. *MHR: Basic Science of Reproductive Medicine*, *2*(7), 527–534. doi:10.1093/molehr/2.7.527 PMID:9239663

Arici, A., Head, J. R., MacDonald, P. C., & Casey, M. L. (1993). Regulation of interleukin-8 gene expression in human endometrial cells in culture. *Molecular and Cellular Endocrinology*, *94*(2), 195–204. doi:10.1016/0303-7207(93)90168-J PMID:8224523

Arici, A., MacDonald, P. C., & Casey, M. L. (1995). Regulation of monocyte chemotactic protein-1 gene expression in human endometrial cells in cultures. *Molecular and Cellular Endocrinology*, *107*(2), 189–197. doi:10.1016/0303-7207(94)03442-V PMID:7768330

Ayrout, M., Simon, V., Bernard, V., Binart, N., Cohen-Tannoudji, J., Lombès, M., & Chauvin, S. (2017). A novel non genomic glucocorticoid signaling mediated by a membrane palmitoylated glucocorticoid receptor cross talks with GnRH in gonadotrope cells. *Scientific Reports*, *7*(1), 1–13. doi:10.103841598-017-01777-2 PMID:28484221

Bamberger, A. M., Minas, V., Kalantaridou, S. N., Radde, J., Sadeghian, H., Löning, T., & Schulte, H. M. (2006). Corticotropin-releasing hormone modulates human trophoblast invasion through carcinoembryonic antigen-related cell adhesion molecule-1 regulation. *American Journal of Pathology*, *168*(1), 141–150. doi:10.2353/ajpath.2006.050167 PMID:16400017

Barbieri, R. L. (2014). The endocrinology of the menstrual cycle. In *Human Fertility* (pp. 145–169). Humana Press. doi:10.1007/978-1-4939-0659-8_7

Bartocci, A., Pollard, J. W., & Stanley, E. R. (1986). Regulation of colony-stimulating factor 1 during pregnancy. *The Journal of Experimental Medicine*, *164*(3), 956–961. doi:10.1084/jem.164.3.956 PMID:3489064

Bell, C. E., Calder, M. D., & Watson, A. J. (2008). Genomic RNA profiling and the programme controlling preimplantation mammalian development. *Molecular Human Reproduction*, *14*(12), 691–701. doi:10.1093/molehr/gan063 PMID:19043080

Birdsall, M. A., Hopkisson, J. F., Grant, K. E., Barlow, D. H., & Mardon, H. J. (1996). Endocrinology and paracrinology: Expression of heparin-binding epidermal growth factor messenger RNA in the human endometrium. *MHR: Basic Science of Reproductive Medicine*, *2*(1), 31–34. doi:10.1093/molehr/2.1.31 PMID:9238654

Bongaerts, E., Bové, H., Ameloot, M., & Nawrot, T. (2019). Ambient Black Carbon Particles Reach the Fetal Side of Human Placenta. *Environmental Epidemiology*, *3*, 34–35. doi:10.1097/01.EE9.0000606004.04305.f0

Boomsma, C. M., Keay, S. D., & Macklon, N. S. (2012). Peri-implantation glucocorticoid administration for assisted reproductive technology cycles. *Cochrane Database of Systematic Reviews*, (6), CD005996. doi:10.1002/14651858.CD005996.pub3 PMID:22696356

Brann, D. W., & Mahesh, V. B. (1991). Role of corticosteroids in female reproduction. *The FASEB Journal*, *5*(12), 2691–2698. doi:10.1096/fasebj.5.12.1655548 PMID:1655548

Bulmer, J. N., Morrison, L., Longfellow, M., Ritson, A., & Pace, D. (1991). Granulated lymphocytes in human endometrium: Histochemical and immunohistochemical studies. *Human Reproduction (Oxford, England)*, *6*(6), 791–798. doi:10.1093/oxfordjournals.humrep.a137430 PMID:1757516

Burden, E. B. (2005). The Pollution in Newborns. A Benchmark Investigation of Industriacl Chemicals, Pollutants and Pesticides in Umbilical Cord Blood. Environmental Working Group (EWG).

Caballero-Campo, P., Domínguez, F., Coloma, J., Meseguer, M., Remohí, J., Pellicer, A., & Simón, C. (2002). Hormonal and embryonic regulation of chemokines IL-8, MCP-1 and RANTES in the human endometrium during the window of implantation. *Molecular Human Reproduction*, *8*(4), 375–384. doi:10.1093/molehr/8.4.375 PMID:11912286

Carson, D. D., Rohde, L. H., & Surveyor, G. (1994). Cell surface glycoconjugates as modulators of embryo attachment to uterine epithelial cells. *The International Journal of Biochemistry*, *26*(10-11), 1269–1277. doi:10.1016/0020-711X(94)90096-5 PMID:7851630

Catalano, R. D., Critchley, H. O., Heikinheimo, O., Baird, D. T., Hapangama, D., Sherwin, J. R. J. R., Charnock-Jones, D. S., Smith, S. K., & Sharkey, A. M. (2007). Mifepristone induced progesterone withdrawal reveals novel regulatory pathways in human endometrium. *Molecular Human Reproduction*, *13*(9), 641–654. doi:10.1093/molehr/gam021 PMID:17584828

Ceelen, M., Van Weissenbruch, M. M., Prein, J., Smit, J. J., Vermeiden, J. P., Spreeuwenberg, M., van Leeuwen, F. E., & Delemarre-van De Waal, H. A. (2009). Growth during infancy and early childhood in relation to blood pressure and body fat measures at age 8–18 years of IVF children and spontaneously conceived controls born to subfertile parents. *Human Reproduction (Oxford, England)*, *24*(11), 2788–2795. doi:10.1093/humrep/dep273 PMID:19648588

Ceelen, M., van Weissenbruch, M. M., Roos, J. C., Vermeiden, J. P., van Leeuwen, F. E., & Delemarre-van de Waal, H. A. (2007). Body composition in children and adolescents born after in vitro fertilization or spontaneous conception. *The Journal of Clinical Endocrinology and Metabolism*, *92*(9), 3417–3423. doi:10.1210/jc.2006-2896 PMID:17595253

Ceelen, M., van Weissenbruch, M. M., Vermeiden, J. P., van Leeuwen, F. E., & Delemarre-van de Waal, H. A. (2008). Cardiometabolic differences in children born after in vitro fertilization: Follow-up study. *The Journal of Clinical Endocrinology and Metabolism*, *93*(5), 1682–1688. doi:10.1210/jc.2007-2432 PMID:18285409

Chatterjee, A., & Harper, M. J. K. (1970). Interruption of implantation and gestation in rats by reserpine, chlorpromazine and ACTH: Possible mode of action. *Endocrinology*, *87*(5), 966–969. doi:10.1210/endo-87-5-966 PMID:4320640

Chawla, M. K., Gutierrez, G. M., Young, W. S. III, McMullen, N. T., & Rance, N. E. (1997). Localization of neurons expressing substance P and neurokinin B gene transcripts in the human hypothalamus and basal forebrain. *The Journal of Comparative Neurology*, *384*(3), 429–442. doi:10.1002/(SICI)1096-9861(19970804)384:3<429::AID-CNE8>3.0.CO;2-5 PMID:9254037

Chen, Q., Peng, H., Zhang, Y., Lei, L., Cao, Y., & Duan, E. (2009). Embryo implantation: A time for recalling and forwarding. *Chinese Science Bulletin*, *54*(22), 4083–4093. doi:10.100711434-009-0631-3

Chini, B., & Fanelli, F. (2000). Molecular basis of ligand binding and receptor activation in the oxytocin and vasopressin receptor family. *Experimental Physiology*, *85*(s1), 59s–66s. doi:10.1111/j.1469-445X.2000.tb00008.x PMID:10795907

Choi, B. H. (1991). Effects of methylmercury on the developing brain. In *Advances in Mercury Toxicology* (pp. 315–337). Springer. doi:10.1007/978-1-4757-9071-9_20

Chou, P. Y., Wu, M. H., Pan, H. A., Hung, K. H., & Chang, F. M. (2011). Use of an oxytocin antagonist in in vitro fertilization–embryo transfer for women with repeated implantation failure: A retrospective study. *Taiwanese Journal of Obstetrics & Gynecology*, *50*(2), 136–140. doi:10.1016/j.tjog.2011.04.003 PMID:21791296

Ciechanowska, M., Łapot, M., Paruszewska, E., Radawiec, W., & Przekop, F. (2018). The influence of dopaminergic system inhibition on biosynthesis of gonadotrophin-releasing hormone (GnRH) and GnRH receptor in anoestrous sheep; hierarchical role of kisspeptin and RFamide-related peptide-3 (RFRP-3). *Reproduction, Fertility, and Development*, *30*(4), 672–680. doi:10.1071/RD16309 PMID:29019791

Colicchia, M., Campagnolo, L., Baldini, E., Ulisse, S., Valensise, H., & Moretti, C. (2014). Molecular basis of thyrotropin and thyroid hormone action during implantation and early development. *Human Reproduction Update*, *20*(6), 884–904. doi:10.1093/humupd/dmu028 PMID:24943836

Costello, A., Abbas, M., Allen, A., Ball, S., Bell, S., Bellamy, R., Friel, S., Groce, N., Johnson, A., Kett, M., Lee, M., Levy, C., Maslin, M., McCoy, D., McGuire, B., Montgomery, H., Napier, D., Pagel, C., Patel, J., ... Patterson, C. (2009). Managing the health effects of climate change: Lancet and University College London Institute for Global Health Commission. *Lancet*, *373*(9676), 1693–1733. doi:10.1016/S0140-6736(09)60935-1 PMID:19447250

Couzinet, B., Strat, N. L., Ulmann, A., Baulieu, E. E., & Schaison, G. (1986). Termination of early pregnancy by the progesterone antagonist RU 486 (mifepristone). *The New England Journal of Medicine*, *315*(25), 1565–1570. doi:10.1056/NEJM198612183152501 PMID:3785321

Crinnion, W. J. (2009). Maternal levels of xenobiotics that affect fetal development and childhood health. *Alternative Medicine Review*, *14*(3). PMID:19803547

Cross, J. C., Werb, Z., & Fisher, S. J. (1994). Implantation and the placenta: Key pieces of the development puzzle. *Science*, *266*(5190), 1508–1518. doi:10.1126cience.7985020 PMID:7985020

Das, S. K., Chakraborty, I., Paria, B. C., Wang, X. N., Plowman, G., & Dey, S. K. (1995). Amphiregulin is an implantation-specific and progesterone-regulated gene in the mouse uterus. *Molecular Endocrinology (Baltimore, Md.)*, *9*(6), 691–705. PMID:8592515

Das, S. K., Wang, X. N., Paria, B. C., Damm, D., Abraham, J. A., Klagsbrun, M., Andrews, G. K., & Dey, S. K. (1994). Heparin-binding EGF-like growth factor gene is induced in the mouse uterus temporally by the blastocyst solely at the site of its apposition: A possible ligand for interaction with blastocyst EGF-receptor in implantation. *Development*, *120*(5), 1071–1083. PMID:8026321

Dey, S. K. (2010). How we are born. *The Journal of Clinical Investigation*, *120*(4), 952–955. doi:10.1172/JCI42380 PMID:20364092

Díaz-Gimeno, P., Ruiz-Alonso, M., Blesa, D., Bosch, N., Martínez-Conejero, J. A., Alamá, P., Garrido, N., Pellicer, A., & Simón, C. (2013). The accuracy and reproducibility of the endometrial receptivity array is superior to histology as a diagnostic method for endometrial receptivity. *Fertility and Sterility*, *99*(2), 508–517. doi:10.1016/j.fertnstert.2012.09.046 PMID:23102856

Donaghay, M., & Lessey, B. A. (2007). Uterine receptivity: Alterations associated with benign gynecological disease. *Seminars in Reproductive Medicine*, *25*(6), 461–475. doi:10.1055-2007-991044 PMID:17960531

Du Ruisseau, P., Tache, Y., Brazeau, P., & Collu, R. (1979). Effects of chronic immobilization stress on pituitary hormone secretion, on hypothalamic factor levels, and on pituitary responsiveness to LHRH and TRH in female rats. *Neuroendocrinology*, *29*(2), 90–99. doi:10.1159/000122910 PMID:116141

Ecker, D. J., Stein, P., Xu, Z., Williams, C. J., Kopf, G. S., Bilker, W. B., Abel, T., & Schultz, R. M. (2004). Stein P, Xu Z, Williams CJ, Kopf GS, Bilker WB, Abel T, Schultz RM. Long-term effects of culture of preimplantation mouse embryos on behavior. *Proceedings of the National Academy of Sciences of the United States of America*, *101*(6), 1595–1600. doi:10.1073/pnas.0306846101 PMID:14747652

Enders, A. C. (1991). Structural responses of the primate endometrium to implantation. *Placenta*, *12*(4), 309–325. doi:10.1016/0143-4004(91)90340-L PMID:1946242

Enders, A. C., Hendrickx, A. G., & Schlafke, S. (1983). Implantation in the rhesus monkey: Initial penetration of endometrium. *The American Journal of Anatomy*, *167*(3), 275–298. doi:10.1002/aja.1001670302 PMID:6881070

Evans, N. P., North, T., Dye, S., & Sweeney, T. (2004). Differential effects of the endocrine-disrupting compounds bisphenol-A and octylphenol on gonadotropin secretion, in prepubertal ewe lambs. *Domestic Animal Endocrinology*, *26*(1), 61–73. doi:10.1016/j.domaniend.2003.09.005 PMID:14732453

Falany, J. L., & Falany, C. N. (1996). Regulation of estrogen sulfotransferase in human endometrial adenocarcinoma cells by progesterone. *Endocrinology*, *137*(4), 1395–1401. doi:10.1210/endo.137.4.8625916 PMID:8625916

Fazleabas, A. T., Bell, S. C., Fleming, S., Sun, J., & Lessey, B. A. (1997). Distribution of integrins and the extracellular matrix proteins in the baboon endometrium during the menstrual cycle and early pregnancy. *Biology of Reproduction*, *56*(2), 348–356. doi:10.1095/biolreprod56.2.348 PMID:9116133

Feinberg, R. F., & Kliman, H. J. (1997). *MAG (mouse ascites Golgi) mucin in the endometrium: a potential marker of endometrial receptivity to implantation. Endometrium and Endometriosis*. Blackwell Science.

Fisher, J. S. (2004). Environmental anti-androgens and male reproductive health: Focus on phthalates and testicular dysgenesis syndrome. *Reproduction (Cambridge, England)*, *127*(3), 305–315. doi:10.1530/rep.1.00025 PMID:15016950

Franceschini, M. D., Custer, C. M., Custer, T. W., Reed, J. M., & Romero, L. M. (2008). Corticosterone stress response in tree swallows nesting near polychlorinated biphenyl-and dioxin-contaminated rivers. *Environmental Toxicology and Chemistry: An International Journal*, *27*(11), 2326–2331. doi:10.1897/07-602.1 PMID:18476749

Fukuda, M. N., Sato, T., Nakayama, J., Klier, G., Mikami, M., Aoki, D., & Nozawa, S. (1995). Trophinin and tastin, a novel cell adhesion molecule complex with potential involvement in embryo implantation. *Genes & Development*, *9*(10), 1199–1210. doi:10.1101/gad.9.10.1199 PMID:7758945

Ghosh, D., De, P., & Sengupta, J. (1994). Luteal phase ovarian oestrogen is not essential for implantation and maintenance of pregnancy from surrogate embryo transfer in the rhesus monkey. *Human Reproduction (Oxford, England)*, *9*(4), 629–637. doi:10.1093/oxfordjournals.humrep.a138561 PMID:8046014

Giudice, L. C. (1999). Potential biochemical markers of uterine receptivity. *Human Reproduction (Oxford, England)*, *14*(suppl 2), 3–16. doi:10.1093/humrep/14.suppl_2.3 PMID:10690796

Glasser, S. R., Mulholland, J., Mani, S. K., & Julian, J. (1991). Blastocyst-endometrial relationships: Reciprocal interactions between uterine epithelial and stromal cells and blastocysts. *Trophoblast Research*, *5*, 229–229.

Gordon, M. (1975). Cyclic changes in the fine structure of the epithelial cells of human endometrium. *International Review of Cytology*, *42*, 127–172. doi:10.1016/S0074-7696(08)60980-8 PMID:172466

Gore, A. C., Attardi, B., & DeFranco, D. B. (2006). Glucocorticoid repression of the reproductive axis: Effects on GnRH and gonadotropin subunit mRNA levels. *Molecular and Cellular Endocrinology*, *256*(1-2), 40–48. doi:10.1016/j.mce.2006.06.002 PMID:16839661

Graham, C. H., Lysiak, J. J., McCrae, K. R., & Lala, P. K. (1992). Localization of transforming growth factor-β at the human fetal-maternal interface: Role in trophoblast growth and differentiation. *Biology of Reproduction*, *46*(4), 561–572. doi:10.1095/biolreprod46.4.561 PMID:1374270

Graham, C. H., McCrae, K. R., & Lala, P. K. (1993). Molecular mechanisms controlling trophoblast invasion of the uterus. *Placenta*, *14*, 237–250. doi:10.1016/S0143-4004(05)80297-8

Graham, J. D., Yeates, C., Balleine, R. L., Harvey, S. S., Milliken, J. S., Bilous, A. M., & Clarke, C. L. (1996). Progesterone receptor A and B protein expression in human breast cancer. *The Journal of Steroid Biochemistry and Molecular Biology*, *56*(1-6), 93–98. doi:10.1016/0960-0760(95)00226-X PMID:8603052

Groll, J. M., Usadi, R. S., Lessey, B. A., Lininger, R., Young, S. L., & Fritz, M. A. (2009). Effects of variations in serum estradiol concentrations on secretory endometrial development and function in experimentally induced cycles in normal women. *Fertility and Sterility*, *92*(6), 2058–2061. doi:10.1016/j.fertnstert.2009.06.018 PMID:19608171

Gurpide, E., Gusberg, S. B., & Tseng, L. (1976). Estradiol binding and metabolism in human endometrial hyperplasia and adenocarcinoma. *Journal of Steroid Biochemistry*, *7*(11-12), 891–896. doi:10.1016/0022-4731(76)90007-8 PMID:1025366

Hamilton, G. S., Lysiak, J. J., Han, V. K. M., & Lala, P. K. (1998). Autocrine-paracrine regulation of human trophoblast invasiveness by insulin-like growth factor (IGF)-II and IGF-binding protein (IGFBP)-1. *Experimental Cell Research*, *244*(1), 147–156. doi:10.1006/excr.1998.4195 PMID:9770358

Han, V. K., Bassett, N., Walton, J., & Challis, J. R. (1996). The expression of insulin-like growth factor (IGF) and IGF-binding protein (IGFBP) genes in the human placenta and membranes: Evidence for IGF-IGFBP interactions at the feto-maternal interface. *The Journal of Clinical Endocrinology and Metabolism*, *81*(7), 2680–2693. PMID:8675597

Hillhouse, E. W., & Grammatopoulos, D. K. (2006). The molecular mechanisms underlying the regulation of the biological activity of corticotropin-releasing hormone receptors: Implications for physiology and pathophysiology. *Endocrine Reviews*, *27*(3), 260–286. doi:10.1210/er.2005-0034 PMID:16484629

Homer, C. S., Hanna, E., & McMichael, A. J. (2009). Climate change threatens the achievement of the millennium development goal for maternal health. *Midwifery*, *25*(6), 606–612. doi:10.1016/j.midw.2009.09.003 PMID:19836867

Huang, F., Wang, H., Zou, Y., Liu, Q., Cao, J., & Yin, T. (2013). Effect of GnRH-II on the ESC proliferation, apoptosis and VEGF secretion in patients with endometriosis in vitro. *International Journal of Clinical and Experimental Pathology*, *6*(11), 2487. PMID:24228111

Hvidtjørn, D., Schieve, L., Schendel, D., Jacobsson, B., Sværke, C., & Thorsen, P. (2009). Cerebral palsy, autism spectrum disorders, and developmental delay in children born after assisted conception: A systematic review and meta-analysis. *Archives of Pediatrics & Adolescent Medicine*, *163*(1), 72–83. doi:10.1001/archpediatrics.2008.507 PMID:19124707

Hyatt, O. M., Lemke, B., & Kjellstrom, T. (2010). Regional maps of occupational heat exposure: Past, present, and potential future. *Global Health Action*, *3*(1), 5715. doi:10.3402/gha.v3i0.5715 PMID:21165172

Irving, J. A., & Lala, P. K. (1995). Functional role of cell surface integrins on human trophoblast cell migration: Regulation by TGF-β, IGF-II, and IGFBP-1. *Experimental Cell Research*, *217*(2), 419–427. doi:10.1006/excr.1995.1105 PMID:7535237

Irwin, J. C., & Giudice, L. C. (1998). IGFBP-1 binds to the α5β1 integrin in human cytotrophoblasts and inhibits invasion into decidualized endometrium stromal cells in vitro. *Growth Hormone & IGF Research*, *8*, 21–23. doi:10.1016/S1096-6374(98)80318-3 PMID:10990441

Kalantaridou, S. N., Zoumakis, E., Makrigiannakis, A., Lavasidis, L. G., Vrekoussis, T., & Chrousos, G. P. (2010). Corticotropin-releasing hormone, stress and human reproduction: An update. *Journal of Reproductive Immunology*, *85*(1), 33–39. doi:10.1016/j.jri.2010.02.005 PMID:20412987

Khairy, M., Dhillon, R. K., Chu, J., Rajkhowa, M., & Coomarasamy, A. (2016). The effect of peri-implantation administration of uterine relaxing agents in assisted reproduction treatment cycles: A systematic review and meta-analysis. *Reproductive Biomedicine Online*, *32*(4), 362–376. doi:10.1016/j.rbmo.2016.01.004 PMID:26936145

Kim, S. K., Han, E. J., Kim, S. M., Lee, J. R., Jee, B. C., Suh, C. S., & Kim, S. H. (2016). Efficacy of oxytocin antagonist infusion in improving in vitro fertilization outcomes on the day of embryo transfer: A meta-analysis. *Clinical and Experimental Reproductive Medicine*, *43*(4), 233. doi:10.5653/cerm.2016.43.4.233 PMID:28090463

King, A., Gardner, L., & Loke, Y. W. (1996). Evaluation of oestrogen and progesterone receptor expression in uterine mucosal lymphocytes. *Human Reproduction (Oxford, England)*, *11*(5), 1079–1082. doi:10.1093/oxfordjournals.humrep.a019300 PMID:8671394

Kinsey-Jones, J. S., Li, X. F., Knox, A. M. I., Wilkinson, E. S., Zhu, X. L., Chaudhary, A. A., Milligan, S. R., Lightman, S. L., O'byrne, K. T., & O'byrne, K. T. (2009). Down-regulation of hypothalamic kisspeptin and its receptor, Kiss1r, mRNA expression is associated with stress-induced suppression of luteinising hormone secretion in the female rat. *Journal of Neuroendocrinology*, *21*(1), 20–29. doi:10.1111/j.1365-2826.2008.01807.x PMID:19094090

Kjellstrom, T. (2009). Climate change, direct heat exposure, health and well-being in low and middle-income countries. *Global Health Action*, 2. PMID:20027264

Knoester, M., Helmerhorst, F. M., van der Westerlaken, L. A., Walther, F. J., & Veen, S. (2007). Matched follow-up study of 5–8-year-old ICSI singletons: Child behaviour, parenting stress and child (health-related) quality of life. *Human Reproduction (Oxford, England)*, *22*(12), 3098–3107. doi:10.1093/humrep/dem261 PMID:17905745

Kobayashi, R., Terakawa, J., Omatsu, T., Hengjan, Y., Mizutani, T., Ohmori, Y., & Hondo, E. (2017). The window of implantation is closed by estrogen via insulin-like growth factor 1 pathway. *Journal of Reproduction & Infertility*, *18*(2), 231. PMID:28868248

Kodaman, P. H., & Taylor, H. S. (2004). Hormonal regulation of implantation. *Obstetrics and Gynecology Clinics*, *31*(4), 745–766. PMID:15550333

Kraus, W. L., & Katzenellenbogen, B. S. (1993). Regulation of progesterone receptor gene expression and growth in the rat uterus: Modulation of estrogen actions by progesterone and sex steroid hormone antagonists. *Endocrinology*, *132*(6), 2371–2379. doi:10.1210/endo.132.6.8504742 PMID:8504742

Krogenaes, A. K., Taubøll, E., Stien, A., Oskam, I. C., Lyche, J. L., Dahl, E., Thomassen, R. F., Sweeney, T., & Ropstad, E. (2008). Valproate affects reproductive endocrine function, testis diameter and some semen variables in non-epileptic adolescent goat bucks. *Theriogenology*, *70*(1), 15–26. doi:10.1016/j.theriogenology.2008.01.029 PMID:18394693

Kurita, T., Young, P., Brody, J. R., Lydon, J. P., O'Malley, B. W., & Cunha, G. R. (1998). Stromal progesterone receptors mediate the inhibitory effects of progesterone on estrogen-induced uterine epithelial cell deoxyribonucleic acid synthesis. *Endocrinology*, *139*(11), 4708–4713. doi:10.1210/endo.139.11.6317 PMID:9794483

Kwong, W. Y., Wild, A. E., Roberts, P., Willis, A. C., & Fleming, T. P. (2000). Maternal undernutrition during the preimplantation period of rat development causes blastocyst abnormalities and programming of postnatal hypertension. *Development*, *127*(19), 4195–4202. PMID:10976051

Lessey, B. A. (2011). Assessment of endometrial receptivity. *Fertility and Sterility*, *96*(3), 522–529. doi:10.1016/j.fertnstert.2011.07.1095 PMID:21880273

Lessey, B. A., & Castelbaum, A. J. (2002). Integrins and implantation in the human. *Reviews in Endocrine & Metabolic Disorders*, *3*(2), 107–117. doi:10.1023/A:1015450727580 PMID:12007287

Lessey, B. A., Gui, Y., Apparao, K. B. C., Young, S. L., & Mulholland, J. (2002). Regulated expression of heparin-binding EGF-like growth factor (HB-EGF) in the human endometrium: A potential paracrine role during implantation. *Molecular Reproduction and Development*, *62*(4), 446–455. doi:10.1002/mrd.10129 PMID:12112577

Lessey, B. A., Killam, A. P., Metzger, D. A., Haney, A. F., Greene, G. L., & McCARTY, K. S. Jr. (1988). Immunohistochemical analysis of human uterine estrogen and progesterone receptors throughout the menstrual cycle. *The Journal of Clinical Endocrinology and Metabolism*, *67*(2), 334–340. doi:10.1210/jcem-67-2-334 PMID:2455728

Lessey, B. A., & Young, S. L. (2014). Homeostasis imbalance in the endometrium of women with implantation defects: The role of estrogen and progesterone. *Seminars in Reproductive Medicine*, *32*(5), 365–375. doi:10.1055-0034-1376355 PMID:24959818

Liang, X. H., Deng, W. B., Li, M., Zhao, Z. A., Wang, T. S., Feng, X. H., & Yang, Z. M. (2014). Egr1 protein acts downstream of estrogen-leukemia inhibitory factor (LIF)-STAT3 pathway and plays a role during implantation through targeting Wnt4. *The Journal of Biological Chemistry*, *289*(34), 23534–23545. doi:10.1074/jbc.M114.588897 PMID:25012664

Lindhard, A., Bentin-Ley, U., Ravn, V., Islin, H., Hviid, T., Rex, S., & Sørensen, S. (2002). Biochemical evaluation of endometrial function at the time of implantation. *Fertility and Sterility*, *78*(2), 221–233. doi:10.1016/S0015-0282(02)03240-5 PMID:12137855

Liu, F., & Li, R. (2019). *The Role of Neuroendocrine in Embryo Implantation*. Embryology - Theory and Practice, Bin Wu and Huai L. Feng, IntechOpen. doi:10.5772/intechopen.87863

Liu, X., Nie, J., & Guo, S. W. (2011). Elevated immunoreactivity to tissue factor and its association with dysmenorrhea severity and the amount of menses in adenomyosis. *Human Reproduction (Oxford, England)*, *26*(2), 337–345. doi:10.1093/humrep/deq311 PMID:21115504

Ludwig, H., & Spornitz, U. M. (1991). Microarchitecture of the human endometrium by scanning electron microscopy: Menstrual desquamation and remodeling. *Annals of the New York Academy of Sciences*, *622*(1), 28–46. doi:10.1111/j.1749-6632.1991.tb37848.x PMID:2064187

Lyche, J. L., Oskam, I. C., Skaare, J. U., Reksen, O., Sweeney, T., Dahl, E., & Ropstad, E. (2004). Effects of gestational and lactational exposure to low doses of PCBs 126 and 153 on anterior pituitary and gonadal hormones and on puberty in female goats. *Reproductive Toxicology (Elmsford, N.Y.)*, *19*(1), 87–95. doi:10.1016/j.reprotox.2004.05.005 PMID:15336716

Ma, L., Benson, G. V., Lim, H., Dey, S. K., & Maas, R. L. (1998). Abdominal B (AbdB) HoxaGenes: Regulation in adult Uterus by estrogen and progesterone and repression in müllerian duct by the synthetic estrogen diethylstilbestrol (DES). *Developmental Biology*, *197*(2), 141–154. doi:10.1006/dbio.1998.8907 PMID:9630742

Maggi, R., Cariboni, A. M., Marelli, M. M., Moretti, R. M., Andre, V., Marzagalli, M., & Limonta, P. (2016). GnRH and GnRH receptors in the pathophysiology of the human female reproductive system. *Human Reproduction Update*, *22*(3), 358–381. doi:10.1093/humupd/dmv059 PMID:26715597

Makrigiannakis, A., Zoumakis, E., Kalantaridou, S., Coutifaris, C., Margioris, A. N., Coukos, G., Rice, K. C., Gravanis, A., & Chrousos, G. P. (2001). Corticotropin-releasing hormone promotes blastocyst implantation and early maternal tolerance. *Nature Immunology*, *2*(11), 1018–1024. doi:10.1038/ni719 PMID:11590404

Manipalviratn, S., DeCherney, A., & Segars, J. (2009). Imprinting disorders and assisted reproductive technology. *Fertility and Sterility*, *91*(2), 305–315. doi:10.1016/j.fertnstert.2009.01.002 PMID:19201275

Margalioth, E. J., Ben-Chetrit, A., Gal, M., & Eldar-Geva, T. (2006). Investigation and treatment of repeated implantation failure following IVF-ET. *Human Reproduction (Oxford, England)*, *21*(12), 3036–3043. doi:10.1093/humrep/del305 PMID:16905766

Martin, K. L., Barlow, D. H., & Sargent, I. L. (1998). Heparin-binding epidermal growth factor significantly improves human blastocyst development and hatching in serum-free medium. *Human Reproduction (Oxford, England)*, *13*(6), 1645–1652. doi:10.1093/humrep/13.6.1645 PMID:9688406

Matsuzaki, S., Fukaya, T., Suzuki, T., Murakami, T., Sasano, H., & Yajima, A. (1999). Oestrogen receptor α and β mRNA expression in human endometrium throughout the menstrual cycle. *Molecular Human Reproduction*, *5*(6), 559–564. doi:10.1093/molehr/5.6.559 PMID:10341004

McGinnis, W., & Krumlauf, R. (1992). Homeobox genes and axial patterning. *Cell*, *68*(2), 283–302. doi:10.1016/0092-8674(92)90471-N PMID:1346368

Michael, A. E., Pester, L. A., Curtis, P., Shaw, R. W., Edwards, C. R. W., & Cooke, B. A. (1993). Direct inhibition of ovarian steroidogenesis by Cortisol and the modulatory role of 11β-hydroxysteroid dehydrogenase. *Horumon To Rinsho*, *38*(6), 641–644. doi:10.1111/j.1365-2265.1993.tb02147.x PMID:8334750

Miles, H. L., Hofman, P. L., Peek, J., Harris, M., Wilson, D., Robinson, E. M., Gluckman, P. D., & Cutfield, W. S. (2007). In vitro fertilization improves childhood growth and metabolism. *The Journal of Clinical Endocrinology and Metabolism*, *92*(9), 3441–3445. doi:10.1210/jc.2006-2465 PMID:17566097

Millar, R., Lowe, S., Conklin, D., Pawson, A., Maudsley, S., Troskie, B., Ott, T., Millar, M., Lincoln, G., Sellar, R., Faurholm, B., Scobie, G., Kuestner, R., Terasawa, E., & Katz, A. (2001). A novel mammalian receptor for the evolutionarily conserved type II GnRH. *Proceedings of the National Academy of Sciences of the United States of America*, *98*(17), 9636–9641. doi:10.1073/pnas.141048498 PMID:11493674

Millar, R. P. (2003). GnRH II and type II GnRH receptors. *Trends in Endocrinology and Metabolism*, *14*(1), 35–43. doi:10.1016/S1043-2760(02)00016-4 PMID:12475610

Miller, M. W. (1993). Migration of cortical neurons is altered by gestational exposure to ethanol. *Alcoholism, Clinical and Experimental Research*, *17*(2), 304–314. doi:10.1111/j.1530-0277.1993.tb00768.x PMID:8488973

Mitchell, M., Schulz, S. L., Armstrong, D. T., & Lane, M. (2009). Metabolic and mitochondrial dysfunction in early mouse embryos following maternal dietary protein intervention. *Biology of Reproduction*, *80*(4), 622–630. doi:10.1095/biolreprod.108.072595 PMID:19129514

Mucio-Ramírez, S., Sánchez-Islas, E., Sánchez-Jaramillo, E., Currás-Collazo, M., Juárez-González, V. R., Álvarez-González, M. Y., Orser, L. E., Hou, B., Pellicer, F., Kodavanti, P. R. S., & León-Olea, M. (2017). Perinatal exposure to organohalogen pollutants decreases vasopressin content and its mRNA expression in magnocellular neuroendocrine cells activated by osmotic stress in adult rats. *Toxicology and Applied Pharmacology*, *329*, 173–189. doi:10.1016/j.taap.2017.05.039 PMID:28579251

Musumeci, G., Castorina, S., Castrogiovanni, P., Loreto, C., Leonardi, R., Aiello, F. C., & Imbesi, R. (2015). A journey through the pituitary gland: Development, structure and function, with emphasis on embryo-foetal and later development. *Acta Histochemica, 117*(4-5), 355–366. doi:10.1016/j.acthis.2015.02.008 PMID:25858531

Nardo, L. G., Nikas, G., Makrigiannakis, A., Sinatra, F., & Nardo, F. (2003). Synchronous expression of pinopodes and alpha v beta 3 and alpha 4 beta 1 integrins in the endometrial surface epithelium of normally menstruating women during the implantation window. *The Journal of Reproductive Medicine, 48*(5), 355–361. PMID:12815909

Naruse, K., Lash, G. E., Innes, B. A., Otun, H. A., Searle, R. F., Robson, S. C., & Bulmer, J. N. (2009). Localization of matrix metalloproteinase (MMP)-2, MMP-9 and tissue inhibitors for MMPs (TIMPs) in uterine natural killer cells in early human pregnancy. *Human Reproduction (Oxford, England), 24*(3), 553–561. doi:10.1093/humrep/den408 PMID:19088110

Nikas, G. (1999). Cell-surface morphological events relevant to human implantation. *Human Reproduction (Oxford, England), 14*(suppl 2), 37–44. doi:10.1093/humrep/14.suppl_2.37 PMID:10690799

Nikas, G., & Aghajanova, L. (2002). Endometrial pinopodes: Some more understanding on human implantation. *Reproductive Biomedicine Online, 4*, 18–23. doi:10.1016/S1472-6483(12)60111-4 PMID:12470560

Nordengren, J., Pilka, R., Noskova, V., Ehinger, A., Domanski, H., Andersson, C., Hoyer-Hansen, G., Hansson, S. R., & Casslen, B. (2004). Differential localization and expression of urokinase plasminogen activator (uPA), its receptor (uPAR), and its inhibitor (PAI-1) mRNA and protein in endometrial tissue during the menstrual cycle. *Molecular Human Reproduction, 10*(9), 655–663. doi:10.1093/molehr/gah081 PMID:15243126

Norwitz, E. R., Schust, D. J., & Fisher, S. J. (2001). Implantation and the survival of early pregnancy. *The New England Journal of Medicine, 345*(19), 1400–1408. doi:10.1056/NEJMra000763 PMID:11794174

Oskam, I. C., Lyche, J. L., Krogenæs, A., Thomassen, R., Skaare, J. U., Wiger, R., Dahl, E., Sweeney, T., Stien, A., Ropstad, E., & Ropstad, E. (2005). Effects of long-term maternal exposure to low doses of PCB126 and PCB153 on the reproductive system and related hormones of young male goats. *Reproduction (Cambridge, England), 130*(5), 731–742. doi:10.1530/rep.1.00690 PMID:16264102

Parry, M. (2009). Climate change is a development issue, and only sustainable development can confront the challenge. *Climate and Development, 1*(1), 5–9. doi:10.3763/cdev.2009.0012

Phumsatitpong, C., & Moenter, S. M. (2018). Estradiol-dependent stimulation and suppression of gonadotropin-releasing hormone neuron firing activity by corticotropin-releasing hormone in female mice. *Endocrinology, 159*(1), 414–425. doi:10.1210/en.2017-00747 PMID:29069304

Piccirilli, D., Baldini, E., Massimiani, M., Camaioni, A., Salustri, A., Bernardini, R., Centanni, M., Ulisse, S., Moretti, C., & Campagnolo, L. (2018). Thyroid hormone regulates protease expression and activation of Notch signaling in implantation and embryo development. *The Journal of Endocrinology, 236*(1), 1–12. doi:10.1530/JOE-17-0436 PMID:28993437

Pierzynski, P. (2011). Oxytocin and vasopressin V1A receptors as new therapeutic targets in assisted reproduction. *Reproductive Biomedicine Online, 22*(1), 9–16. doi:10.1016/j.rbmo.2010.09.015 PMID:21130036

Psychoyos, A. (1986). Uterine Receptivity for Nidation. *Annals of the New York Academy of Sciences*, *476*(1), 36–42. doi:10.1111/j.1749-6632.1986.tb20920.x PMID:3541745

Rance, N. E., Young, W. S. III, & McMullen, N. T. (1994). Topography of neurons expressing luteinizing hormone-releasing hormone gene transcripts in the human hypothalamus and basal forebrain. *The Journal of Comparative Neurology*, *339*(4), 573–586. doi:10.1002/cne.903390408 PMID:8144747

Redmond, G. P. (2004). Thyroid dysfunction and women's reproductive health. *Thyroid, 14*(3), 5-15.

Rhind, S. M., Evans, N. P., Bellingham, M., Sharpe, R. M., Cotinot, C., Mandon-Pepin, B., Centanni, M., Ulisse, S., Moretti, C., Campagnolo, L., & Fischer, B. (2010). Effects of environmental pollutants on the reproduction and welfare of ruminants. *Animal*, *4*(7), 1227–1239. doi:10.1017/S1751731110000595 PMID:20582145

Rice, D., & Barone, S. Jr. (2000). Critical periods of vulnerability for the developing nervous system: Evidence from humans and animal models. *Environmental Health Perspectives*, *108*(suppl 3), 511–533. PMID:10852851

Rodier, P. M., Aschner, M., & Sager, P. R. (1984). Mitotic arrest in the developing CNS after prenatal exposure to methylmercury. *Neurobehavioral Toxicology and Teratology*, *6*(5), 379–385. PMID:6514102

Rosario, G. X., Modi, D. N., Sachdeva, G., Manjramkar, D. D., & Puri, C. P. (2005). Morphological events in the primate endometrium in the presence of a preimplantation embryo, detected by the serum preimplantation factor bioassay. *Human Reproduction (Oxford, England)*, *20*(1), 61–71. doi:10.1093/humrep/deh534 PMID:15576397

Rylander, C., Oyvind Odland, J., & Manning Sandanger, T. (2013). Climate change and the potential effects on maternal and pregnancy outcomes: An assessment of the most vulnerable–the mother, fetus, and newborn child. *Global Health Action*, *6*(1), 19538. doi:10.3402/gha.v6i0.19538 PMID:23481091

Sakka, S. D., Malamitsi-Puchner, A., Loutradis, D., Chrousos, G. P., & Kanaka-Gantenbein, C. (2009). Euthyroid hyperthyrotropinemia in children born after in vitro fertilization. *The Journal of Clinical Endocrinology and Metabolism*, *94*(4), 1338–1341. doi:10.1210/jc.2008-1624 PMID:19190111

Samuelsson, A. M., Matthews, P. A., Argenton, M., Christie, M. R., McConnell, J. M., Jansen, E. H., Piersma, A. H., Ozanne, S. E., Twinn, D. F., Remacle, C., Rowlerson, A., Poston, L., & Taylor, P. D. (2008). Diet-induced obesity in female mice leads to offspring hyperphagia, adiposity, hypertension, and insulin resistance: A novel murine model of developmental programming. *Hypertension*, *51*(2), 383–392. doi:10.1161/HYPERTENSIONAHA.107.101477 PMID:18086952

Savabieasfahani, M., Kannan, K., Astapova, O., Evans, N. P., & Padmanabhan, V. (2006). Developmental programming: Differential effects of prenatal exposure to bisphenol-A or methoxychlor on reproductive function. *Endocrinology*, *147*(12), 5956–5966. doi:10.1210/en.2006-0805 PMID:16946013

Sayem, A. S. M., Giribabu, N., Muniandy, S., & Salleh, N. (2017). Effects of thyroxine on expression of proteins related to thyroid hormone functions (TR-α, TR-β, RXR and ERK1/2) in uterus during peri-implantation period. *Biomedicine and Pharmacotherapy*, *96*, 1016–1021. doi:10.1016/j.biopha.2017.11.128 PMID:29221723

Sinclair, K. D., Allegrucci, C., Singh, R., Gardner, D. S., Sebastian, S., Bispham, J., Thurston, A., Huntley, J. F., Rees, W. D., Maloney, C. A., Lea, R. G., Craigon, J., McEvoy, T. G., & Young, L. E. (2007). DNA methylation, insulin resistance, and blood pressure in offspring determined by maternal periconceptional B vitamin and methionine status. *Proceedings of the National Academy of Sciences of the United States of America*, *104*(49), 19351–19356. doi:10.1073/pnas.0707258104 PMID:18042717

Speroff, L., & Fritz, M. A. (Eds.). (2005). Clinical Gynecologic Endocrinology and Infertility. Lippincott Williams and Wilkins.

Stavreus-Evers, A., Nikas, G., Sahlin, L., Eriksson, H., & Landgren, B. M. (2001). Formation of pinopodes in human endometrium is associated with the concentrations of progesterone and progesterone receptors. *Fertility and Sterility*, *76*(4), 782–791. doi:10.1016/S0015-0282(01)01993-8 PMID:11591414

Su, R. W., & Fazleabas, A. T. (2015). Implantation and establishment of pregnancy in human and nonhuman primates. In *Regulation of Implantation and Establishment of Pregnancy in Mammals* (pp. 189–213). Springer. doi:10.1007/978-3-319-15856-3_10

Sweeney, T., Nicol, L., Roche, J. F., & Brooks, A. N. (2000). Maternal exposure to octylphenol suppresses ovine fetal follicle-stimulating hormone secretion, testis size, and sertoli cell number. *Endocrinology*, *141*(7), 2667–2673. doi:10.1210/endo.141.7.7552 PMID:10875272

Tan, O., & Bukulmez, O. (2011). Biochemistry, molecular biology and cell biology of gonadotropin-releasing hormone antagonists. *Current Opinion in Obstetrics & Gynecology*, *23*(4), 238–244. doi:10.1097/GCO.0b013e328348a3ce PMID:21666463

Tarara, R., Enders, A. C., Hendrickx, A. G., Gulamhusein, N., Hodges, J. K., Hearn, J. P., Eley, R. B., & Else, J. G. (1987). Early implantation and embryonic development of the baboon: Stages 5, 6 and 7. *Anatomy and Embryology*, *176*(3), 267–275. doi:10.1007/BF00310182 PMID:3631530

Taylor, H. S. (2000). The role of HOX genes in the development and function of the female reproductive tract. In Seminars in reproductive medicine (Vol. 18, No. 01, pp. 081-090). Thieme Medical Publishers, Inc. doi:10.1055-2000-13478

Taylor, H. S., Arici, A., Olive, D., & Igarashi, P. (1998). HOXA10 is expressed in response to sex steroids at the time of implantation in the human endometrium. *The Journal of Clinical Investigation*, *101*(7), 1379–1384. doi:10.1172/JCI1597 PMID:9525980

Taylor, H. S., Igarashi, P., Olive, D. L., & Arici, A. (1999). Sex steroids mediate HOXA11 expression in the human peri-implantation endometrium. *The Journal of Clinical Endocrinology and Metabolism*, *84*(3), 1129–1135. doi:10.1210/jc.84.3.1129 PMID:10084606

Taylor, H. S., VandenHeuvel, G. B., & Igarashi, P. (1997). A conserved Hox axis in the mouse and human female reproductive system: Late establishment and persistent adult expression of the Hoxa cluster genes. *Biology of Reproduction*, *57*(6), 1338–1345. doi:10.1095/biolreprod57.6.1338 PMID:9408238

Tei, C., Maruyama, T., Kuji, N., Miyazaki, T., Mikami, M., & Yoshimura, Y. (2003). Reduced expression of αvβ3 integrin in the endometrium of unexplained infertility patients with recurrent IVF-ET failures: Improvement by danazol treatment. *Journal of Assisted Reproduction and Genetics*, *20*(1), 13–20. doi:10.1023/A:1021254620888 PMID:12645863

Thimon, V., Calvo, E., Koukoui, O., Légaré, C., & Sullivan, R. (2008). Effects of vasectomy on gene expression profiling along the human epididymis. *Biology of Reproduction*, *79*(2), 262–273. doi:10.1095/biolreprod.107.066449 PMID:18434627

Troy, P. J., Daftary, G. S., Bagot, C. N., & Taylor, H. S. (2003). Transcriptional repression of peri-implantation EMX2 expression in mammalian reproduction by HOXA10. *Molecular and Cellular Biology*, *23*(1), 1–13. doi:10.1128/MCB.23.1.1-13.2003 PMID:12482956

Tsuboi, T., Osawa, T., Kimura, K., Kubo, M., & Haritani, M. (2011). Experimental infection of early pregnant cows with bovine viral diarrhea virus: Transmission of virus to the reproductive tract and conceptus. *Research in Veterinary Science*, *90*(1), 174–178. doi:10.1016/j.rvsc.2010.04.024 PMID:20553701

Twig, G., Shina, A., Amital, H., & Shoenfeld, Y. (2012). Pathogenesis of infertility and recurrent pregnancy loss in thyroid autoimmunity. *Journal of Autoimmunity*, *38*(2-3), J275–JJ28. doi:10.1016/j.jaut.2011.11.014 PMID:22218218

Van den Boogaard, E., Vissenberg, R., Land, J. A., van Wely, M., van der Post, J. A., Goddijn, M., & Bisschop, P. H. (2011). Significance of (sub) clinical thyroid dysfunction and thyroid autoimmunity before conception and in early pregnancy: A systematic review. *Human Reproduction Update*, *17*(5), 605–619. doi:10.1093/humupd/dmr024 PMID:21622978

Van der Linden, M., Buckingham, K., Farquhar, C., Kremer, J. A., & Metwally, M. (2011). Luteal phase support for assisted reproduction cycles. *Cochrane Database of Systematic Reviews*, 10. PMID:21975790

Vergetaki, A., Jeschke, U., Vrekoussis, T., Taliouri, E., Sabatini, L., Papakonstanti, E. A., & Makrigiannakis, A. (2014). Galectin-1 overexpression in endometriosis and its regulation by neuropeptides (CRH, UCN) indicating its important role in reproduction and inflammation. *PLoS One*, *9*(12), e114229. doi:10.1371/journal.pone.0114229 PMID:25473847

Wang, H., Critchley, H. O., Kelly, R. W., Shen, D., & Baird, D. T. (1998). Progesterone receptor subtype B is differentially regulated in human endometrial stroma. *Molecular Human Reproduction*, *4*(4), 407–412. doi:10.1093/molehr/4.4.407 PMID:9620842

Wang, H., & Dey, S. K. (2006). Roadmap to embryo implantation: Clues from mouse models. *Nature Reviews. Genetics*, *7*(3), 185–199. doi:10.1038/nrg1808 PMID:16485018

Weihua, Z., Saji, S., Mäkinen, S., Cheng, G., Jensen, E. V., Warner, M., & Gustafsson, J. Å. (2000). Estrogen receptor (ER) β, a modulator of ERα in the uterus. *Proceedings of the National Academy of Sciences of the United States of America*, *97*(11), 5936–5941. doi:10.1073/pnas.97.11.5936 PMID:10823946

Whirledge, S., & Cidlowski, J. A. (2013). A role for glucocorticoids in stress-impaired reproduction: Beyond the hypothalamus and pituitary. *Endocrinology*, *154*(12), 4450–4468. doi:10.1210/en.2013-1652 PMID:24064362

Whirledge, S. D., Oakley, R. H., Myers, P. H., Lydon, J. P., DeMayo, F., & Cidlowski, J. A. (2015). Uterine glucocorticoid receptors are critical for fertility in mice through control of embryo implantation and decidualization. *Proceedings of the National Academy of Sciences of the United States of America*, *112*(49), 15166–15171. doi:10.1073/pnas.1508056112 PMID:26598666

Wilcox, A. J., Baird, D. D., & Weinberg, C. R. (1999). Time of implantation of the conceptus and loss of pregnancy. *The New England Journal of Medicine*, *340*(23), 1796–1799. doi:10.1056/NEJM199906103402304 PMID:10362823

Wood, G. W., Hausmann, E., & Choudhuri, R. (1997). Relative role of CSF-1, MCP-1/JE, and RANTES in macrophage recruitment during successful pregnancy. *Molecular Reproduction and Development: Incorporating Gamete Research*, *46*(1), 62–70. doi:10.1002/(SICI)1098-2795(199701)46:1<62::AID-MRD10>3.0.CO;2-5 PMID:8981365

Wu, H. M., Huang, H. Y., Lee, C. L., Soong, Y. K., Leung, P. C., & Wang, H. S. (2015). Gonadotropin-releasing hormone type II (GnRH-II) agonist regulates the motility of human decidual endometrial stromal cells: Possible effect on embryo implantation and pregnancy. *Biology of Reproduction*, *92*(4), 98–1. doi:10.1095/biolreprod.114.127324 PMID:25761596

Yalcin, M., Dyskin, E., Lansing, L., Bharali, D. J., Mousa, S. S., Bridoux, A., Hercbergs, A. H., Lin, H. Y., Davis, F. B., Glinsky, G. V., Glinskii, A., Ma, J., Davis, P. J., & Mousa, S. A. (2010). Tetraiodothyroacetic acid (tetrac) and nanoparticulatetetrac arrest growth of medullary carcinoma of the thyroid. *The Journal of Clinical Endocrinology and Metabolism*, *95*(4), 1972–1980. doi:10.1210/jc.2009-1926 PMID:20133461

Yoo, H. J., Barlow, D. H., & Mardon, H. J. (1997). Temporal and spatial regulation of expression of heparin-binding epidermal growth factor-like growth factor in the human endometrium: A possible role in blastocyst implantation. *Developmental Genetics*, *21*(1), 102–108. doi:10.1002/(SICI)1520-6408(1997)21:1<102::AID-DVG12>3.0.CO;2-C PMID:9291586

Yoshinaga, K. (1989). Receptor concept in implantation research. *Progress in Clinical and Biological Research*, *294*, 379–387. PMID:2542989

Yoshinaga, K. (2014). Progesterone and its downstream molecules as blastocyst implantation essential factors. *American Journal of Reproductive Immunology*, *72*(2), 117–128. doi:10.1111/aji.12253 PMID:24754263

Young, S. L. (2013). Eestrogen and progesterone action on endometrium: A translational approach to understanding endometrial receptivity. *Reproductive Biomedicine Online*, *27*(5), 497–505. doi:10.1016/j.rbmo.2013.06.010 PMID:23933037

Yu, B., Ruman, J., & Christman, G. (2011). The role of peripheral gonadotropin-releasing hormone receptors in female reproduction. *Fertility and Sterility*, *95*(2), 465–473. doi:10.1016/j.fertnstert.2010.08.045 PMID:20888559

Zhou, J., & Bondy, C. (1992). Insulin-like growth factor-II and its binding proteins in placental development. *Endocrinology*, *131*(3), 1230–1240. doi:10.1210/endo.131.3.1380437 PMID:1380437

Chapter 9
The Effect of Temperature on Sperm Motility and Viability

Junaid Ahmad Malik
https://orcid.org/0000-0003-4411-2015
Department of Zoology, Government Degree College, Bijbehara, India

ABSTRACT

Sperm motility is a key factor in permitting us to decide semen quality and capacity to fertilize. Motility in semen is for the most part constrained by K^+ in salmonids, and presumably additionally in sturgeons, and by osmotic pressure in other freshwater and seawater fish species; however, different elements, for example, combination of encircling metabolites and particles (Ca^{2+}, Mg^{2+}, and so on.), pH and temperature likewise impact motility attributes. In the chapter, the author has fundamentally surveyed and abridged the impacts of temperature on the motility of spermatozoa in some fish species (salmonids, cyprinids, and sturgeons) and of humans also. DSper fluxes are initiated by warm temperatures and intercedecation conductance, which shares a pharmacological profile suggestive of TRPV4. Together, these outcomes recommend that TRPV4 actuation triggers early membrane depolarization, encouraging both CatSper and Hv1 gating and, therefore, sperm hyperactivation.

INTRODUCTION

Because of the changing environment of earth, the regular populaces are reacting to unnatural temperature changes by shifting extents, declining and going wiped out (Chen et al., 2011; Parmesan and Yohe, 2003; Raftery et al, 2017; Thomas et al., 2004). With a hotter, increasingly unstable environment, outrageous climatic occasions, for example, heatwaves are anticipated to turn out to be progressively common (Christidis et al., 2015; Meehl and Tebaldi, 2004). Heatwaves create extraordinary warm conditions, with frequently short and stochastic onsets, heatwaves are probably going to be especially problematic for biological capacity and function (Bridle and Vines, 2007). Reproductive sensitivity to increments in temperature that animals frequently experience with their indigenous habitat is notable in mammals, where adjustments that permit testicular cooling of 2°C to 8°C underneath core internal temperature level are basic to permit normal male fecundity (Setchell, 2006). Indeed, even gentle increments in the

DOI: 10.4018/978-1-7998-4480-8.ch009

encompassing thermal condition can disturb male reproductive capacity in endotherms: for instance, uncovering male mice for 24 hr to an air temperature of 32°C brought about fertility regressions of ~75% (Burfening et al., 1970), and various comparable investigations uncover such sensitivities (Hurley et al., 2018; Setchell, 2006).New discoveries published in the journal Nature Communications uncover that heatwaves harm sperm in bugs - with negative effects for fertility across ages. Heatwaves split male fertility, and traded off sperm competitive capacity. Progressive heatwaves exacerbated these impacts, with a second heatwave actuating practically complete sterility in males (Sales et al., 2018). The examination says that male fertility during heatwaves could assist with clarifying why environmental change is having such an effect on species populaces, incorporating atmosphere related eliminations as of late. The investigation shows that sperm function is a particularly sensitive characteristic when the earth warms up, and in a model framework speaking to a colossal measure of worldwide biodiversity. A hotter climate will be progressively unpredictable and risky, with outrageous occasions like heatwaves getting progressively recurrent, strong and extensive. The analysts caution this could add additional strain to populaces previously enduring environmental change after some time (Sales et al., 2018).

The deviation in high encompassing temperature and sogginess is viewed as a significant part causing disparity in semen production and semen quality. A few investigations have indicated that high ambient temperature, heat stress and additionally blistering climate adversely affect semen production (Colenbrander et al., 1993; Suriyasomboon et al., 2004) and semen quality (Cameron and Blackshaw, 1980; Larsson and Einarsson, 1984; Stone, 1982). In many investigations, an expanded extent of unusual spermatozoa has been found after heat treatment, yet the outcomes shift among boars, and the outcomes are likewise identified with the various systems for heat stress that have been utilized (Larsson and Einarsson, 1984; Wettemann et al., 1976). Dynamic sperm motility is significant in assessing the ripeness capability of spermatozoa, and is essential for fertilization in humans (Amelar et al., 1980). Sperm motility relies upon various variables, including the temperature at which the semen is kept between the hour of discharge and the hour of investigation. Temperatures lower than 4°C and higher than 37°C have been appeared to influence contrarily sperm capacities, for example, motility (Appell et al., 1977) and entrance into cervix (Stumpf et al., 1984). A few investigators have demonstrated that the semen ought to be kept at body temperature following discharge, while others have discovered that human, just as rat, spermatozoa lose motility whenever put away at 37°C (Cohen et al., 1985; Dougherty et al., 1975; Freund and Peterson, 1976; Muehleis and Long, 1976; World Health Organization, 1999).Single contrivance that might be answerable for decreased motility and along these lines fertility capability of sperm is an upsurgein assembly of Reactive Oxygen Species (ROS, for example, hydrogen peroxide, superoxide anion and hydroxyl radicals, in semen (Aitken et al., 1989). ROS are expected to manage sperm exertion by animating a redox-directed cAMP-intervened pathway that controls the enlistment of capacitation and the acrosome response (Aitken, 1999). Be that as it may, an expansion in ROS formationmore than average levels bring about oxidative pressure prompting a decrease in sperm motility and a diminished limit with regards to sperm-oocyte combination (Aitken et al., 1989). Moreover, ROS are likewise known to bout DNA, causing strand disruptions and harm to the nuclear constituents of human spermatozoa (Aitken et al., 1998).

Fenkes et al (2017), demonstrated that from the inception in the breeding season, brown trout sperm whirling velocity was influenced by acclimation temperature, initiation temperature and period since sperm actuation. The impact of actuation temperature on sperm swimming velocity fluctuated over the time since sperm initiation, as approved from the critical collaboration between these terms. Sperm swimming velocity declined over the period post-sperm enactment and contrasts in sperm swimming rate between acclimation/initiation temperature lots were obvious in the initial 16 s post-actuation. It has been recommended that an ideal initiation temperature for sperm development exists that compares with species-explicit adjacentthermalsituations (Lahnsteiner and Mansour, 2012; Vladic and Jarvi, 1997). This shows our unfriendly treatment of 8°C is nearer to the thermal ideal for brown trout sperm than the warm treatment of 13°C. Water temperatures of waterways during upstream movement and of reproducing grounds upon appearance can contrast considerably in light of the fact that they intently follow air temperature vacillations (Isaak et al., 2011; van Vliet et al., 2011). The book chapter deals with the exposition of spermatozoa, the usually mediocre class of human discharge must trade off men's fertility. The pressure of the higher temperature in some workplaces and, especially, the higher scrotal temperature created by apparel are frequently referenced as conceivable causative factors right now. The plausible effect of outfit and way of life has regularly been discussed (Mieusset et al., 1987; Thonneau et al., 1997), and this issue has stayed especially in the ''perhaps'' segment, not least as a result of the absence of awareness in men who customarily use almost no attire. For sure, the pleiomorphism of gorilla spermatozoa drove to believe that attire instigated hyperthermia is probably not going to represent the low quality of human discharge (Seuanez, 1980). Besides, the second rate nature of human sperm has in recent times been connected likewise to a nonattendance of the specific powers probably brought by sperm rivalry (Van der Horst and Maree, 2014).

TEMPERATURE STRESS ON FISH SPERMS

As fish cultivating extends and collecting of wild stocks turns out to be increasingly serious, there is a developing requirement for methods of gamete stockpiling to encourage counterfeit reproduction systems. The most well-known parameters utilized in considering sperm science are the structure and motility parameters of the spermatozoa, the natural chemistry of the sperm cells and of the fundamental plasma, the metabolism of spermatozoa (respiration and energetics of motility), the motility that can be actuated after storage and productivity of stockpiled sperm, including freezability for cryopreservation (Billard and Cosson, 1992; Billard et al., 1995a,b). Studies on motility of fish spermatozoa have been confined to an extremely set number of groups, generally species with business esteem in aquaculture or those associated with management programs (Billard and Cosson, 1992). The impact of high temperatures on fish sperm speed is to a great extent species explicit, with there being either undesirable, zero or an ideal impact of temperature on sperm motility (Mehlis and Baker, 2014). Sperm motility is a key factor deciding the quality and impregnating capacity of semen (Billard, 1978; Stoss, 1983), and a few parameters have been utilized to assess motility. The most usually utilized has been the timespan required from actuation to arrive at relative discontinuance of sperm movement (Billard, 1986), the purported 'motility length'. Later examinations allude to the level of motile sperm examined apparently (Billard et al., 1987, 1995a; Cosson et al., 1999; Ingermann et al., 2002). Increasingly quantitative ways to deal with examining sperm motility include utilizing Computer Assisted Semen Analysis (CASA) frameworks (Cosson et al., 1999; Billard et al., 1995a,b; Dreanno et al., 1998).

A few parameters of the swimming medium, for example, ion concentration (K^+, Na^+, Ca^{2+}, Mg^{2+}), osmotic pressure, pH, temperature and dilution, influence the motility term of fish spermatozoa (Stoss, 1983; Billard et al., 1995a; Cosson et al., 1999; Cosson, 2004). In all fish species examined so far, spermatozoa are immotile in the testis and seminal plasma (Stoss, 1983; Billard, 1986). In salmonids, the restraint of motility in semen is for the most part constrained by the convergence of K^+ions (Schlenk and

Kahmann, 1938; Stoss, 1983; Billard, 1986), however osmotic pressure is the primary factor in cyprinids (Stoss, 1983; Billard, 1986; Linhart et al., 1991; Cosson, 2004). Sturgeon spermatozoa, similar to that of teleost fish, are immotile in the seminal plasma (Ginzburg, 1968). The hindrance of motility in sturgeon semen is predominantly because of the convergence of K^+ions (Gallis et al., 1991; Linhart et al., 1991; Billard, 2000; Alavi et al., 2004). Different variables, for example, osmotic pressure, have been appeared (Drabkina, 1961; Linhart et al., 2002) to influence motility of sturgeon spermatozoa.

Further, Lahnsteiner et al. (1996) proposed that sperm speed, a key factor for assessing motility of the overall number of spermatozoa, could be grouped by CASA into the accompanying classes: immotile (velocity <5 µm s^{-1}), locally motile (velocity = 5-20 µm s^{-1}) and completely motile (velocity > 20 µm s^{-1}). Also, motile spermatozoa can be characterized based on their swimming conduct as: straight (linearity index > 0.9), non-linear (linearity index ≤ 0.9) and round (maximal radius for circular tracks=20 µm) (Fig. 1). Cosson et al. (1997) expressed that computer-assisted valuation permits target correlation of the movement of fish spermatozoa presented to different physiological circumstances, prompting examination of their intensity for swimming, capacity to fertilize and to live during freezing performances.The motility span, fertilizing capacity and speed of spermatozoa rely upon temperature of the enactment medium (Ginzburg, 1968; Stoss, 1983; Billard et al., 1995b) and of that of the broodstock holding tank (Williot et al., 2000). Since the vigorous assets of fish spermatozoa are restricted, an expansion in speed brought about by a temperature ascend in the swimming procedure prompts a shorter term of motility, and then again, bringing down the swimming temperature brings about a drawn out length of motility and decreased cell speed (Schlenk and Kahmann, 1938; Ginzburg, 1968; Stoss, 1983).

Figure 1. Classification of motility classes of fish spermatozoa using computer-assisted semen analysis (CASA), according to Lahnsteiner et al. (1996) (Adapted from Alavi and Cosson, 2005)

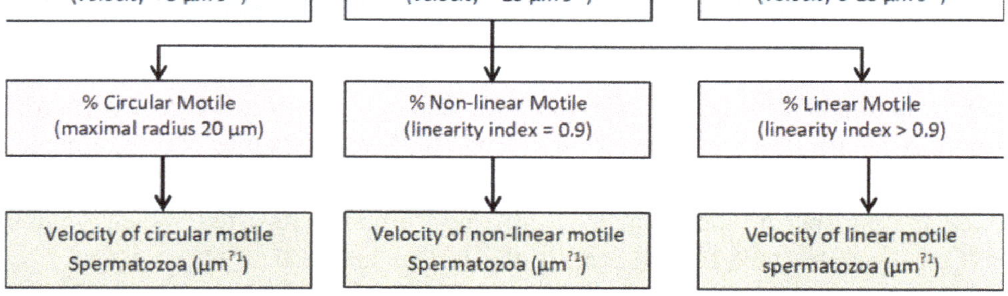

Salmonids

Alavi et al. (2005) presumed that temperature has critical impacts both on all out length and span of dynamic movement in *Salmotruttafario*. Following investigations of Schlenk and Kahmann (1938), apparently, at 12.5⁰C and 16⁰C, the forward speed of spermatozoa of *Salmotruttafario* is high (160-164 μ/s) 4 s after their enactment by freshwater. After 8, 16, and 26 sec, this rate reduces down to 85-91, 24-33 and 2-5 μ/s, respectively. Likewise, forwardeffort of the lake trout spermatozoa, *Salvelinusnamaycush*, stops totally beneath 29 sec at 12.5⁰C to16⁰C and beneath 56 sec at 2.25⁰C (Schlenk and Kahmann, 1938). The beat frequencies of flagella of rainbow trout spermatozoa have been estimated at various temperatures (Billard and Cosson, 1988, 1992): the beating repetition is low at 5⁰C, stays consistent up to 10⁰C, thenrises quickly at around 14⁰C and balances out at values over 21⁰C (Table 1, Fig. 2).The correlation of sperm motility constraints during the enactment of sperm at various temperatures shows that both the proportion of motile spermatozoa and the absolute extent of accelerative measure reduce when the temperature of the swimming medium and the underlying beat recurrence upsurge (Fig. 2) (Alavi et al, 2005).

Table 1.Effects of temperature on initial beat frequencies and total duration of forward movement of rainbow trout live spermatozoa in a swimming solution composed of 80 mMNaCl, 40 mMKCl, 1 mM $CaCl_2$, and 20 mMTris, pH 9.0, at a dilution of 1:10. (A) Time duration during which 100% of the spermatozoa were motile. (B) Elapsed time after which most spermatozoa stopped moving (Adapted from Billard and Cosson, 1992).

Temperature (⁰C)	*Initial beat frequency (Hz)*	*Duration of forward movement (sec)*	
		A	**B**
5	32	140	>180
10	33	70	>180
14	49	60	150
21	55	20	40
26	55	14	20

Figure 2. Changes in the beat frequency of spermatozoa of rainbow trout at different temperatures (Adapted from Billard and Cosson, 1992).

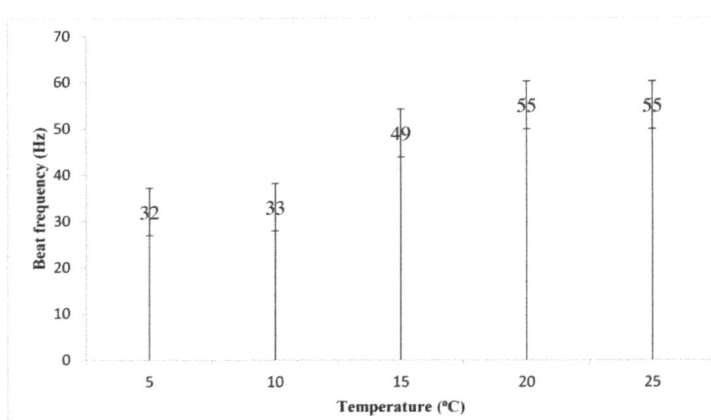

Cyprinids

The absolute span of motility of *Cyprinuscarpio* spermatozoa has been estimated as 70-80 sec post initiation by Jezierska and Witeska (1999) and as 120 sec by Elster and Mann (1952) at 20-21^{0}C. The motility term of grass carp spermatozoa is shorter than the common carp (Belova, 1981; Jezierska and Witeska, 1999). It has been affirmed that spermatozoa are motile for longer at 20^{0}C than at 26, or 30^{0}C in common carp, or 30^{0}C in grass carp (Jezierska and Witeska, 1999). Billard and Cosson (1989) announced that the complete motility term in carp is longer than in trout (more than 1 min versus 20-25 sec). The beat recurrence of most of spermatozoa decreases dynamically after initiation from 80 to 90 sec in carp (Billard and Cosson, 1992; Billard et al., 1995b). As in other fish species, the temperature of the impregnation solution and swimming medium utilized in tests are significant for carp manual semen injection.

Sturgeons

Less data is handy on the impacts of temperature on motility span of sturgeon spermatozoa and these evidences are exceptionally fickle in the inscription. Mims (1991) detailed that the motility time of the spermatozoa of paddlefish, *Polyodonspathula*, can be up to 4.4 min, while Linhart et al. (1995) have revealed that only 1-5% of spermatozoa are motile at 6 min. Williot et al. (2000) found that the motility of Siberian sturgeon, *Acipenserbaeri*, spermatozoa diminishes when the temperature is prolonged from 10^{0}C to 17.5^{0}C; in any case, no critical distinction is seen amongst males kept at 10^{0}C, 12^{0}Cor 15^{0}C.

TEMPERATURE STRESS ON FOWL SPERM

During the normal assessment of smears of new sperm of the domestic fowl it was noticed that the level of movement was distinctly diminished if the temperature of the slide was raised any significant degree above room temperature. Fowl sperm, are impervious to body heat, since the testicles and excretory channels of the male are normally situated inside the peritoneal cavity (Munro, 1938). Besides, they are equipped for enduring and achieving fertility in the female for a time of 2 or 3 weeks (Curtis and Lambert, 1929) and once in a while 4 to 5 weeks (Crew, 1926; Nicolaides, 1934) subsequent to cessation of mating. The motility of fowl sperm in vitro relies on a deal of temperature and medium of suspension. Most kinds of manufactured diluents bolster motility at room temperature however restrain such at bodily heat level. The immobilization isn't lasting and development is again continued when the temperature is brought down. Incredibly low temperatures, for example simply above freezing, likewise immobilize fowl sperm. Rodent and guinea-pig sperm, then again, are generally dynamic at bodily heat levels in similar manufactured diluents and are just immobilized by low temperatures, responses which, as per the literature, have all the earmarks of being commonly typical for mammalian sperm (Munro, 1938).

SCROTAL HEAT STRESS AND MAMMALIAN SPERM

The testicular temperature of mammals vagaries somewhere in the range of 2°C and 8°C beneath the corebodily heat level (Durairajanayagam et al., 2015), and its security is vital to guarantee typical spermatogenesis and sperm development. Scrotal heat stress (SHS) can happen in males and reduce sperm quality. SHS can be prompted by some way of life factors; for instance, during sauna introduction, the scrotal temperature rises to bodily level inside 10 minutes, and this could prompt interruption of spermatogenesis (Jockenhovel et al., 1990). Occupational warmth presentation, for example, that accomplished by professional drivers and culinary professionals, can altogether raise scrotal temperature (Bujan et al., 2000) and likely lead to progressively sub-par sperm quality (Figa-Talamanca et al., 1996).The impacts of SHS on the testis, epididymis, and spermatozoa have been concerted in various types of mammals, including humans (Alves et al., 2016; Banks et al., 2005; Cai et al., 2011; Kanter et al., 2013; Lucio et al., 2016; Rao et al., 2015). Normal warmth treatment in these investigations comprised of the use of transient warmth at 42°C to 43°C for 30 min (Banks et al., 2005; Cai et al., 2011; Perez-Crespo et al., 2008, 2017). Be that as it may, for infertilemales presented to SHS factors, the warmth to which they are uncovered is typically recurrent rather than fleeting, and the temperature isn't incredibly high. To successfully impersonate human SHS conditions, afrequent and mild scrotal heat stress (fmSHS) at a proper temperature is required. Since the whole spermatogenesis pattern of the mouse is roughly five weeks (Nagy, 2006), SHS ought to be instigated for not less than five weeks to guarantee that all spermatozoa removed are presented to warm stress. Further, Wan et al. (2020) presumed that SHS at 39°C and 41°C made mellow debilitation to the sperm, prompting a phenotype like oligoasthenozoospermia in people. Agreed with the fact that most SHS conditions in people are near internal heat level, SHS of 39°C was implemented to devise the fmSHS model. fmSHS diminished the rate of fertilization, hindered on-time progress, and decreased the implantation pace of the embryos in the IVF-ET process, yet it didn't influence the growth or capacity of blastocysts. The fmSHS mice would in general produce more Y-bearing spermatozoa and promulgated extra male posterity.

TEMPERATURE STRESS ON TRPV4 AND ION CHANNELS OF HUMAN SPERM

An ion channel protein that is encoded by the TRPV4 gene in humans is the transient receptor potential cation channel subfamily V member 4. The protein encoded is a Ca^{2+} -permeable, non-specific channel of cation which has been known to be involved in many biological processes, disorders, and disease as well. Osmotic, mechanical and chemical signals activate the channel. It also responds to the temperature changes and warmth. In detecting high temperatures, TRPV1 plays an important function. Two TRPV1-related ion channel proteins, TRPV3 and TRPV4, have recently been documented to be triggered by mild temperature elevations greater than 30-35 ° C (Watanabe et al., 2002). TRPV3 and TRPV4 are present in cultured keratinocytes and play an important role in the in vitro detection of hot temperatures and hypoosmolarity (Chung et al., 2004). When tested on hotplates above 50 °C, mice without TRPV1 demonstrated impairment in the identification of chemically mediated inflammatory pain (Davis et al., 2000).The activation of TRPV4 causes initial membrane depolarization, facilitating both CatSper gating and, consequently, sperm hyperactivation.

The capacity of human spermatozoa to reconnoitre the female regenerative tract and in the end find and prepare the egg is fundamental for reproduction (Okabe, 2013). To achieve these objectives, a spermatozoon must detect milieu and adjust its motility, which is controlled to some degree by ATP formation and flagellar homeostasis (Lishko et al., 2012). Of essential significance is the change from even basal tail twisting into 'hyperactivated motility' – a deviated, high-amplitude, whip-like beating pattern of the flagellum - that empowers sperm to defeat the egg's defensive vestments (Ho et al., 2002;

Ishijima et al., 2006). The steroid hormone progesterone (P4) goes about as a significant trigger of human sperm hyper-actuation (Suarez, 2008). P4 is discharged by cumulus cells encompassing the egg (Schuetz and Dubin, 1981) and causes a vigorous rise of human sperm cytoplasmic [Ca^{2+}] by means of the main Ca^{2+} channel of sperm (Smith et al., 2013). The steroid executes by means of its non-genomic receptor ABHD2, a serine hydrolase that, upon P4 binding, discharges hindrance of human CatSper (Miller et al., 2016).

The Catsper channel is a sperm-specific, pH-dependent, Ca^{2+} -permeable, and low-voltage-dependent channel that is important for sperm flagellum hyperactivity, egg chemotaxis, capacitation, and acrosome reaction. Ca^{2+} entry into the sperm tail by an alkalinization-activated voltage-sensitive Ca^{2+} selective current (CatSper) is necessary for the powerful uneven movement of hyperactivated spermatozoa (Miller et al., 2015). The Ca^{2+}incursionequipped from the opening of CatSper channels is a vital achievement in the process of fertilization and starts hyperactivated motility (Suarez, 2008; Carlson et al., 2005, 2003; Chung et al., 2014; Jin et al., 2007).

Hv1 articulates to a significant segment in the CatSper actuation course, yet it likewise initiates membrane hyperpolarization by trading positive charges out of the cell. Hv1 is a voltage-gated channel and relies upon membrane depolarization to be enacted (Ramsey et al., 2006; Sasaki et al., 2006). In this way, both CatSper and Hv1 must depend on yet unidentified depolarizing ion channels. Sperm hyperactivation, in any case, happens in the fallopian tubes, where P4 fixations are not adequate to haltKSper (Demott and Suarez, 1992). Thus, the present model is feeling the loss of a fourth part – the 'Depolarizing Channel of Sperm' (DSper) (Miller et al., 2015). Enactment of the theoretical DSper particle channel would initiate enduring membrane depolarization and give the fundamental positive net charge inundation for CatSper/Hv1 action.

Utilizing entire cell voltage-clamp estimations, Mundt et al. (2018) recorded a novel non-CatSper conductance in both capacitated and non-capacitated spermatozoa. This unidentified, non-selectivecation conductance displayed outward correction and articulated temperature affectability in a range that coordinates the temperature range of TRPV4 action. Also, the pharmacological profile of DSper looks to some extent like TRPV4. In view of the electrophysiological, biochemical and resistant cytochemical information, the investigators reasoned the molecular character of DSper is TRPV4. The information proposes that TRPV4 action is expanded upon capacitation. Since capacitation envelops changes in the phosphorylation condition of numerous proteins (Visconti et al., 2011), and TRPV4 requires tyrosine phosphorylation to work appropriately (Wegierski et al., 2009), all things considered, TRPV4 phosphorylation is required.TRPV4 –additionally and unequivocally its hyperfunction, might trigger the aversive impact of expanded scrotal temperatures on sperm formation and epididymal safeguarding. As proposed by Bedford et al. (1991), extended scrotal temperatures byclothes and outfits contributes considerably to the second rate nature of human discharge. On the other hand, TRPV4 may signify an alluring objective for male fertility control, since TRPV4 is probably known to lie upstream in the signalling cascades prompting sperm hyperactivation and can be heterologously communicated for high-throughput utilitarian investigations.

CONCLUSION

The regular populations respond to unnatural changes in temperature by shifting edges, decreasing and wiping out due to the changing environment of the earth. The motility of sperm is a crucial factor in helping us to determine the nature and potential of semen to fertilise. Heat waves produce exceptional warm temperatures, heatwaves are possibly going to be extremely troublesome for biological ability and work with sometimes brief and stochastic onsets. The high temperature and soggy deviation is seen as a significant component that causes disparity in the production of semen and semen quality. In assessing the maturity capability of spermatozoa, dynamic sperm motility is important and is essential for human fertilisation. Temperatures below 4°C and above 37°C have been found to have a detrimental effect on sperm ability, such as motility. To a large degree, the effect of high temperatures on the pace of fish sperm is explicit in animals, with either unfavourable, zero or optimal effects of temperature on the motility of sperm. The motility period, fertilising ability and speed of spermatozoa depend upon temperature of the enactment medium. The association of sperm motility restrictions during the enactment of sperm at different temperatures indicates that both the proportion of motile spermatozoa and the total degree of accelerative measure reduce as the temperature of the swimming medium and the underlying beat recurrence upsurge.

Fowl sperm, since the male's testicles and excretory tubes are typically found within the peritoneal cavity, is impervious to body heat. The in vitro motility of fowl sperm depends on the temperature and suspension of the medium. Incredibly low temperatures, such as merely above zero, immobilise fowl sperm as well. The testicular temperature of mammals' vagaries somewhere in the range of 2°C and 8°C under the central bodily heat level. In males, scrotal heat stress (SHS) may occur and reduce the production of sperm. TRPV1 plays an essential role when sensing high temperatures. The Catsper channel, which is important for sperm flagellum hyperactivity, is a sperm-specific, pH-dependent, Ca^{2+}-permeable, and low-voltage-dependent channel. TRPV4 may reflect an appealing target for the regulation of male fertility, because TRPV4 is possibly understood to lie upstream in the signalling cascades triggering hyperactivation of sperm and can be transmitted heterologously for high-throughput utilitarian studies. Overall, the connexion between temperature and sperm motility is species-dependent. This chapter demonstrates an interplay between direct and indirect thermal challenges affecting the quality of sperm in different animal species during the monotonous and reproductive seasons.

AUTHOR CONTRIBUTION

Conceived, designed and wrote the book chapter along with literature survey.

FUND DISCLAIMER

No grant received from any funding agency in public, commercial, or not-for-profit sector.

CORRESPONDING AUTHOR

Junaid Ahmad Malik

REFERENCES

Aitken, R. J. (1999). The Amoroso Lecture. The human spermatozoon-a cell in crisis? *Journal of Reproduction and Fertility*, *115*(1), 1–7. doi:10.1530/jrf.0.1150001 PMID:10341716

Aitken, R. J., Clarkson, J. S., & Fishel, S. (1989). Generation of reactiveoxygen species, lipid peroxidation, and human spermfunction. *Biology of Reproduction*, *41*(1), 183–197. doi:10.1095/biolreprod41.1.183 PMID:2553141

Aitken, R. J., Gordon, E., Harkiss, D., Twigg, J. P., Milne, P., Jennings, Z., & Irvine, D. S. (1998). Relative impact of oxidative stress on the functional competence and genomic integrity of human spermatozoa. *Biology of Reproduction*, *59*(5), 1037–1046. doi:10.1095/biolreprod59.5.1037 PMID:9780307

Alavi, S. M. H., & Cosson, J. (2005). Sperm motility in fishes. I. Effects of temperature and pH: A review. *Cell Biology International*, *29*(2), 101–110. doi:10.1016/j.cellbi.2004.11.021 PMID:15774306

Alavi, S. M. H., Cosson, J., Karami, M., & Abdoulhay, H. (2004). Chemical composition and osmolality of seminal plasma ofAcipenserpersicus, their physiological relationship with spermmotility. *Aquaculture Research*, *35*(13), 1238–1243. doi:10.1111/j.1365-2109.2004.01132.x

Alves, M. B., Andrade, A. F., Arruda, R. P., Batissaco, L., Florez-Rodriguez, S. A., Oliveira, B. M., Torres, M. A., Lanconi, R., Ravagnani, G. M., Prado Filho, R. R., Vellone, V. S., Losano, J. D., Franci, C. R., Nichi, M., & Celeghini, E. C. (2016). Recovery of normal testicular temperature after scrotal heat stress in ramsassessed by infrared thermography and its effects on seminal characteristics and testosterone bloodserum concentration. *Theriogenology*, *86*(3), 795–805. doi:10.1016/j.theriogenology.2016.02.034 PMID:27045627

Amelar, R. D., Dubin, L., & Schoenfeld, C. (1980). Sperm motility. *Fertility and Sterility*, *134*, 197–215. PMID:6250914

Appell, R. A., Evans, P. R., & Blandy, J. P. (1977). The effect oftemperature on the motility and viability of sperm. *British Journal of Urology*, *49*(7), 751–756. doi:10.1111/j.1464-410X.1977.tb04566.x PMID:23196

Banks, S., King, S. A., Irvine, D. S., & Saunders, P. T. (2005). Impact of a mild scrotal heat stress on DNAintegrity in murine spermatozoa. *Reproduction (Cambridge, England)*, *129*(4), 505–514. doi:10.1530/rep.1.00531 PMID:15798026

Bedford, J. M. (1991). Effects of elevated temperature on the epididymis and testis: Experimental studies. *Advances in Experimental Medicine and Biology*, *286*, 19–32. doi:10.1007/978-1-4684-5913-5_3 PMID:2042501

Belova, N. V. (1981). Ecological and physiological characteristics of sperm ofpond Cyprinid fishes. [In Russian]. *VoprIchtiol*, *21*, 525–536.

Billard, R. (1978). Changes in structure and fertilizing ability of marine andfreshwater fish spermatozoa diluted in media of various salinities. *Aquaculture (Amsterdam, Netherlands)*, *14*(3), 187–198. doi:10.1016/0044-8486(78)90094-7

Billard, R. (1986). Spermatogenesis and spermatology of some teleost fishspecies. *Reproduction, Nutrition, Developpement*, *2*(4), 877–920. doi:10.1051/rnd:19860601

Billard, R. (2000). Biology and control of reproduction of sturgeon in fish farm. *Iranian Journal of Fisheries Science*, *2*, 1–20.

Billard, R., Cosson, J., Crim, L. W., & Suquet, M. (1995a). Spermphysiology and quality. In Brood stockmanagement andegg and larval quality (pp. 25-52). Blackwell Science.

Billard, R., Cosson, J., Perchec, G., & Linhart, O. (1995b). Biology of sperm and artificial reproduction in carp. *Aquaculture (Amsterdam, Netherlands)*, *124*(1-4), 95–112. doi:10.1016/0044-8486(94)00231-C

Billard, R., & Cosson, M. P. (1988). Sperm motility in Rainbow trout, Parasalmogairdneri; effects of pH and temperature. In Reproduction in fish basic and applied aspects in endocrinology and genetics (pp. 161-167). Paris, France: Academic Press.

Billard, R., & Cosson, M. P. (1989). Measurement of sperm motility in trout and carp. In N. De Pauw, E. Jaspers, H. Ackefors, & N. Wilkins (Eds.), *Aquaculture, a biotechnology in progress* (pp. 499–503). Bredene.

Billard, R., & Cosson, M. P. (1992). Some problems related to the assessment ofsperm motility in freshwater fish. *The Journal of Experimental Zoology*, *261*(2), 122–131. doi:10.1002/jez.1402610203

Billard, R., Cosson, M. P., & Christen, R. (1987). Some recent data on the biology on trout spermatozoa. In *Proceedings of the 3rd international symposium on reproductive physiology of fish* (pp. 187-190). St John's, Newfoundland, Canada: Academic Press.

Bridle, J. R., & Vines, T. H. (2007). Limits to evolution at range margins: When andwhy does adaptation fail? *Trends in Ecology & Evolution*, *22*(3), 140–147. doi:10.1016/j.tree.2006.11.002 PMID:17113679

Bujan, L., Daudin, M., Charlet, J. P., Thonneau, P., & Mieusset, R. (2000). Increase in scrotal temperaturein car drivers. *Human Reproduction (Oxford, England)*, *15*(6), 1355–1357. doi:10.1093/humrep/15.6.1355 PMID:10831568

Burfening, P. J., Elliott, D. S., Eisen, E., & Ulberg, L. C. (1970). Survival of embryosresulting from spermatozoa produced by mice exposed to elevated ambienttemperature. *Journal of Animal Science*, *30*(4), 578–582. doi:10.2527/jas1970.304578x

Cai, H., Ren, Y., Li, X. X., Yang, J. L., Zhang, C. P., Chen, M., Fan, C. H., Hu, X. Q., Hu, Z. Y., Gao, F., & Liu, Y. X. (2011). Scrotal heat stress causes a transient alteration in tight junctions and induction of TGF-betaexpression. *International Journal of Andrology*, *34*(4pt1), 352–362. doi:10.1111/j.1365-2605.2010.01089.x PMID:20633196

Cameron, R. D., & Blackshaw, A. W. (1980). The effect of elevatedambient temperature on spermatogenesis in the boar. *Journal of Reproduction and Fertility*, *59*(1), 173–179. doi:10.1530/jrf.0.0590173 PMID:7401033

Carlson, A. E., Quill, T. A., Westenbroek, R. E., Schuh, S. M., Hille, B., & Babcock, D. F. (2005). Identical phenotypes of CatSper1and CatSper2 null sperm. *The Journal of Biological Chemistry*, *280*(37), 32238–32244. doi:10.1074/jbc.M501430200 PMID:16036917

Chen, I., Hill, J. K., Ohlemüller, R., Roy, D. B., & Thomas, C. D. (2011). Rapid rangeshifts of species associated with high levels of climate warming. *Science, 333*(6045), 1024–1026. doi:10.1126cience.1206432 PMID:21852500

Christidis, N., Jones, G. S., & Stott, P. A. (2015). Dramatically increasing chance ofextremely hot summers since the 2003 European heatwave. *Nature Climate Change, 5*(1), 3–7. doi:10.1038/nclimate2468

Chung, J. J., Shim, S. H., Everley, R. A., Gygi, S. P., Zhuang, X., & Clapham, D. E. (2014). Structurally distinct Ca(2+) signalling domains of sperm flagella orchestrate tyrosine phosphorylation and motility. *Cell, 157*(4), 808–822. doi:10.1016/j.cell.2014.02.056 PMID:24813608

Chung, M. K., Lee, H., Mizuno, A., Suzuki, M., & Caterina, M. J. (2004). TRPV3 and TRPV4 mediate warmth-evoked currents in primary mouse keratinocytes. *The Journal of Biological Chemistry, 279*(20), 21569–21575. doi:10.1074/jbc.M401872200 PMID:15004014

Cohen, J., Fehilly, C. B., & Waiters, D. E. (1985). Prolonged storage ofhuman spermatozoa at room temperature or in arefrigerator. *Fertility and Sterility, 44*(2), 254–262. doi:10.1016/S0015-0282(16)48747-9 PMID:4018281

Colenbrander, B., Feitsma, H., & Grooten, H. J. (1993). Optimizingsemen production for artificial insemination in swine. *Journal of Reproduction and Fertility. Supplement, 48*, 207–215. PMID:8145205

Cosson, J. (2004). The ionic and osmotic factors controlling motility of fishspermatozoa. *Aquaculture International, 12*(1), 69–85. doi:10.1023/B:AQUI.0000017189.44263.bc

Cosson, J., Billard, R., Cibert, C., Dreanno, C., & Suquet, M. (1999). Ionic factorsregulating the motility of fish sperm. In C. Gagnon (Ed.), *The male gamete: From basic to clinical applications* (pp. 161–186). Cache Rive Press.

Cosson, J., Billard, R., Gibert, C., Dreanno, C., Linhart, O., & Suquet, M. (1997). Movements of fish sperm flagella studied by high speed videomicroscopycoupled to computer assisted image analysis. *Polskie Archiwum Hydrobiologii, 10*, 518–527.

Crew, F. A. E. (1927). XX.—On Fertility in the Domestic Fowl. *Proceedings of the Royal Society of Edinburgh, 46*, 230–238. doi:10.1017/S0370164600022021

Curtis, V., & Lambert, W. V. (1929). A study of fertility in poultry. *Poultry Science, 8*(3), 142–150. doi:10.3382/ps.0080142

Davis, J. B., Gray, J., Gunthorpe, M. J., Hatcher, J. P., Davey, P. T., Overend, P., Harries, M. H., Latcham, J., Clapham, C., Atkinson, K., Hughes, S. A., Rance, K., Grau, E., Harper, A. J., Pugh, P. L., Rogers, D. C., Bingham, S., Randall, A., & Sheardown, S. A. (2000). Vanilloid receptor-1 is essential for inflammatory thermal hyperalgesia. *Nature, 405*(6783), 183–187. doi:10.1038/35012076 PMID:10821274

Demott, R. P., & Suarez, S. S. (1992). Hyperactivated sperm progress in the mouse oviduct. *Biology of Reproduction, 46*(5), 779–785. doi:10.1095/biolreprod46.5.779 PMID:1591334

Dougherty, K. A., Emilson, L. B., Cockett, A. T., & Urry, R. L. (1975). Acomparison of subjective measurements of humansperm motility and viability with two live-dead stainingtechniques. *Fertility and Sterility, 26*, 700–703. doi:10.1016/S0015-0282(16)41238-0 PMID:50237

Drabkina, B. M. (1961). Effect of different water salinities on the survival rate ofspermatozoa, eggs and larvae of sturgeon. *Doklady Akademii Nauk SSSR*, *138*, 492–495.

Dreanno, C., Suquet, M., Desbruyeres, E., Cosson, J., Le Delliou, H., & Billard, R. (1998). Effect of urine on semen quality in Turbot, *Psettamaxima. Aquaculture (Amsterdam, Netherlands)*, *169*(3-4), 247–262. doi:10.1016/S0044-8486(98)00262-2

Durairajanayagam, D., Agarwal, A., & Ong, C. (2015). Causes, effects and molecular mechanisms oft-esticular heat stress. *Reproductive Biomedicine Online*, *30*(1), 14–27. doi:10.1016/j.rbmo.2014.09.018 PMID:25456164

Elster, J., & Mann, H. (1952). Weitereuntersuchungenuber die physiologiederbefruchtung und die zu-ordnug der gametenbeifischen. *Arch HydrobiolSuppl*, *20*, 267–276.

Fenkes, M., Fitzpatrick, J. L., Ozolina, K., Shiels, H. A., & Nudds, R. L. (2017). Sperm in hot water: Direct and indirect thermal challenges interact to impact on brown trout sperm quality. *The Journal of Experimental Biology*, *220*(14), 2513–2520. doi:10.1242/jeb.156018 PMID:28455440

Figa-Talamanca, I., Cini, C., Varricchio, G. C., Dondero, F., Gandini, L., Lenzi, A., Lombardo, F., Angelucci, L., Di Grezia, R., & Patacchioli, F. R. (1996). Effects of prolonged autovehicle driving on malereproduction function: A study among taxi drivers. *American Journal of Industrial Medicine*, *30*, 750–758. doi:10.1002/(SICI)1097-0274(199612)30:6<750::AID-AJIM12>3.0.CO;2-1 PMID:8914722

Freund, M., & Peterson, R. N. (1976). Semen evaluation and fertility. In E. S. Hafez (Ed.), Human Semen and FertilityRegulation in Men (p. 34454). Academic Press.

Gallis, J. L., Fedrigo, E., Jatteau, P., Bonpunt, E., & Billard, R. (1991). Siberiansturgeon spermatozoa: Effects of dilution, pH, osmotic pressure,sodium and potassium ions on motility. In Acipenser (pp. 143-51). Bordeaux, France: Academic Press.

Ginzburg, A. S. (1968). Fertilization of fishes and the problem of polyspermy. New York: Academy of Science USSR.

Ho, H. C., Granish, K. A., & Suarez, S. S. (2002). Hyperactivated motility of bull sperm is triggered at the axoneme by $Ca^{2+}$and not cAMP. *Developmental Biology*, *250*(1), 208–217. doi:10.1006/dbio.2002.0797 PMID:12297107

Hurley, L. L., McDiarmid, C. S., Friesen, C. R., Griffith, S. C., & Rowe, M. (2018). Experimental heat-waves negatively impact sperm quality in the zebra finch. *Proceedings. Biological Sciences*, *285*(1871), 2017–2547. doi:10.1098/rspb.2017.2547 PMID:29343605

Ingermann, R. L., Holcomb, M., Robinson, M. L., & Cloud, J. G. (2002). Carbondioxide and pH affect sperm motility of white sturgeon (*Acipensertransmontanus*). *The Journal of Experimental Biology*, *205*, 2885–2890. PMID:12177152

Isaak, D. J., Wollrab, S., Horan, D., & Chandler, G. (2011). Climate changeeffects on stream and river temperatures across the northwest U.S. from1980–2009 and implications for salmonid fishes. *Climatic Change*, *113*(2), 499–524. doi:10.100710584-011-0326-z

Ishijima, S., Mohri, H., Overstreet, J. W., & Yudin, A. I. (2006). Hyperactivation of monkey spermatozoa is triggered by Ca^{2+} and completed by cAMP. *Molecular Reproduction and Development*, *73*(9), 1129–1139. doi:10.1002/mrd.20420 PMID:16804884

Jezierska, B., & Witeska, M. (1999). The effect of time and temperature on motilityof spermatozoa of common and grass carp. *Electronic Journal of the Polish Agricultural University*, *2*, 1–8.

Jin, J., Jin, N., Zheng, H., Ro, S., Tafolla, D., Sanders, K. M., & Yan, W. (2007). Catsper3 and Catsper4 are essential for spermhyperactivated motility and male fertility in the mouse. *Biology of Reproduction*, *77*(1), 37–44. doi:10.1095/biolreprod.107.060186 PMID:17344468

Jockenhovel, F., Grawe, A., & Nieschlag, E. (1990). A portable digital data recorder for long-termmonitoring of scrotal temperatures. *Fertility and Sterility*, *54*, 694–700. doi:10.1016/S0015-0282(16)53832-1 PMID:2209892

Kanter, M., Aktas, C., & Erboga, M. (2013). Heat stress decreases testicular germ cell proliferation andincreases apoptosis in short term: An immunohistochemical and ultrastructural study. *Toxicology and Industrial Health*, *29*(2), 99–113. doi:10.1177/0748233711425082 PMID:22082826

Lahnsteiner, F., Berger, B., Weismann, T., & Patzner, R. A. (1996). Motility ofspermatozoa of *Alburnusalburnus* (Cyprinidae) and its relationshipto seminal plasma composition and sperm metabolism. *Fish Physiology and Biochemistry*, *15*(2), 167–179. doi:10.1007/BF01875596 PMID:24194090

Lahnsteiner, F., & Mansour, N. (2012). The effect of temperature on spermmotility and enzymatic activity in brown trout *Salmotrutta, burbotLotalota* andgrayling *Thymallusthymallus*. *Journal of Fish Biology*, *81*(1), 197–209. doi:10.1111/j.1095-8649.2012.03323.x PMID:22747813

Larsson, K., & Einarsson, S. (1984). Seminal changes in boarsafter heat stress. *Acta Veterinaria Scandinavica*, *25*, 57–66. PMID:6464925

Linhart, O., Cosson, J., Mims, S. D., Shelton, W. L., & Rodina, M. (2002). Effects ofions on the motility of fresh and demembranated paddlefish(*Polyodonspathula*) spermatozoa. *Reproduction (Cambridge, England)*, *124*, 713–719. doi:10.1530/rep.0.1240713 PMID:12417010

Linhart, O., Mims, S. D., & Shelton, W. L. (1995). Motility of spermatozoa fromShovelnose sturgeon, *Scaphirhynchusplatorynchus*, and Paddlefish,*Polyodonspathula*. *Journal of Fish Biology*, *47*, 902–909. doi:10.1111/j.1095-8649.1995.tb06011.x

Linhart, O., Slechta, V., & Slavik, T. (1991). Fish sperm composition andbiochemistry. *Bulletin of the Institute of Zoology, Academia Sinica*, *16*, 285–311.

Lishko, P. V., Kirichok, Y., Ren, D., Navarro, B., Chung, J. J., & Clapham, D. E. (2012). The control of male fertility byspermatozoan ion channels. *Annual Review of Physiology*, *74*(1), 453–475. doi:10.1146/annurev-physiol-020911-153258 PMID:22017176

Lucio, A. C., Alves, B. G., Alves, K. A., Martins, M. C., Braga, L. S., Miglio, L., Alves, B. G., Silva, T. H., Jacomini, J. O., & Beletti, M. E. (2016). Selected sperm traits are simultaneously altered after scrotal heatstress and play specific roles in in vitro fertilization and embryonic development. *Theriogenology*, *86*(4), 924–933. doi:10.1016/j.theriogenology.2016.03.015 PMID:27087533

Meehl, G. A., & Tebaldi, C. (2004). More intense, more frequent, and longer lastingheat waves in the 21st century. *Science*, *305*(5686), 994–997. doi:10.1126cience.1098704 PMID:15310900

Mehlis, M., & Bakker, T. C. M. (2014). The influence of ambient water temperature on sperm performance and fertilization success in three-spined sticklebacks(*Gasterosteusaculeatus*). *Evolutionary Ecology*, *28*(4), 655–667. doi:10.100710682-014-9707-x

Mieusset, R., Bujan, L., Mondinat, C., Monsat, A., Pontonnier, F., & Grandjean, H. (1987). Association of scrotal hyperthermia with impaired spermatogenesis ininfertile men. *Fertility and Sterility*, *48*(6), 1006–1011. doi:10.1016/S0015-0282(16)59600-9 PMID:3678498

Miller, M. R., Mannowetz, N., Iavarone, A. T., Safavi, R., Gracheva, E. O., Smith, J. F., Hill, R. Z., Bautista, D. M., Kirichok, Y., & Lishko, P. V. (2016). Unconventional endocannabinoidsignalling governs sperm activation via the sex hormoneprogesterone. *Science*, *352*(6285), 555–559. doi:10.1126cience.aad6887 PMID:26989199

Miller, M. R., Mansell, S. A., Meyers, S. A., & Lishko, P. V. (2015). Flagellar ion channels of sperm: Similarities and differencesbetween species. *Cell Calcium*, *58*(1), 105–113. doi:10.1016/j.ceca.2014.10.009 PMID:25465894

Mims, S. D. (1991). Evaluation of activator solutions, motility duration andshort-term storage of paddlefish spermatozoa. *Journal of the World Aquaculture Society*, *22*(4), 224–229. doi:10.1111/j.1749-7345.1991.tb00738.x

Muehleis, P. M., & Long, S. Y. (1976). The effects of altering the pHof seminal fluid on the sex ratio of rabbit offspring. *Fertility and Sterility*, *27*(12), 1438–1445. doi:10.1016/S0015-0282(16)42261-2 PMID:12027

Mundt, N., Spehr, M., & Lishko, P. V. (2018). TRPV4 is the temperature-sensitive ion channel of human sperm. *eLife*, *7*, e35853. doi:10.7554/eLife.35853 PMID:29963982

Munro, S. S. (1938). Fowl sperm immobilization by a temperature-media interaction and its biological significance. *Quarterly Journal of Experimental Physiology: Translation and Integration*, *27*(3), 281–291. doi:10.1113/expphysiol.1938.sp000753

Nagy, A., Gertsenstein, M., Vintersten, K., & Behringer, R. (2006). *A Laboratory Manual Manipulating the Mouse Embryo* (3rd ed.). Chemical Industry Press.

Nicolaides, C. (1934). Fertility studies in poultry. *Poultry Science*, *13*(3), 178–183. doi:10.3382/ps.0130178

Okabe, M. (2013). The cell biology of mammalian fertilization. *Development*, *140*(22), 4471–4479. doi:10.1242/dev.090613 PMID:24194470

Parmesan, C., & Yohe, G. (2003). A globally coherent fingerprint of climate changeimpacts across natural systems. *Nature*, *421*(6918), 37–42. doi:10.1038/nature01286 PMID:12511946

Perez-Cerezales, S., Ramos-Ibeas, P., Lopez-Cardona, A., Pericuesta, E., Fernandez-Gonzalez, R., Pintado, B., & Gutierrez-Adan, A. (2017). Elimination of methylation marks at lysines 4 and 9 of histone 3(H3K4 and H3K9) of spermatozoa alters offspring phenotype. *Reproduction, Fertility, and Development*, *29*(4), 740–746. doi:10.1071/RD15349 PMID:26678259

Perez-Crespo, M., Pintado, B., & Gutierrez-Adan, A. (2008). Scrotal heat stress effects on sperm viability,sperm DNA integrity, and the offspring sex ratio in mice. *Molecular Reproduction and Development*, *75*(1), 40–47. doi:10.1002/mrd.20759 PMID:17474098

Raftery, A. E., Zimmer, A., Frierson, D. M. W., Startz, R., & Liu, P. (2017). Less than 2°C warming by 2100 unlikely. *Nature Climate Change*, *7*(9), 637–645. doi:10.1038/nclimate3352 PMID:30079118

Ramsey, I. S., Moran, M. M., Chong, J. A., & Clapham, D. E. (2006). A voltage-gated proton-selective channel lacking thepore domain. *Nature*, *440*(7088), 1213–1216. doi:10.1038/nature04700 PMID:16554753

Rao, M., Zhao, X. L., Yang, J., Hu, S. F., Lei, H., Xia, W., & Zhu, C. H. (2015). Effect of transient scrotalhyperthermia on sperm parameters, seminal plasma biochemical markers, and oxidative stress inmen. *Asian Journal of Andrology*, *17*(4), 668–675. doi:10.4103/1008-682X.146967 PMID:25652627

Sales, K., Vasudeva, R., Dickinson, M. E., Godwin, J. L., Lumley, A. J., Michalczyk, Ł., Hebberecht, L., Thomas, P., Franco, A., & Gage, M. J. (2018). Experimental heatwaves compromise sperm function and cause transgenerational damage in a model insect. *Nature Communications*, *9*(1), 1–11. doi:10.103841467-018-07273-z PMID:30425248

Sasaki, M., Takagi, M., & Okamura, Y. (2006). A voltage sensor-domain protein is a voltage-gated proton channel. *Science*, *312*(5773), 589–592. doi:10.1126cience.1122352 PMID:16556803

Schlenk, W., & Kahmann, H. (1938). The chemical composition of seminal fluidsand their physiological importance study with trout sperm. *Biochemical Zoology*, *295*, 283–301.

Schuetz, A. W., & Dubin, N. H. (1981). Progesterone and prostaglandin secretion by ovulated rat cumulus cell-oocyte complexes. *Endocrinology*, *108*(2), 457–463. doi:10.1210/endo-108-2-457 PMID:7449734

Setchell, B. (2006). The effects of heat on the testes of mammals. *Animal Reproduction*, *3*, 81–91.

Seuanez, H. N. (1980). Chromosomes and spermatozoa of the African great apes. *Journal of Reproduction and Fertility. Supplement*, *28*, 91–104. PMID:6934315

Smith, J. F., Syritsyna, O., Fellous, M., Serres, C., Mannowetz, N., Kirichok, Y., & Lishko, P. V. (2013). Disruption of theprincipal, progesterone-activated sperm Ca^{2+} channel in a CatSper2-deficient infertile patient. *Proceedings of the National Academy of Sciences of the United States of America*, *110*(17), 6823–6828. doi:10.1073/pnas.1216588110 PMID:23530196

Stone, B. A. (1982). Heat induced infertility of boars: The interrelationshipbetween depressed sperm output and fertility andan estimation of the critical air temperature above which spermoutput is impaired. *Animal Reproduction Science*, *4*(4), 283–299. doi:10.1016/0378-4320(82)90043-4

Stoss, J. (1983). Fish gamete preservation and spermatozoan physiology. In Fish physiology (pp. 305-50). Academic Press. doi:10.1016/S1546-5098(08)60306-4

Stumpf, P. G., Garry, F., & Lloyd, T. (1984). Effects of temperature on*in vitro* penetration of human sperm into bovine cervicalmucus. *The Journal of Reproductive Medicine*, *29*, 833–836. PMID:6520822

Suarez, S. S. (2008). Control of hyperactivation in sperm. *Human Reproduction Update*, *14*(6), 647–657. doi:10.1093/humupd/dmn029 PMID:18653675

Suriyasomboon, A., Kunavongkrit, A., Lundeheim, N., & Einarsson, S. (2004). Effect of temperature and humidity onsperm production in Duroc boars under different housing systemsin Thailand. *Livestock Production Science*, *89*(1), 19–31. doi:10.1016/j.livprodsci.2003.12.008

Thomas, C. D., Cameron, A., Green, R. E., Bakkenes, M., Beaumont, L. J., Collingham, Y. C., & Hughes, L. (2004). Extinction risk from climate change. *Nature*, *427*(6970), 145–148. doi:10.1038/nature02121 PMID:14712274

Thonneau, P., Ducot, B., Bujan, L., Mieusset, R., & Spira, A. (1997). Effect of maleoccupational heat exposure on time to pregnancy. *International Journal of Andrology*, *20*(5), 774–778. doi:10.1046/j.1365-2605.1997.d01-303.x PMID:16130271

van der Horst, G., & Maree, L. (2014). Sperm form and function in the absence of sperm competition. *Molecular Reproduction and Development*, *81*(3), 204–216. doi:10.1002/mrd.22277 PMID:24273026

vanVliet, M. T. H., Ludwig, F., Zwolsman, J. J. G., Weedon, G. P., & Kabat, P. (2011). Global river temperatures and sensitivity to atmospheric warming andchanges in river flow. *Water Resources Research*, *47*, W02544.

Visconti, P. E., Krapf, D., De La Vega-beltrán, J. L., Acevedo, J. J., & Darszon, A. (2011). Ion channels, phosphorylation and mammalian sperm capacitation. *Asian Journal of Andrology*, *13*(3), 395–405. doi:10.1038/aja.2010.69 PMID:21540868

Vladic, T., & Jarvi, T. (1997). Sperm motility and fertilization time span in Atlanticsalmon and brown trout - The effect of water temperature. *Journal of Fish Biology*, *50*, 1088–1093.

Wan, X., He, X., Liu, Q., Duan, P., & Li, H. (2020). Frequent and mild scrotal heat stress impairs embryo development, implantation, and offspring sex ratio in mice. *Reproductive Biomedicine Online*, *40*(5), 617–626. doi:10.1016/j.rbmo.2020.01.020 PMID:32276888

Watanabe, H., Vriens, J., Suh, S. H., Benham, C. D., Droogmans, G., & Nilius, B. (2002). Heat-evoked activation of TRPV4 channels in a HEK293 cell expression system and in native mouse aorta endothelial cells. *The Journal of Biological Chemistry*, *277*(49), 47044–47051. doi:10.1074/jbc.M208277200 PMID:12354759

Wegierski, T., Lewandrowski, U., Muller, B., Sickmann, A., & Walz, G. (2009). Tyrosine phosphorylation modulates theactivity of TRPV4 in response to defined stimuli. *The Journal of Biological Chemistry*, *284*(5), 2923–2933. doi:10.1074/jbc.M805357200 PMID:19033444

Wettemann, R. P., Wells, M. E., Omtvedt, I. T., Pope, C. E., & Turman, E. J. (1976). Influence of elevated ambient temperatureon reproductive performance of boars. *Journal of Animal Science*, *42*(3), 664–669. doi:10.2527/jas1976.423664x PMID:1262278

Williot, P., Kopeika, E. F., & Goncharov, B. F. (2000). Influence of testis state,temperature and delay in semen collection on spermatozoa motilityin the cultured Siberian sturgeon (*Acipenserbaeri* Brandt). *Aquaculture (Amsterdam, Netherlands)*, *189*(1-2), 53–61. doi:10.1016/S0044-8486(00)00358-6

World Health Organisation. (1999). *WHO laboratory manual for the examination of human semen and sperm-cervical mucus interaction*. Cambridge University Press.

Chapter 10
Effects on Fertility and Reproductive Behavior From Environmental Contaminants in Extreme Environments

Irfan Ashraf Badroo
Government Degree College, Kupwara, India

Ashiq Hussain Khanday
Sant Gadge Baba Amravati University, Maharashtra, India

Suriya Ashraf Badroo
Muslim Education Trust, Kashmir, India

Saba Khursheed Khan
C. T. University, Ludhiana, India

Irshad Aziz Malik
Vidya Bharti Mahavidyalaya, India

Hemlata Pradeep Nandurkar
Sant Gadge Baba Amravati University, Maharashtra, India

Syed A. Untoo
College of Science and Commerce, India

ABSTRACT

Human infertility is increasing at an alarming rate and has affected nearly 12% of reproductive aged couples worldwide. Various reasons have been linked to its cause, besides the age of female partner, a huge number of environmental contaminants produced and used by humans throughout the world are one of the reasons for growing the infertility rate. Their exposure is virtually unavoidable. The literature review provides an overview regarding the effect of chemical agents on fertility. The environmental contaminants included in this review are plastics, pesticides, phthalates, metals, industrial chemicals, endocrine disrupters, and nano materials. The goal is to discuss the association between environmental contaminants and reproductive disorders with particular emphasis on infertility.

DOI: 10.4018/978-1-7998-4480-8.ch010

INTRODUCTION

Fertility is the ability to establish a clinical pregnancy (Zegers-Hochschild et al., 2017). The most important health problems are reproductive disorders. World Health Organization (WHO) has recognized infertility as a worldwide public health problem. Infertility is a major problem to the residents of developing countries and worldwide 186 million people have suffered with this disorder (Inhorn and Patrizio, 2015). Infertility is a state of not establishing clinical pregnancy after unprotected sexual intercourse for almost regular 12 months (WHO-ICMART glossary). The term infertility has also been summed up as a disability (Zegers-Hochschild, et al., 2017) in which couples are unsuccessful to conceive (Jenkins, et al., 2004). While infertility has also been used interchangeably with subfertility (Zegers-Hochschild, et al., 2017). Further infertility is a time bound while sterility is a permanent state of infertility (Zegers-Hochschild, et al., 2017). Infertility has affected about 12% of the reproductive couple's worldwide (Ombelet et al, 2008) and almost 20% couples undergo evaluation of infertility. The negative factor responsible for infertility is increasing women's age (Hart, 2016), but there are various other factors that are said to be responsible and play an escalating role in disorder infertility.

Humans are exposed to various environmental pollutants. The route of exposure occurs through multiple routes, it may be via drinking water, food and from work place and raise concern for human health, particularly reproduction (Martínez-Sales et al., 2015). Epidemiologic evidence has indicated that fertility and pregnancy have been profoundly affected by industrial and environmental chemicals (Russ et al., 2015). Chemicals are available everywhere on the globe, hence their exposure is certain and adverse health effects on reproductive system both on females and males have been documented. Even in various organisms environmental chemical are suspected to be responsible for health effects particularly related to reproductive system (Aarab et al., 2004). Hence, low fertility is the main cause of decline in population in few countries. A survey by National Health and Nutrition Examination Survey data from 2003 to 2004 documented at least 43 different chemicals that had affected pregnant women in United States (Woodruff et al., 2011). Further robust scientific evidence also points out the exposure of environmental agents and their effect on reproductive health (Diamanti-Kandarakis et al., 2009; Boekelheide et al., 2012; ACOG, 2013). Various factors are thus responsible for infertility, but in the present chapter we focus mainly on plastics, pesticides, phthalates, metals, industrial chemicals, endocrine disrupters, and nanomaterials. The goal is to discuss the association between chemicals and reproductive disorders with particular emphasis on infertility.

PLASTICS

Plastic plays an imperative role in everyday life and everyone is involved with plastic made products in different ways. Plastic is being used indispensably in our lively hood materials such as health, agriculture, automotive, construction and textiles (Andrady, 2011; Andrady and Neal, 2009; Dris et al., 2015). Plastic reliability is increasing by every single day because of its variety characteristic feature. The potential for diverse applications, including widespread use of disposable items and the durability of plastic is anticipated, but the associated problems of waste management and plastic debris are not (Yarsley and Couzens, 1945). Studies have revealed that the long term use of plastic containers can increase health risks. Plastic containers or bottles are composed of different chemicals such as phthalates, Bisphenol A (BPA), brominated flame retardants, antiminitroxide, and poly-fluorinated chemicals etc., which are a serious risk factor for health (Halden, 2010). BPA is a xenoestrogen, and having the structural similarity with estrogen can therefore bind with the estrogen receptor site. Various evidences have indicated the adverse impact of BPA on human fertility and is also responsible for the reproductive pathologies, e.g., cryptorchidism, cancers, testicular dysgenesis syndrome, follicle loss in female and decreased fertility in male. BPA has been found to bind to estrogen receptors but still it is less potent than estradiol. BPA also binds to the androgen receptor (AR) and act as its antagonist when its level is high (Rochester, 2013; Wisniewski et al., 2015; Lucas et al., 2009; Zhou et al., 2017). European Chemicals Agency 2017, listed BPA as a substance of concern because of its endocrine disruptor properties.

Effect on Females

Many animal studies show that BPA can delay the onset of female puberty, and has influence on estrous cycle (Rochester, 2013) and maturation of the oocyte (Zhou et al., 2017; Santangeli et al., 2017). Fertile women have the lower levels of BPA as compared to the infertile women. Higher level of BPA in the urine of women who undergo in vitro procedure is negatively correlated with oocytes maturation and peak concentrations of estradiol (Mok-Lin et al., 2010). Most likely, the every-day exposure to BPA reduces the ovarian reserve (Richter et al., 2010).

Effect on Males

There are no controversies regarding the disruption of spermatogenesis by BPA (Zhou et al., 2017; Lassen, 2014). BPA acts as an antagonist of the androgen receptor (AR) at higher concentrations (Hejmej et al., 2011). 98% of the couples who need the infertility treatment were observed to have BPA in their urine samples and its level was negatively correlated with sperm count and motility (Meeker et al., 2011). Correlation of increased BPA urine levels with decreased free androgen index was statistically found significant (Mendiola et al., 2010). Evidences of higher urinary concentration of BPA in correlated with the increased percentage of immature sperm (Radwan et al., 2018). Cariati et al., (2019) in a review provide an overview of investigated data on BAP. It is concluded that BAP is hazardous to male fertility. While, Rehfeld et al., (2020) revealed that bisphenols can affect Ca^{2+} signaling in human sperm cells that disrupt human sperm function and can be a contributing factor in human infertility.

PHTHALATES

Phthalates are a group of chemicals used as plasticizer to impart flexibility to polyvinylchloride (PVC) therefore used in different products (ATSDR, 2002). The adverse effects of Phthalates are very certain because of its broad utilization and hence, it is difficult to avoid the exposure. Badroo et al., (2020) has reviewed the multiple health effects of phthalates including the reproductive issues in humans. Phthalates have been shown to affect the male reproductive system in animal studies. Phthalates cause antiandrogenic effects manifested as decreased anogenital distance (AGD), decreased sperm production, decreased testosterone levels, cryptorchidism, and infertility (Fisher et al., 2003; Foster, 2006; Jarfelt et al., 2005; McKee et al., 2004; Moore et al., 2001).

Effect on Females

The involvement of phthalates have been seen to interfere the normal functioning of endocrine system (Damstra et al., 2004; Latini et al., 2004; Sharpe and Irvine, 2004). Endocrine system is responsible for growth and sexual development of both males and females (Sharpe, 2001; Lovekamp-Swan and Davis, 2003). However, Phthalate has been reported to interfere with thyroid function of men and pregnant women (Meeker et al., 2007; Huang et al., 2007).

Phthalates serum level in young girls can cause premature development of breast (thelarche) (Colon et al., 2000) and Phthalate metabolite level is linked to endometrioses (Silva et al., 2005). Phthalate metabolites urinary concentration have been linked to Breast cancer (Lopez-Carillo et al., 2010). Diethyl phthalate may be associated with increased risk of breast cancer (López-Carrillo et al., 2010). Phthalate exposure is also linked to AGD in females (Phillip et al., 1996; Callegari et al., 1987). Female exposure to DEHP is responsible for shorter time to pregnancy (Specht et al., 2015), while as monoethyl phthalate (MEP) is associated with longer time to pregnancy (Thomsen et al., 2016).

DEHP and its metabolite MEHP were found to be toxic to ovary of mature rats (Davis et al., 1994a, b; Lovekamp et al., 2001). Butyl benzyl phthalate (BBzP), di isobutyl phthalate (DIBP), di-n-butyl phthalate (DBP), di ethyl phthalate (DEP) and di-isononyl phthalate (DINP) were reported to be responsible for weak estrogenic activity in vitro Harris et al., (1997), while DBP and DEHP were not shown to be estrogenic (Gray et al., 1984).

Effect on Males

Phthalate exposure can alter the semen characteristics, sperm DNA and reproductive hormones (Akingbemi *et al.*, 2001; Duty *et al.*, 2003) and levels of FSH, SHBG, LH and testosterone in the serum thus showing the connection between serum hormones with that of urinary level of phthalate (Duty *et al.*, 2005). Testosterone production increases with the low dose of phthalates (Foster *et al.*, 2001), whilst high dose inhabit the testosterone levels.

PESTICIDES

Over the past few decade agriculture sector has changed deeply in order to enhance the production and food quality. This was achieved with massive use of pesticides. Pesticides interfere with the vital function of the body and are proven to be toxic. Human exposure is certain; it may be directly or indirectly. Humans get exposed in the workplace, through their environment (Ye et al., 2017). Pesticide toxicity may be acute and chronic, hence can adversely affect the physiology. Several studies have evaluated that exposure of pesticides effects reproduction. It can be of female and male reproductive system.

Effect on Females

Females working in the cotton growing regions exposed to organochlorine pesticides have shown adverse effects in their reproductive system during puberty (Bapayeva et al., 2016). Further a study on 281 women by Fuortes et al. (1997) found that the cause of infertility among women was due to the pesticide exposure and it was in their workplace.

Effect on Males

Pesticide effects male reproductive system. In one of the study pesticide have shown to reduce the semen quality (Tuc et al., 2007; Hossainet al., 2010). It was also shown that pesticides affect the spermatogenesis process through modulating the hormones involved in it (Hossain et al., 2010). Further pesticides interfere spermatogenesis by damaging the testes. The effects may be by modifying the sperm count, motility etc (Bretveldet al., 2007). There are also the studies that pesticides lead the alteration in the morphology of sperms and hence decrease the count per ejaculation and viable sperms (Shojaei Saadi and Abdollahi, 2012a, b). Organophosphorus (OP) pesticides alter sperm morphology (Yucra et al., 2006). OP exposure also affects the seminiferous tubules and spermatogenesis process hence alter the sperm count and it morphology (Recio-Vega et al., 2008). Sperm concertation is the count of sperms per milliliter of semen. Seminiferous tubules have been found to be damaged by the OPs which later alter the sperm concentration (Perry et al., 2011). While a study by Yucra et al., (2006) suggested that pesticide exposure affects the seminal vesicle resulting in the decline of sperm concentration. Hence pesticides affects sperm quality, sperm count, sperm morphology and sperm concentration (Cremonese et al., 2017).

METALS

Metallic element having relatively high atomic weight or atomic number and a relatively high density are called as heavy metals (Hawkes, 1997). Out of total heavy metals known only a few are of concern to human health (Table 1) (Baird and Cann, 2012). Due to their highly toxic nature they can cause damage even at very low concentrations (Wirth and Mijal, 2010). Some heavy metals such as cadmium, lead, and mercury can be harmful to health, especially in high doses or over a long period of time (Llobet et al., 2003). Heavy metals are used in a large number of products that in our daily life use: such as glazed ceramics, nonstick pans, soaps, pesticides, plastic toys, eye shadow, lipstick, plastics, mobile phones, antiseptics, cars, utensils, batteries, paints, solar panels, and particle accelerators (Kozin and Hansen, 2013). It has been identified that human fertility is affecting by heavy metals. Conception chances of

Table 1. List of common toxic heavy metals sorted by atomic number, weight, and density.

Name	*Symbol*	*Atomic Number*	*Atomic Weight*	*Density (g/cm3)*
Chromium	Cr	24	51.9961	7.19
Iron	Fe	26	55.845	7.87
Cobalt	Co	27	58.9332	8.9
Nickel	Ni	28	58.6934	8.9
Copper	Cu	29	63.546	8.96
Zinc	Zn	30	65.39	7.13
Arsenic	As	33	74.9216	5.72
Cadmium	Cd	48	112.411	8.65
Mercury	Hg	80	200.59	13.55
Lead	Pb	82	207.2	11.35

infertile women can be enhanced by decreasing the heavy metal body load (Pant et al., 2003). Exposure of cadmium can delay a woman's chances of getting pregnant (Pant et al., 2003). Recent studies suggest that men exposed to heavy metals effect of heavy metals on male fertility. Evidences have shown that high or even moderate concentrations of lead can cause fertility problems in humans (McGregor and Mason, 1991).

Effect on Females

Environmental factors such as metals may interfere with reproduction in women (Mendola et al., 2008). Some metals including cadmium (Cd), arsenic (As), and lead (Pb) are recognized reproductive toxicants (Centers for Disease Control and Prevention, 2005). Studies have seen extensive toxicological studies reporting the adverse effect of cadmium, lead and mercury on humans as different organ failures, as well as potential implications in impaired fertility (Jang and Hoffman, 2011). Lead or cadmium can affect hormonal levels in women, which leads to conceiving difficulties (Butrimovitz et al., 1983; Hovatta et al., 1998). Bjørklund et al., (2019) in a mini review explained the hazards of Mercury to fertility and pregnancy outcomes.

Effect on Males

Male reproductive functions can be adversely affected by several metals. Lead can reduce sperm mobility and sperm count in men (Butrimovitz et al., 1983; Hovatta et al., 1998). Higher concentrations of cadmium in seminal plasma of men (65 µg/dL) had a significant decline in number of motile sperms and sperm count (Zeng et al., 2002; Mason, 1990). Mercury (40.6 mmol/L) in whole blood resulted in 50% less of progressive sperm motility and 14% less of normal morphology (Pollack et al., 2012). Heavy metals produce cellular level structural and functional impairments in the male reproductive system (Hess, 1998). Even at low levels of exposure Cadmium acts as a male reproductive toxicant by several pathogenetic mechanisms (de Angelis et al., 2017). Zhou, (2016) reported that out 207 urine samples of infertile men thirteen metals were detected such as cadmium (Cd), arsenic (As), cobalt (Co), copper (Cu), chromium (Cr), iron (Fe), manganese (Mn), lead (Pb), molybdenum (Mo), nickel (Ni), mercury (Hg), selenium (Se), and zinc (Zn). The environmental exposure to Mn, Hg, and Ni may be associated with increased sperm DNA damage (Zhou et al., 2016). Lee et al., (2020) reported infertility in the blood samples of US adults in relation with low levels of cadmium and lead. Ali et al., (2020) studies revealed the higher concentration of Cu, Pb and Cd in the serum and blood of infertile men. However, significant decrease in mean Zn levels was observed in the serum of infertile men.

INDUSTRIAL CHEMICALS

Chemicals produced from different industries effects the reproductive health of different organisms including the male and female workers (Rim, 2017). Chemical exposure to any parent can cause genetic mutations or abnormalities in the offspring. The development of the fetus can also be disrupted through heritable changes (epigenetic mechanisms) in gametic cells, causing no change in the underlying DNA sequence of the organism; instead, non-genetic factors cause the organism's genes to express in a different way. There are some evidences of epigenetic effects of paternal exposure that can impact on the pregnancy outcome, but human evidence is still not convincing (Hovatta, 2008). Women's exposed to the high levels of the solvents at the work places may enhance the risks of miscarriage birth defects and low birth weight (Lindbohm, 1995; Ha et el., 2002; Chevrier et al., 2006; Thulstrup and Bonde, 2006). Works in the industries like dry cleaning, manufacturing, painting, pharmaceutical, shoe-making, semi-conductor, audio speaker and laboratory industries were found to have high risks of miscarriage. Some studies have related the neurobehavioural effects, preterm birth and childhood leukemia to the solvent exposures (Wennborg et al., 2002; Infante-Rivard et al., 2005; Julvez and Grandjean, 2009).

The inhalation of chemicals and their intermediates can readily have absorbed in the respiratory and gastrointestinal track. Almost all the chemical solvents can effect the reproductive health adversely. The azo textile dye Disperse Red 1 caused harmful effects in the regeneration and reproduction of newborn and adult specimens (Ribeiro and de Aragão Umbuzeiro, 2014). Human exposure to di-*n*-butyl phthalate (DBP) estimates range from 0.007 mg/kg/d to 0.01 mg/kg/d in the general population and up to 0.233 mg/kg/d in patients taking DBP-coated medications. Because levels of phthalates tend to be higher in women, evaluating ovarian effects of DBP exposure is of great importance, and it was shown that DBP could disrupt ovarian function in mice at doses relevant to humans. Higher doses of di-*n*-butyl phthalate (DBP) were found in the patients taking DBP-coated medications. It is found that DBP could disrupt ovarian function in mice at doses relevant to humans, therefore the evaluation of DBP exposure in women is of great importance (Sen et al., 2015).

Majority of the world's population is employed in agriculture. Since, from the past few decades the use of pesticides has increased numerously. Most of the pesticides used spread in the environment and cause the toxic effects to different organisms including humans specifically to those who work with them. Pesticides besides various health issues can also have adverse effects on the reproduction. Assessment of the health risks posed by the use of pesticides is a great matter of concern for the occupational toxicologists. Minimization of health risks caused by pesticides is technically feasible. Therefore, the risk management should be a precondition for the use of these pesticides (Maroni et al., 2006). Carcinogenic chemical can have the health hazardous effects on the pregnant mother and it's newborn. In the recent years the attention towards the synthetic chemicals have gather the information of reports of increased testicular cancers, male reproductive tract abnormalities and semen quality decrease. A huge number of natural and synthetic chemicals have been reported to disrupt the regular functioning of the reproductive system (Vandenberg et al., 2012). Maske et al., (2018) studied the effect of n-butyl paraben on F1 female rats and drafted the suggestion to avoid exposure to parabens that health monitors need to advice females planning for pregnancy.

Effect on Females

Acute toxicity of organochlorine (OC) pesticides can adversely affect the development of the female reproductive system during puberty (Bapayeva et al., 2016). Chronic exposure of organophosphate compounds can have considerable potential adverse effects on pregnant women, the fetus and young children (De Silva, 2006). Chemical contamination to the breast milk can be an exposure route for the infants that have not been widely recognized (Kacew, 1994).

Effects on Males

Experimental can studies have revealed that polybrominated diphenyl ethers (PBDEs) exposure can adversely affect reproductive hormones in older men (Makey et al., 2016). Pesticide exposure can decrease the sperm motility and concentration seen in the farmers with high pesticide use (Perry, 2008). The chronic exposure of the current use pesticides can affect the reproductive hormones, genital measures a semen quality in men (Cremonese et al., 2017). Exposure of plasticizer di (2-ethylhexyl) phthalate and the organic industrial compounds polychlorinated biphenyls to the mouse revealed complex effects on the male reproductive system (Fiandanese et al., 2016). Dibromochloropropane (DBCP) a previously used pesticide is now considered as the major testicular toxin. It can cause irreversible sterility or oligospermia on the long lasting severe exposure (Lähdetie, 1995). Another banned fumigant dibromide (EDP) can also cause adverse effects on the male reproduction. Moreover, the exposure of some herbicides can adversely affect the semen quality e.g. alachlor, atrazine and 2,4-dichlorophenoxyacetic acid (2,4-D), and three insecticides, carbaryl, chlordecone and diazinon (Jensen et al., 2006; Lähdetie, 1995).

ENDOCRINE DISRUPTING CHEMICALS

Endocrine disrupters (EDCs) are those substances that have potency to interfere with the hormones that are responsible for reproduction, development and homeostasis. The interference is in their synthesis, secretion, transport, metabolism and binding (Kopp et al., 2017). Endocrine disrupters are so widespread in the environment that their exposure is certain and affect human health. EDCs include dioxins (e.g., 2, 3, 7, 8-tetrachlorodibenzo-p-dioxin (TCDD), plasticizers (e.g., bisphenol A (BPA), pesticides (e.g., 1, 1, 1-Trichloro-2, 2-bis (p-chlorophenyl) ethane (DDT)), polycyclic hydrocarbons etc. These EDCs have a negative impact on female reproductive system and interfere with ovarian folliculogenesis and corpus luteum formation ultimately affecting fertility. While other EDCs alter sex steroid hormones. There are other EDCs that target estrogen, androgen receptors and cause ovarian toxicity (Craig et al. 2011). Plasticizers are used in various plastic containers, beverage bottles etc. BPA a plasticizer leaches out from the contained material and affects reproductive health (Craig et al. 2011). It has been found that BPAs largely affect female infertility, function of uterus, estrous cycle and embryo implantation (Ziv-Gal and Flaws 2016). BPA has also been found to decrease sperm quality. Other EDCs like, 2, 3, 7, 8-tetrachlorodibenzo-p-dioxin, 1, 1, 1-Trichloro-2, 2-bis (p-chlorophenyl) ethane and polychlorinated biphenyls (PCBs) alter menstrual and ovarian function (Mendola et al. 2008) and also decrease hormone level of progesterone and estrogen (Windham et al. 2005; Perry et al. 2006). It has also been reported that PCBs alter the structure integrity of human sperm chromatin (Rignell-Hydbom et al. 2005).

NANOPARTICLES

Silver nanoparticles (AgNPs) injection has been found to delay the physical development and largely impaired the behaviour in the offspring when injected in mother. It has also been found that silver (Ag) accumulated in the testes after administration of AgNPs. Furthermore, in early gestation AgNPs presence were seen in yolk sac after early administration in mice. While the radiolabeled AgNPs were detected in the breast milk, placenta, and postnatal offspring after injection during late gestation in rats.

Ag is in the ionic form and is bioavailable, hence route of exposure in humans is required. The studies have confirmed that nanoparticles (NPs) have a negative impact on the pregnancy and fetal development. The NPs may translocate from respiratory tract to placenta and to fetus (Ema *et al.*, 2017). The maternal inhalation of NPs may affect several organs in the offspring but due to limited studies there is an uncertainty about the implication for embryo fetal development. Henceforth due to over production and utilization of NPs, a testing strategy must be established (Hougaard et al., 2015). The maternal exposure to titanium is associated with the neural tube defect (Li et al., 2016). Skovmand et al., (2019) observed that utero exposure to carbon black nanoparticles did not affect male-specific reproductive failure, fertility, gestation and litter parameters in any generation, while Pinho et al., (2020) revealed that exposure to zinc oxide nanoparticles (ZnONPs) for longer durations and higher concentrations can have toxic effects to male germ cells.

CONCLUSION

The chapter includes the recent literature about the environmental contaminants and their effect on the reproductive health particularly fertility. The literature strongly indicated that environmental contaminants interfere with the both female and male reproductive functions and structure as well. The environmental contaminants studied were pesticides, phthalates, metals, industrial chemicals, endocrine disrupters, and nanomaterials. Environmental contaminants studied have different mechanism and pattern of impact in the human fertility. Further research needs to be done to study the effect of environmental contaminants at molecular level. This will help in better understanding the mechanism leading the infertility related outcomes.

AUTHOR CONTRIBUTION

IAB conceived and designed the book chapter. IAB, SAB, SKK, IAM, HPN, SAU performed literature survey. IAB, SAB, SKK, SAU, AHK wrote the book chapter.

FUND DISCLAIMER

No financial support from any funding agency was received.

CORRESPONDING AUTHORS

Irfan Ashraf Badroo

REFERENCES

Aarab, N., Minier, C., Lemaire, S., Unruh, E., Hansen, P. D., Larsen, B. K., Andersen, O.-K., & Narbonne, J.-F. (2004). Biochemical and histological responses in mussel (Mytilus edulis) exposed to North Sea oil and a mixture of North Sea oil and alkylphenols. *Marine Environmental Research*, *58*(2-5), 437–441. doi:10.1016/j.marenvres.2004.03.121

Akingbemi, B. T., Ge, R., Klinefelter, G. R., Zirkin, B. R., & Hardy, M. P. (2004). Phthalate-induced Leydig cell hyperplasia is associated with multiple endocrine disturbances. *Proceedings of the National Academy of Sciences of the United States of America*, *101*(3), 775–780. doi:10.1073/pnas.0305977101

Ali, H. M., Qater, A. N. A., & Selman, M. O. (2020). Relation of Some Heavy Metals with Male Infertility. *Medico-Legal Update*, *20*(1), 1440–1445.

Andrady, A. L. (2011). Microplastics in the marine environment. *Marine Pollution Bulletin*, *62*(8), 1596–1605. doi:10.1016/j.marpolbul.2011.05.030

Andrady, A. L., & Neal, M. A. (2009). Applications and societal benefits of plastics. *Philosophical Transactions of the Royal Society of London. Series B, Biological Sciences*, *364*(1526), 1977–1984. doi:10.1098/rstb.2008.0304

Badroo, I. A., Khanday, A. H., Ali, Z. K., & Badroo, S. A. (2020). Environmental Phthalate Exposure in Relation to Reproductive Outcomes and Other Health Endpoints in Humans. In *Handbook of Research on Environmental and Human Health Impacts of Plastic Pollution* (pp. 288–313). IGI Global. doi:10.4018/978-1-5225-9452-9.ch015

Baird, C., & Cann, M. (2012). *Environmental chemistry. 5th ed.* Toxic heavy metals. W. H. Freeman and Company.

Bapayeva, G., Issayeva, R., Zhumadilova, A., Nurkasimova, R., Kulbayeva, S., & Tleuzhan, R. (2016). Organochlorine pesticides and female puberty in South Kazakhstan. *Reproductive Toxicology (Elmsford, N.Y.)*, *65*, 67–75. doi:10.1016/j.reprotox.2016.06.017

Bjørklund, G., Chirumbolo, S., Dadar, M., Pivina, L., Lindh, U., Butnariu, M., & Aaseth, J. (2019). Mercury exposure and its effects on fertility and pregnancy outcome. *Basic & Clinical Pharmacology & Toxicology*, *125*(4), 317–327. doi:10.1111/bcpt.13264

Boekelheide, K., Darney, S. P., Daston, G. P., David, R. M., Luderer, U., Olshan, A. F., Sanderson, W. T., Willhite, C. C., & Woskie, S. (2004). NTP-CERHR Expert Panel Report on the reproductive and developmental toxicity of 2-bromopropane. *Reproductive Toxicology (Elmsford, N.Y.)*, *18*(2), 189–217. doi:10.1016/j.reprotox.2003.10.003

Bretveld, R., Brouwers, M., Ebisch, I., & Roeleveld, N. (2007). Influence of pesticides on male fertility. *Scandinavian Journal of Work, Environment & Health*, *33*(1), 13–28. doi:10.5271jweh.1060

Butrimovitz, G. P., Sharlip, I., & Lo, R. (1983). Extremely low seminal lead concentrations and male fertility. *Clinica Chimica Acta*, *135*(2), 229–231. doi:10.1016/0009-8981(83)90139-0

Callegari, C., Everett, S., Ross, M., & Brasel, J. A. (1987). Anogenital ratio: Measure of fetal virilization in premature and full-term newborn infants. *The Journal of Pediatrics*, *111*(2), 240–243. doi:10.1016/S0022-3476(87)80075-6

Cariati, F., D'Uonno, N., Borrillo, F., Iervolino, S., Galdiero, G., & Tomaiuolo, R. (2019). Bisphenol a: An emerging threat to male fertility. *Reproductive Biology and Endocrinology*, *17*(1), 6. doi:10.118612958-018-0447-6

Centers for Disease Control and Prevention. (2005). Third National Report on Human Exposure to Environmental Chemicals. National Center for Environmental Health, Division of Laboratory Sciences, Atlanta, GA.

Chevrier, C., Dananché, B., Bahuau, M., Nelva, A., Herman, C., Francannet, C., Robert-Gnansia, E., & Cordier, S. (2006). Occupational exposure to organic solvent mixtures during pregnancy and the risk of non-syndromic oral clefts. *Occupational and Environmental Medicine*, *63*(9), 617–623. doi:10.1136/oem.2005.024067

Colon, I., Caro, D., Bourdony, C. J., & Rosario, O. (2000). Identification of phthalate esters in the serum of young Puerto Rican girls with premature breast development. *Environmental Health Perspectives*, *108*(9), 895.

Cordier, S. (2008). Evidence for a role of paternal exposures in developmental toxicity. *Basic & Clinical Pharmacology & Toxicology*, *102*(2), 176–181. doi:10.1111/j.1742-7843.2007.00162.x

Craig, Z. R., Wang, W., & Flaws, J. A. (2011). Endocrine-disrupting chemicals in ovarian function: Effects on steroidogenesis, metabolism and nuclear receptor signaling. *Reproduction (Cambridge, England)*, *142*(5), 633–646. doi:10.1530/REP-11-0136

Cremonese, C., Piccoli, C., Pasqualotto, F., Clapauch, R., Koifman, R. J., Koifman, S., & Freire, C. (2017). Occupational exposure to pesticides, reproductive hormone levels and sperm quality in young Brazilian men. *Reproductive Toxicology (Elmsford, N.Y.)*, *67*, 174–185. doi:10.1016/j.reprotox.2017.01.001

Damstra, T., Barlow, S., Bergman, A., Kavlock, R., & Van Der Kraak, G. (Eds.). (2004). *Global assessment of the state of the science of endocrine disruptors*. International Programme on Chemical Safety.

Davis, B. J., Maronpot, R. R., & Heindel, J. J. (1994). Di-(2-ethylhexyl) phthalate suppresses estradiol and ovulation in cycling rats. *Toxicology and Applied Pharmacology*, *128*(2), 216–223. doi:10.1006/taap.1994.1200

Davis, B. J., Weaver, R., Gaines, L. J., & Heindel, J. J. (1994). Mono-(2-ethylhexyl) phthalate suppresses estradiol production independent of FSH-cAMP stimulation in rat granulosa cells. *Toxicology and Applied Pharmacology*, *128*(2), 224–228. doi:10.1006/taap.1994.1201

De Angelis, C., Galdiero, M., Pivonello, C., Salzano, C., Gianfrilli, D., Piscitelli, P., Lenzi, A., Colao, A., & Pivonello, R. (2017). The environment and male reproduction: The effect of cadmium exposure on reproductive function and its implication in fertility. *Reproductive Toxicology (Elmsford, N.Y.)*, *73*, 105–127. doi:10.1016/j.reprotox.2017.07.021

De Silva, H. J., Samarawickrema, N. A., & Wickremasinghe, A. R. (2006). Toxicity due to organophosphorus compounds: What about chronic exposure? *Transactions of the Royal Society of Tropical Medicine and Hygiene*, *100*(9), 803–806. doi:10.1016/j.trstmh.2006.05.001

Diamanti-Kandarakis, E., Bourguignon, J.-P., Giudice, L. C., Hauser, R., Prins, G. S., Soto, A. M., Zoeller, R. T., & Gore, A. C. (2009). Endocrine-disrupting chemicals: An Endocrine Society scientific statement. *Endocrine Reviews*, *30*(4), 293–342. doi:10.1210/er.2009-0002

Dris, R., Gasperi, J., Rocher, V., Saad, M., Renault, N., & Tassin, B. (2015). Microplastic contamination in an urban area: A case study in Greater Paris. *Environmental Chemistry*, *12*(5), 592–599. doi:10.1071/EN14167

Duty, S. M., Ackerman, R. M., Calafat, A. M., & Hauser, R. (2005). Personal care product use predicts urinary concentrations of some phthalate monoesters. *Environmental Health Perspectives*, *113*(11), 1530–1535. doi:10.1289/ehp.8083

Duty, S. M., Calafat, A. M., Silva, M. J., Ryan, L., & Hauser, R. (2005). Phthalate exposure and reproductive hormones in adult men. *Human Reproduction (Oxford, England)*, *20*(3), 604–610. doi:10.1093/humrep/deh656

Duty, S. M., Silva, M. J., Barr, D. B., Brock, J. W., Ryan, L., Chen, Z., Herrick, R. F., Christiani, D. C., & Hauser, R. (2003). Phthalate exposure and human semen parameters. *Epidemiology (Cambridge, Mass.)*, *14*(3), 269–277. doi:10.1097/01.EDE.0000059950.11836.16

Ema, M., Okuda, H., Gamo, M., Honda, K. A. (2017). Review of reproductive and developmental toxicity of silver nanoparticles in laboratory animals. *Reproductive Toxicology, 67*, 149-64.

Fiandanese, N., Borromeo, V., Berrini, A., Fischer, B., Schaedlich, K., Schmidt, J.-S., Secchi, C., & Pocar, P. (2016). Maternal exposure to a mixture of di (2-ethylhexyl) phthalate (DEHP) and polychlorinated biphenyls (PCBs) causes reproductive dysfunction in adult male mouse offspring. *Reproductive Toxicology (Elmsford, N.Y.)*, *65*, 123–132. doi:10.1016/j.reprotox.2016.07.004

Fisher, J. S., Macpherson, S., Marchetti, N., & Sharpe, R. M. (2003). Human 'testicular dysgenesis syndrome': A possible model using in-utero exposure of the rat to dibutyl phthalate. *Human Reproduction (Oxford, England)*, *18*(7), 1383–1394. doi:10.1093/humrep/deg273

Foster, P. M. (2006). Disruption of reproductive development in male rat offspring following in utero exposure to phthalate esters. *International Journal of Andrology*, *29*(1), 140–147. doi:10.1111/j.1365-2605.2005.00563.x

Foster, P. M. D., Mylchreest, E., Gaido, K. W., & Sar, M. (2001). Effects of phthalate esters on the developing reproductive tract of male rats. *APMIS*, *109*(S103), S272–S277. doi:10.1111/j.1600-0463.2001.tb05776.x

Fuortes, L., Clark, M. K., Kirchner, H. L., & Smith, E. M. (1997). Association between female infertility and agricultural work history. *American Journal of Industrial Medicine*, *31*(4), 445–451. doi:10.1002/(SICI)1097-0274(199704)31:4<445::AID-AJIM11>3.0.CO;2-#

Gray, T. J. B., & Beamand, J. A. (1984). Effect of some phthalate esters and other testicular toxins on primary cultures of testicular cells. *Food and Chemical Toxicology*, *22*(2), 123–131. doi:10.1016/0278-6915(84)90092-9

Ha, E., Cho, S.-I., Chen, D., Chen, C., Ryan, L., Smith, T. J., Xu, X., & Christiani, D. C. (2002). Parental exposure to organic solvents and reduced birth weight. *Archives of Environmental Health: An International Journal*, *57*(3), 207–214. doi:10.1080/00039890209602938

Habbema, J. D. F., Collins, J., Leridon, H., Evers, J. L., Lunenfeld, B., & teVelde, E. R. (2004). Towards less confusing terminology in reproductive medicine: A proposal. *Human Reproduction (Oxford, England)*, *19*(7), 1497–1501. doi:10.1093/humrep/deh303

Halden, R. U. (2010). Plastics and public health. *Annual Review of Public Health*, *31*(1), 179–194. doi:10.1146/annurev.publhealth.012809.103714

Hart, R. J. (2016). Physiological aspects of female fertility: Role of the environment, modern lifestyle, and genetics. *Physiological Reviews*, *96*(3), 873–909. doi:10.1152/physrev.00023.2015

Hauser, R., Gaskins, A. J., Souter, I., Smith, K. W., Dodge, L. E., Ehrlich, S., Meeker, J. D., Calafat, A. M., & Williams, P. L. (2016). Urinary phthalate metabolite concentrations and reproductive outcomes among women undergoing in vitro fertilization: Results from the EARTH study. *Environmental Health Perspectives*, *124*(6), 831–839. doi:10.1289/ehp.1509760

Hawkes, S. J. (1997). What is a "heavy metal?". *Journal of Chemical Education*, *74*(11), 1374. doi:10.1021/ed074p1374

Hejmej, A., Kotula-Balak, M., & Bilińska, B. (2011). *Antiandrogenic and estrogenic compounds: effect on development and function of male reproductive system*. Steroids–Clinical Aspect.

Hess, R. A. (1998). Effects of environmental toxicants on the efferent ducts, epididymis and fertility. *Journal of Reproduction and Fertility. Supplement*, 247–259.

Hossain, F., Ali, O., D'souza, U. J., & Naing, D. K. S. (2010). Effects of pesticide use on semen quality among farmers in rural areas of Sabah, Malaysia. *Reproductive Toxicology, 56*, 118-140.

Hovatta, O., Venäläinen, E. R., Kuusimäki, L., Heikkilä, J., Hirvi, T., & Reima, I. (1998). Aluminium, lead and cadmium concentrations in seminal plasma and spermatozoa, and semen quality in Finnish men. *Human Reproduction (Oxford, England)*, *13*(1), 115–119. doi:10.1093/humrep/13.1.115

Huang, P. C., Kuo, P. L., Guo, Y. L., Liao, P. C., & Lee, C. C. (2007). Associations between urinary phthalate monoesters and thyroid hormones in pregnant women. *Human Reproduction (Oxford, England)*, *22*(10), 2715–2722. doi:10.1093/humrep/dem205

Infante-Rivard, C., Siemiatycki, J., Lakhani, R., & Nadon, L. (2005). Maternal exposure to occupational solvents and childhood leukemia. *Environmental Health Perspectives*, *113*(6), 787–792. doi:10.1289/ehp.7707

Inhorn, M. C., & Patrizio, P. (2015). Infertility around the globe: New thinking on gender, reproductive technologies and global movements in the 21st century. *Human Reproduction Update*, *21*(4), 411–426. doi:10.1093/humupd/dmv016

Jang, D. H., & Hoffman, R. S. (2011). Heavy metal chelation in neurotoxic exposures. *Neurologic Clinics*, *29*(3), 607–622. doi:10.1016/j.ncl.2011.05.002

Jarfelt, K., Dalgaard, M., Hass, U., Borch, J., Jacobsen, H., & Ladefoged, O. (2005). Antiandrogenic effects in male rats perinatally exposed to a mixture of di (2-ethylhexyl) phthalate and di (2-ethylhexyl) adipate. *Reproductive Toxicology (Elmsford, N.Y.)*, *19*(4), 505–515. doi:10.1016/j.reprotox.2004.11.005

Jensen, T. K., Bonde, J. P., & Joffe, M. (2006). The influence of occupational exposure on male reproductive function. *Occupational Medicine*, *56*(8), 544–553. doi:10.1093/occmed/kql116

Julvez, J., & Grandjean, P. (2009). Neurodevelopmental toxicity risks due to occupational exposure to industrial chemicals during pregnancy. *Industrial Health*, *47*(5), 459–468. doi:10.2486/indhealth.47.459

Kacew, S. (1994). Current issues in lactation: Advantages, environment, silicone. Biomedical and environmental sciences. *BES*, *7*(4), 307–319.

Kopp, R., Martínez, I. O., Legradi, J., & Legler, J. (2017). Exposure to endocrine disrupting chemicals perturbs lipid metabolism and circadian rhythms. *Journal of Environmental Sciences (China)*, *62*, 133–137. doi:10.1016/j.jes.2017.10.013

Kozin, L. F., & Hansen, S. C. (2013). *Mercury handbook: chemistry, applications and environmental impact*. Royal Society of Chemistry.

Lähdetie, J. (1995). Occupation-and exposure-related studies on human sperm. *Journal of Occupational and Environmental Medicine*, *37*(8), 922–930. doi:10.1097/00043764-199508000-00006

Lassen, T. H., Frederiksen, H., Jensen, T. K., Petersen, J. H., Joensen, U. N., Main, K. M., Skakkebaek, N. E., Juul, A., Jørgensen, N., & Andersson, A.-M. (2014). Urinary bisphenol A levels in young men: Association with reproductive hormones and semen quality. *Environmental Health Perspectives*, *122*(5), 478–484. doi:10.1289/ehp.1307309

Latini, G., Gallo, F., & De Felice, C. (2004). Birth characteristics and hepatoblastoma risk in young children. Cancer. *Interdisciplinary International Journal of the American Cancer Society*, *101*(1), 210–210.

Lee, S., Min, J. Y., & Min, K. B. (2020). Female Infertility Associated with Blood Lead and Cadmium Levels. *International Journal of Environmental Research and Public Health*, *17*(5), 1794. doi:10.3390/ijerph17051794

Li, Z., Huo, W., Li, Z., Wang, B., Zhang, J., & Ren, A. (2016). Association between titanium and silver concentrations in maternal hair and risk of neural tube defects in offspring: A case-control study in north China. *Reproductive Toxicology (Elmsford, N.Y.)*, *66*, 115–121. doi:10.1016/j.reprotox.2016.10.006

Lindbohm, M. L. (1995). Effects of parental exposure to solvents on pregnancy outcome. *Journal of Occupational and Environmental Medicine*, *37*(8), 908–914. doi:10.1097/00043764-199508000-00004

Llobet, J. M., Falco, G., Casas, C., Teixido, A., & Domingo, J. L. (2003). Concentrations of arsenic, cadmium, mercury, and lead in common foods and estimated daily intake by children, adolescents, adults, and seniors of Catalonia, Spain. *Journal of Agricultural and Food Chemistry*, *51*(3), 838–842. doi:10.1021/jf020734q

Lopez-Carrillo, L., Hernandez-Ramirez, R. U., Calafat, A. M., Torres-Sanchez, L., Galvan Portillo, M., Needham, L. L., & Cebrian, M. E. (2010). Ex- posure to phthalates and breast cancer risk in Northen Mexico. *Environmental Health Perspectives*, *118*(4), 539–544. doi:10.1289/ehp.0901091

Lovekamp-Swan, T., & Davis, B. J. (2003). Mechanisms of phthalate ester toxicity in the female reproductive system. *Environmental Health Perspectives*, *111*(2), 139–145. doi:10.1289/ehp.5658

Lucas, B., Fields, C., & Hofmann, M. C. (2009). Signaling pathways in spermatogonial stem cells and their disruption by toxicants. *Birth Defects Research. Part C, Embryo Today*, *87*(1), 35–42. doi:10.1002/bdrc.20145

Makey, C. M., McClean, M. D., Braverman, L. E., Pearce, E. N., Sjödin, A., Weinberg, J., & Webster, T. F. (2016). Polybrominated diphenyl ether exposure and reproductive hormones in North American men. *Reproductive Toxicology (Elmsford, N.Y.)*, *62*, 46–52. doi:10.1016/j.reprotox.2016.04.009

Maroni, M., Fanetti, A. C., & Metruccio, F. (2006). Risk assessment and management of occupational exposure to pesticides in agriculture. *La Medicina del Lavoro*, *97*(2), 430–437.

Martínez-Sales, M., García-Ximénez, F., & Espinós, F. J. (2015). Zebrafish as a possible bioindicator of organic pollutants with effects on reproduction in drinking waters. *Journal of Environmental Sciences (China)*, *33*, 254–260. doi:10.1016/j.jes.2014.11.012

Maske, P., Dighe, V., & Vanage, G. (2018). n-butylparaben exposure during perinatal period impairs fertility of the F1 generation female rats. *Chemosphere*, *213*, 114–123. doi:10.1016/j.chemosphere.2018.08.130

Mason, H. J. (1990). Occupational cadmium exposure and testicular endocrine function. *Human and Experimental Toxicology*, *9*(2), 91–94. doi:10.1177/096032719000900205

McElgunn, B. (1998). Reproductive and developmental hazards in the workplace. *Clinical Excellence for Nurse Practitioners*, *2*, 140–145.

McGregor, A. J., & Mason, H. J. (1991). The effects of occupational exposure to cadmium, lead and mercury vapour on male reproductive endocrine function. *Heavy Metals in the Environment*. Edinburgh. *CEP Consultants*, *1*, 375–378.

McKee, R. H., Butala, J. H., David, R. M., & Gans, G. (2004). NTP center for the evaluation of risks to human reproduction reports on phthalates: Addressing the data gaps. *Reproductive Toxicology (Elmsford, N.Y.)*, *18*(1), 1–22. doi:10.1016/j.reprotox.2003.09.002

Meeker, J. D., Calafat, A. M., & Hauser, R. (2007). Di (2-ethylhexyl) phthalate metabolites may alter thyroid hormone levels in men. *Environmental Health Perspectives*, *115*(7), 1029–1034. doi:10.1289/ehp.9852

Meeker, J. D., Rossano, M. G., Protas, B., Diamond, M. P., Puscheck, E., Daly, D., Paneth, N., & Wirth, J. J. (2008). Cadmium, lead, and other metals in relation to semen quality: Human evidence for molybdenum as a male reproductive toxicant. *Environmental Health Perspectives*, *116*(11), 1473–1479. doi:10.1289/ehp.11490

Meeker, J. D., Yang, T., Ye, X., Calafat, A. M., & Hauser, R. (2011). Urinary concentrations of parabens and serum hormone levels, semen quality parameters, and sperm DNA damage. *Environmental Health Perspectives*, *119*(2), 252–257. doi:10.1289/ehp.1002238

Mendiola, J., Jørgensen, N., Andersson, A.-M., Calafat, A. M., Ye, X., Redmon, J. B., Drobnis, E. Z., Wang, C., Sparks, A., Thurston, S. W., Liu, F., & Swan, S. H. (2010). Are environmental levels of bisphenol A associated with reproductive function in fertile men? *Environmental Health Perspectives*, *118*(9), 1286–1291. doi:10.1289/ehp.1002037

Mendola, P., Messer, L. C., & Rappazzo, K. (2008). Science linking environmental contaminant exposures with fertility and reproductive health impacts in the adult female. *Fertility and Sterility*, *89*(2), 81–94. doi:10.1016/j.fertnstert.2007.12.036

Mok-Lin, E., Ehrlich, S., Williams, P. L., Petrozza, J., Wright, D. L., Calafat, A. M., Ye, X., & Hauser, R. (2010). Urinary bisphenol A concentrations and ovarian response among women undergoing IVF. *International Journal of Andrology*, *33*(2), 385–393. doi:10.1111/j.1365-2605.2009.01014.x

Moore, R. W., Rudy, T. A., Lin, T. M., Ko, K., & Peterson, R. E. (2001). Abnormalities of sexual development in male rats with in utero and lactational exposure to the antiandrogenic plasticizer Di (2-ethylhexyl) phthalate. *Environmental Health Perspectives*, *109*(3), 229–237. doi:10.1289/ehp.01109229

Ng, T. P., Goh, H. H., Ng, Y. L., Ong, H. Y., Ong, C. N., Chia, K. S., Chia, S. E., & Jeyaratnam, J. (1991). Male endocrine functions in workers with moderate exposure to lead. *Occupational and Environmental Medicine*, *48*(7), 485–491. doi:10.1136/oem.48.7.485

No, A. C. O. (2013). Exposure to toxic environmental agents. *Fertility and Sterility*, *100*(4), 931–934. doi:10.1016/j.fertnstert.2013.08.043

Ombelet, W., Cooke, I., Dyer, S., Serour, G., & Devroey, P. (2008). Infertility and the provision of infertility medical services in developing countries. *Human Reproduction Update*, *14*(6), 605–621. doi:10.1093/humupd/dmn042

Pant, N., Upadhyay, G., Pandey, S., Mathur, N., Saxena, D. K., & Srivastava, S. P. (2003). Lead and cadmium concentration in the seminal plasma of men in the general population: Correlation with sperm quality. *Reproductive Toxicology (Elmsford, N.Y.)*, *17*(4), 447–450. doi:10.1016/S0890-6238(03)00036-4

Perry, M. J. (2008). Effects of environmental and occupational pesticide exposure on human sperm: A systematic review. *Human Reproduction Update*, *14*(3), 233–242. doi:10.1093/humupd/dmm039

Perry, M. J., Ouyang, F., Korrick, S. A., Venners, S. A., Chen, C., Xu, X., Lasley, B. L., & Wang, X. (2006). A prospective study of serum DDT and progesterone and estrogen levels across the menstrual cycle in nulliparous women of reproductive age. *American Journal of Epidemiology*, *164*(11), 1056–1064. doi:10.1093/aje/kwj329

Perry, M. J., Venners, S. A., Chen, X., Liu, X., Tang, G., Xing, H., Barr, D. B., & Xu, X. (2011). Organophosphorous pesticide exposures and sperm quality. *Reproductive Toxicology (Elmsford, N.Y.)*, *31*(1), 75–79. doi:10.1016/j.reprotox.2010.08.006

Phillip, M., De, C. B., Pilpel, D., Karplus, M., & Sofer, S. (1996). Clitoral and penile sizes of full term newborns in two different ethnic groups. Journal of pediatric endocrinology & metabolism. *Journal of Pediatric Endocrinology & Metabolism*, *9*(2), 175–179.

Pollack, A. Z., Schisterman, E. F., Goldman, L. R., Mumford, S. L., Perkins, N. J., Bloom, M. S., Rudra, C. B., Browne, R. W., & Wactawski-Wende, J. (2012). Relation of blood cadmium, lead, and mercury levels to biomarkers of lipid peroxidation in premenopausal women. *American Journal of Epidemiology*, *175*(7), 645–652. doi:10.1093/aje/kwr375

Radwan, M., Wielgomas, B., Dziewirska, E., Radwan, P., Kałużny, P., Klimowska, A., Hanke, W., & Jurewicz, J. (2018). Urinary bisphenol A levels and male fertility. *American Journal of Men's Health*, *12*(6), 2144–2151. doi:10.1177/1557988318799163

Recio-Vega, R., Ocampo-Gómez, G., Borja-Aburto, V. H., Moran-Martínez, J., & Cebrian-Garcia, M. E. (2008). Organophosphorus pesticide exposure decreases sperm quality: Association between sperm parameters and urinary pesticide levels. *Journal of Applied Toxicology*, *28*(5), 674–680. doi:10.1002/jat.1321

Rehfeld, A., Andersson, A. M., & Skakkebæk, N. E. (2020). Bisphenol A Diglycidyl Ether (BADGE) and Bisphenol Analogs, but Not Bisphenol A (BPA), Activate the CatSper Ca^{2+} Channel in Human Sperm. *Frontiers in Endocrinology*, *11*, 324. doi:10.3389/fendo.2020.00324

Ribeiro, A. R., & de Aragão Umbuzeiro, G. (2014). Effects of a textile azo dye on mortality, regeneration, and reproductive performance of the planarian, Girardiatigrina. *Environmental Sciences Europe*, *26*(1), 22. doi:10.118612302-014-0022-5

Richter, C. A., Birnbaum, L. S., Farabollini, F., Newbold, R. R., Rubin, B. S., Talsness, C. E., Vandenbergh, J. G., Walser-Kuntz, D. R., & vom Saal, F. S. (2007). In vivo effects of bisphenol A in laboratory rodent studies. *Reproductive Toxicology (Elmsford, N.Y.)*, *24*(2), 199–224. doi:10.1016/j.reprotox.2007.06.004

Rignell-Hydbom, A., Rylander, L., Giwercman, A., Jönsson, B. A. G., Lindh, C., Eleuteri, P., Rescia, M., Leter, G., Cordelli, E., Spano, M., & Hagmar, L. (2005). Exposure to PCBs and p, p¢-DDE and human sperm chromatin integrity. *Environmental Health Perspectives*, *113*(2), 175–179. doi:10.1289/ehp.7252

Rim, K. T. (2017). Reproductive Toxic chemicals at work and efforts to protect workers' health: A literature review. *Safety and Health at Work*, *8*(2), 143–150. doi:10.1016/j.shaw.2017.04.003

Rochester, J. R. (2013). Bisphenol A and human health: A review of the literature. *Reproductive Toxicology (Elmsford, N.Y.)*, *42*, 132–155. doi:10.1016/j.reprotox.2013.08.008

Russ, H. (2016). Urinary phthalate metabolite concentrations and reproductive outcomes among women undergoing in vitro fertilization: Results from the EARTH study. *Environmental Health Perspectives*, *124*(6), 831–839. doi:10.1289/ehp.1509760

Santangeli, S., Maradonna, F., Olivotto, I., Piccinetti, C. C., Gioacchini, G., & Carnevali, O. (2017). Efects of BPA on female reproductive function: Te involvement of epigenetic mechanism. *General and Comparative Endocrinology*, *245*, 122–126. doi:10.1016/j.ygcen.2016.08.010

Sen, N., Liu, X., & Craig, Z. R. (2015). Short term exposure to di-n-butyl phthalate (DBP) disrupts ovarian function in young CD-1 mice. *Reproductive Toxicology (Elmsford, N.Y.)*, *53*, 15–22. doi:10.1016/j.reprotox.2015.02.012

Sharpe, R. M. (2001). Hormones and testis development and the possible adverse effects of environmental chemicals. *Toxicology Letters*, *120*(1-3), 221–232. doi:10.1016/S0378-4274(01)00298-3

Sharpe, R. M., & Irvine, D. S. (2004). How strong is the evidence of a link between environmental chemicals and adverse effects on human reproductive health? *BMJ (Clinical Research Ed.)*, *328*(7437), 447–451. doi:10.1136/bmj.328.7437.447

Skovmand, A., Jensen, A. C., Maurice, C., Marchetti, F., Lauvås, A. J., Koponen, I. K., Jensen, K. A., Goericke-Pesch, S., Vogel, U., & Hougaard, K. S. (2019). Effects of maternal inhalation of carbon black nanoparticles on reproductive and fertility parameters in a four-generation study of male mice. *Particle and Fibre Toxicology*, *16*(1), 13. doi:10.118612989-019-0295-3

Specht, I. O., Bonde, J. P., Toft, G., Lindh, C. H., Jonsson, B. A., & Jorgensen, K. T. (2015). Serum phthalate levels and time to pregnancy in couples from Greenland, Poland and Ukraine. *PLoS One*, *10*(3), e0120070. doi:10.1371/journal.pone.0120070

Thulstrup, A. M., & Bonde, J. P. (2006). Maternal occupational exposure and risk of specific birth defects. *Occupational Medicine*, *56*(8), 532–543. doi:10.1093/occmed/kql115

Tuc, V. P., Wangsuphachart, V., Tasanapradit, P., Fungladda, W., Van Trong, P., & Nhung, N. T. (2007). Impacts of pesticide use on semen characteristics among rice famers in Kienxuong District, Thaibinh Province, Vietnam. *The Southeast Asian Journal of Tropical Medicine and Public Health*, *38*(3), 569.

Vandenberg, L. N., Colborn, T., Hayes, T. B., Heindel, J. J., Jacobs, D. R. Jr, Lee, D. H., Shioda, T., Soto, A. M., vom Saal, F. S., Welshons, W. V., Zoeller, R. T., & Myers, J. P. (2012). Hormones and endocrine-disrupting chemicals: Low-dose effects and nonmonotonic dose responses. *Endocrine Reviews*, *33*(3), 378–455. doi:10.1210/er.2011-1050

Wennborg, H., Bonde, J. P., Stenbeck, M., & Olsen, J. (2002). Adverse reproduction outcomes among employees working inbiomedical research laboratories. *Scandinavian Journal of Work, Environment & Health*, *28*(1), 5–11. doi:10.5271jweh.640

Windham, G. C., Lee, D., Mitchell, P., Anderson, M., Petreas, M., & Lasley, B. (2005). Exposure to organochlorine compounds and effects on ovarian function. *Epidemiology (Cambridge, Mass.)*, *16*(2), 182–190. doi:10.1097/01.ede.0000152527.24339.17

Wirth, J. J., & Mijal, R. S. (2010). Adverse effects of low level heavy metal exposure on male reproductive function. *Systems Biology in Reproductive Medicine*, *56*(2), 147–167. doi:10.3109/19396360903582216

Wisniewski, P., Romano, R. M., Kizys, M. M., Oliveira, K. C., Kasamatsu, T., Giannocco, G., Chiamolera, M. I., Dias-da-Silva, M. R., & Romano, M. A. (2015). Adult exposure to bisphenol A (BPA) in Wistar rats reduces sperm quality with disruption of the hypothalamic–pituitary–testicular axis. *Toxicology*, *329*, 1–9. doi:10.1016/j.tox.2015.01.002

Woodruff, T. J., Zota, A. R., & Schwartz, J. M. (2011). Environmental chemicals in pregnant women in the United States: NHANES 2003–2004. *Environmental Health Perspectives*, *119*(6), 878–885. doi:10.1289/ehp.1002727

World Health Organization (WHO). (2018). *Global Prevalence of Infertility, Infecundity and Childlessness*. http://www.who.int/reproductivehealth/ topics/infertility/burden/en

Yarsley, V. E., & Couzens, E. G. (1945). *Plastics Middlesex*. Penguin Books Limited.

Ye, M., Beach, J., Martin, J. W., & Senthilselvan, A. (2017). Pesticide exposures and respiratory health in general populations. *Journal of Environmental Sciences (China)*, *51*, 361–370. doi:10.1016/j.jes.2016.11.012

Youness, E. R., Mohammed, N. A., & Morsy, F. A. (2012). Cadmium impact and osteoporosis: Mechanism of action. *Toxicology Mechanisms and Methods*, *22*(7), 560–567. doi:10.3109/15376516.2012.702796

Yucra, S., Rubio, J., Gasco, M., Gonzales, C., Steenland, K., & Gonzales, G. F. (2006). Semen quality and reproductive sex hormone levels in Peruvian pesticide sprayers. *International Journal of Occupational and Environmental Health*, *12*(4), 355–361. doi:10.1179/oeh.2006.12.4.355

Zegers-Hochschild, F., Adamson, G. D., Dyer, S., Racowsky, C., De Mouzon, J., Sokol, R., Rienzi, L., Sunde, A., Schmidt, L., Cooke, I. D., Simpson, J. L., & van der Poel, S. (2017). The international glossary on infertility and fertility care, 2017. *Human Reproduction (Oxford, England)*, *32*(9), 1786–1801. doi:10.1093/humrep/dex234

Zeng, X., Jin, T., Zhou, Y., & Kong, Q. (2002). Alterations of serum hormone levels in male workers occupationally exposed to cadmium. *Journal of Toxicology and Environmental Health. Part A.*, *65*(7), 513–521. doi:10.1080/15287390252807975

Zhou, W., Fang, F., Zhu, W., Chen, Z. J., Du, Y., & Zhang, J. (2017). Bisphenol A and ovarian reserve among infertile women with polycystic ovarian syndrome. *International Journal of Environmental Research and Public Health*, *14*(1), 18. doi:10.3390/ijerph14010018

Zhou, Y., Fu, X. M., He, D. L., Zou, X. M., Wu, C. Q., Guo, W. Z., & Feng, W. (2016). Evaluation of urinary metal concentrations and sperm DNA damage in infertile men from an infertility clinic. *Environmental Toxicology and Pharmacology*, *45*, 68–73. doi:10.1016/j.etap.2016.05.020

Ziv-Gal, A., & Flaws, J. A. (2016). Evidence for bisphenol A-induced female infertility: A review (2007–2016). *Fertility and Sterility*, *106*(4), 827–856. doi:10.1016/j.fertnstert.2016.06.027

KEY TERMS AND DEFINITIONS

Cryptorchidism: A condition in which one or both testes hasn't moved into the scrotum before birth.

Epidemiology: The study of the distribution and determinants of health-related states and events in specified populations and the application of this study to the control of health problems.

Infertility: It is the inability of a couple to reproduce naturally or we can say not able to get pregnant.

Pesticides: A group of chemical substances that are used to kill, repel, or control the pests including rodents, insects, certain forms of plants and fungi.

Phthalates: A family of chemical compounds used in hundreds of products mainly as plasticizers, increase the flexibility, transparency, durability, and longevity to plastics.

Spermatogenesis: A process of formation of male germ cells in the testis.

Sterility: It's the physiological inability to produce offspring or to conceive a child.

Testicular Dysgenesis Syndrome: The malformations of the male reproductive organs beside damaged semen quality, lower testosterone production and testicular germ cell cancer, constitute a syndrome of testicular dysgenesis (TDS) with a common origin in fetal life.

Xenoestrogen: Substances that are close enough in molecular structure to estrogen also called as "foreign" estrogens, can bind to estrogen receptor sites with potentially hazardous outcomes.

Chapter 11
Impact of Climate Change on Animal Fertility

Urfeya Mirza
SKUAST, India

Uiase Bin Farooq
SKUAST, India

Shahnaz Anjum
University of Kashmir, Srinagar, India

ABSTRACT

The fertility and reproductive efficiency of both male and female animals is very sensitive to climatic disturbances, particularly hyperthermia. The most evident consequences of heat stress are decreased quantity and quality of sperm production in males and compromised fertility in females. In this chapter, the effects of climate change on male and female reproduction have been separately and thoroughly elaborated. The effect of heat stress on embryo and fetal development has also been described. In the end, various approaches for mitigating the effects of heat stress on animals have been brought to light.

INTRODUCTION

Animals play an exceptional role in the human population chiefly in the form of the products they produce, and this role has grown progressively during the last 40 years (McLeod, 2011). Despite rapid development in management and genetics, the reproductive performance of high producing animals has suffered a remarkable decrease (Royal *et al.*, 2000; Lucy, 2001; López-Gatius, 2003).The blend of elements including temperature, humidity, rainfall, radiation, air movement, ionization, and atmospheric pressure constitutes the climate (Johnson, 1987). Every homoeothermic animal has an optimal ambient temperature zone (Thermo neutral Zone) within which the metabolic rate is at minimum (Yousef, 1985; Johnson, 1987) (Fig. 1). The comfortable range of environmental temperatures for the production or reproduction of most of the animals is between -0.5° and 20°C (Johnson, 1987; Verga*et al.*, 2007).

DOI: 10.4018/978-1-7998-4480-8.ch011

However, over the past several decades due to anthropical and natural factors, the earth has been changing rapidly and there are predicted temperature increments of over 0.1-0.2°C per decade (MacCracken *et al.*, 2008), which has led toa significant increase in heat stress on human, animal and plant population (Lecha, 2007). One of the principal causes of low fertility in livestock is heat stress (Wolfenson*et al.*, 2000) and it has been mostly studied dairy cattle (López-Gatius, 2003; Collier *et al.*, 2006; Roth, 2008). Environmental temperature, relative humidity, wind speed, and radiation all contribute to the level of heat stress (De Rensis and Scaramuzzi, 2003). Elevated temperatures have negative effects on reproductive processes in both female and male animals (Das *et al.*, 2016). In females, it leads to premature aging of the oocyte (Payton *et al.*, 2004), reduced conception rate (Bridges *et al.*, 2005) while in males it may lead to temporary or permanent infertility (Yaeram*et al.*, 2006; Tusell*et al.*, 2011). High ambient temperatures affect the sexual behaviour in animals (Bolocan, 2013; Graves *et al.*, 2018; Boni, 2019). Heat stress may also lead to premature embryo loss in animals (Ju*et al.*, 2005). Consequently, heat stress is a general worldwide problem distressing 60% of the livestock population, which causes huge economic losses (Wolfenson*et al.*, 2000). This is the reason why research in the last decades has been focused on ways to mitigate heat stress in animals (Ealy*et al.*, 1993; West, 2003; Wolfenson*et al.*, 1995).

Figure 1. Thermal zones in animals (adapted from Yousef, 1985).

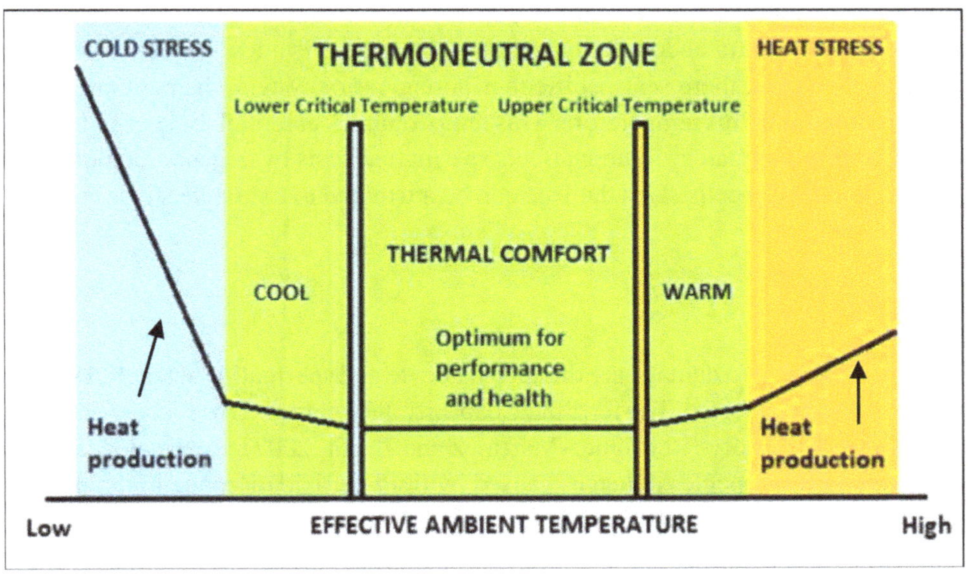

Impact of Climate Change on Animal Reproduction

In both the male and female animals, the processes reproductive are very sensitive to disturbance by hyperthermia, with the most evident consequences being decreased quantity and quality of sperm production in males and compromised fertility in females. Each aspect of reproductive function is disrupted either by the significant change in body temperature caused by high environmental temperature or by the physiological adaptations encountered by the animal body to reduce the abnormal increase in temperature. Heat stress considerably reduces the fertility of female animals by poor expression of estrus due to hormonal imbalance or by the bodily lethargy felt by heat-stressed animals. It addition to this, it affects the ovarian follicle development and normal follicular dynamics pattern; thereby compromising thegrowth of the female gamete. During pregnancy, heat stress inhibits the growth of the fetus thereby augmenting the chances of fetal loss and affects the feed intake of animals negatively affecting the weight of the conceptus (Sejian,2012). In males, heat stress affects spermatogenesis by damaging thespermatogonal germ cells, Sertoli cells, orLeydig cells. The stress-related damage to testes is due to hypoxia leading to oxidative stress and germ cell destruction. This causes analteration in the semen attributes.Furthermore, there is also an effect on sexual behavior in male animals. Heat stress leads to impairment of development and increased mortality of embryos due to a rise in intracellular reactive oxygen species and fall in glutathione levels, which is unfavorable to embryo survivability(Hansen, 2007). Highest incidence of embryonic motility due to heat stress is experienced in first week of pregnancy.

Effects on Male Animal Fertility

- ***Testes***

Various parameters like scrotal circumference, testicular consistency, tone, size, and weight determine spermatogenic functions. Stress due to extremes of temperature destroys the germinal epithelium and atrophy of seminiferous tubules, thereby, altering the testicular measurements. In most mammals, testes are located in the scrotum extra abdominally and the temperature is a few degrees lower than standard body temperature. Moreover, there is a specific thermoregulatory system, the pampini form plexus and scrotal muscles, maintaining this temperature. This temperature is essential for spermatogenesis and any disturbance to the thermoregulatory system affects spermatogenesis by degeneration of spermatogonial germ cells, sertoli and leydig cells. This damage can be attributed to oxidative stress caused by hypoxia (Paul*et al.,* 2009).

- ***Male Gametes***

The cells most vulnerable to damage are the spermatocyte and spermatid (Setchell, 1998); however, B spermatogonia can also be affected. The fertilizing capacity and reproductive success of the spermatozoa are directly related to its quality (Rodriguez-Martinez and Barth, 2007). Spermatozoa are particularly sensitive to reactive oxygen species damages caused by lipid peroxidation due to the presence of polyunsaturated fatty acids in higher amounts in spermatozoa (Alvarez *et al.*, 1987). Reactive oxygen species reduce the viability and motility of sperms and induce lipid peroxidation and acrosome reaction (Fatma *et al.*, 2009) which leads to infertility (Tremellen, 2008).

- ***Semen Characteristics***

The semen characteristics are not instantaneouslyinfluenced by alterations in temperature of testes because for some time after temperature stress,the destroyedspermatogenic cells do not go intothe ejaculates. The semen attributes like sperm motility, sperm morphology, sperm concentration, sperm viability, and acrosome integrity are negatively affected which may, in turn, lead to infertility or decreased fertility. The other components of seminal plasma like total lipid, total protein, cholesterol, globulin, alkaline phosphatase, and creatinine are not affected by heat stress, while acid phosphatase and albumin concentrations in the seminal plasma are significantly reduced as seen in bulls (Vogleraet *al.,* 1993).

- ***Libido***

As in females, reproductive behavior in males is also influenced by high temperatures. During summer, a decrease in libido is shown by male animals; however, normal mating can occur during cooler parts of the day (Winfield *et al.*, 1981). In severe heat stress, the decrease in libido is accompanied by degradation of semen quality which can be seen for about two months after the relapse of heat stress in bulls (Hansen, 2005) and rabbits (Zeidanet *al.*, 1997; Daader and Seleem, 1999). In other species like boars, the decrease in libido is compensated by increase in ejaculation time, fertility remaining unaffected (De Alba and Riera, 1966; Egbunike and Dede,1979).

- ***Hormones***

Plasma testosterone concentration decreases following heat stress and afterward in the acclimatization process, serum testosterone levels return to the normal levels (Wettemann and Desjardins, 1979). Thus, heat stress leads to a temporary decrease inplasma testosterone concentrations. It has been seen in bulls and boars (Rhynes and Ewing, 1973; Wettemann and Desjardins, 1979) that high temperature decreases thecapability of the testes to synthesize testosterone. However, in sheep, heat stress has no significant effect on testosterone concentration (Rasoolìet *al.,* 2010). As heat stress alters LH production in females, similarly it can compromise LH secretion in males as well.

Effects on Female Animal Fertility

- ***Estrus***

High ambient temperature can influence the onset (Sejian *et al.,* 2011), length (Abilayet *al.*, 1975), intensity (Gangwaret *al.*, 1965) and expression (Nobel *et al.*, 1997) of estrus and also increases the length of anestrous and silent ovulation (Gwazdauskaset *al.*, 1981). The possible reason to this is the bodily lethargy created by high temperatures, as an adaptation to limit heat production. In case of ewes, modifications in the pulsatile release of luteinizing hormone (LH) and decrease in estrogen levelsafter heat stress are possible reasons for the delay of onset of estrus (Dobson and Smith, 2000). Variations in estrus lead to significant decrease in the number of mounts in hot climate (Pennington *et al.*, 1985).

- ***Ovarian Follicles***

Heat stress ceases ovarian follicular development resulting in a decreased reproductive competence of domestic animals during summer. Heat stress affects follicle development (Roth *et al.*, 2000) by causing metabolic and hormonal alterations. All the changes are evident in the dominant follicular fluid (Shehab-El-Deenet *al.*, 2010) and consequently influence the development of oocyte lying within (Leroy *et al.*, 2005). Follicular growth is influenced by high temperatures due to the decrease in the dimension of the first-wave of dominant follicles and the prior appearance of the second-wave of dominant preovulatory follicles (Wolfensonet *al.*, 1995; Roth *et al.*, 2000).

- ***Oocytes***

The mechanism by which heat stress disrupts oocyte function is most probably due to alterations in follicular function. The preovulatory oocyte is very sensitive to increased temperatures as shown in cows (Putney *et al.*, 1989) and mice (Roth *et al.*, 2008). In addition, the process of oocyte maturation is influenced by high temperatures (Wang *et al.*, 2009). During the preovulatory period, heat stress involves the production of reactive oxygen species, as the effects of heat stress are reduced by the administration of antioxidants (Roth *et al.*, 2008). While as, for maturing oocytes, apoptosis plays a significant role in the effects of thermal stress (Soto and Smith, 2009). After heat stress is subsided, a recovery period of two to three estrous cycles is required for the appearance of normal oocytes again (Roth *et al.*, 2001).

- ***Sexual Behavior***

Stress due to high temperature, decreases the usual expression of various sexual behaviors, leading to decreased productivity of animals. There is an abnormality in various forms of estrus sexual behaviorsin ewes due to heat stress e.g., tail fanning, standing, circling, head-turning, and approaching males (Lynch*et al.*,1992). Heat stress also affects sexual behavior in heifers (Bolocan, 2013), gilts (Graves *et al.*, 2018), buffaloes (Marai and Haeeb, 2010) and mice (Politch and Herrenkohl, 1984).

- ***Fertilization, Embryo- and Fetal-development***

Heat stress reduces fertilization rate following in vitro fertilization (IVF) of bovine oocytes, decreases conception rates, and affects early cleavage of embryos (Edwards and Hansen, 1996; Rivera and Hansen, 2001). The effect of heat stress relapses as the embryo develops to form morula or blastocyst from a 2-cell stage (Ealy*et al.,* 1993). The effect is seen in sheep (Dutt, 1964) and pigs (Tompkins *et al.*, 1967) but not on rats (Arechiga and Hansen, 1998). High temperature leads to increased incidences of abnormalities in bovine embryos (Putney *et al.,* 1989). This may again be attributed to the production of reactive oxygen species after exposure to high temperatures. Cows if subjected to heat stress, decrease the intake of dry matter which in turn reduces their energy at the cost of the energy required to maintain pregnancy (Al-Katanani *et al.,* 2002). Heat stress also affects the placental function by disturbing blood perfusion of the placental vascular bed which results in the reduced weight of the fetus in ewes(Wallace *et al.*, 2005). As angiogenesis is more extensive in the mid-gestation phase as compared to the late gestation phase, heat stress affect smore during the in mid-gestation. The effect of maternal heat stress on fetal development occurs in most of the animals. Heat stress reduces the release of placental hormones and the effect of this can be seen on the milk yield (Wolfenson*et al.*, 1988), which indirectly affects the nutrition of the neonate. Maternal heat stress may also increase the incidence of fetal abnormalities (Graham *et al.*, 1998).

- ***Hypothalamo-pituitary-Ovarian (HPO) Axis***

The main factors regulating ovarian activity are gonadotrophin-releasing hormone from the hypothalamus and the gonadotrophins, LH and FSH from the anterior pituitary gland. The effects of heat stress on these hormones and their concentrations in peripheral blood may be observed to be inconsistent due to variable reasons related to physiology and state of health of the animal. The levels remain unaltered

(Gauthier, 1986) while others report increase in concentrations (Roman-Ponce *et al.*, 1981), and some report a decline in concentrations (Gilad *et al.*, 1993; Lee, 1993) due to heat stress.

Further, heat stress may lead to an endocrine imbalance which in turn affects the folliculogenesis (De Rensis and Scaramuzzi, 2003) by destroying the luteal and follicular cells, decreasing hormone production and interfering with oocyte development (Torres-Junior *et al.*, 2008). There is decrease in plasma estradiol and inhibin concentration but no effect on follicle-stimulating hormone (FSH) secretion in heat-stressed animals (Wolfenson *et al.*, 1995). In goats, heat stress reduced plasma concentrations of estradiol, aromatase activity and delayed ovulation (Ozawa *et al.*, 2005) while in the case of rats, heat stress reduced the levels of gonadotropin receptors (Shimizu *et al.*, 2005). There is reduced concentration of progesterone (Wolfenson*et al.*, 1993) and LH (Rosenberg *et al.*, 1982) during heat stress in animals. Decreased concentration of progesterone in plasma has a negative effect on fertility, particularly during luteal and pre-conception phase of estrous cycle (Ahmad *et al.*, 1995). High temperature also affects endometrial prostaglandin production which may result in premature luteolysis and embryo loss (Vasconcelos*et al.*, 1998).

Fig. 2 provides the description of the mechanisms generated by heat stress which may affect reproduction in the lactating dairy cow. Heat stress can act in more than one way to reduce fertility in lactating dairy cow. Heat stress can reduce dry matter intake to indirectly inhibit GnRH and LH secretion from the hypothalamo-pituitary system (dashed red lines). However, it is not clear if heat stress can also directly influence the hypothalamo-pituitary system (dashed blue line) to reduce GnRH and LH secretion. Heat stress can directly compromise the uterine environment (solid lines) to cause embryo loss and infertility (De Rensis and Scaramuzzi, 2003).

Figure 2. Effect of heat stress on hypothalamo-pituitary-ovarian system (adapted from De Rensis and Scaramuzzi, 2003).

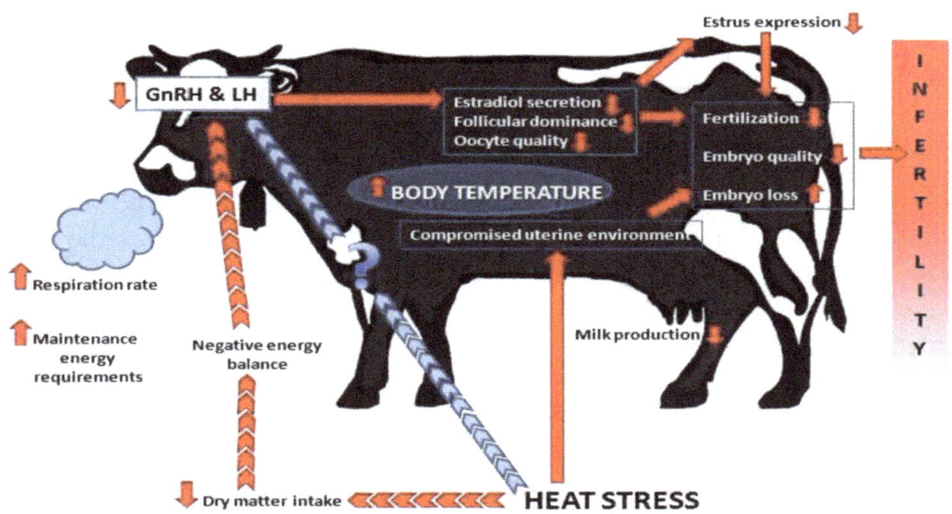

APPROACHES FOR MITIGATION OF HEAT STRESS EFFECTS

To improve fertility during summers or under other conditions of heat stress, some approaches can be developed:

Management Procedures

Various management procedures like the use of shade and evaporative cooling systems can be used to lessen the effect of increased temperatures. The high quality feed at a regular time and the cool drinking water could be made available all the times to counteract heat stress. Use of estrus detection aids and hormones can be brought into practice to improve summer fertility by eliminating the need for estrus detection. Moreover, the use of novel methods like embryo transfer technology can improve pregnancy rates under heat stress. Most importantly, the selection of breeds that are genetically adapted to the hot environment would improve fertility. Enhanced reproductive performance comes under three broad categories. The first category includes the practices associated with the day-to-day management of a dairy herd. Rate of detection of estrus, rate of conception, prevalence of some common reproductive disorders such as retained fetal membranes, dystocia, ovarian cystic follicles, and uterine infections, and management-related decisions such as interval of postpartum delay before the first insemination, application of artificial insemination contrary to natural service, and decisions about culling are among the important traits in this category. The second category involves pharmacological control of reproductive processes and systems for relating this control to a herd. Here primary emphasis is laid on the precise control of the ovulation time so that inseminations can be programmed at preset times. The third category involves biotechnical procedures that have the latency to increase the rate of genetic changes in a population. The practices in this area include storage and transfer of embryo, control of sex ratios, cloning of genetically upgraded individuals, and insertion of the reproductive performance enhancing nuclear and mitochondrial. Also, among and within herds, the conception rates vary considerably and several factors contribute to this variation. The effects of milk yield on conception rates appear marginal among the controlled studies in well-managed herds (Britt, 1982; Fonseca *et al.*, 1983; Hillers *et al.*, 1984). The conception rates are affected by parity affects (Fonseca *et al.*, 1983; Oltenacu*et al.*, 1983; Freeman, 1984; Hillers *et al.*, 1984; Ron *et al.*, 1984) and it has been estimated that the postpartum interval to first insemination has little effect on fertility beyond about 50 to 60 days postpartum (Hillers *et al.*, 1984; Ron *et al.*, 1984). Environment has a great impact on conception rate of bovine, such as the environmental temperature affecting the rate of conception in lactating cows (Ron *et al.*, 1984).

Use of Antioxidant Agents

Nowadays, the antioxidants are frequently used as a pharmacological aide to limit infertility. Certainly, people rightly perceive oxidative stress and antioxidant treatment as significant modulators of infertility. While the direct effects of this treatment on the quality of semen and oocytes are still under investigation, a momentous figure of indications points over to the loss of vascular tone as a major cause of erectile dysfunction and, probably, alterations to female reproduction. Administration of antioxidants improves fertility in heat-stressed animals (Cetica *et al.*, 2001). Both enzymatic antioxidants (like Superoxide dismutase, Catalase, Glutathione peroxidase) and non-enzymatic antioxidants (like Glutathione, Bovine serum albumin, Taurine, Carotenoids, Fatty acids, Vitamins A, E, and C) can be incorporated into the daily diet of animals to counteract heat stress.Regardless of the multifactorial nature of human infertility, there is a rising awareness that reactive oxygen and nitrogen species (ROS/RNS) and the concomitant oxidative damage may prove as potent modulators of reproductive health (Lanzafameet *al.*, 2009; Desai *et al.*, 2010). Oxidative and nitrosative stress is related to both the risk factors for infertility (e.g. diabetes, hypertension, aging, and smoking), and directly in reproductive disorders as sundry as endometriosis,pre-eclampsia, and oocyte implantationin women,and erectile dysfunction, motility, and sperm damage in men (Sikka, 2004; Desai *et al.*, 2010).There is a convincing hypothetical justification to limit oxidative stress associated with these conditions by the use of antioxidant therapies or pharmacological agents, which prompt endogenous antioxidant defenses. Majority of the evidences showing therapeutic efficiency of antioxidants are from experimental animal models or in vitro studies (Lanzafame *et al.*, 2009).

CONCLUSION

The climate in general and hyperthermia in particular has a pronounced effect on fertility and reproductive efficiency of animals.The most evident consequences of hyperthermia are decreased quantity and quality of sperms in males and compromised fertility in females.Moreover in females, heat stress affects the onset, length, intensity and expression of estrus, development of ovarian follicles, and disruption of preovulatory oocyte function. In males, it affects the libido, testicular morphology, seminal attributes, and spermatozoan characteristics. Steroidogenesis and sexual behavior are affected in both male and female animals. Heat stress also affects embryo and fetal development. Heat stress involves the production of reactive oxygen species that are responsible for most of the damage caused. To counteract the damage caused and to improve fertility during the conditions of heat stress, some approaches can be developed which include various managemental practices, nutritional management, use of hormones, use of estrus detection aids, use of embryo transfer or selection of breeds that are genetically adapted to the hot environment. Enzymatic and non-enzymatic antioxidant agents can be administrated to the heat-stressed animals which decrease the reactive oxygen species and can improve the reproductive performance.

AUTHOR CONTRIBUTION

UM designed and wrote the book chapter. UBF performed literature survey. SA performed manuscript editing.

FUND DISCLAIMER

No grant received from any funding agency in public, commercial, or not-for-profit sector.

CORRESPONDING AUTHOR

Urfeya Mirza

REFERENCES

Abilay, T. A., Johnson, H. D., & Madam, L. M. (1975). Influence of environmental heath on peripheral plasma progesterone and cortisol during the bovine estrus cycle. *Journal of Dairy Science*, *58*(12), 1836–1840. doi:10.3168/jds.S0022-0302(75)84795-3 PMID:1239464

Ahmad, N., Schrick, F. N., Butcher, R. L., & Inskeep, E. K. (1995). Effect of persistent follicles on early embryonic losses in beef cows. *Biology of Reproduction*, *52*(5), 1129–1135. doi:10.1095/biolreprod52.5.1129 PMID:7626713

Al-Katanani, Y. M., Paula-Lopes, F. F., & Hansen, P. J. (2002). Effect of season and exposure to heat stress on oocyte competence in Holstein cows. *Journal of Dairy Science*, *85*(2), 390–396. doi:10.3168/jds.S0022-0302(02)74086-1 PMID:11913699

Alvarez, J. G., Touchstone, J. C., Blasco, L., & Storey, B. (1987). Spontaneous lipid peroxidation and production of hydrogen peroxide and superoxide in human spermatozoa: Superoxide dismutase as a major enzyme protectant against oxygen toxicity. *Journal of Andrology*, *8*(5), 338–348. doi:10.1002/j.1939-4640.1987.tb00973.x PMID:2822642

Arechiga, C. F., & Hansen, P. J. (1998). Response of preimplantation murine embryos to heat shock as modified by developmental stage and glutathione status. *In Vitro Cellular & Developmental Biology. Animal*, *34*(8), 655–659. doi:10.100711626-996-0016-8 PMID:9769152

Bolocan, E. (2013). Effects of heat stress on sexual behavior in heifers. *Lucrari Stiintifice. Zootehnie si Biotehnologii*, *42*, 141–148.

Boni, R. (2019). Heat stress, a serious threat to reproductive function in animals and humans. *Molecular Reproduction and Development*, *86*(10), 1307–1323. doi:10.1002/mrd.23123 PMID:30767310

Bridges, P. J., Brusie, M. A., & Fortune, J. E. (2005). Elevated temperature (heat stress) in vitro reduces androstenedione and estradiol and increases progesterone secretion by follicular cells from bovine dominant follicles. *Domestic Animal Endocrinology*, *29*(3), 508–522. doi:10.1016/j.domaniend.2005.02.017 PMID:16153500

Britt, J. H. (1982). Phenotypic relationships between reproductive performance and milk yield in dairy cattle. In *Procceedings of the National Invitational Dairy Cattle Reproduction Workshop.*

Cetica, P. D., Pintos, L. N., Dalvit, G. C., & Beconi, M. T. (2001). Antioxidant enzyme activity and oxidative stress in bovine oocyte in vitro maturation. *IUBMB Life*, *51*(1), 57–64. doi:10.1080/15216540152035073 PMID:11419698

Collier, R. J., Dahl, G. E., & VanBaale, M. J. (2006). Major advances associated with environmental effects on dairy cattle. *Journal of Dairy Science*, *89*(4), 1244–1253. doi:10.3168/jds.S0022-0302(06)72193-2 PMID:16537957

Daader, A. H., & Seleem, T. S. T. (1999). Recent trends in rabbit production. *Proceedings of 1st International Conference on Indigenous Versus Acclimatized Rabbits,* 23-50.

Das, R., Sailo, L., Verma, N., Bharti, P., Saikia, J., Imtiwati, & Kumar, R. (2016). Impact of heat stress on health and performance of dairy animals: A review. *Veterinary World*, *9*(3), 260–268. doi:10.14202/vetworld.2016.260-268 PMID:27057109

De Alba, & Riera, S. (1966). Sexual maturity and spermatogenesis under heat stress in the bovine. *Animal Production*, *8*, 137–144.

De Rensis, F., & Scaramuzzi, R. J. (2003). Heat stress and seasonal effects on reproduction in the dairy cow – a review. *Theriogenology*, *60*(6), 1139–1151. doi:10.1016/S0093-691X(03)00126-2 PMID:12935853

Desai, N., Sabanegh, E. Jr, Kim, T., & Agarwal, A. (2010). Free radical theory of aging: Implications in male infertility. *Urology*, *75*(1), 14–19. doi:10.1016/j.urology.2009.05.025 PMID:19616285

Dobson, H., & Smith, R. F. (2000). What is stress and how does itaffects fertility. *Animal Reproduction Science*, *60*, 743–752. doi:10.1016/S0378-4320(00)00080-4 PMID:10844239

Dutt, R. H. (1964). Detrimental effects of high ambient temperatureon fertility and early embryo survival in sheep. *International Journal of Biometeorology*, *8*(1), 47–56. doi:10.1007/BF02186927 PMID:5888798

Ealy, A. D., Drost, M., & Hansen, P. J. (1993). Developmental changes in embryonic resistance to adverse effects of maternal heat stress in cows. *Journal of Dairy Science*, *76*(10), 2899–2905. doi:10.3168/jds.S0022-0302(93)77629-8 PMID:8227617

Edwards, J. L., & Hansen, P. J. (1996). Elevated temperature increases heat shock protein 70 synthesis in bovine two-cell embryos and compromises function of maturing oocyte. *Biology of Reproduction*, *55*(2), 340–346. doi:10.1095/biolreprod55.2.341

Egbunike, G. N., & Dede, T. I. (1979). The influence of short-term exposure to tropical sunlight on boar seminal characteristics. *Biology of Reproduction*, *20*, 235–241.

Fatma, B. A., Nozha, C. F., Ines, D., Hamadi, A., Basma, H., & Leila, A. K. (2009). Sperm quality improvement after date seed oil in vitro supplementation in spontaneous and induced oxidative stress. *Asian Journal of Andrology*, *11*(3), 393–398. doi:10.1038/aja.2008.6 PMID:19151733

Fonseca, F. A., Britt, J. H., McDaniel, B. T., Wilk, J. C., & Rakes, A. H. (1983). Reproductive traits of Holsteins and Jerseys. Effects of age, milk yield, and clinical abnormalities on involution of cervix and uterus, ovulation, estrous cycles, detection ofestrus, conception rate, and days open. *Journal of Dairy Science*, *66*(5), 1128–1147. doi:10.3168/jds.S0022-0302(83)81910-9 PMID:6683729

Freeman, A. E. (1984). Secondary traits: Sire evaluation and the reproductive complex. *Journal of Dairy Science*, *67*(2), 449–458. doi:10.3168/jds.S0022-0302(84)81324-7 PMID:6371077

Gangwar, P. C., Branton, C., & Evans, D. L. (1965). Reproductive and physiological response of Holstein heifers to controlled and natural climatic conditions. *Journal of Dairy Science*, *48*(2), 222–227. doi:10.3168/jds.S0022-0302(65)88200-5 PMID:14277425

Gauthier, D., Coulaud, G., & Varo, H. (1986). The influence of season and shade on estrous behaviour, timing of preovulatory LH surge and the pattern of progesterone secretion in FFPN and Creole heifers in a tropical climate. *Reproduction, Nutrition, Developpement*, *26*(3), 767–775. doi:10.1051/rnd:19860502 PMID:3749595

Gilad, E., Meidan, R., Berman, A., Graber, Y., & Wolfenson, D. (1993). Effect of heat stress on tonic and GnRH induced gonadotrophin secretion in relation to concentration of oestradiol in plasma of cyclic cows. *Journal of Reproduction and Fertility*, *99*(2), 315–321. doi:10.1530/jrf.0.0990315 PMID:8107012

Graham, J. M. Jr, Edwards, M. J., & Edwards, M. J. (1998). Teratogen update: Gestational effects of maternalhyperthermia due to febrile illnesses and resultant patterns of defects in humans. *Teratology*, *58*(5), 209–221. doi:10.1002/(SICI)1096-9926(199811)58:5<209::AID-TERA8>3.0.CO;2-Q PMID:9839360

Graves, K., Seibert, J., Keating, A., Baumgard, L., & Ross, J. (2018). Characterizing the acute heat stress response in gilts: II.Assessing repeatability and association with fertility. *Journal of Animal Science*, *96*(6), 10. doi:10.1093/jaskx037 PMID:29788126

Gwazdauskas, F. C., Thatcher, W. W., Kiddy, C. A., Pape, M. J., & Wilcox, C. J. (1981). Hormonal pattern during heat stress following PGF2alpha-tham salt induced luteal regression in heifers. *Theriogenology*, *16*(3), 271–285. doi:10.1016/0093-691X(81)90012-1 PMID:16725640

Hansen, P. J. (2005). Managing the heat-stressed cow to improve reproduction. *Proceedings of the 7th Western Dairy Management Conference*, 9-11.

Hansen, P. J. (2007). To be or not to be - Determinants of embryonic survival following heat shock. *Theriogenology*, *68S*, S40–S48. doi:10.1016/j.theriogenology.2007.03.013 PMID:17467047

Hillers, J. K., Senger, P. L., Darlington, R. L., & Fleming, W. N. (1984). Effects of production, season, age of cow, days dry and days in milk on conception to first service in large commercial dairy herds. *Journal of Dairy Science*, *67*(4), 861–867. doi:10.3168/jds.S0022-0302(84)81378-8 PMID:6539346

Johnson, H. D. (1987). Bioclimates and livestock. In Bioclimatology and adaptation of livestock. Elsevier.

Ju, J. C., Jiang, S., Tseng, J. K., Parks, J. E., & Yang, X. (2005). Heat shock reduces developmental competence and alters spindle configuration of bovine oocytes. *Theriogenology*, *64*(8), 1677–1689. doi:10.1016/j.theriogenology.2005.03.025 PMID:15951010

Lanzafame, F. M., La Vignera, S., Vicari, E., & Calogero, A. E. (2009). Oxidative stress and medical antioxidant treatment in male infertility. *Reproductive Biomedicine Online*, *19*(5), 638–659. doi:10.1016/j.rbmo.2009.09.014 PMID:20021713

Lecha, L. B. (2007). *Fourth Assessment Report*. Cambridge University Press.

Lee, N. (1993). Environmental stress effect on bovine reproduction. *The Veterinary Clinics of North America*, *9*, 263–273. PMID:8348371

Leroy, J. L. M. R., Vanholder, T., Mateusen, B., Christophe, A., Opsomer, G., de Kruif, A., Genicot, G., & Van Soom, A. (2005). Non-esterified fatty acids in follicular fluid of dairy cows and their effect on developmental capacity of bovine oocytes in vitro. *Reproduction (Cambridge, England)*, *130*(4), 485–495. doi:10.1530/rep.1.00735 PMID:16183866

López-Gatius, F. (2003). Is fertility declining in dairy cattle? A retrospective study in northeastern Spain. *Theriogenology*, *60*(1), 89–99. doi:10.1016/S0093-691X(02)01359-6 PMID:12620583

Lucy, M. C. (2001). Reproductive loss in high-producing dairy cattle: Where will it end? *Journal of Dairy Science*, *84*(6), 1277–1293. doi:10.3168/jds.S0022-0302(01)70158-0 PMID:11417685

Lynch, J. J., Hinch, G. N., & Adams, D. B. (1992). *The behavior of sheep:biological principles and implications for production*. C.A.B. International.

MacCracken, M. C., Moore, F., & Topping, J. C. (Eds.). (2008). *Sudden and disruptive climate change: exploring the real risks and how we can avoid them*. Earthscan.

Marai, F. M. I., & Haeeb, A. A. M. (2010). Buffaloes' reproductive and productive traits as affected by heat stress. *Tropical and Subtropical Agroecosystems*, *12*(2), 193–217.

McLeod, A. (2011). *World livestock 2011-livestock in food security*. Food and Agriculture Organization of the United Nations (FAO). http://www.fao.org/docrep/014/i2373e/i2372e00.htm

Nobel, R. L., Jobst, S. M., Dransfield, M. B. G., Pandolfi, S. M., & Balley, T. L. (1997). Use of radio frequency data communication system, Heat Watch, to describe behavioural estrus in dairy cattle. *Journal of Dairy Science*, 179.

Oltenacu, P. A., Britt, J. H., Braun, R. K., & Mellenberger, R. W. (1983). Relationships among type of parturition, type of discharge from genital tract, involution of cervix, and subsequent reproductive performance in Holstein cows. *Journal of Dairy Science*, *66*(3), 612–619. doi:10.3168/jds.S0022-0302(83)81832-3 PMID:6841757

Ozawa, M., Tabayashi, D., Latief, T. A., Shimizu, T., Oshima, I., & Kanai, Y. (2005). Alterations in folliculardynamics and steroidogenic abilities induced by heatstress during follicular recruitment in goats. *Reproduction (Cambridge, England)*, *129*(5), 621–630. doi:10.1530/rep.1.00456 PMID:15855625

Paul, C., Teng, S., & Saunders, P. T. (2009). A single, mild, transientscrotal heat stress causes hypoxia and oxidative stress in mousetestes, which induces germ cell death. *Biology of Reproduction*, *80*(5), 913–919. doi:10.1095/biolreprod.108.071779 PMID:19144962

Payton, R. R., Romar, R., Coy, P., Saxton, A. M., Lawrence, J. L., & Edwards, J. L. (2004). Susceptibility of bovine germinal vesicle-stage oocytes from antral follicles to direct effects of heat stress in vitro. *Biology of Reproduction*, *71*(4), 1303–1308. doi:10.1095/biolreprod.104.029892 PMID:15201201

Pennington, J. A., Albright, J. L., Diekman, M. A., & Callahan, C. J. (1985). Sexual activity of Holstein cows: Seasonal effects. *Journal of Dairy Science*, *68*(11), 3023–3030. doi:10.3168/jds.S0022-0302(85)81197-8 PMID:4078128

Politch, J. A., & Herrenkohl, L. R. (1984). Effects of prenatal stress on reproduction in male and female mice. *Physiology & Behavior*, *32*(1), 95–99. doi:10.1016/0031-9384(84)90077-5 PMID:6718542

Putney, D. J., Mullins, S., Thatcher, W. W., Drost, M., & Gross, T. S. (1989). Embryonic development in superovulated dairy cattle exposed to elevated ambient temperatures between the onset of estrus and insemination. *Animal Reproduction Science*, *19*(1-2), 37–51. doi:10.1016/0378-4320(89)90045-6

Rasooli, A., Jalali, M. T., Nouri, M., Mohammadian, B., & Barati, F. (2010). Effects of chronic heat stress on testicular structures, serum testosterone and cortisol concentrations in developing lambs. *Animal Reproduction Science*, *117*(1-2), 55–59. doi:10.1016/j.anireprosci.2009.03.012 PMID:19428197

Rhynes, W. E., & Ewing, L. L. (1973). Testicular endocrine function in Hereford bulls exposed to high ambient temperature. *Endocrinology*, *92*(2), 509–515. doi:10.1210/endo-92-2-509 PMID:4682863

Rivera, R. M., & Hansen, P. J. (2001). Development of cultured bovine embryos after exposure to high temperatures in the physiological range. *Reproduction (Cambridge, England)*, *121*, 107–115. doi:10.1530/rep.0.1210107 PMID:11226033

Rodriguez-Martinez, H., & Barth, A. D. (2007). In vitro evaluation of sperm quality related to in vivo function and fertility. *Society of Reproduction and Fertility Supplement*, *64*(1), 39–54. doi:10.5661/RDR-VI-39 PMID:17494215

Roman-Ponce, H., Thatcher, W. W., & Wilcox, C. J. (1981). Hormonal interrelationships and physiological responses of lactating dairy cows to shade management system in a tropical environment. *Theriogenology*, *16*(2), 139–154. doi:10.1016/0093-691X(81)90097-2 PMID:16725628

Ron, M., Bar-Anan, R., & Wiggans, G. R. (1984). Factors affecting conception rate of Israeli Holstein cattle. *Journal of Dairy Science*, *67*(4), 861. doi:10.3168/jds.S0022-0302(84)81377-6 PMID:6725731

Rosenberg, M., Folman, Y., Herz, Z., Flamenbaum, I., Berman, A., & Kaim, M. (1982). Effect of climatic conditions on peripheral concentrations of LH, progesterone and estradiol-17b in high milk-yielding cows. *Journal of Reproduction and Fertility*, *66*(1), 139–146. doi:10.1530/jrf.0.0660139 PMID:7120177

Roth, Z. (2008). Heat stress, the follicle, and its enclosed oocyte: Mechanism and potential strategies to improve fertility in dairy cows. *Reproduction in Domestic Animals*, *43*(Suppl. 2), 238–244. doi:10.1111/j.1439-0531.2008.01168.x PMID:18638130

Roth, Z., Arav, A., Bor, A., Zeron, Y., Braw-Tal, R., & Wolfenson, D. (2001). Improvement of quality of oocytes collected in the autumn by enhanced removal of impaired follicles from previously heat- stressed cows. *Reproduction (Cambridge, England)*, *122*, 737–744. doi:10.1530/rep.0.1220737 PMID:11690534

Roth, Z., Aroyo, A., Yavin, S., & Arav, A. (2008). The antioxidantepigallocatechin gallate (EGCG) moderates thedeleterious effects of maternal hyperthermia on follicle enclosedoocytes in mice. *Theriogenology*, *70*(6), 887–897. doi:10.1016/j.theriogenology.2008.05.053 PMID:18585774

Roth, Z., Meidan, R., Braw-Tal, R., & Wolfenson, D. (2000). Immediate and delayed effects of heat stress on follicular development and its association with plasma FSH and inhibin concentration in cows. *Journal of Reproduction and Fertility*, *120*(1), 83–90. doi:10.1530/jrf.0.1200083 PMID:11006149

Royal, M. D., Darwash, A. O., Flint, A. P. F., Webb, R., Woolliams, J. A., & Lamming, G. E. (2000). Declining fertility in dairy cattle: Changes in traditional and endocrine parameters of fertility. *Animal Science (Penicuik, Scotland)*, *70*(3), 487–501. doi:10.1017/S1357729800051845

Sejian, V. (2012). Climate change, green house gas emission and sheepproduction. In Status Papers on Future Research in Sheep Production and Production Development, 50 Years Research Contributions 1962-2012. Central Sheep and Wool Research Institute.

Sejian, V., Lakritz, J., Ezeji, T., & Lal, R. (2011). Forage and Flaxseed impact on enteric methane emission in dairy cows. *ResearchJournal of Veterinary Sciences*, *4*(1), 1–8. doi:10.3923/rjvs.2011.1.8

Setchell, B. P. (1998). The Parkes Lecture.Heat and the testis. *Journal of Reproduction and Fertility*, *114*(2), 179–194. doi:10.1530/jrf.0.1140179 PMID:10070346

Shehab-El-Deen, M. A. M. M., Leroy, J. L. M. R., Fadel, M., Saleh, S., Maes, D., & Van Soom, A. (2010). Biochemical changes in the follicular fluid of the dominant follicle of high producing dairy cows exposed to heat stress early post partum. *Animal Reproduction Science*, *117*(3-4), 189–200. doi:10.1016/j.anireprosci.2009.04.013 PMID:19481380

Shimizu, T., Ohshima, I., Ozawa, M., Takahashi, S., Tajima, A., Shiota, M., Miyazaki, H., & Kanai, Y. (2005). Heat stress diminishes gonadotropin receptor expression and enhances susceptibility to apoptosis of rat granulose cells. *Reproduction (Cambridge, England)*, *129*(4), 463–472. doi:10.1530/rep.1.00502 PMID:15798021

Sikka, S. C. (2004). Role of oxidative stress and antioxidants in andrology and assisted reproductive technology. *Journal of Andrology*, *25*(1), 5–18. doi:10.1002/j.1939-4640.2004.tb02751.x PMID:14662779

Soto, P., & Smith, L. C. (2009). BH4 Peptide Derived From Bcl-xL and Bax-Inhibitor Peptide Suppresses Apoptotic Mitochondrial Changes in Heat Stressed Bovine Oocytes. *Molecular Reproduction and Development*, *76*(7), 637–646. doi:10.1002/mrd.20986 PMID:19062170

Tompkins, E. C., Heidenreich, C. J., & Stob, M. (1967). Effect of after-breeding thermal stress on embryonic mortality of swine. *Journal of Animal Science*, *26*(2), 377–380. doi:10.2527/jas1967.262377x

Torres-Junior, J. R. D., Pires, M. D. A., De Sa, W. F., Feffeira, A. D., Viana, J. H. M., Camargo, L. S. A., Ramos, A. A., Folhadella, I. M., Polisseni, J., de Freitas, C., Clemente, C. A. A., De Sa, M. F., Paula-Lopes, F. F., & Baruselli, P. S. (2008). Effect of maternal heat stress on follicular growth and oocyte competence in Bosindicus cattle. *Theriogenology*, *69*(2), 155–166. doi:10.1016/j.theriogenology.2007.06.023 PMID:17953981

Tremellen, K. (2008). Oxidative stress and male infertility-A clinical perspective. *Human Reproduction Update*, *14*(3), 243–258. doi:10.1093/humupd/dmn004 PMID:18281241

Tusell, L., David, I., Bodin, L., Legarra, A., Rafel, O., López-Bejar, M., & Piles, M. (2011). Using the product threshold model for estimating separately the effect of temperature on male and female fertility. *Journal of Animal Science*, *89*(12), 3983–3995. doi:10.2527/jas.2011-3924 PMID:21764834

Vasconcelos, J. L. M., Silcox, R. W., Lacerda, J. A., Pursley, G. R., & Wiltbank, M. C. (1998). Pregnancy rate, pregnancy loss, and response to heat stress after AI at 2 different times from ovulation in dairy cows. *Biology of Reproduction*, *56*(Suppl 1), 140.

Verga, M., Luzi, F., & Carenzi, C. (2007). Effects of husbandry and management systems on physiology and behaviour of farmed and laboratory rabbits. *Hormones and Behavior*, *52*(1), 122–129. doi:10.1016/j.yhbeh.2007.03.024 PMID:17482618

Voglera, C. J., Bamea, J. H., DeJarnettea, J. M., McGilliarda, M. L., & Saacke, R. G. (1993). Effects of elevated testicular temperature onmorphology characteristics of ejaculated spermatozoa in the bovine. *Theriogenology*, *40*(6), 1207–1219. doi:10.1016/0093-691X(93)90291-C

Wallace, J. M., Regnault, T. R., Limesand, S. W., Hay, W. W. Jr, & Anthony, R. V. (2005). Investigating the causes of low birth weight in contrasting ovine paradigms. *The Journal of Physiology*, *565*(1), 19–26. doi:10.1113/jphysiol.2004.082032 PMID:15774527

Wang, J. Z., Sui, H. S., Miao, D. Q., Liu, N., Zhou, P., Ge, L., & Tan, J. (2009). Effects of heat stress during in vitro maturation on cytoplasmic versus nuclear components of mouse oocytes. *Reproduction (Cambridge, England)*, *137*(2), 181–189. doi:10.1530/REP-08-0339 PMID:19029342

West, J. W. (2003). Effects of heat stress on production in dairy cattle. *Journal of Dairy Science*, *86*(6), 2131–2144. doi:10.3168/jds.S0022-0302(03)73803-X PMID:12836950

Wettemann, R. P., & Desjardins, C. (1979). Testicular function in boars exposed to elevated ambient temperature. *Biology of Reproduction*, *20*(2), 235–241. doi:10.1095/biolreprod20.2.235 PMID:454734

Winfield, C. G., Hemsworth, P. H., Galloway, D. B., & Makin, A. W. (1981). Sexual behaviour and semen characteristics of boars: Effects of high temperature. *Australian Journal of Experimental Agriculture and Animal Husbandry*, *21*(108), 39–45. doi:10.1071/EA9810039

Wolfenson, D., Flamenbaum, I., & Berman, A. (1988). Hyperthermia and body energy store effects on estrous behavior, conception rate, and corpus luteum function in dairy cows. *Journal of Dairy Science*, *71*(12), 3497–3504. doi:10.3168/jds.S0022-0302(88)79956-7 PMID:3235739

Wolfenson, D., Luft, O., Berman, A., & Meidan, R. (1993). Effects of season, incubation temperature and cell age on progesterone and prostaglandin F2α production in bovine luteal cells. *Animal Reproduction Science*, *32*(1-2), 27–40. doi:10.1016/0378-4320(93)90055-V

Wolfenson, D., Roth, Z., & Meidan, R. (2000). Impaired reproduction in heat-stressed cattle: Basic and applied aspects. *Animal Reproduction Science*, *60/61*, 535–547. doi:10.1016/S0378-4320(00)00102-0 PMID:10844222

Wolfenson, D., Thatcher, W. W., & Badinga, L. (1995). Effect of heat stress on follicular development during the estrous cycle in lactating dairy cattle. *Biology of Reproduction*, *52*(5), 1106–1113. doi:10.1095/biolreprod52.5.1106 PMID:7626710

Yaeram, J., Setchell, B. P., & Maddocks, S. (2006). Effect of heat stress on the fertility of male mice in vivo and in vitro. *Reproduction, Fertility, and Development*, *18*(6), 647–653. doi:10.1071/RD05022 PMID:16930511

Yousef, M. K. (1985). *Stress physiology in livestock*. CRC Press.

Zeidan, A. E. B., Marai, I. F. M., & Abd El-Kariem, Z. A. (1997). Effects of intratesticular injection of gonadotropin-releasing hormone on reproductive performance of low fertile male rabbits under Egyptian summer conditions. *Proceedings of 1st International Conference on Animal Production and Health*, 557–566.

Chapter 12
Climate Change and Livestock Fertility

Vishakha Shrimali
ICMR-National Institute of Occupational Health (NIOH), India

Nibedita Naha
ICMR-National Institute of Occupational Health (NIOH), Ahmedabad, India

Sukanta Mondal
ICAR-National Institute of Animal Nutrition and Physiology (ICAR-NIANP), India

ABSTRACT

Climate change is a global threat to livestock sector to so many species and ecosystem in different parts of the world. Climate change, heat stress, and nutritional stress are the major intriguing factors responsible for reduced fertility in farm animals in tropical countries. Heat and nutritional stresses affect the reproductive performance by decreasing the expression of estrous behavior, altering ovarian follicular development and hormonal profiles, compromising oocyte competence, and inhibiting embryonic development in livestock. Climate is changed by greenhouse gases that released into atmosphere through man-made activities. Livestock contribute 18% of the production of greenhouse gases itself and causes climate change including heat stress, which has direct and indirect impact on fertility of the animals as well as reduce milk production. Adaptation to climate change and lowering its negative effect by alteration of animal micro-environment using different essential technologies are the main mitigation strategies to recover heat stress damage in this respect.

INTRODUCTION

Livestock production is cover almost 45% of earth's land and one of the dominant lands use for business purpose; however, after use the land become variably different depending upon the environmental condition, and unsuitable for other purposes (Baumgard et al, 2012). Developing country like India,

DOI: 10.4018/978-1-7998-4480-8.ch012

agriculture is an integral part of livestock and most of the people depends on that with various reason of economical support (Das, 2017). The world population is increased approximately 1 billion inhabitants during the past 12 years and make the count till 7.6 billion in 2017 according to the United Nations. However, these population growths are slower than last 10 years (1.24% and 1.10% respectively, per year) and predict that it will increase 8.6 billion in 2030 and 9.8 billion in 2050. This population growth and demand of people to get the income and to maintain the economic balance give rise to the demand for livestock products in developing countries (UN, 2017).

In the 21st century, there is huge increase in the demand of animal products that also expected in the next decade. The food and water are one of the primary needs for human. At the same time the increased population due to their miscellaneous activities causes local and global climate changes (i.e., global warming), leading to the major alteration of the livestock production. Although the adverse effect of global warming is not being same everywhere but it is accepted that it relatively affects the livestock and other crop production. In general, hot environment or rather heat stress is related with different factors like growth, quality and quantity of meat and milk production, quality of egg and its weight etc., which are the indirect indicators of reproductive performance, metabolic and health status, and immune response of an animal (Nardone et al, 2010). In the contrary, all over worlds, at some points global warming has positive effect at forecast: it shows the more productivity at the mid and high latitude as An increased temperature of 1-3°C. Also as a need of warm temperature in these areas, frost, heat waves or heavy rainfall decrease the advantage of hot temperature and a lowering of temperature also negatively affects the crop and other cereal production at lower latitude (Thornton et al, 2007). This high temperature between 2–3°C in any country for prolong time is sufficient to cause climate change of the country and simultaneously increases stressful condition to the dairy animal; and as a result of which animal shows reduced growth, meat and milk production. The global population is increased so rapidly that the demand of milk and meat is increasing day by day. The developing county like India stands at a second largest producer in cattle after Brazil (13% of world population), largest number of buffaloes (56% of world population) in the world (FAO, 2014), which is also a source of revenue; hence any impact of climate change on these livestock poses a serious threat to the country's economy.

Developed countries are also not an exception in this regard. For example, USA has been reported the economic loss of 2.4 billion USD due to change in climate, and mainly heat stress give so many opportunities to other factors that cumulatively restricted the growth of livestock. This is also related with several diseases that affect the livestock production directly via restriction on vectors, environmental habitants, and disease causing agents (Stem et al, 1989). The developing countries mainly face this negative effect of climate change where population growth, food need and security, and other socioeconomic factors suffers from this vulnerable impact (Bishaw et al, 2013). As a result of this some important live hood factors like agriculture activity, average health due to food resources, water quality and other global warming affected factors like shrinking glaciers, river floods, disease prevalence, erratic weathers and disrupt ecosystem leads the whole country of natural hazards and disasters (Younas et al, 2012) that have an impact on global livestock sector.

IMPACT OF CLIMATE CHANGE ON LIVESTOCK

The worldwide increased population of human and all different kinds of man-made activities increase the risk of climate change, which forces the world to think about the overall threat of other species, ecosystem as well as the livestock production. First, change in climate depending mainly on temperature that increases by greenhouse gases release in the atmosphere naturally and other agriculture-based anthropogenic activities. The greenhouse gases mainly include methane, and livestock production itself emits 18% of the global methane emission. In livestock production, specifically ruminant livestock plays an important role in the contribution of total agricultural emission of methane. Methane emissions from livestock in India is much less than the other developed countries around 7.26–10.4 metric ton/year due to its livestock population (Chauhan and Ghosh, 2014). The climate change mostly due to extreme weather condition, affect the daily life of different communities all over the world. There are many other factors related with type of climate change issue (Nunes et al, 2019). The most expected impact of climate change is increasing the chances of vulnerability of livestock system and their essential factors that play a crucial role in the overall livestock production sector. Also climate change directly from air, temperature, humidity, wind speed and so on, affects the animal performance such as, growth, milk production, wool production and reproductive potential (Rust and Rust, 2013).

Moreover, livestock production system will alternatively affect as a result of slow but sustainable changes in people's live hood related to climate change issue such as, global warming. Besides, livestock food chain plays a major role in greenhouse emissions. Climate change is having powerful impact on global ecosystems and hence, on global livestock sector in many ways like productivity and other essentials. Further, agriculture production is a main part of livestock sector so farming sector requires various mitigation and adaptive technology, which might have direct effect on the productivity of livestock and reduce the negative impact of climate change (Thornton and Gerber, 2010). As discussed above, the ruminant livestock play an important role in greenhouse gas emission mainly methane, but other than this nitrous oxide is another important greenhouse gas raised from animal agriculture. Methane, mainly produce by enteric fermentation and manure storage, has 28 times higher than the carbon dioxide for contributing global warming. Nitrous oxide arising by organic or inorganic fertilizers, and one single molecule has a potential to create 265 times higher global warming effect than the carbon dioxide. Besides, carbon dioxide used as a standard unit to represent the contribution of other gases in the global warming effect (IPCC, 2013). Feed production additionally distributes in global production of carbon dioxide and nitrous oxide; hence livestock reproduction.

Agriculture sector by decomposing plant residues, mineralization of soil organic matter, use of lands etc. increase the emission of carbon dioxide in soil. Other manufacturing products due to anthropogenic activities such as, synthetic fertilizers, pesticides and crude oil are also attributed a lot on farm and agricultural activities. Nitrous oxide emissions are increased by inorganic fertilizer in soil (Grossi et al, 2018). The research showed increase global warming of an average of +0.74°C causes melting of snow and iceberg; thereby rising of the global sea surface by 3.0 mm annually (IPCC, 2008a), that might have negative influence on the livestock sector. The agriculture livestock sector is the highest domestic producer of the country like India, and contribute between 70%–90% of the labour work. The agriculture supply also produces 50% of the household income that mainly involve animals like beef, cattle, goats, sheep and chicken (FAO, 2007).

Further, climate change affects the livestock in two ways: directly by temperature, humidity, wind speed, rain and due to this direct change it indirectly influences the reproductive performances of the animals reflected by growth, milk/meat production and so on (Houghton, 2001). Hence, global climate change leads to reduce the overall agriculture livestock production system. As increased demand for animal protein for human need as population become more altered with change in eating habit; therefore, animal productivity plays a significant role in maintaining the food chain. High demand of animal product offers market opportunities and economical support for small and marginal farmers but now-a-days faces serious concern due to global warming effect on livestock sector. This negative change in climate specifically due to greenhouse gas emissions is more reflected in the developing countries as they still suffer from the lack of facilities and resources, necessary for mitigation strategies. Subsequently, heat stress is one of the major factors to reduce the animal growth and also a limiting factor for small farmers to maintain the optimum temperature for growth, which directly affect the production of animal proteins and other livestock activities mostly in the developing countries; they have to compromise for food security and other limiting factors of climate change (Baumgard et al, 2012). Study also evident that the global climate change significantly destroys the grassland ecosystem through productivity, composition and carbon exchange (Baumgard et al, 2012). Seasonal effect further increases the consequences of temperature and water availability under doubled atmospheric carbon dioxide equilibrium conditions due to climate change (Baumgard et al, 2012).

The major effect of climate change on the ruminant livestock is to make favourable condition and proliferation of many diseases like malaria and zoonotic parasitic diseases such as, leishmaniosis, and cryptosporidiosis, giardiasis, trypanosomiasis, schistosomiasis, filariasis, onchocerciasis, and loiasis (Patz et al, 2000). Thus the overall effect of climate change on livestock in present as well as in future is uncertain and varies from area to area, and depending on the different combination of temperature, rainfall and plant responses to elevated carbon dioxide concentration in the atmosphere (McKeon et al, 2009). Livestock production sector mainly agricultural as well as ruminant livestock plays an important role in the world as about 40% of the world agriculture production contributes to the live hood and need of food security of around one billion people in the world. On global platform livestock also contribute 15% of the total animal food need and 25% of the dairy product and animal protein (FAO, 2009), which is very important as a protein supplement and provides micronutrients that cannot be obtained from the vegetable or any plant products (FAO, 2009).

EFFECT OF CLIMATE CHANGE ON LIVESTOCK REPRODUCTION

Livestock sector plays a crucial role in handling the economical balance of the country. Climate change directly affects the livestock production and reproductive capacity in both male and female animals, where duration and intensity of heat are the main determinant of animal performance (Lacetera, 2019). Livestock animal is homoeothermic animal; so capable to maintain their standard body temperature, and regulate heat production by metabolic activity and physiological behaviour of heat loss in the surrounding as an immediate effect of environmental temperature rise (DeShazer et al, 2009). Generally, animal maintain their body temperature by increasing evaporation rate only when the surrounding temperature is above than their body temperature. There are three mechanisms for maintaining the thermal gradient like conduction, convention and radiation. So when the ambient temperature rises, animal despite of heat try to adapt with the adverse effect by increasing evaporation rate through sweating and panting though not always succeed under the present condition of global warming. The increase global temperature significantly increases the relative humidity as a primary change that directly affects the body and metabolic activity of the animal. This combine effect of heat and humidity ultimately affects the different physiological and biochemical condition of the body, reproductive traits and indirectly influences the overall food and water intake (Sinha et al, 2017).

Heat stress on animal leads to 20–30% decrease of pregnancy rate (Schuller et al, 2014), suggesting the reduction of fertility rate in such heat stressed animal. Further, damage of both male and female gametes at different developmental stages due to climate change are also reported that include spermatogenesis in males and oocyte complication, maturation and implantation defects, embryo and fatal growth retardation as well as defects in lactation in females (Casey et al, 2019; Benyei et al, 2001). According to research heat stress directly affects food intake of dairy cattle, which reduces milk production (Silanikove, 2000; DeShazer et al, 2009). Recent experiment shows disparate slops in feed intake and milk yield of the heat stressed animal (Shwartz et al, 2009). Further, heat stress causes inadequate feed intake due to insufficient nutrient supply, manifested by reduction in egg production, egg mass and quality of shell (Das, 2017).

Climate change, mainly high temperature cause decrease of some essential metabolic activities and biochemical synthesis of specific hormones, disturbance of cellular microenvironment, protein synthesis, microfilament and microtubule architecture, and reduce the oocyte competence, manifested by programmed cell death or apoptosis (Edwards and Hansen, 1997). Further, the high temperature country possesses more livestock fertility issue in summer because heat directly affects the mechanism of oocyte and sperm formation. Therefore, livestock based industries have increase their success ratio of reproductive capacity of animals by understanding the mechanism behind the successful reproduction strategies in different climatic conditions because it's a key driver of maintaining the livestock productivity to overcome the adverse effect of climate change (Devi et al, 2019).

Male Reproduction

Climate change leads to the several changes in atmosphere but global warming due to rise in temperature is the most important and vast distributed problem for livestock sector all over the world because like female, it directly affects the fertility potential of the male reproductive system. Spermatogenesis requires a lower temperature to produce viable sperms, specifically 1-8°C lower than the normal body temperature of 36^{0}C (98.6^{0}F) (Dash et al, 2016). Male reproductive system is sensitive for high temperature, solar radiation, relative humanity and wind speed that may reduce the quality as well as quantity of mature spermatozoa. High temperature for long- or short-term exposure directly affects the spermetogonial cells in the seminiferous tubules, manifested by low sperm density (Rasooli et al, 2010).

Seasonal variation plays a key role in the male animal by deteriorating semen quality and fertility potential. In summer, thyroxin hormone, which is mainly responsible for food intake, decreases due to low metabolic activity and subsequent reduction in food intake that indirectly responsible for low sperm count and ejaculation volume (Zafar et al, 1988). Another study on bulls reported that ejaculate volume, sperm concentration, sperm motility and production of mature spermatozoa are reduced in high temperature than that of the low temperature (Das, 2017). Further, metabolic disturbance not only lowers the sperm concentration in fall but also induces the morphological abnormality of spermatozoa that significantly reduces the percentage of fertility in male animal (Phillips et al, 1943).

Moreover, there are different developmental stages of gamete in male reproductive system, which are affected by various factors at different stages of development including heat stress and adverse climatic condition resulting in production of abnormal and poorer quality of sperm that affects the overall fertility. Heat stress exposure also alters normal sperm cell to abnormal sperm cell as result of which cell itself entre the apoptotic pathway (Sinha Hikim et al, 2003) and loose DNA integrity (Banks et al, 2005). Sertoli cell, which nurturing the spermatogenesis, is the most and primary victim of climate change issue in testis that have an additive effect on overall abnormality of germ cell development process (Setchell, 1998). Apart from sertoli cell, the second most affected part of spermatozoa is acrosome as the percentage of acrosome integrity is reduced during summers in comparison with the winter (Bhakat et al, 2014). Again this reduced acrosome activity decreases the average fertility of male animal in summer i.e., due to heat stress (Birck et al, 2010). According to study on heat stressed bull, the reaction time and total time taken for successful ejaculation are significantly rises, and accompanied with low libido score (Mandal et al, 2000).

Female Reproduction

The effect of global warming and heat stress on female reproduction directly affects the oocyte quality and production as well as estrus cycle, regulated by several sex steroid hormones including progesterone. Change in duration of estrus cycle is observed upon alteration of plasma progesterone level due to cyclic change in corpus luteum. Heat stress-induce malproduction of ovarian hormone and reduce expression of follicle and ovarian cells due to this hormonal regulation significantly affects the reproduction process. High temperature also alters estrogen production in the cell and directly influencing oestrous cycle (Sirotkin and Kacaniova, 2010). The concentration of progesterone, estradiol and luteinizing hormone (LH) shows the reduction in the expression level of estrus especially in summer, responsible for low reproductive potential (Bhat et al, 1983). Buffaloes have lower rate of reproduction mainly due to decrease efficiency and poor expression of estrus symptoms more likely in hot season and it also shows the long calving interval (Tao et al, 2012). The combination of high heat and high ambient humanity cause the impairment of hormonal regulation in buffaloes (Heba et al, 2016). Also a silent estrus remarkably responsible as a limiting factor that representing the reproductive capacity in buffalos (Singh et al, 2013). Heat exposure also increases estrus cycle which results in the aged ovulation with low efficiency for further development process. Oocyte contains antral follicles, which is mostly sensitive to do the damage of oocyte due to heat stress (Roth, 2008).

Moreover, ovarian pool is more to the climate change and mostly by the heat stress in this regard. According to research heat stress decreases the developmental competence by altering the growth and maturation of oocyte. Heat causes production of oxidative stress-induced apoptotic cell death in the animal by the irreversible change in meiotic cycle and spindle distribution (Lin et al, 2006; Hansen, 2009). Animal exposure to heat stress for short- or long-term slow down the process of foetal growth and increases the risk of foetal loss. This is happened because over exposure of heart activate alternative mechanism that changes the foetal body temperature; thereby causing foetal death when mother is thermally stressed (Habeeb et al, 2018). Some studies reveal that cows are negatively affected by heat stress (Nardone et al,2010) as it reduces the oocyte growth by altering the level of progesterone, LH and follicle stimulating hormone (FSH) during estrus cycle (Ronchi et al, 2001). Further, previous study reported that the first wave of dominant follicle in heat exposed cows and also in lactating cows is found with less diameter and loss of biological fluids, which is a pre-requisite for better growth and development of the embryo on 8th day of the estrus cycle (Putney et al, 1988a;b). Moreover, it is also evident that heat stress adversely affects the biochemical activity, inhibits follicular growth, decreases prestrus and increases the level of estradiol-17β, and promote the smaller size of the second wave of dominant follicle (Hooda et al, 2011)

In India, Gir cow exhibits the negative effect of heat stress on reproductive functions. It showed delay in oocyte developmental capacity by an increase in the occurrence of large follicles, reduction in the oestrous cycle length and duration, intensity of estrus, and reduces the progesterone level in the body that directly affect the reproduction of female animal (Torres-junior et al, 2008). The total time required for recovering the effect of heat stress-induced damage due to climate change is two or three estrus cycles; thereby suggesting direct effect of heat stress on the ovarian pool of oocyte might be due to lowering the concentration of LH (Roth, 2017). LH also plays a crucial role in the process of follicular growth, estradiol production, progesterone concentration and hence, ovulation. The lowering of LH concentration further indicates low nutrient consumption in the animal and due to inadequate nutrient intake, it finally affects the other consequences of several hormonal activities like insulin, ghrelin, leptin etc. as a direct prove of climate change (heat stress) issue on livestock reproductive potential though variation experience from species to species (Szymanski et al, 2007).

Oocyte Completion

Different reproductive factors like blood flow and hormonal concentrations in uterus affects the reproductive potential of the animal associated with temperature. Each of this factor can directly influence on the oocyte completion process leading to impairment of embryonic development and induce cell death (Putney et al, 1988a;b). Estradiol and progesterone in serum also play pivotal role for follicular development in this regard (Luo et al, 2005). The high concentrations of estradiol increases the follicle size due to increase in the number of granulosa cells while growth and development process of the follicles. The process of ovulation requires specific size of the follicle, generally equal or greater than the initial size depending on the species, which is again governed by LH upon stimulation of estradiol and the whole process is strongly controlled by the seasonal variation or climatic condition. In summer and autumn, the result shows decline in the production of follicular fluid estradiol and progesterone in different species of buffalos as compared with the winter and spring (Khan et al, 2012).

The low secretion of LH is responsible for dominant follicle formation but in summer low concentration of LH decreases estradiol level, resulting in reduction of oocyte maturation process (Wolfenson et al, 2000) and poor expression of estrus ultimately leading to low fertility potential of the livestock (Heba et al, 2016; Roth et al, 2000). The decreased concentration of steroid hormone during hot season also alters the vascular responses (Roth et al, 2001) as observed in heat stressed cows that express low progesterone level in summer as compared to winter (Ronchi et al, 2001). The overall effect is manifested by decrease length of estrus cycle and oocyte maturation, as well as low conception rate (Ronchi et al, 2001). Moreover, high temperature causes delay in follicle selection and decrease in the probability of dominance of the dominant follicle, leading to impairment of oocyte completion, a pre-requisite of fertilization.

Fertilization Potential

Higher temperature during summer reduces fertility potential of the cattle mostly cow, buffalo and sheep (Hansen and Arechiga, 1999). Heat stress-induced low fertility potential of bull causing more commercial issue than cow because one bull can fertilize 30–40 cows with natural matting or thousands of cows through artificial fertilization (Kastelic, 2013); thereby influences the pregnancy rate of the cow. Under heat stress, the secretion of estrogen and estradiol decreases that cause suppression of estrus, ovulation, transport of gametes and conception rate as well as directly affects the fertilization potential of the animal (Wolfenson et al, 2000).

Hypothalamic-pituitary-gonadal (ovarian) (HPG) axis is most important factor that directly affects by the high temperature and high intensity of radiant heat (Ahmed et al, 2015). High temperature not only decreases embryo development process at higher magnitude but also affects the pre-attachment stage of embryo (Ealy et al, 1993). Heat stress alters the concentration of gonadotrophin, inhibin and prostaglandin F2 alpha (PGF2α). The length of luteal phase in heat exposed cows is longer than the cows, which are kept in thermally stable atmosphere. It also shows that the secretion of PGF2α by uterine endometrial cells alters due to low concentration of estradiol because of elevated temperature. The production of estradiol directly affects the secretion of prostaglandin and luteolysis in uterine endometrium wall, necessary for implantation (Samal, 2013). Hence, successful fertilization and reproduction requires the proper maintenance of HPG axis in their optimum limits, which is associated with internal state and external surroundings (Kriegsfeld et al, 2010). Low concentration of progesterone decreases the optimum functional capacity of uterine endometrial wall and subsequently affects fertility potential of the livestock.

Additionally, increase concentration of PGF2α threatens the chances of successful fertilization as well as pregnancy rate of an animal due to activation of luteolysis (Krishnan et al, 2010; Upadhyay et al, 2009). Heat stressed animal has impaired conception rate following exposure to high temperature (heat) either during breeding period or 42 days before/on the day/20 days after insemination (Greer, 1983; Schuller et al, 2014). Follicular dynamics reduces due to heat stress leading to poor quality of oocyte or extended upon removal of heat stress, which might be a reason for lower fertility of dairy cattle in cooler autumn season (Collier et al, 2006).

Embryo Development

Climate change and global warming are mainly associated with heat stress, which damage the embryo development process (Wolfenson et al, 2000) and increase the chance of embryo mortality in ruminant livestock (Hansen, 2007). Also it remarkably affects utero-embryo cross-talk among all the dairy cattles (Hansen and Arechiga, 1999). High temperature increases the number of cell death in 2nd developing stage of embryos, indicating that heat stress directly alters embryonic stage and reduces developmental competence (Fear and Hansen, 2011). Long term exposure to high heat affects the production embryo by superovulation and reduces the early embryonic development in summer (Zeron et al, 2001). The livestock sector is also facing a problem of collection and storage of gametes in hot environment. Further, the expression of maternal transcripts i.e., single **c**opy gene **mo**loney **s**arcoma oncogene product, serine/threonine kinase (C-MOS), growth differentiation factor 9 (GDF9), POU (**p**ituitary-specific Pit-1; **o**ctamer transcription factor proteins Oct-1 and Oct-2; neural **U**nc-86 transcription factor) domain class 5 transcription factor 1 (POU5F1) and glyceraldehyde 3-phosphate dehydrogenase (GAPDH), which are responsible for oocyte composition, maturation and early embryo development varies during summer and winter seasons (Gendelman et al, 2010). Moreover, pre implantation of embryo at an early stage of development is damaged by heat stress (Hansen, 2013). Heat stress reduces the endometrial prostaglandin secretion that leads to production of premature luteolysis and embryo loss; according to research mostly embryonic loss occur in heat stressed cattle within 42 days of insemination (Singh et al, 2019). In heat stressed cow, early embryonic death is also observed (Sakai et al, 2018).

Heat stress also affects uterine blood flow that changes the uterine micro-environment leading to damage in utero-embryo cross-talk. Heat stress may not only affect the cleavage rate but also decreases the blastocyst development as well as foetal growth (Torres-junior et al, 2008) might be due to reduce blood flow in uterus, which causes placental insufficiency to provide nutrients to foetus (Biggers et al, 1987); even foetus born with low body weight and size (Dunlap et al, 2015). Heat stress also modulates livestock gestation period; it causes reduction in the length of period resulted in a lowering of body weight at the time of birth and weaning. The foetal growth and development and their immune system are also affected by heat stress as an indicator of increasing global warming (Tao et al, 2012). Previous study further revealed that heat stress decreases blood progesterone concentration, oocyte maturation, embryo implantation and early embryonic death in ruminant livestock regulated by alteration of hormonal stimulation (Khodaei-Motlagh et al, 2011). Heat stressed swine exposed for 5 days reduces the number of viable embryo at an average of 38% after 27th day of gestation period (Tompkins et al, 1967) further confirming the adverse effect of heat stress is not limited to the oocyte and estrus cycle only. It also has impact on the timed embryo transfer, conception rates etc. that reduce in hot season as compared to the cooler season; hence prove the direct effects of heat stress on fertility potential as well as uterine micro-environment (Putney et al, 1988a;b).

Hormonal Imbalance

Climate change alters the physiology of reproductive tract via hormonal imbalance that affects the fertility and reproductive performance of livestock. Like reduce semen quality in males, heat stress also decreases oocyte quality and embryo development in females through lowering of LH and estradiol leading to reduction of intensity and length of estrus expression, increase incidence of anestrus and silent heat (Krishnan et al, 2017). Hypothalamic-pituitary-adrenal (HPA) axis secretes corticotrophin type hormone, adrenocorticotropic hormone, cortisol and aldosterone that further alter by thermal stress (Einarsson et al, 2008). Corticotrophin type hormone stimulates secretion of somatostatin and thyroid stimulating hormone from hypothalamus via pituitary, which is inhibited by heat stress leading to down-regulation of the effects of both these hormones (Quinteiro-Filho et al, 2012).

Moreover, gonadotrophin releasing hormone system regulates reproductive suppression of both male and female livestock during heat exposure for more extended duration (Denver, 2009; Gilad et al, 1993) as this hormonal system controls the secretion of gonadotrophins, LH, FSH etc. via anterior pituitary that regulates steroidogenesis and gametogenesis (oogenesis and spermatogenesis) (Tsutsui et al, 2000). Glucocorticoids secretion also low during acclimatization at 35°C and much lower in heat-stressed dairy cattle as compared to thermally balanced animal (Casey et al, 2019). Heat stress reduces fertility in heat-stressed cattle before insemination (Al-Katanani et al, 2002). According to research, in tropical climate condition fertility ratio is decreased in buffalos when it exposed to temperature humidity index (THI) higher than 75 as compared with cattle, suggesting buffalos are more sensitive to heat stress (Dash et al, 2016). Uterine temperature also plays a key role in implantation as increase of uterine temperature by 0.5°C during fertilization or implantation (Wilson et al, 1998). In case of cows, most damage occur in between estrus and day 7 of pregnancy exposed to climatic change (heat stress) (Ealy et al, 1993), which in buffalos is about 7 to 21 days (Amundson et al, 2006).

The overall detrimental effect of climate change on livestock reproduction and the general economy of the country is shown below (Fig. 1).

Figure 1. Climate change (heat stress) response on livestock reproduction and economy

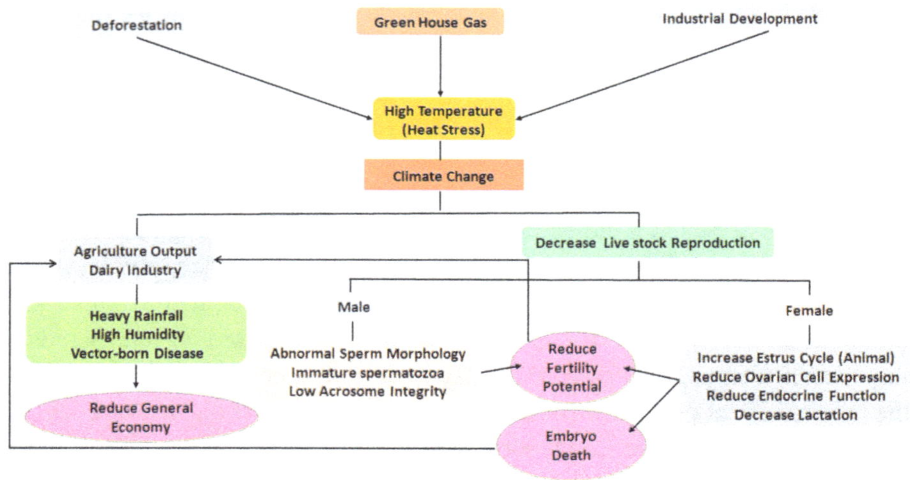

MITIGATION STRATEGIES TO OVERCOME CLIMATE CHANGE ISSUES

General Approach

Climate change is mostly occurred and current issue in livestock industry. Livestock sector needs to solve this problem or need some strategies to overcome the adverse effect as discussed above. The first target of mitigation strategy should limit or reduce the production of greenhouse gases that causes rise in global temperature and directly affect livestock at different stages of reproduction along with several vector borne diseases. Livestock play an important role in greenhouse gas production. Methane is mainly produced by ruminate livestock by enteric fermentation, which can be reduced by providing good quality food including supplements like bovine somatotropins, feed antibiotics etc.; thereby improving nutrition level (Boadi et al, 2004). Vaccines are also available to suppress methane emission but study reveals that vaccines are not much effective to reduce methane production initially as reduction of fermentation is temporary (Rowan et al, 2013).

Secondly, adaptation with reference to climate change especially continuous rise of global temperature. It needs preparation for the adjustment of livestock with human activity or into the newly changed environment to reduce damage and ensure beneficial opportunity for maintaining the natural lifestyle of the livestock. Adaptation can reduce the high risk of climate change (Das, 2017). It is also important to maintain a balance between nutrients and energy uptake (through quality and availability of food) as both are indirectly associated with fertility potential and reproductive performance of the livestock at high temperate zone of arid or semi-arid areas (Sejian et al, 2015). Further, essential macronutrients like sodium and potassium in feed increase the production performance of animal (Chauhan et al, 2015). Inclusion of ascorbic acid in feed can also reduce heat stress-induced problems that affected fertility potential like weight gain, oxidative stress, semen quality, poor immunity and high body temperature (Abdin and Khatoon, 2013; Kumar et al, 2010).

Thirdly, there are several management techniques to reduce the effect of heat stress with own positive and negative side effects (Lawrence, 2008). Housing also provides the control over heat stress inducer but it has relatively high investment cost for each head (Das, 2017); so not always feasible for low- and middle-income countries. Therefore, orientation of shade, rod size and roof material etc. with low maintenance cost are encouraged to reduce heat load in micro-environment of the shade. Hay is cheap and suitable roof material for dairy cattle shade in developing countries including India. Corrugated steel sheet is also most popular shade material for relative longevity and low maintaining requirements (Das, 2017). As heat stress reduces milk production in dairy cattle so it is essential to provide economically suitable structure i.e., protective shade for animal from exposure of heat and recover efficiency by avoiding a loss of milk production. Covering shade should be large with maximum ventilation for proper circulation of air mostly in hot and highly humid climate, and in accordance with the type of dairy cattle that increases the comfort level of the animal and recovery from damage of heat. The minimum standard size of shade is 50-60 square per feet per animal (Blogs.nrcs.usda.gov., 2020). According to study a proper ventilated shade can reduce 30–40% of heat stress of the animal (Kendall et al, 2006).

Further, orientation of shade has significant effect on milk production of the crossbreed cow in the coastal area where shade is designed and oriented in such a way that cattle are more exposed to cool sky, which act as cold sink (Berman and Horovitz, 2012). In extremely hot climate with low rainfall (10–12 cm), east-west shade orientation is most preferable because of cooler in nature. Second preferable orientation with the north-south long axis to expose more sunlight in the morning and afternoon, and keep the area dry and clean, specifically applicable for hilly region (Das et al, 2015). Humidity is another factor that affects the lactating dairy cattle. Evaporative cooling system can provide a comfortable climate zone in low rain or dry area. This system works on high pressure, fine mist and large volume of air, which evaporate huge quantity of moisture to keep the shade's micro-environment cool (Collier et al, 2006). This evaporating system can efficiently work only in semi dry area where relative humanity is >70% and the potential recovery rate in THI is ≤10% (Collier et al, 2006).

Also modern technology like fully air conditioned animal house to get complete solution of heat problem is common in most of the developed countries, which indirectly increases the production cost of milk as well as the quantity of milk. Research showed cooled zone cows i.e., exposure of head and neck to cooled air, produces 19% more milk than the normal conditioned climate cow (Bianca, 1964) though not economically valid for low- and middle-income farmers. The important factor is movement of air can relief the heat stress, although it reduces humidity and evaporation rate, which encourage to build open side constructed places for maintaining natural ventilation (Bucklin et al, 1991). All these approaches including orientation of shades and maintaining standard THI as discussed above, are the most economically convincing mitigation strategies that overcome the damage of extreme climate and improve milk and meat production in the livestock sector.

Hormone and Gene-based Approach

Hormonal treatment has a potential to minimize the effect of extreme climate including heat stress in dairy animal. Gonadotropin receptor hormone therapy at the early stages of estrus can recover the production of endogenous LH and improves the rate of conception (AE-96. Extension.puedue.edu., 2020). Gonadotrophic receptor hormone or human chorionic gonadotrophin (HCG) dose can improve the progesterone level by injected on 5th day of estrus cycle in case of chronic heat exposure, resulting in ovulation or production of dominant follicles that reduces corpus luteum, responsible for progesterone secretion under normal condition (Samal, 2013); hence this kind of artificial insemination program is effective when it associates with summer fertility. Same hormone-based mitigation technique also follows for PGF2α injection after 7 days of 1st dose to regress the quality of corpus luteum for successful maturation of follicles. In cows, the second dose of gonadotrophic receptor hormone after 48 hr and PGF2α may induce ovulation and artificial insemination at 16 hr can ensure the successful conception (Pursley et al, 1998). Moreover, inclusion of antioxidant in these hormonal replacement therapies regulates the overall damage due to heat stress (Benmrad and Stevenson, 1986).

Genetic manipulation develops more resistant to heat exposure. In animal there are some proteins responsible for regulation of body temperature; thus for better cellular mechanism under extreme climatic condition identification of that particular genes responsible for resistance to heat may useful to develop more susceptible breed where gene administration can regulate body temperature leading to recover reproduction and fertility rate of the heat sensitive species (Krishnan et al, 2017). As acclimatization of animal can change hormonal signalling and alters target tissues such as, male and female gonads; therefore, to improve our understanding towards successful mitigation of climate change issues by genetic manipulation through genetic selection of heat stress-resistance breed is a promising area of mitigation. Hormones like thyroid, prolactin, somatotropin, glucocorticoids and mineralocorticoids also known as homeostatic regulators to avoid heat stress-induce damage in livestock (Collier et al, 2006).

In *B. Taurus* of tropical environment, one phenotype is characterized which has dominant gene responsible for profuse sweating, lower rectal temperature and lower respiratory rate (Mariasegaram et al, 2007). Further, genes are responsible to produce heat shock proteins (HSPs) that associate with thermal tolerance and regulation of body temperature. Some of them are used as markers while selecting specific breeding program; for example, heat tolerant gene HSP90AB1 in Thai native cattle (Deb et al, 2014), HSP70A1A and HSBP1 in Chinese Holstein cattle (Charoensook et al, 2001). Further, there is also HSPs thermos-tolerant genes present in ruminant livestock that undergoes changes of expression pattern upon heat exposure (Wang et al, 2013). Administration of HSPs into embryo culture shows increase rate of hatching as compared with embryo chronically exposed to heat, may be due to induction of heat shock protein (Rayan et al, 1992). Therefore, based on these observations hormone and genetic approach of

Figure 2. Common mitigation strategies to overcome the climate change issue on livestock reproduction

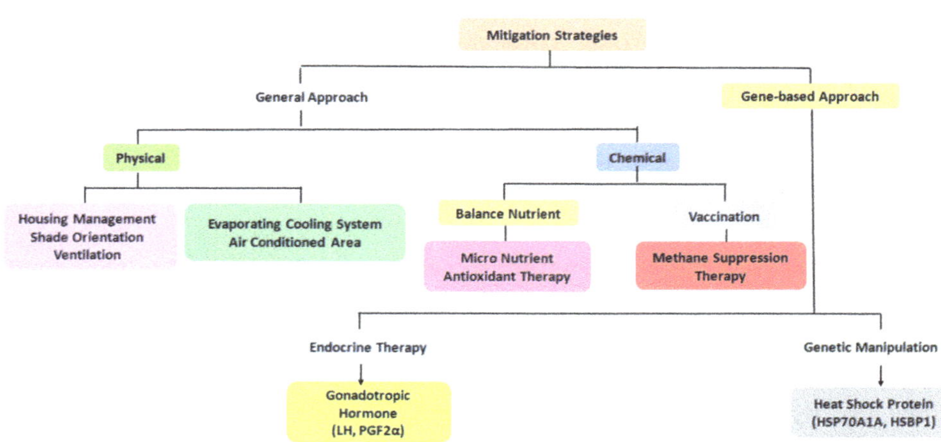

mitigation strategies using HSPs gain importance to recover from damage caused by climate change including maintenance of reproductivity in the livestock sector.

The common mitigation strategies for overcoming the detrimental effects of climatic changes on livestock sector is shown in Fig. 2, as mentioned below.

CONCLUSION

Climate change is a major threat to animal agriculture through its potential effects on heat stress, food and water security, extreme weather events, vulnerable shelter and population migration. Environmental stress is an important economic problem because of heat induced reductions in fertility of livestock. Therefore, increasing reproductive efficiency under elevated ambient temperatures is a top priority for the animal agriculture in tropical and subtropical areas. A vast number of studies illustrate the very diverse immediate effects of environmental factors and hence climate change on reproductive efficiency leading to alterations on genetic, phenotypic and behavioral levels. Decreased development of oocytes and/or embryos subjected to heat stress could be caused, at least in part, by heat shock-induced apoptosis and alterations of the chromatin and spindle microtubules. Enhancement of thermotolerance of oocytes and embryos appears to be one of the solutions to this problem. Using functional genomics to identify genes that are up- or downregulated during a stressful event can lead to the identification of animals that are genetically superior for coping with climactic stress and toward the creation of therapeutic drugs and treatments that target affected genes.

ABBREVIATIONS

C-MOS, Single copy gene Moloney sarcoma oncogene product, serine/threonine kinase; **FSH**, Follicle stimulating hormone; **GAPDH**, Glyceraldehyde 3-phosphate dehydrogenase; **GDF9**, Growth differentiation factor 9; **HCG**, Human chorionic gonadotrophin; **HSPs**, Heat shock proteins; **HPA**, Hypothalamic-pituitary-adrenal axis; **HPG**, Hypothalamic-pituitary-gonadal axis; **LH**, Luteinizing hormone; **PGF2α**, Prostaglandin F2 alpha; **POU5F1**, POU (Pituitary-specific Pit-1; Octamer transcription factor proteins Oct-1 and Oct-2; neural Unc-86 transcription factor) domain class 5 transcription factor 1; **THI**, Temperature humidity index.

AUTHOR CONTRIBUTION

SM, NN conceived and designed the book chapter. VS, NN, SM performed literature survey and wrote the book chapter. SM, NN shared equal contribution.

FUND DISCLAIMER

No grant received from any funding agency in public, commercial, or not-for-profit sector.

ACKNOWLEDGMENT

Gujarat University for Ph.D. registration of Ms. Vishakha Shrimali.

CORRESPONDING AUTHORS

Nibedita Naha, Sukanta Mondal

REFERENCES

Abdin, Z., & Khatoon, A. (2013). Heat stress in poultry and the beneficial effects of ascorbic acid (vitamin c) supplementation during periods of heat stress. *World's Poultry Science Journal*, *69*(1), 135–152. doi:10.1017/S0043933913000123

AE-96. (n.d.). Available from: https://www.extension.purdue.edu/extmedia/AE/AE-96.html

Ahmed, A., Tiwari, R., Mishra, G., Jena, B., Dar, M., & Bhat, A. (2015). Effect of Environmental Heat Stress on Reproduction Performance of Dairy Cows- A Review. *International Journal of Livestock Research*, *5*(4), 10. doi:10.5455/ijlr.20150421122704

Al-Katanani, Y. M., Paula-Lopes, F. F., & Hansen, P. J. (2002). Effect of season and exposure to heat stress on oocyte competence in Holstein cows. *Journal of Dairy Science*, *85*(2), 390–396. doi:10.3168/jds.S0022-0302(02)74086-1

Amundson, J. L., Mader, T. L., Rasby, R. J., & Hu, Q. S. (2006). Environmental effects on pregnancy rate in beef cattle. *Journal of Animal Science*, *84*(12), 3415–3420. doi:10.2527/jas.2005-611

Banks, S., King, S. A., Irvine, D. S., & Saunders, P. T. K. (2005). Impact of a mild scrotal heat stress on DNA integrity in murine spermatozoa. *Reproduction (Cambridge, England)*, *2005*(129), 505–514. doi:10.1530/rep.1.00531

Baumgard, L., Rhoads, R., Rhoads, M., Gabler, N., Ross, J., & Keating, A. (2012). Impact of Climate Change on Livestock Production. *Environmental Stress and Amelioration in Livestock Production*, 413-468.

Benmrad, M., & Stevenson, J. (1986). Gonadotropin-Releasing Hormone and Prostaglandin F2α for Postpartum Dairy Cows: Estrous, Ovulation, and Fertility Traits. *Journal of Dairy Science*, *69*(3), 800–811. doi:10.3168/jds.S0022-0302(86)80469-6

Benyei, B., Gaspard, A., & Barros, C. W. C. (2001). Changes in embryo production results and ovarian recrudescence during the acclimation to the semiarid tropics of embryo donor Holstein–Frisian cows raised in a temperate climate. *Animal Reproduction Science*, *68*(1-2), 57–68. doi:10.1016/S0378-4320(01)00144-0

Berman, A., & Horovitz, T. (2012). Radiant heat loss, an unexploited path for heat stress reduction in shaded cattle. *Journal of Dairy Science*, *95*(6), 3021–3031. doi:10.3168/jds.2011-4844

Bhakat, M., Mohanty, T. K., & Gupta, A. K. (2014). Effect of season on semen quality of crossbred (Karan Fries) bulls. *Advances in Animal and Veterinary Science*, *2*(11), 632–637. doi:10.14737/journal.aavs/2014/2.11.632.637

Bhat, P. N., Kumar, R., & Raheja, K. L. (1983). Breeding behavior of murrah buffalo. *Proceedings of the Fourth International Congress of SABRAO*, 91-97.

Bianca, W. (1964). Animal Housing in Hot Countries. *International Journal of Biometeorology*, *8*(2), 99–101. doi:10.1007/BF02187573

Biggers, B., Geisert, R., Wetteman, R., & Buchanan, D. (1987). Effect of Heat Stress on Early Embryonic Development in the Beef Cow. *Journal of Animal Science*, *64*(5), 1512–1518. doi:10.2527/jas1987.6451512x

Birck, A., Christensen, P., Labouriau, R., Pedersen, J., & Borchersen, S. (2010). In vitro induction of the acrosome reaction in bull sperm and the relationship to field fertility using low-dose inseminations. *Theriogenology*, *73*(9), 1180–119. doi:10.1016/j.theriogenology.2009.10.010

Bishaw, B., Henry, N., Jeremias, M., Abdu, A., Jonathan, M., & Gemedo, D. et al. (2013). *Farmers' Strategies for Adapting to and Mitigating Climate Variability and Change through Agroforestry in Ethiopia and Kenya*. Academic Press.

Boadi, D., Wittenberg, K., Scott, S., Burton, D., Buckley, K., Small, J., & Ominski, K. H. (2004). Effect of low and high forage diet on enteric and manure pack greenhouse gas emissions from a feedlot. *Canadian Journal of Animal Science*, *84*(3), 445–453. doi:10.4141/A03-079

Bucklin, R. A., Turner, L. W., Beede, D. K., Bray, D. R., & Hemken, R. W. (1991). Methods to relieve heat stress for dairy cows in hot, humid climates. *Applied Engineering in Agriculture*, *7*(2), 241–247. doi:10.13031/2013.26218

Casey, G., Shayegh, S., Moreno-Cruz, J., Bunzl, M., Galor, O., & Caldeira, K. (2019). The impact of climate change on fertility. *Environmental Research Letters*, *14*(5), 054007. doi:10.1088/1748-9326/ab0843

Charoensook, R., Gatphayak, K., Sharifi, A., Chaisongkram, C., Brenig, B., & Knorr, C. (2011). Polymorphisms in the bovine HSP90AB1 gene are associated with heat tolerance in Thai indigenous cattle. *Tropical Animal Health and Production*, *44*(4), 921–928. doi:10.100711250-011-9989-8

Chauhan, D., & Ghosh, N. (2014). Impact of Climate Change on Livestock Production: A Review. *Journal of Animal Research*, *4*(2), 223. doi:10.5958/2277-940X.2014.00009.6

Chauhan, S. S., Celi, P., Leury, B., Liu, F., & Dunshea, F. R. (2015). Exhaled breath condensate hydrogen peroxide concentration, a novel biomarker for assessment of oxidative stress in sheep during heat stress. *Animal Production Science*, *56*(7), 1105–1112. doi:10.1071/AN14070

Collier, R., Dahl, G., & VanBaale, M. (2006). Major Advances Associated with Environmental Effects on Dairy Cattle. *Journal of Dairy Science*, *89*(4), 1244–1253. doi:10.3168/jds.S0022-0302(06)72193-2

Das, S. (2017). Impact of Climate Change on Livestock, Various Adaptive and Mitigative Measures for Sustainable Livestock Production. *Approaches in Poultry. Dairy and Veterinary Sciences*, *1*(4), 33–42.

Das, S. (2017). Impact of Climate Change on Livestock, Various Adaptive and Mitigative Measures for Sustainable Livestock Production. *Approaches in Poultry, Dairy & Veterinary Sciences*, *1*(4).

Das, S. K., Karunakaran, M., Barbuddhe, S. B., & Singh, N. P. (2015). Effect of Orientation, Ventilation, Floor Space Allowance and Cooling Arrangement on Milk Yield and Microclimate of Dairy Shed in Goa. *Journal of Animal Research*, *5*(2), 231–235. doi:10.5958/2277-940X.2015.00040.6

Dash, S., Chakravarty, A. K., Singh, A., Upadhyay, A., Singh, M., & Yousuf, S. (2016). Effect of heat stress on reproductive performances of dairy cattle and buffaloes. *Veterinary World*, *9*(3), 235–244. doi:10.14202/vetworld.2016.235-244

Dash, S., Chakravarty, A. K., Singh, A., Upadhyay, A., Singh, M., & Yousuf, S. (2016). Effect of heat stress on reproductive performances of dairy cattle and buffaloes. *Veterinary World*, *9*(3), 235–244. doi:10.14202/vetworld.2016.235-244

Deb, R., Sajjanar, B., Singh, U., Kumar, S., Singh, R., Sengar, G., & Sharma, A. (2014). Effect of heat stress on the expression profile of Hsp90 among Sahiwal (Bos indicus) and Frieswal (Bos indicus × Bos taurus) breed of cattle. *Gene*, *536*(2), 435–440. doi:10.1016/j.gene.2013.11.086

Denver, R. (2009). Stress hormones mediate environment-genotype interactions during amphibian development. *General and Comparative Endocrinology*, *164*(1), 20–31. doi:10.1016/j.ygcen.2009.04.016

DeShazer, J. A., Hahn, G. L., & Xin, H. (2009). Basic principals of the thermal environment and livestock energetics. In J. A. DeShazer (Ed.), *Livestock energetics and thermal environment management*. ASABE.

DeShazer, J. A., Hahn, G. L., & Xin, H. (2009). Basic principles of the thermal environment and livestock energetics. In *Livestock energetics and thermal environmental management*. American Society of Agricultural and Biological Engineers (ASABE).

Devi, P., Debbarma, S., Kumar, G., & Thakur, P. (2019). Effect of heat stress on reproduction in farm animals and its mitigation: A review. *Journal of Entomology and Zoology Studies*, *7*(2), 342–345.

Dunlap, K., Brown, J., Keith, A., & Satterfield, M. (2015). Factors controlling nutrient availability to the developing fetus in ruminants. *Journal of Animal Science and Biotechnology*, *6*(1), 16. doi:10.118640104-015-0012-5

Ealy, A. D., Drost, M., Robinson, O. W., & Britt, J. H. (1993). Developmental changes in embryonic resistance to adverse effects of maternal heat stress in cows. *Journal of Dairy Science*, *76*(10), 2899–2905. doi:10.3168/jds.S0022-0302(93)77629-8

Edwards, J. L., & Hansen, P. J. (1997). Differential responses of bovine oocytes and pre- implantation embryos to heat shock. *Molecular Reproduction and Development*, *46*(2), 138–145. doi:10.1002/(SICI)1098-2795(199702)46:2<138::AID-MRD4>3.0.CO;2-R

Einarsson, S., Brandt, Y., Lundeheim, N., & Madej, A. (2008). Stress and its influence on reproduction in pigs: A review. *Acta Veterinaria Scandinavica*, *50*(1), 48. doi:10.1186/1751-0147-50-48

FAO. (2007). *Adaptation to climate change in agriculture, forestry, and fisheries: perspective, framework and priorities.* http://www.fao.org/3/a-au030e.pdf

FAO. (2009). *The state of food and agriculture: Livestock in the balance*. http://www.fao.org/3/a-i0680e.pdf

Fear, J. M., & Hansen, P. J. (2011). Developmental changes in expression of genes involved in regulation of apoptosis in the bovine pre-implantation embryo. *Biology of Reproduction*, *84*(1), 43–51. doi:10.1095/biolreprod.110.086249

Gendelman, M., Aroyo, A., Yavin, S., & Roth, Z. (2010). Seasonal effects on gene expression, cleavage timing, and developmental competence of bovine preimplantation embryos. *Reproduction (Cambridge, England)*, *140*(1), 73–82. doi:10.1530/REP-10-0055

Gilad, E., Meidan, R., Berman, A., Graber, Y., & Wolfenson, D. (1993). Effect of heat stress on tonic and GnRH-induced gonadotrophin secretion in relation to concentration of oestradiol in plasma of cyclic cows. *Reproduction (Cambridge, England)*, *99*(2), 315–321. doi:10.1530/jrf.0.0990315

Greer, E. (1983). Heat stress and reproduction in pigs: Its role in seasonal infertility. *Recent Advances in Animal Nutrition in Australia*, *7*, 216–223.

Grossi, G., Goglio, P., Vitali, A., & Williams, A. (2018). Livestock and climate change: Impact of livestock on climate and mitigation strategies. *Animal Frontiers*, *9*(1), 69–76. doi:10.1093/af/vfy034

Habeeb, A., Gad, A., & Atta, M. (2018). Temperature-Humidity Indices as Indicators to Heat Stress of Climatic Conditions with Relation to Production and Reproduction of Farm Animals. *International Journal of Biotechnology and Recent Advances*, *1*(1), 35–50. doi:10.18689/ijbr-1000107

Hansen, P. J. (2007). Exploitation of genetic and physiological determinants of embryonic resistance to elevated temperature to improve embryonic survival in dairy cattle during heat stress. *Theriogenology*, *68S*, S242–S249. doi:10.1016/j.theriogenology.2007.04.008

Hansen, P. J. (2009). Effects of heat stress on mammalian reproduction. *Philosophical Transactions of the Royal Society of London. Series B, Biological Sciences*, *364*(1534), 3341–3350. doi:10.1098/rstb.2009.0131

Hansen, P. J. (2013). Cellular and molecular basis of therapies to ameliorate effects of heat stress on embryonic development in cattle. *Animal Reproduction*, *10*, 322–333.

Hansen, P. J., & Arechiga, C. F. (1999). Strategies for managing reproduction in the heat-stressed dairy cow. *Journal of Animal Science*, *77*(2), 36–50.

Heba, F., Hozen Hodallah, H. A., Shalaby, S. I. A., & Essawy, G. E. S. (2016). Seasonal Heat Stress Effect on Cholesterol, Estradiol and Progesterone during Follicular Development in Egyptian Buffalo. *Wayamba Journal of Animal Science*. http://www.wayambajournal.com/documents/1304195534.pdf

Houghton, J. T. (2001). *Climate Change 2001: The Scientific Basis. Contribution of Working Group I to the third assessment report of the Intergovernmental Panel on Climate Change*. Cambridge University Press.

IPCC. (2007a). Summary for Policymakers. In *Climate change 2007: Impacts, adaptation and vulnerability*. Contribution of Working Group II to the Fourth Assessment Report of the Intergovernmental Panel on Climate Change. Cambridge University Press.

IPCC. (2013). Summary for policymakers. In *Climate change 2013: the physical science basis*. Contribution of Working Group I to the Fifth Assessment Report of the Intergovernmental Panel on Climate Change. Cambridge University Press.

Kastelic, J. P. (2013). Male involvement in fertility and factors affecting semen quality in bulls. *Animal Frontiers*, *3*(4), 20–25. doi:10.2527/af.2013-0029

Kendall, P., Nielsen, P., Webster, J., Verkerk, G., Littlejohn, R., & Matthews, L. (2006). The effects of providing shade to lactating dairy cows in a temperate climate. *Livestock Science*, *103*(1-2), 148–157. doi:10.1016/j.livsci.2006.02.004

Khan, F. A., Das, G. K., Pande, M., Sarkar, M., Mahapatra, R. K., & Shankar, U. (2012). Alterations in follicular fluid estradiol, progesterone and insulin concentrations during ovarian acyclicity in water buffalo (Bubalus bubalis). *Animal Reproduction Science*, *130*(1-2), 27–32. doi:10.1016/j.anireprosci.2011.12.020

Khodaei-Motlagh, M., Shahneh, A. Z., Masoumi, R., & Derensis, F. (2011). Alterations in reproductive hormones during heat stress in dairy cattle. *African Journal of Biotechnology*, *10*(29), 5552–5558.

Kriegsfeld, L., Gibson, E., Williams, W. III, Zhao, S., Mason, A., Bentley, G., & Tsutsui, K. (2010). The Roles of RFamide-Related Peptide-3 in Mammalian Reproduc tive Function and Behaviour. *Journal of Neuroendocrinology*, *22*(7), 692–700. doi:10.1111/j.1365-2826.2010.02031.x

Krishnan, G., Bagath, M., Pragna, P., Vidya, M., Aleena, J., & Archana, P. (2017). Mitigation of the Heat Stress Impact in Livestock Reproduction. *Theriogenology*, 63–87.

Kumar, B., Singh, G., & Meur, S. (2010). Effects of Addition of Electrolyte and Ascorbic Acid in Feed during Heat Stress in Buffaloes. *Asian-Australasian Journal of Animal Sciences*, *23*(7), 880–888. doi:10.5713/ajas.2010.90053

Lacetera, N. (2019). Impact of climate change on animal health and welfare. *Animal Frontiers*, *9*(1), 26–31. doi:10.1093/af/vfy030

Lawrence, A. B. (2008). Applied animal behaviour science: Past, present and future prospects. *Applied Animal Behaviour Science*, *115*(1-2), 1–24. doi:10.1016/j.applanim.2008.06.003

Lin, H., Decuypere, E., & Buyse, J. (2006). Acute heat stress induces oxidative stress in broiler chickens. *Comparative Biochemistry and Physiology. Part A, Molecular & Integrative Physiology*, *144*(1), 11–17. doi:10.1016/j.cbpa.2006.01.032

Luo, M. J., Han, J. B., Li, W., Sui, H. S., & Tan, J. H. (2005). Serum and follicular fluid steroid levels as related to follicular development and granulosa cell apoptosis during the estrous cycle of goats. *Small Ruminant Research*, *57*(1), 57–65. doi:10.1016/j.smallrumres.2004.06.015

Mandal, D. K., Nagpaul, P. K., & Gupta, A. K. (2000). Seasonal variation in seminal attributes and sexual behaviour of Murrah buffalo bulls. *Indian Journal of Dairy Science*, *53*(4), 278–283.

Mariasegaram, R., Chase, C. C., Jr Chaparro, J. X., Olson, T. A., Brenneman, R. A., & Niedz, R. P. (2007). The slick air coat locus maps to chromosome 20 in Senepol-derived cattle. *Animal Genetics*, *38*(1), 54–59. doi:10.1111/j.1365-2052.2007.01560.x

McKeon, G. M., Stone, G. S., Syktus, J. I., Carter, J. O., Flood, N. R., Ahrens, D. G., Bruget, D. N., Chilcott, C. R., Cobon, D. H., Cowley, R. A., Crimp, S. J., Fraser, G. W., Howden, S. M., Johnston, P. W., Ryan, J. G., Stokes, C. J., & Day, K. A. (2009). Climate change impacts on northern Australian rangeland livestock carrying capacity: A review of issues. *The Rangeland Journal*, *31*(1), 1–29. doi:10.1071/RJ08068

Nardone, A., Ronchi, B., Lacetera, N., Ranieri, M., & Bernabucci, U. (2010). Effects of climate changes on animal production and sustainability of livestock systems. *Livestock Science*, *130*(1-3), 57–69. doi:10.1016/j.livsci.2010.02.011

Nunes, L., Meireles, C., Pinto Gomes, C., de Almeida Ribeiro, N. (2019). Global Climate Change Outlook. *Springer Briefs in Environmental Science*, 9-17.

Patz, J. A., Graczyk, T. K., Geller, N., & Vitto, Y. A. (2000). Effects of environmental change on emerging parasitic diseases. *International Journal for Parasitology*, *30*(12-13), 1395–1405. doi:10.1016/S0020-7519(00)00141-7

Phillips, R. W., Knapp, B. Jr, Hemstra, L. C., & Eaton, O. N. (1943). Seasonal variation in the semen of bulls. *American Journal of Veterinary Research*, *4*, 115.

Pursley, J., Silcox, R. W., & Wiltbank, M. C. (1998). Effect of time of artificial insemination on pregnancy rates, calving rates, pregnancy loss and gender ratio after synchronization of ovulation in lactating dairy cows. *Journal of Dairy Science*, *81*(8), 2139–2144. doi:10.3168/jds.S0022-0302(98)75790-X

Putney, D., Drost, M., & Thatcher, W. (1988a). Embryonic development in super ovulated dairy cattle exposed to elevated ambient temperatures between Days 1 to 7 post insemination. *Theriogenology*, *30*(2), 195–209. doi:10.1016/0093-691X(88)90169-0

Putney, D. J., Gross, T. S., & Thatcher, W. W. (1988b). Prostaglandin secretion by endometrium of pregnant and cyclic cattle at day 17 after estrus in response to in vitro heat stress. *Journal of Reproduction and Fertility*, *84*(2), 475–483. doi:10.1530/jrf.0.0840475

Quinteiro-Filho, W., Rodrigues, M., Ribeiro, A., Ferraz-de-Paula, V., Pinheiro, M., Sá, L., Ferreira, A. J. P., & Palermo-Neto, J. (2012). Acute heat stress impairs performance parameters and induces mild intestinal enteritis in broiler chickens: Role of acute hypothalamic-pituitary-adrenal axis activation1. *Journal of Animal Science*, *90*(6), 1986–1994. doi:10.2527/jas.2011-3949

Rascoli, A., Taha Jalali, M., Nouri, M., Mohammadian, B., & Barati, F. (2010). Effects of chronic heat stress on testicular structures, serum testosterone and cortisol concentrations in developing lambs. *Animal Reproduction Science*, *117*(1-2), 55–59. doi:10.1016/j.anireprosci.2009.03.012

Ronchi, B., Stradaioli, G., Verini Supplizi, A., Bernabucci, U., Lacetera, N., Accorsi, P. A., Nardone, A., & Seren, E. (2001). Influence of heat stress and feed restriction on plasma progesterone, estradiol-17β LH, FSH, prolactin and cortisol in Holstein heifers. *Livestock Production Science*, *68*(2-3), 231–241. doi:10.1016/S0301-6226(00)00232-3

Ronchi, B., Stradaioli, G., Verini Supplizi, A., Bernabucci, U., Lacetera, N., Accorsi, P. A., Nardone, A., & Seren, E. (2001). Influence of heat stress or feed restriction on plasma progesterone, oestradiol-17β, LH, FSH, prolactin and cortisol in Holstein heifers. *Livestock Production Science*, *68*(2-3), 231–241. doi:10.1016/S0301-6226(00)00232-3

Roth, Z. (2008). Heat stress, the follicle, and its enclosed oocyte: Mechanisms and potential strategies to improve fertility in dairy cows. *Reproduction in Domestic Animals*, *43*(2), 238–244. doi:10.1111/j.1439-0531.2008.01168.x

Roth, Z. (2017). Effect of heat stress on reproduction in dairy cows: Insights into the cellular and molecular responses of the oocyte. *Annual Review of Animal Biosciences*, *5*(1), 151–170. doi:10.1146/annurev-animal-022516-022849

Roth, Z., Meidan, R., Braw-Tal, R., & Wolfenson, D. (2000). Immediate and delayed effect of heat stress on follicular development and its association with plasma FSH and inhibin concentration in cows. *Journal of Reproduction and Fertility*, *120*(1), 83–90. doi:10.1530/jrf.0.1200083

Roth, Z., Meidan, R., Shaham-Albalancy, A., Braw-Tal, R., & Wolfenson, D. (2001). Delayed effect of heat stress on steroid production in medium-sized and preovulatory bovine follicles. *Reproduction (Cambridge, England)*, *121*, 745–751. doi:10.1530/rep.0.1210745

Rowan, E., Evans, C., Riley-Gilbert, M., Hyman, R., Kafalenos, R., Beucler, B., Rodehorst, B., Choate, A., & Schultz, P. (2013). Assessing the Sensitivity of Transportation Assets to Extreme Weather Events and Climate Change. *Transportation Research Record: Journal of the Transportation Research Board*, *2326*(1), 16–23. doi:10.3141/2326-03

Rust, J., & Rust, T. (2013). Climate change and livestock production: A review with emphasis on Africa. *South African Journal of Animal Science*, *43*(3), 255. doi:10.4314ajas.v43i3.3

Ryan, D. P., Blakewood, E. G., Lynn, J. W., Munyakazi, L., & Godke, R. A. (1992). Effect of heat-stress on bovine embryo development in vitro. *Journal of Animal Science*, *70*(11), 3490–3497. doi:10.2527/1992.70113490x

Sakai, S., Hagihara, N., Kuse, M., Kimura, K., & Okuda, K. (2018). Heat stress affects prostaglandin synthesis in bovine endometrial cells. *The Journal of Reproduction and Development*, *64*(4), 311–317. doi:10.1262/jrd.2018-051

Samal, L. (2013). Heat Stress in Dairy Cows - Reproductive Problems and Control Measures. *International Journal of Livestock Research*, *3*, 14–23.

Schuller, L. K., Burfeind, O., & Heuwieser, W. (2014). Impact of heat stress on conception rate of dairy cows in the moderate climate considering different temperature humidity index thresholds, periods relative to breeding, and heat load indices. *Theriogenology*, *81*(8), 1050–1057. doi:10.1016/j.theriogenology.2014.01.029

Sejian, V., Samal, L., Haque, N., Bagath, M., Hyder, I., Maurya, V. P., Bhatta, R., Ravindra, J. P., Prasad, C. S., & Lal, R. (2015). Overview on adaptation, mitigation and amelioration strategies to improve livestock production under the changing climatic scenario. *Sheep Production Adapting to Climate Change*, 359-398.

Setchell, B. P. (1998). The Parkes lecture Heat and the testis. *Journal of Reproduction and Fertility*, *114*(2), 179–194. doi:10.1530/jrf.0.1140179

Shwartz, G., Rhoads, M. L., VanBaale, M. J., Rhoads, R. P., & Baumgard, L. H. (2009). Effects of a supplemental yeast culture on heat-stressed lactating holstein cows. *Journal of Dairy Science*, *92*(3), 935–942. doi:10.3168/jds.2008-1496

Silanikove, N. (2000). Effects of heat stress on the welfare of extensively managed domestic ruminants. *Livestock Production Science*, *67*(1-2), 1–18. doi:10.1016/S0301-6226(00)00162-7

Singh, A. K., Chandra, G., Ojha, B. K., & Rajoriya, J. S. (2019). Effect of heat stress on livestock reproduction. *Haryana Veterinarian*, *58*, 16–24.

Singh, M., Chaudhari, B. K., Singh, J. K., Singh, A. K., & Maurya, P. K. (2013). Effects of thermal load on buffalo reproductive performance during summer season. *The Journal of Biological Sciences*, *1*(1), 1–8.

Sinha, R., Ranjan, A., Lone, S., Rahim, A., Devi, I., & Tiwari, S. (2017). The impact of climate change on livestock production and reproduction: Ameliorative management. *International Journal of Livestock Research*, *7*(6), 1–8. doi:10.5455/ijlr.20170417042102

Sinha Hikim, A. P., Lue, Y., Yamamoto, C., Vera, Y., Rodriguez, S., Yen, P. H., & … . (2003). Key apoptotic pathways for heat-induced programmed germ cell death in the testis. *Endocrinology*, *144*(7), 3167–3175. doi:10.1210/en.2003-0175

Sirotkin, A. V., & Kacaniova, M. (2010). The effect of high temperature on swine ovarian function in vitro. *Veterinarni Medicina*, *55*(8), 377–382. doi:10.17221/2953-VETMED

Stem, E., Mertz, G. A., Stryker, J. D., & Huppi, M. (1989). Changing animal disease patterns induced by the greenhouse effect. In *The potential effects of global change on the United States*. US Environmental Protection Agency.

Szymanski, L., Schneider, J., Friedman, M., Ji, H., Kurose, Y., Blache, D., Rao, A., Dunshea, F. R., & Clarke, I. J. (2007). Changes in Insulin, Glucose and Ketone Bodies, But Not Leptin or Body Fat Content Precede Restoration of Luteinising Hormone Secretion in Ewes. *Journal of Neuroendocrinology*, *19*(6), 449–460. doi:10.1111/j.1365-2826.2007.01551.x

Tao, S., Monteiro, A. P., Thompson, I. M., Hayen, M. J., & Dahl, G. E. (2012). Effect of late gestation maternal heat stress on growth and immune function of dairy calves. *Journal of Dairy Science*, *95*(12), 7128–7136. doi:10.3168/jds.2012-5697

Thornton, P., & Gerber, P. (2010). Climate change and the growth of the livestock sector in developing countries. *Mitigation and Adaptation Strategies for Global Change*, *15*(2), 169–184. doi:10.100711027-009-9210-9

Thornton, P., Herrero, M., Freeman, A., Mwai, O., Rege, E., Jones, P., & McDermott, J. (2007). Vulnerability, climate change and livestock research opportunities and challenges for poverty alleviation. *Journal of SAT Agricultural Research*, *4*(1), 1–23.

Tompkins, E. C., Heidenreich, C. J., & Stob, M. (1967). Effects of post-breeding thermal stress on embryonic mortality in swine. *Journal of Animal Science*, *26*(2), 4. doi:10.2527/jas1967.262377x

Torres-Junior, J. R. D. S., Pires, M., & de. (2008). Effect of maternal heat stress on follicular growth and oocyte competence in Bos indicus cattle. *Theriogenology*, *69*(2), 155–166. doi:10.1016/j.theriogenology.2007.06.023

Tsutsui, K., Saigoh, E., Ukena, K., Teranishi, H., Fujisawa, Y., Kikuchi, M., Ishii, S., & Sharp, P. J. (2000). A novel avian hypothalamic peptide inhibiting gonadotropin release. *Biochemical and Biophysical Research Communications*, *275*(2), 661–667. doi:10.1006/bbrc.2000.3350

UN. (2017). *United Nations, Department of Economic and Social Affairs, Population Division 2017. World population prospects: the 2017 revision, key findings and advance tables*. Working Paper No. ESA/P/WP/248. Available from https://esa.un.org/unpd/wpp/publications/files/wpp2017_keyfindings.pdf

Upadhyay, R. C., & Ashutosh, S.V. (2009). Impact of climate change on reproductive functions of cattle and buffalo. In *Global Climate Change and Indian Agriculture*. New Delhi: ICAR.

Wang, Y., Huang, J., Xia, P., He, J., Wang, C., Ju, Z., Li, J., Li, R., Zhong, J., & Li, Q. (2013). Genetic variations of HSBP1 gene and its effect on thermal performance traits in Chinese Holstein cattle. *Molecular Biology Reports*, *40*(6), 3877–3882. doi:10.100711033-012-1977-1

Wilson, S. J., Marion, R. S., Spain, J. N., Spiers, D. E., Keisler, D. H., & Lucy, M. C. (1998). Effects of controlled heat stress on ovarian function of dairy cattle. *Journal of Dairy Science*, *81*, 2139–2144.

Wolfenson, D., Roth, Z., & Meidan, R. (2000). Impaired reproduction in heat-stressed cattle: Basic and applied aspects. *Animal Reproduction Science*, *60-61*, 535–547. doi:10.1016/S0378-4320(00)00102-0

Younas, M., Ishaq, M., & Ali, I. (2012). *Effect of Climate Change on Livestock Production in Pakistan*. Academic Press.

Zafar, A. H., Ahmed, N., & Shah, S. K. (1988). Effect of seasonal variation on semen production of Nili-Ravi buffalo bulls. *Buffalo Journal*, *4*(1), 61–68.

Zeron, Y., Ocheretny, A., Kedar, O., Borochov, A., Sklan, D., & Arav, A. (2001). Seasonal changes in bovine fertility: Relation to developmental competence of oocytes, membrane properties and fatty acid composition of follicles. *Reproduction (Cambridge, England)*, 447–454. doi:10.1530/rep.0.1210447

Chapter 13
Climate Change on Fertility and Reproductive Processes of Female Livestock

Joan Nyika
https://orcid.org/0000-0001-8300-6990
Technical University of Kenya, Kenya

ABSTRACT

The effects of climate change continues to be a growing modern-day challenge. Climate change-induced heat stress disrupts reproductive and fertility systems in livestock. In males, it modifies the physiology of the spermatogenic cycle resulting to poor quality semen and high prevalence of secondary sperm defects. In female livestock, heat stress decreases the production of gonadotrophins, results to hormonal imbalance, decreases the quality of oocytes, and lengthens the oestrous period leading to infertility. These effects can be reversed through genetic modifications, nutritive supplementation, physical cooling mechanisms, and hormonal therapies. The successful implementation of the ameliorative strategies is pegged on improved research and their combined administration. Ultimately, climate change mitigation and adaptation are indispensable to overcome fertility problems in livestock among other environmental effects of the climate variations.

INTRODUCTION

Climate change is a growing concern of the contemporary world particularly in developing countries with the highest vulnerability yet, these countries are the least prepared to deal with the resultant effects, are financially limited due to high poverty levels and rely on natural resources heavily. Nyika (2020) confirmed that the effects of climate change are multifaceted and affect natural resources' (land and water) sustainability, ecology, human and animal health with the situation in developing countries being dire due to their vulnerability and unpreparedness. Rising urbanisation trends and a growing population further exacerbate this trend. With climate change, the temperatures rise according to projections by Sheikh et al. (2017) who stated that by the year 2100 global temperatures will be 1.1 to 6.4°C higher

DOI: 10.4018/978-1-7998-4480-8.ch013

than the state at 2010 while rainfall, a main requirement for the fodder production will become more variable. Climate change trends are exacerbated by uncertainty, which makes it difficult to assess and predict its expected effects (Rust and Rust, 2013).

Globally, the livestock sector has been pressured by climate change events and effects in that the demand for animal products is high though optimisation of production capacities is hampered by several environmental conditions. According to Rojas-Downing et al. (2017), there is a growing livestock revolution whereby the preference to meat and meat products will rise from 258 to 455 million tonnes while that of milk and milk products will rise from 664 to 1077 million tonnes between 2006 and 2050. The effects of climate change, which will interfere with production capacity of livestock and will spill over these effects to the economy. This scenario is expected because 17% of the kilocalorie consumption and 33% protein intake is from livestock worldwide. Additionally, the livestock sector employs more than 1.1 billion people and most of them dwell in low income developing countries (Escarcha et al., 2018). There is a rising demand for livestock keeping and its indirect effects such as rising demand for agricultural land to grow fodder and high demand for consumptive water, which are consumers of natural resources (land and water). Consequently, the climate change situation is expected to worsen as the sector currently contributes to 14.5% of global greenhouse gas (GHG) emissions (Gerber et al., 2013). Consequently, climate change will affect livestock production by influencing competition with other natural resources, heat loss, biodiversity loss, vulnerability to livestock diseases and the quality and quantity of produced feeds (Sejian, 2013).

Of growing concern is the suggestion that climate change could interfere with the reproduction capacity of livestock. Kumar et al. (2017) reported that extreme climatic conditions induce stress on animals, which negatively influence reproduction and their production. Among the reported stresses include: reduced pasture and crop yield, frequent droughts and low rainfall leading to nutritional stress as well as heat stress due to increased soil radiation. The effects are more pronounced in animals reared under extensive rather than intensive production systems. Reproduction is an intricate process governed by environmental signals, metabolites and hormones in timed psychological and physiological procedures. The procedures include post-partum recovery, parturition, gestation, motherly recognition of pregnancy, embryonic development, puberty, transportation of gametes, fertilization and gametogenesis. Environmental factors such as nutrition, water access and quality, photoperiod, humidity and temperature affect the procedures when acting alone or in combination. The environmental factors usually aggravated by climate change adversely affect reproduction by interfering with the functions of hormonal systems and reproductive organs (Kumar et al., 2017). Effects such as dysfunctional sperm production and its reduced mobility to the egg and female hormonal abnormalities have been reported in sheep as a result of climate change (Sawyer and Jitik Narayan, 2019). Similarly, reduced egg production and interruption of ovulation in hens is documented as a result of extreme heat (Rojas-Downing et al., 2017). Reproductive roles of bulls, pigs and poultry have been disrupted as a result of climate change (Escarcha et al., 2018). Other general effects of climate change on reproduction include retarded foetal growth, abortion, still/ pre-mature births, embryonic deaths, failed fertilization, indifferent sexual behaviours and delayed puberty among others (Indu et al., 2014). This book chapter explores the effects of climate change on livestock fertility and reproductive systems and suggests a number of measures to mitigate and adapt to these changes with the cognition that climate change and its effects are inevitable.

EFFECTS OF CLIMATE CHANGE ON REPRODUCTION AND FERTILITY

Climate change effects on livestock occur in four categories: (1) accelerated stress and disease due to temperature rises, (2) limited availability of forage due to unfavourable weather, (3) increased costsof adaptation aimed at increasing livestock production/fertility and (4) extreme weather conditions (flooding and drought) (Savsani et al., 2015). Of these four effects, the increase in temperatures characterised by frequent heat waves and hot days that increase the intensity of radiation and duration of air temperature interfere with fertility and reproductive systems of livestock (Bernabucci, 2019). Such thermal stress often leading to hyperthermia impairs and modifies cellular function in tissues of livestock such as cow, sheep, pigs, rabbits and horses. Consequently, blood flow redistribution occurs in these animals reducing their production performance and food intake due to stress as Wolfenson and Roth (2019) noted. In the following sections, the effects of climate change on male and female livestock productive processes that culminate to reduced fertility as summarised in Fig. 1 will be discussed.

Figure 1. Comparative aspects of the climate change-induced heat stress on the reproductive/fertility systems of male and female livestock.

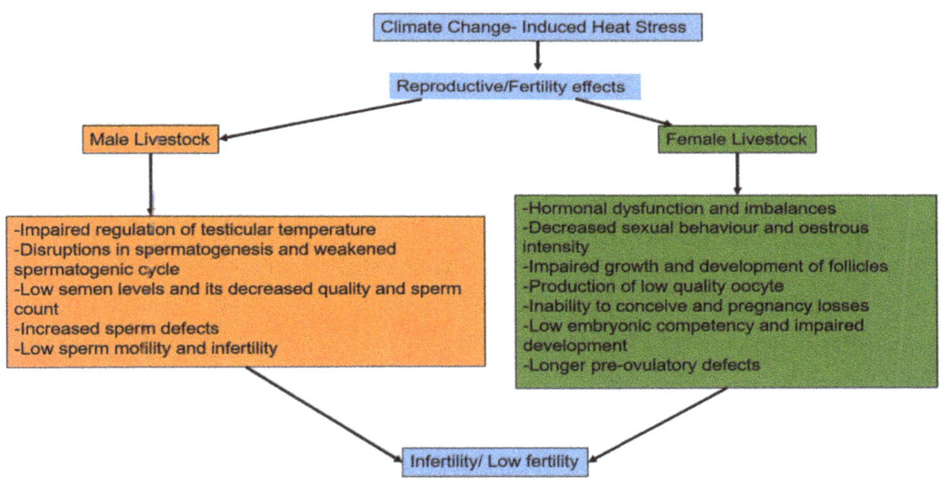

Climate Change Effects on Male Livestock Reproductive Processes

The reproductive system of male livestock consists of the sexual organ (penis), ducts, testis (testicles) and associated glands. Its main function is to generate and release sperms that fuse with female gametes for reproduction. Additionally, the male reproductive system acts as an endocrine gland, which releases male hormones including testosterone, which are attributable to secondary sexual characteristics of male livestock. Male livestock have inherent physiological mechanisms aimed at preventing hyperthermia in the testicles to maintain reproductive integrity even in adverse environmental conditions (Takahashi, 2012). Additionally, they have enlarged sweat glands in their scrotums to maintain normal local temperature. For instance, the testicular temperature of bulls should be about 5°C below the temperature of the rectum for effective spermatogenesis. Optimum temperature for spermatogenesis ranges between 15 and 20°C

. Climate change effects such as high solar radiation, high relative humidity and ambient temperature characteristic of climate variation and change make male livestock vulnerable to reduced fertility due to low sperm production. This phenomenon is not only in bulls but has been reported in male horse, sheep and pigs (Takahashi, 2012). High temperatures also enhance sperm defects by interfering with metabolic functions of sperm cells resulting to increased lipid peroxidation and production of reactive oxygen (Krishnan et al., 2017). In male ruminant animals, hyperthermia increases the temperature of the scrotal skin leading to testicular degeneration, low sperm production and poor quality semen release, which causes infertility and subfertility (Brito et al., 2004).

High temperatures are causes of secondary defects of sperms in livestock, which exhibits as low sperm motility. A study by Ruediger et al. (2016) positively correlated high scrotal temperatures to production of abnormal and/or dead sperms in livestock but negatively correlated these climatic effects to low libido, sperm concentration and motility. These findings are inconsistent with those from other publications such as Kadokawa et al. (2012) who reported that high temperatures influence the quality of ejaculate; and Nardon et al. (2010) who suggested that these climatic modifications result to abnormal sperm morphology, their reduced concentration, viability and motility in bucks and bulls causing infertility. The following sections examine these effects in details under two categories: quality of semen and sperm production process.

Quality of Semen Produced

Semen is a slightly yellowish to viscous substance made of sperms and seminal fluid. The quality of semen is affected by animal illness, fever or stress. Stress interferes with the spermatogenic cycle and consequently the semen quality. Heat stress in male livestock triggers endocrinological and biochemical modifications including the decreased production of glutathione peroxidase[1] and increased production of oxidative markers such as thiobarbituric acid reactive substances (TBARs)[2] (Takahashi, 2012). These modifications affect the quality of semen produced by bovines, rams and boars. In high temperatures, semen of these animals was low in volume and its sperm count was low leading to a high prevalence of sperm defects (Nichi et al., 2006). The quality and quantity of semen produced by boars during high temperatures was found to deteriorate prior to and during collection owing to low maturation rates and stereroidgenesis[3] (Krishnan et al., 2017). Heat effects on semen quality is more predominant in male pigs since they have a reduced sweating capacity. Kunavongkrita et al (2005), made these observations reporting that boars had low volumes of semen whose sperm concentration significantly reduced at summer compared to winter. Additionally, the biochemical constituents of semen such as its total phosphorus, potassium, sodium, citric acid and fructose decreased, an effect resulting from increased lipid peroxidation. The quantity and quality of semen are essential factors in determining male livestock productivity and sexual behaviour.

Spermatogenesis

Spermatogenesis is the process of developing haploid spermatozoa in the testicles particularly within the seminiferous tubules. The process entails mitotic division at the basement membrane of the seminiferous tubules. Factors that influence sperm production such as weight, size, tone and consistency of testicles as well as scrotal circumference are indirectly proportional to high temperatures. Menegassi et al. (2015) noted about this trend highlighting that the temperature range for effective spermatogenesis is between 15.6 and 29.4°C . Temperatures above this range result to degenerated sperms with low fertilisation capacity. The high relative humidity and ionizing irradiation associated with hyperthermia can destroy sperms at their mitotic stage. According to Pinto et al. (2001), rams and bulls exposed to such heat stress are prone to secondary sperm defects and their sperms have low mobility. Furthermore, their production of dead and dysfunctional sperms increased, which reduced their breeding efficiency, reduced the number of spermatids and spermatocytes, the ratio of sertoli cells[4] in comparisons to other cells and seminiferous tube diameter (Krishnan et al., 2017). High temperatures modify the physiology of the scrotal distal and proximal poles increasing sperm necrosis and apoptosis (Brito et al., 2004).

Climate Change Effects on Female Livestock Reproductive Processes

Climate change effects of high temperatures affect female livestock fertility at various stages of development, during conception and have embryoniceffects. A number of mechanisms, which propagate these effects, are shown in Fig. 2. Maternal hyperthermia affects livestock fertility and reproductive system where it acts on both the hypothalamus and pituitary glands exciting the release of adrenocorticotrophic and corticotropic hormones. These hormones stimulate the release of cortisol, catecholamines and glucocorticoids that are stress-relieving hormones. These hormones eventually inhibit the reproductive axis by shortening the oestrus period and intensity, promoting pre-antral and antral follicular apoptosis and modifying the development of follicles. Maternal heat stress is reported to delay growth of female livestock to pubertal stages, disturbs the growth of follicles and meiotic maturation of oocytes and inhibits normal uterine function leading to an inhibitory reproductive process and subfertility (Wolfenson and Roth, 2019). Heat stress also suppresses the role of gonadotrophins[5] including the luteinising hormone (LH).

Figure 2. Mechanisms that propagate climate change-induced infertility in female livestock (Adopted from Wolfenson and Roth, 2019). GnRH; represents gonadotrophin releasing hormone, LH; luteinising hormone, CRH; corticotropic releasing hormone, ACTH; adrenocorticotrophic hormone, An; androstenedione, E_2; follicular oestradiol and CL; corpus luteum.

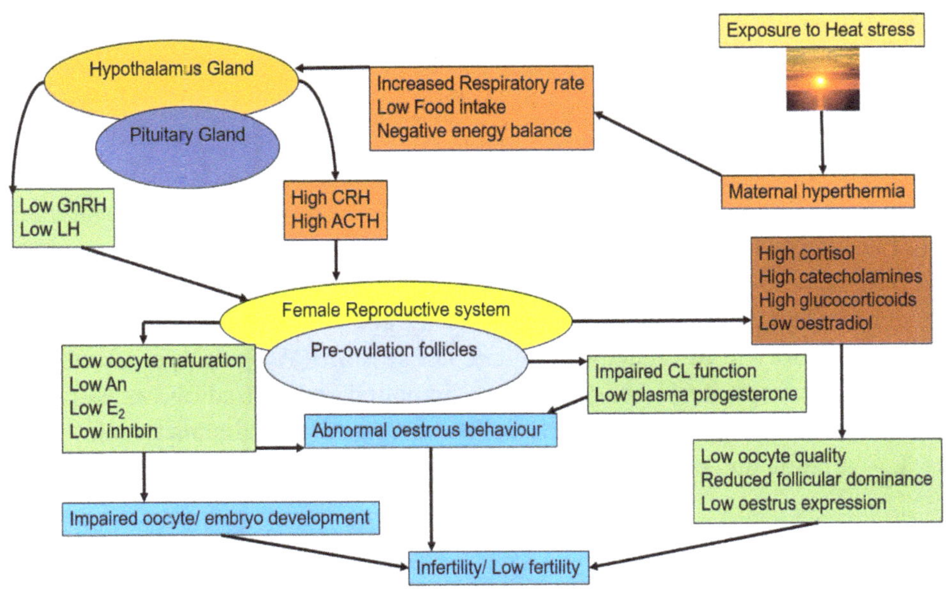

Hormonal Modifications of Female Livestock

Reproductive hormones have an essential role in propagating development of livestock. In female animals, numerous hormones secreted by endocrine glands regulate reproduction. Oestrogen is produced in the Graafian follicle while progesterone is produced in the corpus luteum. The production of these two ovarian hormones is influenced by gonadotropins of the anterior pituitary, which are follicle stimulating (FSH) and luteinizing hormones (LH). The LH enables follicle rapture during ovulation leading to the growth of the corpus luteum while the FSH enhances growth and physiology of the follicles. In regulating ovarian activity, the GnRH, FSH and LH are most important. A number of studies have shown that climate change stimulated rises in temperature disrupt the physiological effectiveness of reproductive hormones compromising productivity of livestock (Takahashi, 2012; Rust and Rust, 2013; Savsani et al., 2015; Krishnan et al., 2017; Sawyer et al., 2019; Wolfenson and Roth, 2019). On the twelfth day of the oestrous cycle, a 16-hour exposure of heifers to temperatures above 40°C resulted to reduced GnRH-stimulated reduction of FSH and LH (Ozawa et al., 2005). The frequency and pulse amplitude of LH decreases significantly at high temperatures hindering the development and ovulation of dominant follicles because of low oestradiol[6] production. In does, hyperthermia causes modified sexual behaviour in the form of lagged ovulation resulting from reductions of LH receptors and plasma concentrations of progesterone according to Wolfenson et al. (2000). Androstenedione and oestradiol reductions in addition to low plasma progesterone levels disrupt endometrial function interfering with embryo development, which results to infertility (Singh et al., 2013). Specific studies focusing on climate change effects on hormonal levels and functions in female livestock have reported consistent results. In heifers, excess heat exposure reduce LH and FSH levels (Takahashi, 2017) while in bucks and rams, E2 and progesterone, respectively (Ozawa et al., 2005).

Climate Change on Ovarian Follicle Development

Follicular development is stimulated by FSH production and after their ripening, oestrogen levels rise. The rise of oestrogen suppresses FSH production from the pituitary via a negative feedback. Follicles also release inhibits to suppress FSH levels. As the oestrogen levels rise, the production of LH at the mid cycle is prompted to enable ovulation. (Wolfenson and Roth, 2019). Mammalian livestock usually have two follicular waves in a 3-week oestrous cycle. In the two, one follicle enlarges to become dominant while the remaining one disappear and become atretic. At the second wave, the dominant follicle grows further to a pre-ovulatory follicle as the cycle ends and if the endocrine conditions are favourable, ovulation can occur. In events of maternal hyperthermia, the process is modified so that follicular dominance is attenuated in two forms. The first one is increased number of dominant follicles occurring in a given follicular wave and the second, is a lengthened period of pre-ovulatory follicle dominance as a result of low levels of inhibin[7]. The latter phenomenon explains the rise of depressed fertility in cows while the former, the prevalence of twin inseminations during hot seasons (Wolfenson and Roth, 2019). Under heat stress, steroidogenic capacity of follicular cells is impaired and oestradiol concentrations decrease. According to Wolfenson et al. (2000), these events lead to low oestrous intensity, inhibit production of LH, alter the CL, ovulation and enhance ovarian cyst development.

Climate Change on Corpus Luteum

The luteinising of the follicles following their rupture leads to the formation of corpus luteum (CP), which is a transitory endocrine gland. This gland secrets oestrogen and progesterone. The physiological function of the CL is to produce progesterone whose role is in embryonic development. Under heat stress, mammalian livestock suffer from luteal insufficiency, which is a deficiency of progesterone that is associated with the animals' inability to support pregnancy. The duration at which animals are exposed to heat stress influences the produced levels of progesterone. Wolfenson et al. (2002) associated short-term heat exposure of heifers to increased progesterone release by adrenal glands while the situation was reversed in chronic hyperthermia. These observations were associated to low progesterone levels, impaired CL and pre-ovulatory follicle formation. According to Inskeep and Dailey (2005), heat stressed cows have non-persistent CL, which inhibits interferon tau (ruminants' pregnancy recognition signal) production and ultimately impairs embryonic growth. This phenomenon was attributed to the loss of pregnancy during the first 8 to 17 days of pregnancy in a retrospective study of dairy cattle at Spain's north-eastern region (Lopez-Gatius, 2003).

Climate Change on Oocyte Maturation

Oocytes are immature eggs that develop in the follicle within the ovarian outer layer. At each cycle, only one oocyte matures to an ovum following ovulation. Oocytes development capacity occurs in various stages during follicular development and as such, heat stress results to their perturbations during follicular growth. Hyperthermia impairs the developmental competence of oocytes by inducing apoptosis, irreversible damage of their meiotic spindle and cytoskeleton and through oxidative damage, which retards their maturation (Singh et al., 2013). The prolonged follicular dominance resulting from climate change-induced heat leads to the aging of oocytes and their low developmental prospect due to premature meiosis.

Many in vivo and in vitro studies have reported the effects of hyperthermia to the quality of resultant oocytes. In Holstein heifers, heat stress results to reduced number of oocytes that do not develop to ovum (Takahashi, 2017). This could be a result of induced cell death, disorganised meiosis, impaired architecture of microtubules and microfilament and decreased protein synthesis. Heat stress after fertilisation resulted to super-ovulation in heifers (Ozawa et al., 2005), retarded foetal growth in sows (Wettemann et al., 1988) and apoptosis in rabbits (Makarevich et al., 2007). The heat-induced oocyte impairment is long-term and can take 2-3 oestrous cycles to recover as Roth et al. (2001) suggested. This phenomenon could explain events where some heifers even during autumn and winter fail to conceive as a sub-population of follicular cells that are damaged are in the process of recovering spontaneously.

Climate Change on Embryo Development

Embryogenesis refers to the preliminary stages of prenatal growth where foetal development occurs. It starts with ovum fertilization by a sperm cell to form a diploid zygote. The zygote undergoes mitotic division and cellular differentiation without considerable growth to form a multicellular embryo. Embryo development is sensitive to heat stress and this sensitivity is indirectly proportional to age. According to Hansen (2007), embryos at early developmental stages are sensitive to heat compared to their resistance at later stages. Thermo-tolerance is associated to antioxidant protection and free radical generation for older embryos. Heat stress induces impaired embryo development, luteal regression and inhibits implantation. These post-conception and placentation defects occur due to accelerated production of prostaglandin F2α (PGF2α)[8] at the uterus. These proteins induced pregnancy losses due to slow embryonic development in rabbits (Scenna et al., 2004) and cows (Soto et al., 2003).

AMELIORATION STRATEGIES OF CLIMATE CHANGE-INDUCED HEAT STRESS IN LIVESTOCK

Climate change-induced heat stress leads to agricultural losses though the challenge is manageable. Collier, Dahl and Van Baale (2006) proposed three approaches to manage heat stress including: 1) nutritional management, 2) genetic development of breeds and 3) physical modifications. To optimize productivity, these strategies can be used individually or as a combination. Wolfenson and Roth (2019) proposed advanced technologies aimed at improving livestock fertility including timed artificial insemination, embryo transfer and gonadotrophin treatment. The following sections discusses these modifications in details.

Physical Modifications

Physical modifications aim at providing suitable microclimate to optimise survival and productive capacity of animals through provision of shade and cooling/evaporative techniques. Cooling techniques and shade maintain reproductive performance of livestock and are critical in tropical climates (Brantly, 2013). They occur naturally in the form of shade or by painting shade roof white colour, installation of isolating material to reduce solar radiation and increasing the shade height to more than three metres (Slimen et al., 2014). Direct water sprays on animals, use of cooling pads and installation of fans, misting and fogging systems can also be done to reduce animal temperatures and has been reported to reduce 30% of incoming solar radiation in addition to providing productive benefits in dairy cattle (Slimen et al., 2014). Cooling approaches are short-term and involve several cooling cycles daily, which last 30-50 minutes. These evaporative cooling strategies of maintaining reproductive soundness inlivestock are expensive though they have 68-98% effectiveness in retaining conception soundness of livestock according to Wolfenson and Roth (2019). Dash et al. (2016) noted that dairy cattle fertility significantly improved following the introduction of cooling systems during summer. In sows, cooling systems reduced the rate of foetal death according to Johnson (2018) who cautioned against implementing it rapidly to prevent negative effects on pig metabolism.

Nutrition Management

Livestock dietary trends influence their productive capacities and heat production ability. In pigs for instance, the intake of fibre enhanced heat production and worsened heat stress during summer (Johnson, 2018). Dash et al. (2016) also pointed that protein energy ratio and ability of cows to digest dry matter influenced their reproductive trends. In cognition of these predispositions, livestock keepers have formulated energy-dense but low fibre feeds for animals to reduce heat productivity without compromising on their fertility (Johnson, 2018). Feeding cows with low-degradable protein of high quality, supplementation with vitamins such as niacin and antioxidants was reported to improve their associated fertility (Dash et al., 2016). Supplementation of livestock feed with minerals such as zinc and betaine and amino acids such as L-glutamate was found to enhance immunity in cows and rabbits and reduce pregnancy losses in sows (Johnson, 2018). According to Staples, Burke and Thatcher (1998), the use of fat supplements on heat stressed cows was shown to improve follicular growth, pregnancy survival, metabolism of prostaglandins and plasma progesterone, which are essential in their fertility. Chauhan et al. (2015) noted that macro-mineral (sodium and potassium) supplementation increased livestock reproductive performance while Abdin and Khattom (2013) reported that L-vitamin C supplementation in mammalian livestock enhanced semen quality and fertility during heat stress. Wolfenson and Roth (2019) noted that supplementation of heat stressed livestock with insulin growth factor (IGF-1) enhanced embryonic survival and was anti-apoptotic to follicles while vitamin A introduction to dairy cows prior to calving improved their reproductive performance.

Hormonal Therapy

The introduction of gonadotropins in livestock has been used as a promising technique to enhance fertility during climate change-induced heat stress. A case example is the induction of GnRH during the preliminary stages of the oestrous cycle that resulted to enhanced production of LH that improves the success rate of conception (Krishnan et al., 2017). Wolfenson et al. (2000) noted that injection of human chorionic gonadotropin (hCG)[9] on the fifth oestrous day enhanced ovulation of dominant follicles to form a CL and increased plasma progesterone levels during heat stress. The most successful hormone therapy used to manage climate change-induced heat stress is the timed artificial insemination (TAI) (Dash et al., 2016; Krishnan et al., 2017). The procedure involves injection of GnRH in livestock to introduce ovulatory follicles followed by injection of PGF2α a week later to prevent CL development and enhance follicular maturation. Afterwards, a second GnRH dose is administered to induce ovulation and insemination of dairy cattle after 16 hours to enhance conception. Other hormonal treatments include controlled internal-drug release (CIDR) synchronized with ovsynch systems that were found to improve conception rates by increasing sensitivity of the oestrous period and through enhanced pregnancy survival in heat stressed dairy cattle (El-Tarabany and El-Tarabany, 2015).

Genetic Modifications

With the inevitability of climate change, long-term strategies to manage heat stress and prevent its effects on animal fertility are essential. Previous strategies focused on selective breeding of livestock to improve productivity though this is antagonistic to heat production (Dash et al., 2016). However, recent modifications are focusing on temperature humidity index (THI)[10] covariate characteristics for better genetic competence aimed at improved reproduction and fertility of livestock (Johnson, 2018). This has been done through 1) introgressive hybridisation of heat tolerant genes to commercial from local breeds and 2) selection of breeds in heat tolerant conditions. In the latter, breeds with slick hair, long hair length and coat colour genes have been selected due to their inherent tolerance to heat. In the former, livestock from hotter areas have been induced with slick hair and heat shock protein (HSP) genes that confer heat tolerance without disrupting fertility (Dash et al., 2016).

In marker-facilitated breed selection, HSP genes are used as markers to enhance heat tolerance in various breeds. Examples are HSBP1 and HSP70 A1 A genes in Chinese dairy cattle (Wang et al., 2013) and HSP90AB1 gene in Thai dairy cattle (Deb et al., 2014). Other important genes that have been used to enhance livestock fertility include glycosyl transferase, fibroblast growth factor and ATP1B2 thyroid hormone receptor among others (Wang et al., 2013). The expression mechanisms of the named genes and their relationship to enhanced fertility in heat stressed livestock require extensive research prior to their use as biological markers (Johnson, 2018).

CONCLUSION

With the predominance of climate change, high relative humidity and solar radiation will induce heat stress in livestock and negatively affect their fertility and reproductive capacity. This chapter explored the various effects of this heat in male and female livestock fertility and reproductive systems using named examples and at various developmental stages. Ovarian components including CL, oocytes, embryos and follicles as well as the spermatogenic cycle were found to be sensitive to high temperatures. The chapter also suggested some ameliorative strategies to manage the effects. These include hormonal therapies to enhance conception and pregnancy survival, genetic modifications to enhance heat-stress adaptation and heat-tolerance, cooling/evaporative adjustments to maintain normal livestock temperature and nutritive adjustments to retain reproductive soundness in such animals. A combination of these strategies is essential for optimal results in various livestock considering that their individual effect is influenced by multiple factors.

AUTHOR CONTRIBUTION

JMN conceived, designed and wrote the book chapter along with literature survey.

FUND DISCLAIMER

No grant received from any funding agency in public, commercial, or not-for-profit sector.

CORRESPONDING AUTHOR

Joan Nyika

REFERENCES

Abdin, Z., & Khatoon, A. (2013). Heat stress in poultry and the beneficial effects of ascorbic acid(vitamin c) supplementation during periods of heat stress. *World's Poultry Science Journal*, *69*(1), 135–152. doi:10.1017/S0043933913000123

Bernabucci, U. (2019). Climate change: Impact on livestock and how we can adopt. *Animal Frontiers*, *9*(1), 3–5. doi:10.1093/af/vfy039 PMID:32002232

Brantly, S. (2013). *Mitigating heat stress in cattle*. Working Trees Info Sheet. USDA National Agroforestry Service. Retrieved from: http://nac.unl.edu/Working_Trees/infosheets.htm

Brito, L., Silva, A., Barbosa, R., & Kastelic, J. (2004). Testicular thermoregulation in Bos indicus, crossbred and Bostaurus bulls: Relationship with scrotal, testicular vascular cone and testicular morphology and effects on semen quality and sperm production. *Theriogenology*, *61*(3), 511–528. doi:10.1016/S0093-691X(03)00231-0 PMID:14662148

Chauhan, S., Celi, P., Leury, B., Liu, F., & Dunshea, F. (2015). Exhaled breath condensate hydrogen peroxide concentration, a novel biomarker for assessment of oxidative stress in sheep during heat stress. *Animal Production Science*, *56*(7), 1105–1112. doi:10.1071/AN14070

Collier, R. J., Dahl, G. E., & VanBaale, M. J. (2006). Major Advances Associated with Environmental Effects on Dairy Cattle. *Journal of Dairy Science*, *89*(4), 1244–1253. doi:10.3168/jds.S0022-0302(06)72193-2 PMID:16537957

Dash, S., Chakravarty, A. K., Singh, A., Upadhyay, A., Singh, M., & Yousuf, S. (2016). Effect of heat stress on reproductive performances of dairy cattle and buffaloes: A review. *Veterinary World*, *9*(3), 235–244. doi:10.14202/vetworld.2016.235-244 PMID:27057105

Deb, R., Sajjanar, B., Singh, U., Kumar, S., Singh, R., Sengar, G., & Sharma, A. (2014). Effect of heat stress on the expression profile of Hsp90 among Sahiwal (Bos indicus) and Frieswal (Bos indicus × Bos taurus) breed of cattle. *Gene*, *536*(2), 435–440. doi:10.1016/j.gene.2013.11.086 PMID:24333856

El-Tarabany, M. S., & El-Tarabany, A. A. (2015). Impact of thermal stress on the efficiency of ovulation synchronization protocols in Holstein cows. *Animal Reproduction Science*, *160*, 138–145. doi:10.1016/j.anireprosci.2015.08.002 PMID:26297623

Escarcha, J., Lassa, J., & Zander, K. (2018). Livestock under Climate Change: A Systematic Review of Impacts and Adaptation. *Climate (Basel)*, *6*(3), 54. doi:10.3390/cli6030054

Gerber, P. J., Steinfeld, H., Henderson, B., Mottet, A., Opio, C., Dijkman, J., Falcucci, A., & Tempio, G. (2013). *Tackling Climate Change through Livestock: A Global Assessment of Emissions and Mitigation Opportunities*. FAO.

Hansen, P. J. (2007). Exploitation of genetic and physiological determinants of embryonic resistance to elevated temperature to improve embryonic survival in dairy cattle during heat stress. *Theriogenology*, *68*(suppl. 1), S242–S249. doi:10.1016/j.theriogenology.2007.04.008 PMID:17482669

Indu, S., Sejian, V., & Naqvi, S. (2014). Impact of simulated heat stress on growth, physiological adaptability, blood metabolites and endocrine responses in Malpura ewes under semiarid tropical environment. *Animal Production Science*, *55*(6), 766. doi:10.1071/AN14085

Inskeep, E., & Dailey, R. (2005). Embryonic death in cattle. *Veterinary Clinics: Food Animal Practice*, *21*, 437–461. PMID:15955439

Johnson, J. S. (2018). Heat stress: Impact on livestock well-being and productivity and mitigation strategies to alleviate the negative effects. *Animal Production Science*, *58*(8), 1404. doi:10.1071/AN17725

Kadokawa, H., Sakatani, M., & Hansen, P. (2012). Perspectives on improvement of reproduction incattle during heat stress in a future Japan. *Animal Science Journal*, *83*(6), 439–445. doi:10.1111/j.1740-0929.2012.01011.x PMID:22694326

Krishnan, G., Bagath, M., Pragna, P., Vidya, M. K., Aleena, J., Archana, P. R., Sejian, V., & Bhatta, R. (2017). *Mitigation of the Heat Stress Impact in Livestock Reproduction in Theriogenology*. InTech Publishers.

Kumar, D., De, K., Sejian, V., & Naqvi, S. (2017). Impact of Climate Change on Sheep Reproduction. *Sheep Production Adapting to Climate Change*, 71–93.

Kunavongkrita, A., Suriyasomboonb, A., Lundeheimc, N., & Hearda, T. (2005). Management and sperm production of boars under differing environmental conditions. *Theriogenology*, *63*(2), 657–667. doi:10.1016/j.theriogenology.2004.09.039 PMID:15626423

Lopez-Gatius, F. (2003). Is fertility declining in dairy cattle? A retrospectivestudy in northeastern Spain. *Theriogenology*, *60*(1), 89–99. doi:10.1016/S0093-691X(02)01359-6 PMID:12620583

Makarevich, A., Olexikova, L., Chrenek, P., Kubovicova, E., Freharova, K., & Pivko, J. (2007). The effect of hyperthermia in vitro on vitality of rabbit preimplantation embryos. *Physiological Research*, *56*, 789–796. PMID:17087597

Menegassi, S., Barcellos, J., Dias, E., Koetz, C. Jr, Pereira, G., Peripolli, V., McManus, C., Canozzi, M. E. A., & Lopes, F. G. (2015). Scrotal infrared digital thermography as a predictor of seasonal effects on sperm traits in Braford bulls. *International Journal of Biometeorology*, *19*(3), 357–364. doi:10.100700484-014-0847-z PMID:24848445

Nardone, A., Ronchi, B., Lacetera, N., Ranieri, M., & Bernabucci, U. (2010). Effects of climate changes on animal production and sustainability of livestock systems. *Livestock Science*, *130*(1), 57–69. doi:10.1016/j.livsci.2010.02.011

Nichi, M., Bols, P., Zuche, R., Barnabe, V., Goovaerts, I., Barnabe, R., & Cortada, C. (2006). Seasonal variation in semen quality in Bos indicus and Bos taurus bulls raised under tropical conditions. *Theriogenology*, *66*(4), 822–828. doi:10.1016/j.theriogenology.2006.01.056 PMID:16529802

Nyika, J. (2020). Climate change situation in Kenya and measures towards adaptive management in the water sector. *International Journal of Environmental Sustainability and Green Technologies*, *11*(2), 34–47. doi:10.4018/IJESGT.2020070103

Ozawa, M., Tabayashi, D., Latief, T., Shimizu, T., Oshima, I., & Kanai, Y. (2005). Alterations in follicular dynamics and steroidogenic abilities induced by heat stress during follicular recruitment in goats. *Reproduction (Cambridge, England)*, *129*(5), 621–630. doi:10.1530/rep.1.00456 PMID:15855625

Pinto, M., de-Alencar-Araripe, M., & Alencar, A. (2001). Effects of scrotal insulation on testissize and semen criteria in Santa Ines hairy sheep raised in the State of Ceara. *Northeast of Brazil Rev. Brasileira Zootec*, *30*(6), 1704–1711.

Rojas-Downing, M. M., Nejadhashemi, A. P., Harrigan, T., & Woznicki, S. A. (2017). Climate change and livestock: Impacts, adaptation, and mitigation. *Climate Risk Management*, *16*, 145–163. doi:10.1016/j.crm.2017.02.001

Roth, Z., Arav, A., Bor, A., Zeron, Y., Braw-Tal, R., & Wolfenson, D. (2001). Improvement of quality of oocytes collected in the autumn by enhanced removal of impaired follicles from previously heat-stressed cows. *Reproduction (Cambridge, England)*, *122*, 737–744. doi:10.1530/rep.0.1220737 PMID:11690534

Ruediger, F., Chacur, M., Alves, F., Oba, E., & Ramos, A. (2016). Digital infrared thermography of the scrotum, semen quality, serumtestosterone levels in Nellore bulls (Bostaurusindicus) and theircorrelation with climatic factors. *CiênciasA grárias Londrina, 37*(1), 221-232.

Rust, J., & Rust, T. (2013). Climate change and livestock production: A review with emphasis on Africa. *South African Journal of Animal Science*, *43*(3), 256–267. doi:10.4314ajas.v43i3.3

Savsani, H., Padodara, R., Bhadaniya, A., Kalariya, V., Javia, B., Ghodasara, S., & Ribadiya, N. (2015). Impact of climate on feeding, production and reproduction of animals-A Review. *Agricultural Reviews (Karnal)*, *36*(1), 26–36. doi:10.5958/0976-0741.2015.00003.3

Sawyer, G., & Jitik Narayan, E. (2019). A Review on the Influence of Climate Change on Sheep Reproduction. In E. Narayan (Ed.), *Comparative Endocrinology of Animals*. doi:10.5772/intechopen.86799

Scenna, F., Edwards, J., Rohrbach, N., Hockett, M., Saxton, A., & Schrick, F. (2004). Detrimental effects of prostaglandin F2alpha on preimplantation bovine embryos. *Prostaglandins Other Lipid Meditation*, *73*(3-4), 215–226. doi:10.1016/j.prostaglandins.2004.02.001 PMID:15287153

Sejian, V. (2013). Climate Change: Impact on Production And Reproduction, Adaptation Mechanisms And Mitigation Strategies In Small Ruminants: A Review. *Indian Journal of Small Ruminants*, *19*(1), 1–22.

Sheikh, A., Bhagat, R., Islam, S., Dar, R., Sheikh, S., Wani, J., & Dogra, P. (2013). Effect of climate change on reproduction and milk production performance of livestock: A review. *Journal of Pharmacognosy and Phytochemistry*, *6*(6), 2062–2064.

Singh, M., Chaudhari, B., Singh, J., Singh, A., & Maurya, P. (2013). Effects of thermal load on buffalo reproductive performance during summer season. *The Journal of Biological Sciences*, *1*(1), 1–8.

Slimen, I., Najar, T., Ghram, A., Dabbebi, H., & Ben Mrad, M. (2014). Reactive oxygen species, heat stress and oxidative-induced mitochondrial damage. *Internal Journal of Hyperthermia*, *30*(7), 513–523. doi:10 3109/02656736.2014.971446 PMID:25354680

Soto, P., Natzke, R., & Hansen, P. (2003). Identification of possible mediators of embryonic mortality caused by mastitis: Actions oflipopolysaccharide, prostaglandin F2alpha, and the nitric oxidegenerator, sodium nitroprusside dihydrate, on oocyte maturationand embryonic development in cattle. *American Journal of Reproductive Immunology*, *50*(3), 263–272. doi:10.1034/j.1600-0897.2003.00085.x PMID:14629032

Staples, C., Burke, J., & Thatcher, W. (1998). Influence of supplemental fats on reproductive tissues and performance of lactating cows. *Journal of Dairy Science*, *81*(3), 856–871. doi:10.3168/jds.S0022-0302(98)75644-9 PMID:9565891

Takahashi, M. (2012). Heat stress on reproductive function and fertility in mammals. *Reproductive Medicine and Biology*, *11*(1), 37–47. doi:10.100712522-011-0105-6 PMID:29699104

Wang, Y., Huang, J., Xia, P., He, J., Wang, C., Ju, Z., Li, J., Li, R., Zhong, J., & Li, Q. (2013). Genetic variations of HSBP1 gene and its effect on thermal performance traits in Chinese Holstein cattle. *Molecular Biology Reports*, *40*(6), 3877–3882. doi:10.100711033-012-1977-1 PMID:23572288

Wettemann, R., Bazer, F., Thatcher, W., Caton, D., & Roberts, R. (1988). Conceptus development, uterine response, blood gases and endocrine function of gilts exposed to increased ambient temperature during early pregnancy. *Theriogenology*, *30*(1), 57–74. doi:10.1016/0093-691X(88)90263-4 PMID:16726449

Wolfenson, D., & Roth, Z. (2019). Impact of heat stress on cow reproduction and fertility. *Animal Frontiers*, *9*(1), 32–39. doi:10.1093/af/vfy027 PMID:32002237

Wolfenson, D., Roth, Z., & Meidan, R. (2000). Impaired reproduction in heat-stressed cattle: Basic and applied aspects. *Animal Reproduction Science*, *60*, 535–547. doi:10.1016/S0378-4320(00)00102-0 PMID:10844222

Wolfenson, D., Sonego, H., Bloch, A., Shaham-Albalancy, A., Kaim, M., Folman, Y., & Meidan, R. (2002). Seasonal differences in progesterone production by luteinized bovine thecal and granulosa cells. *Domestic Animal Endocrinology*, *22*(2), 81–90. doi:10.1016/S0739-7240(01)00127-8 PMID:11900966

KEY TERMS AND DEFINITIONS

CIDR and Ovsynch: A form of TAI, which is independent of the uterus and ovaries.

Endocrinological: Processes related to endocrine glands and the various hormones they produce.

Estrous Cycle: A period of recurring sexual behavior induced by gonadotropin production.

Follicular Apoptosis: The normal death of follicles that occur as a regulated part of growth and development in females.

Lipid Peroxidation: Oxidative metabolism of lipids stimulated by removal of electrons from cell membrane lipids.

Luteal Insufficiency: Insufficient production of progesterone in females, which results to failed implantation and inability to maintain pregnancy.

Spermatogenesis: The process by which spermatozoa develop in the testis' seminiferous tubules.

ENDNOTES

1 An enzyme produced in the cytoplasm to reduce hydrogen peroxide to oxygen and water.

2 A measure for lipid peroxidation levels in body fluids and cell tissues.

3 The process of synthesizing steroids including sex hormones such as testosterone.

4 Cells produced in the testis and are important in spermatogenesis.

5 Hormones produced at the anterior pituitary and whose release affects gonads to stimulate production of sex hormones at ova or sperm.

6 A steroid hormone found in females and helps in maturation of female reproductive systems.

7 A hormone secreted at the gonads that inhibits FSH release at the pituitary gland.

8 A naturally occurring protein that induces labor in events of failed implantation during the luteal phase.

9 A hormone produced after conception to maintain CL at the early stages of pregnancy.

10 A metric to assess the environmental effects of relative humidity and temperature resulting from heat stress.

Chapter 14
Multidimensional Impact of Climate Change on Human Reproduction and Fertility:
A Medical Perspective on Changing Dynamics

Ranjana Hanumant Choudhari
ICMR-National Institute of Occupational Health (NIOH), India

ABSTRACT

Research has identified a multidimensional, interactive relationship between climate change variables and an adverse (mostly) impact on the mammalian reproductive systems, reproductive organs, and fertility in animals, but direct evidence establishing the impact of climate change on reproductive health and fertility in humans is limited. Climate change has established direct or indirect linkages with re-emergence, geospatial redistribution of pathogens of likely reproductive health significance in humans. Similarly, alterations in growth, mortality rate, reproduction, and spatiotemporal distribution of vectors (e.g., zika virus – Aedes aegypti) and intermediate hosts (e.g., Schistosomiasis – snail) of certain infectious diseases of reproductive health importance are influenced by climate change variables like temperature, precipitation, and humidity. The exposure channels or effect pathways, through which the regional and global climate change can directly or indirectly influence the human reproductive ability, health, fertility, progeny, and thus, ultimately, demography can broadly be classified as physical variables, chemical hazards, biological agents, factors related to psycho-socio-behavior, and economy. The chapter is an overall account of how each of these factors, as an inherent component of climate change has the potential to cause a variable degree of impact on human reproduction from a medical point of view. As human reproductive systems are highly vulnerable to diseases and other post-catastrophic effects of extreme climate change events, so it is high time to understand the adversity and resort to proper and sustainable control measures for a healthy reproductive life of future generations.

DOI: 10.4018/978-1-7998-4480-8.ch014

INTRODUCTION

Climate change signifies a continuous, gradual but significant variation in regional or global climate parameters over a considerable prolonged period, with an inherent potential to impact health and economy. Although, the changing parameters of climate vary based on the specific location in the globe, the most universal and sensitive climate variables are mean and peak atmospheric temperatures, rainfall pattern, humidity, salinity and wind velocity (McMichael,2013). Recently, the changes in the global mean temperature and other climatic variables, which were relatively stable over last century, have acquired a visible momentum in changes. (Solomon et al.,2007). The European Environment Agency had documented a continuous rise in world mean temperature in the last century along with a continuous expansion of sea level during previous half century, and a dramatic increase in the melting of Arctic snow, at an alarming rate of about3% per decade (Christiansen et al.,2008) A remarkable rise up to 6°C in global mean temperature was predicted by the Intergovernmental Panel on Climate Change in the 21st century, accompanied by more severe, extreme weather events dominated by floods, droughts and hurricanes(Watson,2001; Hansen et al.,2006).Meteorology scientists have also predicted the dramatic increase in the frequency and severity of extreme weather events due to climate change. Increases in land/mudslides and avalanches may not be solely due to climate change, but has partial roots in climate change (Keiler et al,2010). The broad definition of the anticipated events and consequences due to climate change also encompasses, but is not limited to scarcity of potable water, increasing sea levels causing submerging of coastal densely populated land regions, reduced fertile land and decrease in agricultural production, decrease in biodiversity, increased salinity of coastal areas, food scarcity, release of environmental toxic contaminants, natural catastrophic events, rapid arctic snow melting etc Each one of these aftermath effects of climate change may have multifaceted potential impact on human fertility and reproduction.

Reproductive health is defined as a state of complete physical, mental and social well-being and not merely the absence of reproductive disease or infirmity, in all matters relating to the reproductive system and its functions and processes. In general perspective, reproductive health here implies that people are able to have the capability to reproduce and the liberty to control the self-reproduction (if, when and how many to give births), leading ultimately to healthy progeny. Many of impacts of climate change, including those on migration, geo-spatial redistribution of communicable diseases, physical, mental health and mortality are predictable. (Amegah et al.2016; Mboera et al.2011; Sheffield and Landrigan 2011; Wu et al.2016 ; Bunker et al.2016; Matte et al.2016). Although the effects of climate change on human health are widely evident, studies on the impact of climate change and reproductive health, reproductive outcomes and fertility are limited (Basu et al.2010; Van Zutphen et al.2012) and the magnitude of the risk and its effect on pregnancies are ill recognized (Poursafa et al.2015; Strand et al.2011).

MULTIPLE, COMPLEX, INTER-ACTIVE LINKAGES BETWEEN CLIMATE CHANGE AND HUMAN REPRODUCTIVE OUTCOME

Whether, how and to what extent the climate change can impact demography, through its effect on human reproductive functions, reproductive ability and fertility is a million-dollar question? This under-rated nexus of climate change with human reproduction and fertility, has recently gained much-needed scientific attention and concerns due to its potential economic consequences. Although the climate change variables have potential to mediate their ultimate impact on human reproduction through a variety of complex and interlinked pathways including economical and psychological factors, how various climate change variables ultimately affect the physical aspect of the reproductive health and fertility is discussed at length in this chapter, as illustrated in Fig. 1. Effects of changes in the climate change variables regulating the socio-economic and psycho-social(measured as behavioural changes)determinants of human reproductive health have been addressed in brief, as they are based on less of empirical evidences and more on probable, predictive causation theories like The Climate Change-Economy-Demography Theory. Although, there are multiple interactive linkages among various climate change parameters having potential to influence the human reproductive health and fertility, for ease of description and comprehension purpose, climate change variables can be categorised into five distinctive groups, as under (1) physical, (2) chemical, (3) biological, (4) psycho-social behavioural, and (5) economical variables, discussed below.

Figure 1. Multidimentional impact of climate change on reproductive health and fertility through various interactive linkages.

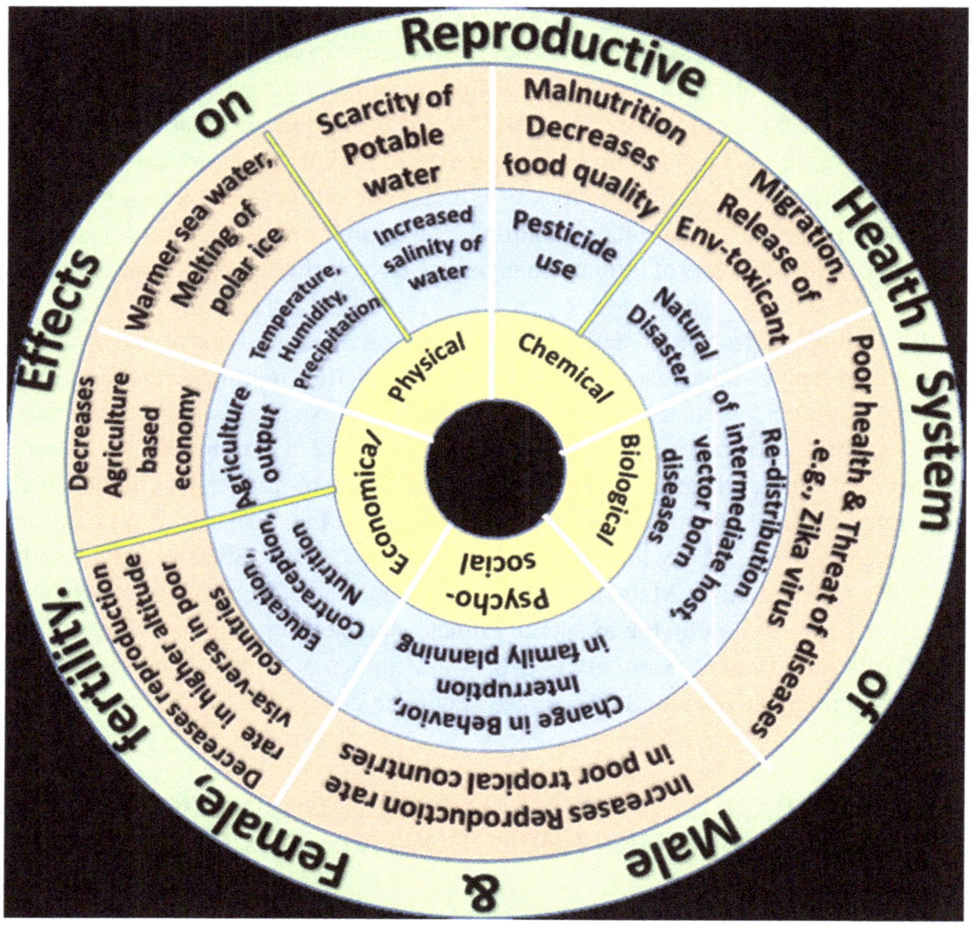

PHYSICAL VARIABLES OF CLIMATE CHANGE

Average and Peak Atmospheric Temperature

i. Physiological thermal homeostasis and reproduction in humans

Humans due to an evolutionary advantage, can maintain a constant core body temperature in the presence of a variable environmental temperature through the effectiveness of thermoregulatory mechanisms (Ziskin and Morrissey,2011). Only for short duration, humans can sustain core temperatures below 35°C or above 41°C. Thus, when a person's core temperature due to warm environment exceeds a critical limit, the body exercises its thermostatic control to stabilize its core temperature through cooling mechanisms such as increased dermal blood flow and evaporation of sweat. When the core temperature falls below the critical limit, the reverse reaction occurs, and the heat generation mechanisms set in to maintain a relatively stable thermal control of the body under normal physiological conditions. Climate change's accelerating impact on global temperatures will make heat a more serious threat to maternal and reproductive health in near future. Exposure to unusually hot weather can lead to changes in length of gestation, birth weight, stillbirth rates, and neonatal stress. Although, thermal fertility limit, means the critical temperature at or above which the functional impairment of fertility, reproductive ability occurs is uncertain, but the dreaded effect of heat stress on fertility is well documented.

ii. Atmospheric temperature and mammalian reproduction: Hypothalamo-pituitary-gonadal axis

Heat induced stress can have large effects on most aspects of reproductive function in mammals, which include disruptions in spermatogenesis and oocyte development, oocyte maturation, early embryonic development, fetal and placental growth. Many effects of elevated temperature on gametes and the early embryo involve increased production of reactive oxygen species and increases oxidative stress, which is involved in pathogenesis of infertility(Jana et al.,2010; Tusell et al.,2014).

iii. Temperature and male reproductive system in mammals

Most mammals have testes covered in the scrotum, outside the body cavity, which maintains intra testicular temperature lower than the core body temperature, with the help of the thermoregulatory system, an arterio-venous plexus called the pampiniform plexus of testes. High testicular temperature in mammals leads to reduced sperm output, decreased sperm motility and an increased proportion of morphologically abnormal spermatozoa (Setchell, 1998; Hansen, 2009; Belhadj-Silmen et al, 2015). Chronic exposure to heat decreases libido, sperm density, semen volume, and anatomical structure of the rabbit testes (Yangli et al, 2012), also may reduce levels of reproductive hormones (Davidson et al., 1996). On exposure to heat stress, synthesis of heat shock proteins is accelerated, which protects other physiological proteins from denaturation (Kwok-Keung et al, 2013). However, above a critical threshold temperature, an inhibition of heat shock protein synthesis occurs, resulting in premature cell death (Kregel, 2002). Weights of reproductive organs including testis, epididymis, vas deferens, and accessory sex glands decreased in animals on exposure to hot temperature (Garrigue, 2017). Decrease in testicular weight, which is a good indicator of dwindling sperm counts is secondary to the decrease in the number of Sertoli and germ cells within the seminiferous epithelium (Nicolino et al, 2001). Impairment of testicular function by heat induced oxidative stress results in reduction in testicular space of the Leydig cells, which could release steroid hormones and other factors involved in the regulation of the hypothalamus–pituitary–gonadal axis. This leads to decrease in testosterone, the most important androgen in males mainly produced by testicular Leydig cells under the control of the pituitary gland (Yangli et al, 2012). Heat induced stress generally led to azoospermia or oligospermia in animal species (Marai et al., 2002; Ngoula et al, 2020).

iv. Temperature and female reproductive system in mammals

High atmospheric temperature can increase uterine temperatures, delay menarche (Kurowicka et al., 2006), ovulation (Rozenboim et al., 2007), ovarian rhythmicity, oogenesis and embryogenesis (Edwards and Hansen, 1997; Beere, 2004; Lawrence et al., 2004), and reduce conception and pregnancy rates (Putney et al., 1989) in different animal species. Most of the ovarian cell functions are under the control of various hormones, growth factors and prostaglandins (Berisha and Schams, 2005; Sirotkin et al., 2005).A high ambient temperatures can either reduce or increase secretion of prostaglandins, steroids, plasma follicle-stimulating hormone (FSH), prolactin, progesterone, testosterone and estradiol levels (Kurowicka et al., 2006; Rozenboim et al., 2007), reduce androstenedione and estradiol, and increase progesterone production by ovarian cells (Bridges et al., 2005; Kurowicka et al., 2006). Thus, high temperatures can suppress ovarian functions through changes in the release of reproductive hormones (Narayansingh et al., 2004; Salvetti et al., 2008). The inhibition of ovarian cell proliferation and/or turnover and oversecretion of ovarian hormones can be causes of adverse effects of heat stress on reproductive functions (Edwards and Hansen, 1997; Beere, 2004; Lawrence et al., 2004; Kunavongkrit et al., 2005; Rozenboim et al., 2007; Sirotkin,2010).

v. Temperature shocks and dynamics of human fertility

Global warming might directly affect human fertility in two key ways.

- Hot weather could adversely affect the frequency and intensity of physically demanding activities such as sexual behaviour.
- Temperature could negatively influence reproductive health factors such as sperm motility, viability in men and menstruation, ovulation in women. There is significant evidence that occupational heat exposure is a significant risk factor for male infertility, affecting sperm morphology and resulting in delayed conception (Thonneau et al.,1998; Hamerezaee et al., 2018).

The human natural experiment regarding climate change, temperature shocks and dynamics of human fertility was carried out with the hypothesis that natural temperature spikes can influence dynamics of human fertility were tested using data on temperature and birth rates from the United States across the years 1931 to 2010. After controlling the socio-economical confounders, including economic opportunities for women and access to birth control, the attributable and quantifiable effect of atmospheric temperature on human fertility was measured. Abnormal hot days (more than 32°C) in the United States, resulted in a reduction in birth rates eight to ten months later, with a maximum effect of 0.4% at nine months from the hot day, and only a partial rebound in birth rates in the three months after the initial decline, possibly due to carry on, residual health effects of hot spikes. This suggests that hot waves could reduce the number of children a woman may have over the course of her reproductive life Use of air conditioning played a decisive role in negating the impact of hot days on conception chances (Barreca et al, 2018). The evidence suggests that reproductive health is especially more vulnerable than sexual health. If the cause was only regarding hot days making decreased in sexual activity, then there would be sharp decline in the birth rates only at 9 months after hot days. Instead, the fall in birth rates continued 10 months later, suggesting that hot days have lasting adverse reproductive health effects. However, more scientific research studies are needed to verify and approve the cause-effect relationship. Many tropical countries, like India, already experience hotter climates than the United States, are more prone to experience the effects of climate change, including adverse fertility outcomes.

vi. Relation between atmospheric temperature and reproductive outcomes-Pregnancy: A physiological compromised thermoregulatory state

As such females as compared to males, report a higher metabolic rate during work with decreased heat dissipation ability through sweating, and reduced heat loss through radiation (Duncan,2006). Altered hormone status during pregnancy and the increased circulatory dependence of the fetus on the mother may negatively influence mother's thermoregulatory ability. Sickness induced severe maternal hyperthermia may increase the incidence of congenital malformation in fetus, but substantial evidence of a similar effect from occupational heat exposure is lacking (ILO, 1998). The human fetus can be susceptible to elevated temperatures during certain critical stages of embryo development such as cell proliferation, cell migration, cell differentiation, and programmed cell death for different organogenesis. The embryo and developing fetus depend mostly on the mother's thermoregulatory capacity in order to control its temperature. Smaller levels of heat stress in mother, although asymptomatic in nature, could result in increased blood flow to the skin at the cost of placental and umbilical blood perfusion, thus eventually increasing temperature of fetus (Ziskin and Morrissey, 2011). Significant associations between the atmospheric temperature variables (due to climate change) and the pregnancy-related outcomes such as eclampsia, preeclampsia, low birth weight, preterm birth (Basu et al.2010; Strand et al.2012) skewed sex ratio, and length of pregnancy (Poursafa et al.2015) are reported.

Association with Pre-eclampsia and Eclampsia

High ambient temperatures are also linked to higher incidences of pregnancy induced hypertension and preeclampsia (Makhseed et al., 1999). Predominant association of eclampsia and preeclampsia with cold and humid temperatures is reported. A study had also reported elevated risks of preeclampsia and eclampsia in women, who had conceived during the warmest months (Beltran et.al.,2013). In fact the association of heat stress with pregnancy induced hypertension can be bidirectional as per a recent study, which showed that elevated averaged temperature during the 1st or 3rd trimesters of pregnancies with maximum exposure to hot climate, confer an increased risk for the development of preeclampsia. Whereas the same relationship was reversed in pregnancies with maximum months of pregnancy in cold seasons. This suggests temperature might be associated with perturbations in maternal heat homeostasis resulting in reallocation of energy resources and their availability to the fetus that may increase the risk for preeclampsia. This bidirectional dynamic association especially highlights how temperature variability can act as a significant physical determinant of maternal reproductive health and subsequent fertility (Shashar et al.,2020). Two spectrums of heat namely hot and cold, different seasons of the year, sunlight intensity and season of fertilization are all associated with higher rates of pregnancy induced hypertension, eclampsia, preeclampsia.

Association with Preterm Births, low Birth Weight and Length of Gestation

Seasonal patterns of pregnancy outcomes, such as low birth weight, preterm birth, and stillbirth, in association with prenatal exposure to ambient temperature are observed. Most of the studies found high risk of preterm birth, stillbirth, and low birth weight in winter, summer or both, which indicates that the extremes of temperature may be a risk factor for poor pregnancy outcomes (Rylander et al,2013). Definition of Preterm birth varies a lot, still broadly births occurring at less than 37 weeks of gestation with a variable lowest margin up to 20 weeks has been adopted by maximum studies. Heat was significantly correlated with increased risk of preterm birth in majority of the studies (Strand et al.,2012; Wang et al,2013.; Schifano et al, 2016; Vicedo-Cabrera et al, 2014; Basu et al.,2010; Kent et al,2014; Arroyo et al., 2016; Schifano et al 2013; Vicedo-Cabrera et al,2015;Cox et al 2016; He et al.,2016; Liang et al 2016; Bruckner et al., 2014; Mathew et al, 2017; Kloog et al.,2015); studies reporting noassociation (Auger et al., 2014; Wolfand Armstrong.,2012). A protective effect was also demonstrated in Shenzhen, China for extreme heat exposure and preterm birth (Liang et al,2016) Excessive heat exposure was correlated with elevated risk of early term birth (Strand et al.,2012; Schifano et al, 2016, Cox et al 2016). Previous studies have suggested that the factors related to preterm births are the secretion of oxytocin and heat-shock proteins secondary to heat stress (Dreiling et al.,1991; Poursafa et al,2015) along with dehydration that results from heat exposure (Shoghag et al.,1997).

Low birth weight at the time of delivery relative to gestational age was found to be significantly associated with increased heat exposure (Kloog et al.,2015; Dadvand et al.,2014; Molinaand Saldarriaga,2017). A contradictory finding coming from Israel.is that lower ambient temperature may increase the risk of low birth weight and Small for gestational age births, and higher pregnancy temperature may decrease the risk of small for gestational age in singleton term infants (Kloog et al.,2018). Exposure to hot climate was associated with 5 days decrease in the average pregnancy duration (Dadvand et al., 2011). Scientific evidence exists for varied and crucial adverse effects of different aspects of climate change including the temperature, season, rain (Yu et al., 2018) and wind on human reproduction particularly on perinatal period. Approximately all studies have reported a significant relationship between heat exposure and intended pregnancy outcomes including eclampsia, preeclampsia, low birth weight, preterm birth, pregnancy induced hypertension, skewed sex ratio and reduced length of pregnancy (Poursafa et al.,2015).

Association with Congenital Malformations and Stillbirths

Prolonged fetal exposure to chronic sickness induced maternal hyperthermia in crucial stages of embryo development has proven teratogenic effect, which can result in severe fetal malformations (Edwards 2006; Dreier et al.2014; Moretti et al 2005;Pei et al,2015; Van Zutphen et al,2012).Thus excessive heat induced unsustainable malformation in structure or function, embryo growth retardation, or fetus death may culminate into spontaneous miscarriage or stillbirth (Gilbert-Barness 2010). Increasing risk of stillbirth with increasing ambient temperatures are indeed reported. (Strand et al., 2012; Bekkar et al., 2020). The pathophysiological mechanism through which excessive maternal heat exposure can reduce placental blood flow is by increase in production of vasoactive substances, elevated blood viscosity, and affecting endothelial cell function, which in turn may increase risk for hypertensive crises and fetal growth retardation (Ha et al., 2017).

The teratogenicity of heat is dependent on the additive or synergistic effect of the two independent factors, namely degree elevation of ambient heat and the critical stage of embryo development; whereas the type of fetal malformation depends on the particular stage of organogenesis at the time of exposure (Edwards,2006; Graham2005; Konkel, 2019;Van Zutphen et al, 2012). Long duration of unseasonable heat exposure is associated with the increased risks for cardiovascular anomalies such as ventricular septal defects and atrial septal defects (Lin et al, 2018). Serious congenital anomalies, not sustainable for life, may culminate into miscarriage or intrauterine fetal death (Parker et al., 2010). Ambient heat in hot regions of the world could therefore induce congenital malformations in humans chronically exposed to heatwaves, resulting in an increased prevalence of spontaneous abortion or stillbirth than would be expected from any other singleton cause, but only a few empirical investigations in real-world settings have been reported. Whereas in an epidemiological study, the linkage between ambient heat exposure and pregnancy outcome was studied by estimating maternal heat exposure level using the wet-bulb globe temperature (WBGT), which measures collective influence of temperature, humidity, heat radiation, and air movement. Finding of increase in the odds of experiencing miscarriage or stillbirth with every degree increase in WBGT, lost its significance after adjusting for confounders. The real-life scenario of maximum ambient heat exposure below 30 degrees may not be sufficient to result into attributable risk of miscarriage or stillbirths (Asamoah et al,2018). In summary, prolonged exposure to high atmospheric temperatures, normally observed in tropical desert countries might cause stillbirth, congenital birth defects, and preterm delivery regardless of maternal ethnicity or age, with younger mothers having an even higher risk of negative outcomes (Basu et al.,2016; Ha et al.,2017; Kuehn et al,2017; Van Zutphen et al.,2012).

Association with Skewed sex Ratio at Birth

The linkage of atmospheric temperature is observed with a lower male: female gender ratio of newborns probably mediated through increased male fetal deaths in utero. Male embryo appears to be especially susceptible to outside environmental stressors, including climate changes (Fukudaet al, 2014).

vii. Developmental temperature and epigenetics pathways of transgenerational damage to fertility

How the developmental temperature can impact reproductive behaviour and reproductive physiology and thus ultimately fertility through epigenetic inheritance is explored in an evolutionary ecological study on Indian meal moth. The sperm production and sperm viability, along with its copulation behaviour are damaged by high temperatures. Longevity of life, the reproductive performance and fertile potential of next generation males, produced by heat exposed males were found to have deteriorated. Thus the heat induced damage to spermatogenesis and the anti-fertile reproductive behaviour can potentially transcend to next generation males through inheritance (Sales et al,2018), which is a warning sign for humans, as the very existence of species depends upon its ability to reproduce.

The effects of global warming on the local climate vary between regions, countries, and specific locations. Although, predicted increase in temperature will be largest in the polar region, in the areas with originally low atmospheric temperatures, the reproductive health impact secondary to hot climate will most likely be subclinical in nature. In urban areas in the tropical region, local hot spots occur due to the 'urban heat island effect'. Increasing humidity will enhance the heat stress of tropical, subtropical regions, causing substantial reproductive health risks in future.

Humidity and Precipitation

A person's capacity to reduce excessive heat by sweat evaporation to maintain core body temperature is strongly influenced by the surrounding temperature, humidity and wind. The rate of sweat evaporation is reduced with increasing humidity and increased by air movement Hence, heat stress indicators incorporate humidity and wind speed, as well as temperature. Climate change is expected to increase moisture levels in North and South America, northern Europe, and northern and central Asia. These regions will experience more rainfall, flooding and storms. Moisture levels are expected to decrease in southern Africa, Central America and the Mediterranean potentially causing substantial droughts. On a global scale, the ozone layer can be destroyed by high moisture and high temperature. Low moisture conditions will produce toxic air particles, while high moisture will cause chemicals in the air to drop to the ground via acid rain or snow. The combination of drought with extreme precipitation may also impact reproduction by affecting the health of pregnant women by water, food- and vector- borne diseases.

Changes in temperature, humidity and precipitation would alter the geo-spatial distribution of many pathogens and vectors of diseases. Pregnant women are especially at greater risk for contact with vectors due to physiologic changes of pregnancy (Sorensen et al, 2018). Increased CO2 production during pregnancy, a chemoattractant for mosquitos, and increased peripheral blood flow and skin temperature in all increase vulnerability for mosquito bites. Additionally, pregnancy driven hormonal changes might induce corresponding changes in immunity, which in turn may suppress host defenses resulting in a higher degree of micro-oraganisms in blood circulation (Kourtis et al.,2014; Mbonye et al.,2005). Studies have documented that the risk of severe malaria in pregnant women, is three times that of non-pregnant women (Rijken et al.,2012). Anemia and reduced transplacental nutrient supply due to the presence of parasite in the placenta, secondary to Malaria infection during pregnancy results in intrauterine growth retardation and increased risk of hemorrhagic complications of delivery (Steketee et al.,1996). About a quarter of maternal deaths between 2004 and 2006 in Indian districts were attributable to malaria, making it the foremost cause of maternal death during pregnancy (Kourtis et al.,2014). Dengue is spreading in geographical distribution and severity in India due to climate change (Mutheneni et al.,2017), has association with increased risk of cesarean section, preeclampsia and intrauterine growth retardation (Pouliot et al., 2010). Zika virus infection, having a vector as Aedes mosquito, is an emergent climate associated infectious disease with severe fetal impacts including microcephaly, central nervous system malformations (Petersen et al.,2016).

Salinity and Reproduction

Climate change may lead to scarcity of potable water in already drought prone, areas, or areas where the surface water is highly contaminated, making the population to resort to overexploitation of highly saline groundwater for drinking purpose. In coastal areas, drought can cause contamination of groundwater, drinking water sources by the influx of saline water. The mixing of salt water with fresh groundwater that occurs in drought regions can advance inland affecting freshwater tubewells, further exacerbated by climate change and sea-level rise (Linzey,2011). High salt intake through drinking water might result in hypertension, particularly risky during pregnancy. Hypertension may cause severe pregnancy-related complications, triggering late pregnancy loss or miscarriage if not controlled. Vineis et al (Vineis et al., 2011) has established a link between high salt intakes with gestational hypertension and preeclampsia, and incidents of miscarriage (Hossain, 2020). In addition, lack of access to clean water and proper sanitation, especially during menstruation and pregnancy when hygiene need is more frequent, creates additional reproductive health challenges to women (Birch et al.,2012).

Knowledge Limitations

To summarise, even though the most extensively studied variable of climate change is atmospheric temperature and its effect on human reproduction, there are certain critical research gaps related to temperature, still unaddressed or understudied. There is probable evidence that extreme temperature can influence many aspects of human reproduction and fertility. However more substantial evidence exists for extremes of environmental temperature can adversely impact pregnancy outcomes, including, but not limited to: changes in length of gestation, birth weight, rates of stillbirth, gender ratio at birth. Since research lacks the adaptation of uniform definitions of heat stress and outcome variables for assessing the effects of heat on reproduction and fetal health, there is need to establish uniform definitions for further research on the ways that different climate change variables, and heat in particular, may affect reproduction and pregnancy outcomes. Especially, humidity and temperature measures such as apparent temperature and heat index (HI) were not universally analyzed within various studies of climate change. Scientific literature has not established a critical period of maternal sensitivity to hot temperatures, which even may vary from region to region. There is limited evidence for critical heat-related window periods. Heat exposures may have immediate effects on birth outcomes, as well as delayed or cumulative effects on pregnancy outcomes.

Medical Perspective

Heat exposure is mostly correlated with a decrease in birth weight, decreased gestational length, increased risk of stillbirth, and increases in neonatal mortality. While stillbirths established a linear relationship, birthweight and gestational length and temperature were nonlinear. Overall, excessive or peak environmental heat exposure, notably in critical window periods of embryo development, in isolation or congruence with other variables of climate change, may cause adverse pregnancy outcomes, particularly in malnourished and already dehydrated pregnant women. Also notable is that early developmental exposure to the effects of climate change can cause immediate harm or damage to the reproductive ability that becomes evident later in life, resulting in lasting effects like infertility and even transcending to the next generation

Health Education

Clinicians and obstetricians should pay attention to the history of excessive exposure to heat, particularly of occupational origin, in investigating cases with history of bad pregnancy outcomes. The awareness of clinicians and expectant mothers on the role of peak environmental or occupational heat exposure during the gestation period will help in addressing the issue to a great extent. Simple measures like the provision of an air cooling facility during critical periods of embryo development may prove as a primordial prevention tool in reducing the adverse impact of atmospheric heat exposure on pregnancy.

Strengthening of Antenatal Services

Expectant mothers with existing morbidities or thermoregulatory compromised states such as hypo or hyperthyroidism, and proven placental insufficiency may be identified as most vulnerable for development of temperature related adverse effects, and accordingly appropriate measures of prevention and control of excessive heat exposure should be advised during antenatal visits, for which establishing the comprehensive antenatal services providing network is mandatory for the women in high risk occupations. In tropical countries like India, majority of poor, daily labour, workforce in unorganised sectors like farming is of women, who are occupationally exposed to extreme environmental temperatures without any legal provision to safeguard their interests. Strengthening antenatal services for such a vulnerable workforce is the key element in the prevention and control of the adverse effects of the myriad of climate change variables.

Policy Implications

Clinicians and policy makers should understand that the risk for adverse outcomes at a specific temperature will not be uniform in all climatic zones, as mean temperature and humidity may vary dramatically between climates. The genetic adaptation to the temperatures typical of their respective climates may also determine the final outcome. The issue of acclimatization is further exaggerated by the timing of heat exposures, wherein heat waves earlier in the warm season may trigger adverse outcomes more readily than at the end of the season after acclimatization has taken place. Occupational policy should be especially designed for the outdoor occupations, to minimize the adverse impact of extreme climate change events such as heatwaves on human reproduction Provision of clean potable water and proper sanitation infrastructure to the women of reproductive age is also invaluable.

CHEMICAL HAZARDS OF CLIMATE CHANGE

Intersection of Climate, Toxicants and Reproductive Toxicity

Climate change can influence the formation, presence, movement, and toxicity of chemicals in the environment through alterations in the temperature and moisture. Increased temperature can vaporize some chemicals, releasing them from soil or water, can also cause chemicals to dissolve in drinking water. Melting polar snow can also release stored chemicals. All of these climate change related activities will increase the quantity and variety of chemicals humans are exposed to. Extreme weather events like cyclones, hurricanes can increase the risk of release of toxicants of reproductive health significance into the environment, which in turn can alter parameters related to human reproduction. As temperature influences the toxicity of chemicals, so does chemical exposure influence the heat tolerance of human beings. This vicious reciprocal relationship potentiates the detrimental consequences of climate change on human reproductive ability and health.

Direct Impact of Environmental Toxicants on Human Reproductive Health

Heavy metals like lead, cadmium, may contribute to reproductive disorders and infertility (Lei et al,2015). Global warming can alter the ecotoxicity of heavy metals like lead, a known reproductive toxicant. Environmental heavy metal exposure even in trace quantity, can cause reproductive dysfunction in women (Bloom et al,2010; Gardella, 2001). Toxic metals may induce hormonal changes affecting the menstrual cycle, ovulation, and female fertility (Sengupta et al, 2015). Epidemiologic studies on menstruation have documented the effect of metal on reproductive hormone levels (Krieg, 2007). Lead and Cadmium levels are measured in human follicular fluid (Paksy et al,2001). Women with blood Pb levels higher than 25 μg/L had a 3-fold increased risk of infertility compared with women whose blood Pb levels were less than 25 μg/L (Chang et al,2010 ;Bloom et al,2010). Lead, cadmium, mercury, and arsenic, are toxic for the reproduction of humans. Scientific literature on adverse effects of exposure to low levels of these metals on measures of male fertility: semen quality and reproductive hormone levels are vastly conclusive and also there is supporting evidence from experimental and occupational studies. (Wirth et al,2010)

Mercury is a highly toxic chemical, have a negative impact on the reproductive and endocrine systems in both male and female. Large depots of mercury have been found in the permafrost of polar regions, which can be released in the atmosphere and sea water due to climate change thawing of permafrost. This may cause biotransformation of mercury into highly toxic chemical form methyl mercury, leading to biomagnification and bioaccumulation of methyl mercury. Exposure to toxic mercury, irrespective of the route of exposure, is mainly associated with infertility or subfertility status. Mercury exposure causes cellular deformation of the Leydig cells and the seminiferous tubules, and testicular degeneration giving rise to abnormal sperm morphology and motility through DNA damage. In females, mercury is related with higher incidence of menstrual abnormalities and hormonal disorders and increased rates of adverse reproductive outcomes including spontaneous abortions and congenital abnormalities (Carvalho Henriques et al,2019; Bjørklund, et al. 2019; Henriques et al,2019). Recent epidemiological evidence shows that, mercury even at low biological exposure levels, considered at par with the climate change induced mercury exposure level, can induce increase in sex(testosterone) and thyroid hormone in humans would have serious future reproductive health implications for the general populations (Castiello et al,2020).

Re-suspension and Re-exposure to old Environmental Toxicants due to Catastrophic Extreme Climate Change Events

Flooding due to Hurricane Katrina resulted in many oil spills and sewage breakdowns and other waste disposal issues in the region of the United States. The receding floodwater left behind the sediments, having the potential to result in dangerous air pollution containing many environmental toxins, such as endocrine-disrupting chemicals (EDCs), which have been implicated in early onset of menstruation Hurricanes can also release harmful chemicals, such as polybrominated diphenyl ethers (PBDEs) through precipitation. In utero exposure to PBDEs (Harley et al,2017) was associated with late menarche. This could thus affect the timing of menarche, resulting in late menarche for girls that are exposed in utero to the PBDE containing rainfall during an extreme climate change event. Maternal preconception levels of certain PBDEs may increase the hazards of incident pregnancy losses (Choi et al,2019) and can disturb maternal glucose homeostasis which increases the risk of Gestational Diabetes Mellitus and its adverse pregnancy effects (Liu et al,2018). The teratogenic effect of PBDE exposure during pregnancy may cause reproductive organ anomalies such as hypospadias in the newborns (Korenet al,2019). PBDEs as an endocrine-active compound interfere with sexual development and sexually dimorphic behaviour in male rats, however the similar evidence on male reproduction function and behaviour in humans is lacking. (Lilienthal et al, 2006; Kuriyama et al,2005). Certain rivers that are severely contaminated with polychlorinated biphenyls (PCBs), and other chemicals, the exposure increases when the sediment is re-suspended in water post extreme climatic events like hurricane and flooding (Schneider et al,2007) PCB exposure is implicated in causation of late puberty in females. Prenatal exposure to persistent organic pollutants, e.g. polychlorinated biphenyls (PCBs) and polybrominated diphenyl ethers (PBDEs) also have an influence on neonatal weight (Lignell et al,2013). Environmental pesticides of historical occurrence, which are banned from use since decades were detected in humans, subsequent to resurfacing of toxins during extreme climate change events This indicates the potential for re-exposure to old contaminated sediment following hurricanes Many chemicals released through contaminated precipitation or soil sediment re-release can result in either early or late onset of puberty and negative reproductive health effects in a female.

Environmental Toxicants and Male Infertility

Exposure to environmental contaminants poses a serious risk to human reproductive health. As compared to animals, a relatively smaller reduction in sperm production in men may have severe consequences for human reproduction Because the normal human sperm count is near the threshold for the number of sperm needed to ensure reproductive competence, hence sperm count is a sensitive and validated endpoint for reproductive toxicology assessment. In rodents, on perinatal exposure to low doses of PBDE-99, 2,3,7,8-tetrachlorodibenzo-p-dioxin (Faqi et al. 1998), PCBs (Kuriyama and Chahoud 2004;Hsu et al. 2003), and the pesticides deltamethrin (Andrade et al. 2002) and lindane (Dalsenter et al. 1997), the defective spermatogenesis is reported in adulthood.

Evidence of Ovarian Toxicity and Female Infertility

A wide range of environmental contaminants from multiple chemical classes has been measured in human ovarian follicular fluid illustrating that the ovary is a potential target for adverse effects of chemical contaminants. Epidemiological studies link exposure to environmental contaminants with adverse effects on reproductive health suggesting potential effects on the ovary. However, inconsistency across studies, potential confounding, and type I error arising from multiple comparisons suggests only limited evidence that exposure to environmental contaminants is associated with impaired ovarian function and fertility. Experimental animal and tissue culture studies provide support for the biological plausibility of the associations noted in epidemiological studies and elucidate potentially important mechanistic pathways. Taken together, the literature suggests that the human ovary is potentially at risk for toxic effects arising from exposure to environmental contaminants (Foster et al,2019). While the epidemiological studies provide limited evidence for the relationship between exposure to environmental contaminants and adverse effects on human fertility and ovarian function, animal studies elucidate the cause-effect relationship of the above association. Experimental studies in animals reveal that environmental contaminants from all chemical classes, at some dose, can induce ovarian toxicity and ultimately may lead to infertility.

Effect of Released Chemicals on Crop, Food Security, Nutrition and Reproductive System

Climate change accelerated release of environmental toxins like pesticides can have potential impact on the both the quality and quantity of crops and food production. This would impact food availability and food intake and disturb onset of puberty in female. Weather events could alter crop availability and intake of vegetables and protein, which is known to affect timing of menarche and nutritional status of the female. These perturbations in timing of menstruations could then result in increased infertility-related conditions such as increased antral follicle count, premature menopause, hysterectomy, fetal loss, longer time to first baby. Effects of the 1944–1965 Dutch famine on reproductive function found that girls exposed to severe famine before puberty, were 1.5 times more likely to have irregular menstrual patterns relative to those unexposed. Those exposed to severe famine twice or more years after menarche were almost 9 times more likely to experience prolonged menstrual irregularity than those unexposed (Elias et al,2006). Adequate food intake and balanced energy expenditure, both critical factors in reproductive function, can be limited by crop availability and, therefore, by climate conditions This interdependence creates a dynamic relationship, where seasonal changes in climate conditions cause seasonal changes in nutrition intake and food-related energy expenditure, leading to seasonal variations in ovulation (Bleiberg et al., 1980; Bantje et al, 1988) specifically calorie-based amenorrhea. Surprisingly, Indian women of the Himalayan ranges, who do not face seasonal changes in food availability, still show a seasonal variation in ovulation as a result of the seasonal variation in the expenditure of energy. (Panter-Brick et al,1993). In summary, the ratios of these dietary food stuffs can result in either early or late initiation of reproduction function, which have both been linked with fertility. Availability and access to adequate food quantity and quality, which is required for appropriate functioning of the reproductive system and hormones, is largely influenced by climate changes. During and post extreme climate change events, food security of the affected population needs to be insured.

Medical Perspective

Most of the studies examine concurrent exposure to multiple chemicals and the association between these chemicals and numerous reproductive outcomes, thus, test multiple hypotheses at a time without correcting for multiple comparisons, increasing the risk of chance discovery. The strength of the associations found is generally weak and inconsistent across studies for individual compounds. Issues relating to exposure for example, age at exposure, dose, duration of exposure, and temporal relationship with the outcome measures are frequently unknown, which further makes interpretation of study results difficult. Further, control for potential confounders such as comorbidities and underlying cause of infertility are challenging issues. Consequently, extrapolation of results from a population undergoing medical treatment for infertility to the general population is not logical. While many more studies need to be done before we can confidently make a conclusion, its highly probable that climate change will increase human exposure to harmful chemicals. The combined stress of acclimatization to a changing climate and dealing with a chemical hazard, may make humans more susceptible to the toxicant's adverse reproductive effects. Mattison (Mattison, 2010) indicated that the impact of environmental toxicants on pregnancy outcomes might have no threshold level, and the only reasonable solution is the prevention and control of toxic exposures. Awareness in medical professionals and infertility experts about the additional, possible climate change related potential chemical hazards, leading to secondary infertility in humans, will aid in proper investigations and treatment of the infertility.

Policy Implications

Disaster management protocols during extreme climate change events should incorporate prevention and control measures to address the possibility of chemical exposure to affected population. Provision of safe potable water and uncontaminated food to the affected population during catastrophic climate change events should be considered as a high priority.

BIOLOGICAL HAZARDS OF CLIMATE CHANGE

The complex interactions between climate change and water-borne diseases -Aftermaths of climate change events like drought, floods, cyclones can negatively impact the quality of potable water The contamination of potable water with flood water and sewage can deteriorate the water quality which in turn, can lead to water-related disease (Cann et al.,2013;.Effler et al,2001; Frumkin et al,2008) The hotter temperatures, in congruence with decreased water during drought conditions may promote the growth of pathogens in water. The author here discuss how the pathogen contaminated water can be responsible for some of the diseases of reproductive significance

E. Coli Bacteria

A negative correlation has been identified between *Escherichia coli* growth and water temperatures Contamination of water can cause epididymitis through urinary tract infection, which may lead to male infertility in long term. About 10% of men who have had acute epididymitis develop persistent azoospermia thereafter, and 30% have oligozoospermia, both are markers of infertility in men (Kay and McDonald.,1983)

Cholera and Diarrhoeal Disease

Climate change is predicted to increase the risk and severity of cholera-like diarrhoeal diseases, with each degree of increase in ambient temperature. Diarrhea and cholera often lead to severe dehydration in expectant mothers, influencing the pregnancy outcome and the neonatal health condition (Rylander et al,2013).

Leptospirosis

The survival and multiplication of leptospirochaete bacteria may get altered by temperature and pH changes in water resulting from the stagnation associated with drought Leptospirosis is a zoonotic disease with global distribution, transmitted from an infected host to the human by direct or indirect contact with infected urine. Although Leptospiral infection has been reported as one cause of reproductive disorders in production animals (Fornazari et al,2012), there is no direct conclusive evidence of it causing reproductive disorders or infertility in humans.

Vector-borne Diseases

Increased temperatures with stagnant water favour mosquito reproduction, thereby increasing density, transmission and spreading of vector-borne diseases Malaria, dengue fever are expected to spread in geospatial distribution, increasing health risks for humans with no previous immunity. Alsoschistosomiasis infection due to widespread redistribution of its intermediate host (snails), may affect reproductive outcomes of a larger number of people in the future.

Malaria

Globally, the deadliest vector-borne disease, may alter the incidence due to climate change. In congruence with raised global temperature, secondary effects of climate change in the form of changes in population density, precipitation pattern, and ecological changes of both humans and the vectors influence on the incidence of malaria. (Rylander et al,2013) For example, as a consequence of the rising rate of warming, significant spatiotemporal distribution of Anapheles mosquito was observed in Nepal. Malaria, which was earlier confined to forests near Tarai lowlands has expanded to further high-altitude districts. Apart from large number of maternal deaths, maternal malaria infection might lead to IUGR and prematurity among newborns, which in turn may lead to low birth weight. (Rylander et al,2013)

Dengue Fever

With increasing temperatures, the geographical area of dengue transmission is expected to increase. Due to factors like population density and lack of prevention strategies, dengue fever is a large health problem in many developing, tropical countries. There are varied adverse health effects and pregnancy outcomes such as placental transmission to the fetus, preterm birth, low birth weight, pre-eclampsia and eclampsia, Caesarean delivery, and fetal/perinatal death also maternal death that are linked to maternal dengue infection during pregnancy (Pouliot et al, 2010). Currently, scientific literature is unaware of the long term effects of dengue on reproductive health and fertility. Over the years, the rate at which vector-borne diseases and their transmission dynamics modify and develop has been shown to be highly dependent to a great extent on changes in temperature and geographical distribution.

Climate Change and Spacio-temporal Distribution of Pathogens Affecting Reproductive Health

Impact of climate Change on Schistosomiasis

The exact impact of climate change on schistosomiasis, a disease that affects more than 250 million people mainly in tropical and subtropical countries, is currently unpredictable, but likely to vary with the snail-parasite species' specific ecologies on the spatio-temporal scale. The evidence associated with climate change responses is variable across geographical regions and snail-parasite taxonomic groups. Indeed, the strongest evidence stems from the People's Republic of China pertaining to *Schistosoma japonicum*. Some evidence is also available from eastern Africa, mainly for *Schistosoma mansoni*. While studies focused on the northern and southern range margins for schistosomiasis, indicate an increase in transmission range as the most likely outcome. Climate change is more likely to shift than to expand the geographic ranges of schistosomiasis. A comparison between the current geographical distributions and the thermo-physiological limitations of the two main African schistosome species (*Schistosoma haematobium and S. mansoni*) offered additional insights, and showed that both species already exist near their thermo-physiological limits. The African species can move considerably out of their "thermal comfort zone" in a future, warmer Africa, but *S. haematobium* in particular is likely to experience less favourable climatic temperatures. The consequences for schistosomiasis transmission will, to a large extent, depend on the adaptive abilities of parasites and geospacial movement of snails in response to climate change (Stensgaard et al, 2019).

Each climate change variable viz temperature, precipitation, heat waves, drought, and flooding could influence snail and schistosome population dynamics (McCreesh et al, 2013). As the parasite relies on specific species of snails to complete their life cycle; the ecology of the snails becomes a key element in transmission of the parasites. More than 200 million people are infected with schistosomes of which 95% live in sub-Saharan Africa and many more are living in areas where transmission is on-going. Climate change driven increased temperature in Zimbabwe may cause a reduced spatial distribution of suitable habitat of host snails with a probable exception of *Bi. pfeifferi*, the intermediate host for intestinal schistosomiasis that may increase around 2055 before declining towards the end of the century. (Pedersen et al,2014). The majority of infected people (85%) live in sub-Saharan Africa. Climate change will affect water resources and millions of people are estimated to become climate refugees. As a consequence, more people will lack access to safe drinking water and proper sanitation. This will facilitate the emergence of schistosomiasis, which will increase the risk of the disease becoming endemic in areas that currently have low endemicity. In Brazil and Eastern Africa, refugee movements and migrations have already demonstrated the spreading of schistosomiasis into previously uninfected areas (Aagaard-Hansen et al,2010; Adekiya et al, 2019).

There are several case reports and case series that report the effect of schistosomiasis on pregnancy health and birth outcome, but there exists limited literature on comparative studies, which investigate the effects of schistosomiasis on pregnancy health, and therefore to establish a causal relationship between the disease and adverse pregnancy outcomes is complicated. Rodent models suggest that schistosomiasis has a strong negative impact on pregnancy outcome. A significant link between being heavily infected by *S. mansoni* and increased risk of anemia among pregnant women is there and Anemia is a risk factor for low birth weight and maternal mortality. In a recent observational study maternal *S. japonicum* was associated with maternal, placental, and fetal inflammation, and placental inflammation is linked with low birth weight (Kurtis et al,2011). World health organisation considers schistosomiasis as one of the three most devastating tropical diseases and has warned against, *S. haematobium*, a urogenital schistosomiasis pathogen, which is known to induce genital lesions, vaginal bleeding, pain during sexual intercourse, vulva nodules, and pathology of the prostate, seminal vesicles and other organs, along with infertility as a potential long-term consequence(WHO). Pathology series suggest associations of female genital schistosomiasis (*Schistosoma haematobium*) with infertility and ectopic pregnancy. Infertility appears to be associated spatially with *S. haematobium* (Woodall and Kramer, 2018).

REMEDIAL MEASURES

Country Specific Health Programmes are Required to Reduce Prevalence of Schistosomiasis

Impact of Climate Change on Geospatial Re-distribution of Zika Virus

Studies have shown that climatic factors strongly influence the ecology, development, behavior, and survival of mosquitoes and the transmission dynamics of the diseases they transmit (Reiter, 2001). A less common vector-borne disease came into limelight, when in February 2016, the World Health Organization declared the mosquito-borne Zika virus as a "public health emergency of international concern" because the disease linked to thousands of birth defects in Brazil spread rapidly. The distribution of the Aedes mosquitos has drastically increased over the past few decades, due to climate change (Yang et al,2016). Asia and its Himalayan region are particularly vulnerable to changes in climate and land use, moreover, Asia has been a hotspot of dengue fever and chikungunya mainly due to its dense human population, unplanned urbanization and poverty. Dengue virus and chikungunya virus are transmitted by Aedes aegypti and Aedes. albopictus mosquitoes which are also eligible vectors of Zika virus Over the last decade, dengue and chikungunya transmissions by Ae. aegypti have extended to the Himalayan countries of Bhutan and Nepal and Zika virus could follow the same path in sub-Himalayan regions The climate change established wide distribution of human-biting Aedes mosquito vectors and a susceptible population with lack of immunity against Zika virus places the sub-himalayan region at a higher risk of zika virus. Some of the countries in the region have already reported Zika virus infection (Dhimal et al,2018).

i. Zika virusin fection and male reproductive system

Zika virus exhibits high tropism to the testes and female genital tracts. An animal study demonstrates a correlation between Zika virus and testis damage and suggests that Zika virus infection, can eventually lead to male infertility. Zika virus persists in the semen of male patients and can induce inflammation in the testis and epididymis, but not in the prostate or seminal vesicle of mice. (Ma et al,2017) and Zika virus is considered as an emerging, climate change driven potential threat to human reproduction system (Li et al,2018).

ii. Zika virus infection and reproduction in female

Perinatal transmission and vertical transmission of Zika virus have been documented in humans (Oliveria Melo et al,2016). The virus is thought to be neurotrophic causing destruction in various parts of the fetal brain resulting in calcifications in the eyes, micro-ophthalmia and microcephaly. It is also linked to hydrops fetalis, hydranencephaly, fetal demise (Sarno et al,2016), placental insufficiency and IUGR (Brasil et al,2016).

Impact of Climate Change on Leishmaniasis

In a case report, subclinical systemic leishmaniasis was found to be the cause of infertility in dog suffering from semen alterations. (Assis et al,2010; Mir et al, 2012) Although no such adverse effect is reported in humans till date, there exists a biological plausibility.

Association of Global Warming and Incidence of Syphilis

A novel relation between atmospheric temperature, humidity and acquisition of syphilis infection has been established. Overall, the relative risk of primary and secondary syphilis increased with humidity and temperature. It slightly increased when the human perceived temperature was around 15 and clearly peaked around 30. When humidity index increased to 38, relative risk started to increase again. Normally, conditions of temperature between 20-25°C are considered to be the most comfortable for outdoor activities. In subtropical areas, the most comfortable range of truly perceived temperature is around 30 Therefore, humans tend to socialize more during such days. Additionally, increased relative risk may be associated with increased testosterone levels, leading to more sexual activities. (Zhang et al.,2017)

Association of Climate Change with Sero-prevalence of Toxoplasmosis Gondii and its Intermediate Host

Toxoplasma gondii is an intracellular parasite, depends on an intermediate host called cat of the family Felidae to complete its life cycle. Occurrence of T. gondii has been discovered in animals like polar bear and arctic fox, even though there is complete absence of felids in the Arctic Archipelago (Sandströmet al,2013). With climate change, northward migration of felids (the intermediate host) may increase the prevalence of T. gondii in Arctic wildlife. (Reiling et al,2019). There is high plausibility that T. gondii may enter world ecosystem via migratory birds due to climate change. Wild birds, being reservoir hosts and vectors of zoonotic pathogen T. gondii, are important in the epidemiology of toxoplasmosis (Verma et al, 2016). Global warming has been postulated to cause serious impact on geographical dispersion of an infectious disease of reproductive implications like toxoplasmosis. As the frequency of T. gondii largely depends on climatic conditions, the climate change over the last few decades has impacted the seroprevalence of Toxoplasmosis in 21 states of America. A significant, increasing trend of higher prevalence of T. gondii infection was noted in recent health surveys compared to two decades old survey, with a geographical re-distribution after year 2000 along with a positive correlation with temperature(Caballero-Ortega et al, 2012) A specific Global Climate Change model has predicted a potential increase of T. gondii prevalence in humans in some regions of Europe as a result of changing environmental conditions (Meerburg et al,2009). Toxoplasmosis infection in women is a known cause of miscarriages and infertility in women

Impact of Climatic change on Mumps

Over the past decade, there have been resurgences and large-scale outbreaks of viral disease mumps, worldwide. Mumps virus causes an acute febrile illness with a nonspecific prodrome followed by painful swelling of the parotid and less commonly, other salivary glands. The complications of mumps in post-pubertal age include, mastitis, oophoritis, and orchitis, which might affect the sperm quality, causing rarely infertility. Mumps infection results in progressive degeneration of the seminiferous epithelium and, occasionally, Sertoli cell-only syndrome. Thus, the remaining Sertoli cells may be critical to spermatogenesis recovery after orchitis healing (Wang et al, 2013). Evidence is available on the relationship between meteorological factors and the resurgence of mumps. Mean temperature, relative humidity, wind velocity, and atmospheric pressure might be important predictors of the mumps incidence. Tropical cyclones may cause a higher increase in mumps cases (Yang et al,2014; Zha et al,2020; Hu et al,2017; Lu et al,2020). Maternal mumps infection during early pregnancy might cause spontaneous abortions, but no clear association between congenital malformations and maternal mumps has been demonstrated. The mumps virus has been implicated in potential damage to the ovaries and causation of premature menopause.

Impact of Climatic Change on Epididymo-orchitis

A common infectious disease, affecting young men. The associations of various meteorological indicators on the incidence of epididymo-orchitis are reported in a national level Taiwanese study. Monthly meteorological indicators, including average temperatures, humidity, rainfall, total rain days, and sunshine hours, were found well-correlated with the incidence of epididymo-orchitis. The average monthly temperature had significant positive linear correlation with the incidence of epididymo-orchitis (Liu et al, 2017). January month reports a significantly higher incidence rate of torsion. In addition, the incidence rate of testicular torsion was negatively associated with temperature (Chiu et al,2012). Increased incidences of testicular torsion during low temperature periods in Japan, the United States, Greece, and Nigeria are reported. (Lyronis et al,2009; Mabogunje,1986; Hoshino et al, 1993; Srinivasan et al, 2007)

PSYCHO-SOCIO-BEHAVIOURAL HAZARDS ON CLIMATE CHANGE: BEHAVIOUR MODULATION - FERTILITY THEORY

Linkages of Climate Change and Reproductive Outcome, via Psycho-behavioural Pathway

Even not so extreme forms of weather, may also indirectly impact reproductive health through its effects on human behaviour. The constraint of psychological restoration as one behavioural pathway through which weather at higher latitudes, can negatively affect reproductive outcome such as very low birthweight, which has very serious long term implications on the survival and general health of the infant, was investigated in a Swedish study. Using time-series analytic methods with national level data across the years 1973–2010, the association between mean monthly temperature and the odds of very low birth weight for each of the summer months was established. Unpleasant weather, being a source of stress may figure in stress-related adverse health outcomes, but also act as a constraint on restoration. Eliasson et al. (2007) found that temperature as compared to cloud cover and wind speed, had a stronger association with the number of people present in different urban outdoor spaces. Those additional variables do nonetheless contribute to the explanation of presence outdoors, whether they have an ability to harm health via this behavioral pathway, either alone or in interaction with temperature is less understood. Further research considering other meteorological conditions that interfere with outdoor activities and human behaviour in different climatic zones, geographical locations are needed for advancement of the

Behavioural Pathway/ modulation theory. In some geographical regions, increasing summer temperatures due to global warming might restrict outdoor activities of pregnant women (Hartig and Catalano, 2013).

Linkages of Climate Change with Reproductive Behaviour, Responses and Reproductive Goals: A Socio-behavioural Pathway

There are numerous ways through which climate variability and its consequential sociobehavioural impacts may influence reproductive behavior and population growth rates Couples do modify their reproductive behaviors and reproductive goals in response to changing climatic conditions. Changes in temperature and precipitation may influence reproductive goals (e.g., the desire for another child, ideal family size) and behaviors (e.g. libido, contraceptive use, childbearing) through multiple pathways (Eissler et al,2019). Use of family planning tends to decrease, when monsoon onset is delayed; but that family planning use increases in response to high temperatures However, these effects often vary significantly across different groups, which suggests that women's ability and likelihood to change their reproductive behavior in response to climate varies (Sellers and Gray 2019). Climatic changes may also stimulate fertility by decreasing child survival, which may lead women to plan additional children to compensate for anticipated child mortality risks or in response to actual child death. However, in many cases climate change is likely to increase fecundity indirectly, outside of intentional decision-making processes about how many children to have. For example, climate-induced changes in labor migration may lead to increased spousal separation, which is inversely related to fertility (Agadjanian et al,2011). Likewise, climate-induced changes in the demand for labor and corresponding time allocation may influence the availability of spare time, which may be positively associated with risk of conception and fertility (Burlando et al,2014).

Acute events associated with climate change, like cyclones, floods may also interrupt access to contraceptives, cause disruptions in public health family planning services. Changes in precipitation can also influence reproductive goals but in a slightly more complex and inconsistent way. There is a dynamic relationship, in which women opt for lesser number of desired children in response to short spells of above-average rainfall, probably due to immediate demands for their own physical work, while multi-year spells of above-average precipitation may lead to increased ideal family size. Assuming such extended periods of above-average precipitation are associated with improved agricultural and economic outcomes, the increases in ideal family size may reflect conscious behaviour and decision to increase household labor force (Simon, 2017). Some populations may experience seasonal changes in food intake behaviour due to economy-or culture-driven variations in food availability that influence reproductive events such as onset of puberty, which can indirectly get affected by changes in climate conditions. Peak incidences of attainment of menarche in girls of the United States (Matchocket al,2004) and European countries (Gueresi,1997) have been recorded in January and July months, whereas in April and October months among girls in Asian girls.

ECONOMICAL HAZARDS ON CLIMATE CHANGE: CLIMATE CHANGE-ECONOMY-DEMOGRAPHY THEORY

Climate change through its substantial impact on the economy, has potential to make a larger, ultimate impact on human fertility. There are various economic channels through which climate change might affect human reproductive ability, including labour reallocation, the gender wage gap, longevity, and child mortality. A quantitative model that takes into account existing estimates of the economic consequences of climate change, predicts that, increases in global temperature differentially affect agricultural and non-agricultural sectors. In equatorial poor countries, climate change has a larger negative effect on agriculture. The resulting scarcity in agricultural goods leads to higher agricultural prices and wages, causing labour reallocation into this sector, which ultimately acts as a driving force towards increase in human fecundity. These patterns are reversed at higher latitudes, suggesting that climate change may exacerbate inequities by reducing fertility and increasing education in richer northern high latitude countries, while increasing fertility and reducing education in poorer tropical countries. Likewise, climate-induced debilitated economic conditions over the longer period may affect individuals' ability to afford contraceptives. Also, climate-induced changes in family resources may lead to intentional decisions to affect family size. If children are considered as net utilisers of resources, climate-induced resource constraints, such as diminished agricultural output or income, may lead to smaller family size. The opposite may occur if children are viewed as income generators. While the model only examines the role of economy mediated mechanism, it suggests that climate change can have an impact on fertility and fertility patterns through socioeconomic pathways.

Migration

Fig. 2 depicts how due to the heightened frequency of extreme weather events, such as droughts and cyclones,millions of people have become climate refugees and migrants. Catastrophic climate events cause crop failure, livestock mortality, and increased foodprices, food and potable water shortage and malnutrition. The probability of the food being contaminated or of poor quality will also increase the migration in search of food and employment. Increased migration will further increase the risk of the emergence of epidemics in new areas since the spreading and control of many tropical diseases have been linked to population movements. It is however evident that people that are forced to resettle due to natural disasters are stressed due to traumatized experiences. Traumatic experiences will also affect the mental health of people Postpartum depression is more common among immigrant mothers (Morrow et al, 2008). In regards to maternal health, extreme events will reduce access to contraceptive and antenatal care. Lack of access to proper pregnancy care will further increase the risk of complications during delivery and also increase the risk of maternal and infant mortality.

Malnutrition: Fertility, Reproductive Health, and Pregnancy Outcome

A well-nourished woman is essential for healthy reproduction. Malnourished women have more chances of having intrauterine growth restriction and higher incidences of low birth weight babies and perinatal mortality. In the long-term perspective, undernutrition among mother could affect reproductive ability. Granulosa cells of ovary are affected by malnutrition. Malnutrition can affect reproductive processes through changes in the secretion of metabolic hormones. Food insecurity reduces the release of hormone leptin from adipose tissues, which can affect reproduction through influence on the hypothalamo-hypophysial system and by direct action on gonads (Sirotkin, 2010). Chronic malnutrition has been associated with delaying menarche, in India (Gopalan and Naidu.1972). The indirect effect of increased natural calamities such as famines, droughts and floods (Fig. 2), which are secondary to climate change may lead to reduced food security, affecting both, the quantity and nutritional quality of food. This may lead to gross mass undernutrition of women particularly affecting poor countries. The reproductive axis is closely linked to nutritional status, especially undernutrition in the female, and inhibitory pathways involving detectors in the hind brain suppress ovulation in subjects with weight loss. Recovery may occur after minimal reacquisition of weight because energy balance is more important than body fat mass. Severe cachexia in women of reproductive age, observed in eating disorders like anorexia nervosa may lead to amenorrhoea, infertility and, in those who do conceive, an increased likelihood of miscarriage and adverse reproductive outcomes (ESHRE Capri Workshop Group, 2006).

Figure 2. Pathways through which acute and subacute climate change events impact human reproduction and fertility output.

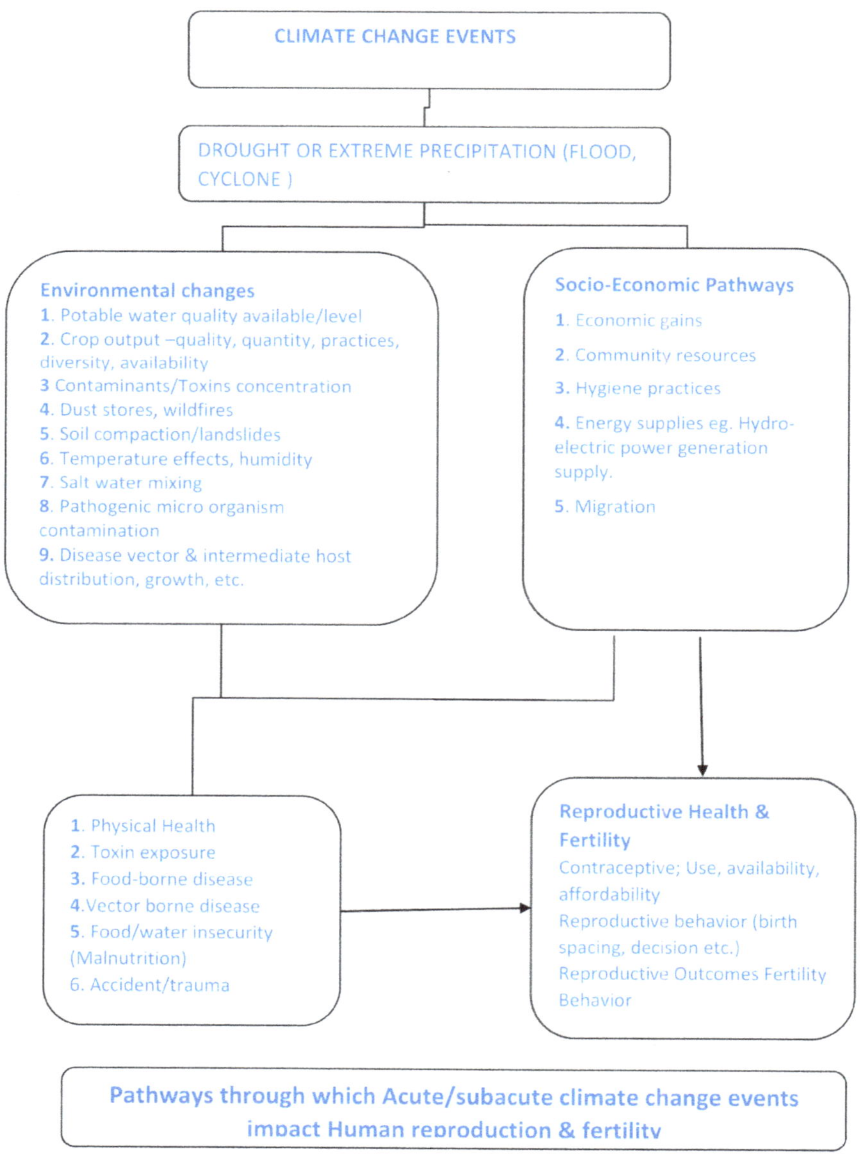

RESEARCH GAPS

Most of the scientific literature focuses exclusively on the antifertility effects of heat stress and does not offer major insights into impact of other biotic and abiotic variables including major social changes resulting from climate change on human reproduction and fertility. More research is needed to improve our understanding of how other climate change variables can impact reproductive health, but acquiring more data will require a uniform method for collectionMost of the conclusive scientific evidence, establishing relationship of abiotic type of climate change variables with the damage to mammalian reproductive organs, systems, functions, and fertility is generated through experimental studies on the suitable animal models, with very limited evidence coming from human epidemiological studies. Due to ethical and other technical issues, such type of scientific evidence generation through intervention and, or experimental studies on humanbeings is not feasible. Hence the findings and conclusions of mammalian studies are generally extrapolated to human beings. But the caution should be exercised in translating such conclusions to humans, as the animal models differ in sensitivities to chemicals, physical hazards and infectivity of biological pathogens.

There is limited knowledge about which regions will likely be heavily affected. Research efforts should focus to fill knowledge gaps by identifying the most vulnerable groups to develop the harmful effects on reproduction and coordinate efforts to control the parameters causing negative reproductive health consequences. The effects of malnutrition, infectious diseases, environmental problems, and direct heat exposure on expectant mothers and pregnancy outcomes will lead to severe health risks. Increased focus on antenatal care is recommended to prevent worsening maternal health and perinatal mortality and morbidity. Interventions to reduce the negative reproductive health impacts caused by climate change are also crucial. Protective effect of some hormones like leptin on the adverse ovarian outcomes needs to be studied. Every effort should be made to develop and maintain good antenatal care during catastrophic climatic conditions through adequate disaster management planning measures.

CONCLUSION

This chapter has focused on how climate change variables can directly or indirectly influences human reproduction and fertility by altering various closely interlinked and interactive factors, biotic and abiotic determinants of reproductive health and outcome. Climate change can also act through malnutrition, safe potable water, proper sanitation; population migrations, changing patterns of disease and morbidity, access to health facility and economical factors. A significant increase of reproductive age women contracting malaria, dengue fever, and schistosomiasis before the end of this century is expected, as a result of changing disease pattern and population migration. Increased incidence of malaria, dengue fever, and schistosomiasis will clearly increase the risk of begetting a child with low birth weight. The risk for birth complications and preterm deliveries will also increase. Thus, these diseases will lead to increased infant and maternal mortality, poorer reproductive health. Increased incidence of all the reproductive diseases will demand better health care systems and better access to medicines for millions of populations at risk in the years to come. The populations in developing countries, in tropical areas are likely to suffer most from climate change due to various pre-existing adversities. The strength of health care systems to respond to an increase in the burden of climate-sensitive reproductive outcomes will play a critical role in determining the reproductive health impact of climate change.

Most of the literature demonstrates the acute or chronic effect of individual climate change variable of interest on human reproduction and fertility; but in real world scenario, humans face threat of concurrent exposure to multiple, adverse climatic parameters, which may potentiate, augument negative effects of each other, causing cumulative ultimate, measurable damage to reproduction and fertility, such type of evidence although foreseeable, is limited. In the absence of such direct evidence, clinicians, infertility experts should advice the precautionary principle of avoiding the harmful exposure to prevent the damage to human reproduction in long term. Appropriate control strategies should be applied through policy changes inorder to mitigate the adverse impact. For eg provision of extra resting, mid day hours, ensuring comfortable environment to the expectant mothers employed in outdoor occupations like construction industry, to minimize ill effects of excessive heat during summers of tropical countries.

Additionally, humans due to evolutionary advantage, might possess and exhibit more or less genetic acclimatization to the changing climate than the other animals. Also the availability of the artificial means of dealing with the adverse effects of climate change, such as access to air cooling system in high atmospheric temperature desert areas, can also negatively affect the human acclimatization to extreme temperature weather, at the same time minimising the adverse impact on the human reproduction to a great extent Moreover, along with biotic factors, abiotic factors of climate change such as social, psychological behavioural, and economical consequences, have potential to alter, modify the effect of physical or chemical climate change variables on human reproduction. Thus, it's the chronic, accumulative impact of individual variable along with the concurrent effect of multiple exposures, that determines the intensity of impact on the reproductive parameters. All these multiple linkages between the climate change variables may cause multidimentional, adverse(mainly) impact on the human reproduction and fertility. The worldwide, increasing rates, trends of infertility, perhaps are visible and measurable refection of the potential damage that climate change and industrialization caused to the human reproduction over the last century.

In future, Climate change will disturb the global physical systems (eg global precipitation patterns) and ecosystems (eg biological agents, vector habitats); these perturbations, in turn, may pose direct and indirect hazards to human reproductive health and fertility. Direct risks involve climatic variables, such as biotic factorsthat directly causes changes in human reproductive physiology, structure and functions. Indirect risks, mainly mediated through abiotic factors, although lack direct causal relations between climatic parameters and human reproduction, but can act through multiple linkagessuch as psycho-social and economic pathways. Conversely, factors related to fertility such as population size, composition and growth ratemay also influence a community's ability get to adapt to a changing climate and must be duly considered in predicting future scenarios.

AUTHOR CONTRIBUTION

RHC conceived, designed and wrote the book chapter along with literature survey.

FUND DISCLAIMER

No grant received from any funding agency in public, commercial, or not-for-profit sector.

CORRESPONDING AUTHOR

Ranjana Hanumant Choudhari

ACKNOWLEDGMENT

Ms. Vishakha Shrimali, Ph.D. student, Biochemistry Div., ICMR-NIOH for digitalization of the illustration.

REFERENCES

Aagaard-Hansen, J., Nombela, N., & Alvar, J. (2010). Population movement: A key factor in the epidemiology of neglected tropical diseases. *Tropical Medicine & International Health, 15*(11), 1281–1288. doi:10.1111/j.1365-3156.2010.02629.x PMID:20976871

Adekiya, T. A., Aruleba, R. T., Oyinloye, B. E., Okosun, K. O., & Kappo, A. P. (2019, December 26). The Effect of Climate Change and the Snail-Schistosome Cycle in Transmission and Bio-Control of Schistosomiasis in Sub-Saharan Africa. *International Journal of Environmental Research and Public Health, 17*(1), 181. doi:10.3390/ijerph17010181 PMID:31887979

Amegah, A. K., Rezza, G., & Jaakkola, J. J. K. (2016). Review article: temperature-related morbidity and mortality in sub-Saharan Africa: a systematic review of the empirical evidence. *Environment International, 91*, 133–149. doi:10.1016/j.envint.2016.02.027 PMID:26949867

Arroyo, V., Díaz, J., Ortiz, C., Carmona, R., Sáez, M., & Linares, C. (2016). Short term effect of air pollution, noise and heat waves on preterm births in Madrid (Spain). *Environmental Research, 145*, 162–168. doi:10.1016/j.envres.2015.11.034 PMID:26706568

Asamoah, B., Kjellstrom, T., & Östergren, P. (2018). Is ambient heat exposure levels associated with miscarriage or stillbirths in hot regions? A cross-sectional study using survey data from the Ghana Maternal Health Survey 2007. *International Journal of Biometeorology, 62*(3), 319–330. doi:10.100700484-017-1402-5 PMID:28748383

Assis, V. P., Ribeiro, V. M., Rachid, M. A., Castro, A. C. S., & Valle, G. R. (2010). Dogs with Leishmania chagasi infection have semen abnormalities that partially revert during 150 days of allopurinol and amphotericin B therapy. *Animal Reproduction Science, 117*(1-2), 183–186. doi:10.1016/j.anireprosci.2009.03.003 PMID:19345025

Auger, N., Naimi, A. I., Smargiassi, A., Lo, E., & Kosatsky, T. (2014). Extreme heat and risk of early delivery among preterm and term pregnancies. *Epidemiology (Cambridge, Mass.), 25*(3), 344–350. doi:10.1097/EDE.0000000000000074 PMID:24595396

Bantje, H. F. W. (1988). Female stress and birth seasonality in tanzania. *Journal of Biosocial Science, 20*(2), 195–202. doi:10.1017/S0021932000017429 PMID:3384835

Barreca, A., Deschenes, O., & Guldi, M. (2018). Maybe Next Month? Temperature Shocks and Dynamic Adjustments in Birth Rates. *Demography*, *55*(4), 1269–1293. doi:10.100713524-018-0690-7 PMID:29968058

Basu, R., Malig, B., & Ostro, B. (2010). High ambient temperature and the risk of preterm delivery. *American Journal of Epidemiology*, *172*(10), 1108–1117. doi:10.1093/aje/kwq170 PMID:20889619

Basu, R., Sarovar, V., & Malig, B. J. (2016). Association Between High Ambient Temperature and Risk of Stillbirth in California. *American Journal of Epidemiology*, *183*(10), 894–901. doi:10.1093/aje/kwv295 PMID:27037268

Beere, H. M. (2004). 'The stress of dying': The role of heat shock proteins in the regulation of apoptosis. *Journal of Cell Science*, *117*(13), 2641–2651. doi:10.1242/jcs.01284 PMID:15169835

Bekkar, B., Pacheco, S., Basu, R., & DeNicola, N. (2020). Association of Air Pollution and Heat Exposure With Preterm Birth, Low Birth Weight, and Stillbirth in the US: A Systematic Review. *JAMA Netw Open.*, *3*(6).

Belhadj-Silmen, I., Najar, T., Ghram, A., & Abdrrabba, M. (2015). Heat stress effects on livestock: Molecular, cellular and metabolic aspects, a review. *Journal of Animal Physiology and Animal Nutrition*, *100*(3), 401–412. doi:10.1111/jpn.12379 PMID:26250521

Beltran, A. J., Wu, J., & Laurent, O. (2013). Associations of meteorology with adverse pregnancy outcomes: A systematic review of preeclampsia, preterm birth and birth weight. *International Journal of Environmental Research and Public Health*, *11*(1), 91–172. doi:10.3390/ijerph110100091 PMID:24362545

Berisha, B., & Schams, D. (2005). Ovarian function in ruminants. *Domestic Animal Endocrinology*, *29*(2), 305–317. doi:10.1016/j.domaniend.2005.02.035 PMID:15998502

Birch, E. L., Meleis, A., & Wachter, S. (2012). The urban water transition: Why we must address the new reality of urbanization, women, Water, and sanitation in sustainable development. *wH2O: The Journal of Gender and Water, 1*. Available at: https://repository.upenn.edu/wh2ojournal/vol1/iss1/1

Bleiberg, F. M., Brun, T. A., Goihman, S., & Gouba, E. (1980). Duration of activities and energy expenditure of female farmers in dry and rainy seasons in Upper-Volta. *British Journal of Nutrition*, *43*(1), 71–82. doi:10.1079/BJN19800065 PMID:7189407

Brasil, P., Pereira, J. P., Gabaglia, C. R., Damasceno, L., Wakimoto, M., & … . (2016). Zika Virus Infection in Pregnant Women in Rio de Janeiro-Preliminary Report. *The New England Journal of Medicine*, *375*(24), 2321–2334. doi:10.1056/NEJMoa1602412 PMID:26943629

Bridges, P. J., Brusie, M. A., & Fortune, J. E. (2005). Elevated temperature (heat stress) in vitro reduces androstenedione and estradiol and increases progesterone secretion by follicular cells from bovine dominant follicles. *Domestic Animal Endocrinology*, *29*(3), 508–522. doi:10.1016/j.domaniend.2005.02.017 PMID:16153500

Bruckner, T. A., Modin, B., & Vågerö, D. (2014). Cold ambient temperature in utero and birth outcomes in Uppsala, Sweden, 1915-1929. *Annals of Epidemiology*, *24*(2), 116–121. doi:10.1016/j.annepidem.2013.11.005 PMID:24332864

Bunker, A., Wildenhain, J., Vandenbergh, A., Henschke, N., Rocklöv, J., Hajat, S., & Sauerborn, R. (2016). Research paper: Effects of air temperature on climate-sensitive mortality and morbidity outcomes in the elderly; a systematic review and meta-analysis of epidemiological evidence. *EBioMedicine*, *6*, 258–268. doi:10.1016/j.ebiom.2016.02.034 PMID:27211569

Caballero-Ortega, H., Uribe-Salas, F. J., Conde-Glez, C. J., Cedillo-Pelaez, C., Vargas-Villavicencio, J. A., Luna-Pastén, H., Cañedo-Solares, I., Ortiz-Alegría, L. B., & Correa, D. (2012). Seroprevalence and national distribution of human toxoplasmosis in Mexico: Analysis of the 2000 and 2006 National Health Surveys. *Transactions of the Royal Society of Tropical Medicine and Hygiene*, *106*(11), 653–659. doi:10.1016/j.trstmh.2012.08.004 PMID:22998951

Cann, K. F., Thomas, D. R., Salmon, R. L., Wyn-Jones, A. P., & Kay, D. (2013). Extreme water-related weather events and waterborne disease. *Epidemiology and Infection*, *141*(4), 671–686. doi:10.1017/S0950268812001653 PMID:22877498

Chiu, B., Chen, C.-S., Keller, J. J., Lin, C.-C., & Lin, H.-C. (2012). Seasonality of testicular torsion: A 10-year nationwide population based study. *The Journal of Urology*, *187*(5), 1781–1785. doi:10.1016/j.juro.2011.12.104 PMID:22425116

Christiansen, T., & Voigt, T. (2008). *Impact of Europe's Changing Climate—2008 Indicator-Based Assessment*. https://ec.europa.eu/jrc/sites/jrcsh/files/jrc_reference_report_2008_09_climate_change.pdf

Cox, B., Vicedo-Cabrera, A. M., Gasparrini, A., Roels, H. A., Martens, E., Vangronsveld, J., Forsberg, B., & Nawrot, T. S. (2016). Ambient temperature as a trigger of preterm delivery in a temperate climate. *Journal of Epidemiology and Community Health*, *70*(12), 1191–1199. doi:10.1136/jech-2015-206384 PMID:27261529

Dadvand, P., Basagaña, X., Sartini, C., Figueras, F., Vrijheid, M., de Nazelle, A., Sunyer, J., & Nieuwenhuijsen, M. J. (2011). Climate extremes and the length of gestation. National Institute of Environmental Health Sciences. *Environmental Health Perspectives*, *119*(10), 1449–1453. doi:10.1289/ehp.1003241 PMID:21659038

Dadvand, P., Ostro, B., Figueras, F., Foraster, M., Basagaña, X., Valentín, A., Martinez, D., Beelen, R., Cirach, M., Hoek, G., Jerrett, M., Brunekreef, B., & Nieuwenhuijsen, M. J. (2014). Residential proximity to major roads and term low birth weight: The roles of air pollution, heat, noise, and road-adjacent trees. *Epidemiology (Cambridge, Mass.)*, *25*(4), 518–525. doi:10.1097/EDE.0000000000000107 PMID:24787556

Davidson, J. F., Whyte, B., Bissinger, P. H., & Schiestl, R. H. (1996). Oxidative stress is involved in heat-induced cell death in Saccharomyces cerevisiae. *Proceedings of the National Academy of Sciences of the United States of America*, *93*(10), 5116–5121. doi:10.1073/pnas.93.10.5116 PMID:8643537

Dhimal, M., Dahal, S., Dhimal, M.L., Mishra, S.R., Karki, K.B., Aryal, K.K., Haque, U., Kabir, M.I., Guin, P., Butt, A.M., Harapan, H., Liu, Q.Y., Chu, C., Montag, D., Groneberg, D.A., Pandey, B.D., Kuch, U., & Müller, R. (2018). Threats of Zika virus transmission for Asia and its Hindu-Kush Himalayan region. *Infect Dis Poverty, 15*(7), 40.

Dreier, J. W., Andersen, A.-M. N., & Berg-Beckhoff, G. (2014). Systematic review and meta-analyses: Fever in pregnancy and health impacts in the offspring. *Pediatrics*, *133*(3), e674–e688. doi:10.1542/peds.2013-3205 PMID:24567014

Dreiling, C. E., Carman, F. S. III, & Brown, D. E. (1991). Maternal endocrine and fetal metabolic responses to heat stress. *Journal of Dairy Science*, *74*(1), 312–327. doi:10.3168/jds.S0022-0302(91)78175-7 PMID:2030175

Duncan, K. (2006). Global climate change, air pollution, and women's health. *WIT Trans Ecol and the Envir.*, *99*, 633–643. doi:10.2495/RAV060611

Edwards, J. L., & Hansen, P. J. (1997). Differential responses of bovine oocytes and preimplantation embryos to heat shock. *Molecular Reproduction and Development*, *46*(2), 138–145. doi:10.1002/(SICI)1098-2795(199702)46:2<138::AID-MRD4>3.0.CO;2-R PMID:9021745

Edwards, M. J. (2006). Review: Hyperthermia and fever during pregnancy. *Birth Defects Research. Part A, Clinical and Molecular Teratology*, *76*(7), 507–516. doi:10.1002/bdra.20277 PMID:16933304

Edwards, M. J. (2006). Review: Hyperthermia and fever during pregnancy. *Birth Defects Research. Part A, Clinical and Molecular Teratology*, *76*(7), 507–516. doi:10.1002/bdra.20277 PMID:16933304

Effler, E., Isaacson, M., Arntzen, L., Heenan, R., Canter, P., Barrett, T., Lee, L., Mambo, C., Levine, W., Zaidi, A., & Griffin, P. M. (2001). Factors contributing to the emergence of Escherichia Coli O157 in Africa. *Emerging Infectious Diseases*, *7*(5), 812–819. doi:10.3201/eid0705.017507 PMID:11747693

Elias, S. G., van Noord, P. A., Peeters, P. H., den Tonkelaar, I., Kaaks, R., & Grobbee, D. E. (2007). Menstruation during and after caloric restriction: The 1944–1945 Dutch famine. *Fertility and Sterility*, *88*(4), 1101–1107. doi:10.1016/j.fertnstert.2006.12.043 PMID:17544416

Fornazari, F., da Silva, R. C., Richini-Pereira, V. B., Beserra, H. E., Luvizotto, M. C., & Langoni, H. (2012). Comparison of conventional PCR, quantitative PCR, bacteriological culture and the Warthin Starry technique to detect Leptospira spp. in kidney and liver samples from naturally infected sheep from Brazil. *Journal of Microbiological Methods*, *90*(3), 321–326. doi:10.1016/j.mimet.2012.06.005 PMID:22713608

Frumkin, H., Hess, J., Luber, G., Malilay, J., & McGeehin, M. (2008). Climate change: The public health response. *American Journal of Public Health*, *98*(3), 435–445. doi:10.2105/AJPH.2007.119362 PMID:18235058

Fukuda, M., Fukuda, K., Shimizu, T., Nobunaga, M., Mamsen, L. S., & Yding Andersen, C. (2014). Climate change is associated with male:female ratios of fetal deaths and newborn infants in Japan. *Fertility and Sterility*, *102*(5), 1364–1370. doi:10.1016/j.fertnstert.2014.07.1213 PMID:25226855

Garrigue, M. (2017). *Effet du stress thermique sur les paramètres séminologiques de taureaux de centre d'insémination* (Thèse d'exercice, Médecine vétérinaire). Ecole Nationale Vétérinaire de Toulouse, Toulouse, France.

Gilbert-Barness, E. (2010). Review: Teratogenic causes of malformations. *Annals of Clinical and Laboratory Science*, *40*, 99–114. PMID:20421621

Graham, J. M. Jr. (2005). Marshall J. Edwards: Discoverer of maternal hyperthermia as a human teratogen. *Birth Defects Research. Part A, Clinical and Molecular Teratology*, *73*(11), 857–864. doi:10.1002/bdra.20185 PMID:16265640

Ha, S., Liu, D., Zhu, Y., Soo Kim, S., Sherman, S., Grantz, K. L., & Mendola, P. (2017). Ambient Temperature and Stillbirth: A Multi-Center Retrospective Cohort Study. *Environmental Health Perspectives*, *125*(6), 067011. doi:10.1289/EHP945 PMID:28650842

Hamerezaee, M., Dehghan, S. F., Golbabaei, F., Fathi, A., Barzegar, L., & Heidarnejad, N. (2018, June). Assessment of Semen Quality among Workers Exposed to Heat Stress: A Cross-Sectional Study in a Steel Industry. *Safety and Health at Work*, *9*(2), 232–235. doi:10.1016/j.shaw.2017.07.003 PMID:29928539

Hansen, J., Sato, M., Ruedy, R., Lo, K., Lea, D. W., & Medina-Elizade, M. (2006). Global temperature change. *Proceedings of the National Academy of Sciences of the United States of America*, *103*(39), 14288–14293. doi:10.1073/pnas.0606291103 PMID:17001018

Hansen, P. J. (2009). Effects of heat stress on mammalian reproduction. Philos. *Trans R Soc.*, *364*(1534), 3341–3350. doi:10.1098/rstb.2009.0131 PMID:19833646

Hartig, T., & Catalano, R. (2013). Cold summer weather, constrained restoration, and very low birth weight in Sweden. *Health & Place*, *22*, 68–74. doi:10.1016/j.healthplace.2013.03.008 PMID:23603428

He, J. R., Liu, Y., Xia, X. Y., Ma, W. J., Lin, H. L., Kan, H. D., Lu, J. H., Feng, Q., Mo, W. J., Wang, P., Xia, H. M., Qiu, X., & Muglia, L. J. (2016). Ambient Temperature and the Risk of Preterm Birth in Guangzhou, China (2001-2011). *Environmental Health Perspectives*, *124*(7), 1100–1106. doi:10.1289/ehp.1509778 PMID:26672059

Hoshino, H., Abe, T., Watanabe, H., Katsuoka, Y., & Kawamura, N. (1993). Correlation between atmospheric temperature and testicular torsion. *Acta Urologica Japonica.*, *39*(11), 1031–1033. PMID:8266870

Hossain, S. (2020). Salinity and Miscarriage: Is There a Link? Impact of Climate Change in Coastal Areas of Bangladesh - A Systematic Review. *European Journal of Environment and Public Health*, *4*(1), em0036.

Hu, W., Li, Y., Han, W., Xue, L., Zhang, W., Ma, W., & Bi, P. (2018, April 1). Meteorological factors and the incidence of mumps in Fujian Province, China, 2005-2013: Non-linear effects. *The Science of the Total Environment*, *619-620*, 1286–1298. doi:10.1016/j.scitotenv.2017.11.108 PMID:29734606

International Labour Office. (1998). *General Hazards: Heat and Cold website*. https://www.iloencyclopaedia.org/part-vi-16255/heat-and-cold

Jana, K., Samanta, P. K., & De, D. K. (2010). Nicotine diminishes testicular gametogenesis, steroidogenesis and steroidogenic acute regulatory protein expression in adult albino rats: Possible influence on pituitary gonadotropins and alteration of testicular antioxidant status. *Toxicological Sciences*, *116*(2), 647–159. doi:10.1093/toxsci/kfq149 PMID:20498001

Kay, D., & McDonald, A. (1983, September). Predicting coliform concentrations in upland impoundments: Design and calibration of a multivariate model. *Applied and Environmental Microbiology*, *46*(3), 611–618. doi:10.1128/AEM.46.3.611-618.1983 PMID:6639016

Keiler, M., Knight, J., & Harrison, S. (2010). Climate change and geomorphological hazards in the eastern European Alps. Philosophical transactions. *Series A, Mathematical, Physical, and Engineering Sciences, 368*(1919), 2461–2479.

Kent, S. T., McClure, L. A., Zaitchik, B. F., Smith, T. T., & Gohlke, J. M. (2014). Heat waves and health outcomes in Alabama (USA): The importance of heat wave definition. *Environmental Health Perspectives*, *S122*(2), 151–158. doi:10.1289/ehp.1307262 PMID:24273236

Kloog, I., Melly, S. J., Coull, B. A., Nordio, F., & Schwartz, J. D. (2015). Using Satellite-Based Spatiotemporal Resolved Air Temperature Exposure to Study the Association between Ambient Air Temperature and Birth Outcomes in Massachusetts. *Environmental Health Perspectives*, *123*(10), 1053–1058. doi:10.1289/ehp.1308075 PMID:25850104

Kloog, I., Novack, L., Erez, O., Just, A. C., & Raz, R. (2018). Associations between ambient air temperature, low birth weight and small for gestational age in term neonates in southern Israel. *Environmental Health: A Global Access Science Source, 17*(1), 76.

Konkel, L. (2019). Taking the Heat: Potential Fetal Health Effects of Hot Temperatures. *Environmental Health Perspectives*, *127*(10), 102002. doi:10.1289/EHP6221 PMID:31652107

Kourtis, A. P., Read, J. S., & Jamieson, D. J. (2014). Pregnancy and infection. *The New England Journal of Medicine*, *370*(23), 2211–2218. doi:10.1056/NEJMra1213566 PMID:24897084

Kregel, K. C. (2002). Heat shock proteins: Modifying factors in physiological stress response and acquired thermotolerance. *Journal of Applied Physiology (Bethesda, Md.)*, *92*(5), 2177–2186. doi:10.1152/japplphysiol.01267.2001 PMID:11960972

Kuehn, L., & McCormick, S. (2017). Heat Exposure and Maternal Health in the Face of Climate Change. *Int J Environ Res Public Health, 14*(8), 853.

Kunavongkrit, A., Suriyasomboon, A., Lundeheim, N., Heard, T. W., & Einarsson, S. (2005). Management and sperm production of boars under differing environmental conditions. *Theriogenology*, *63*(2), 657–667. doi:10.1016/j.theriogenology.2004.09.039 PMID:15626423

Kurowicka, B., Gajewska, A., & Franczak, A. (2006). Effect of early thermal experience on pituitary-gonadal axis in female rats. *Reproductive Biology*, *6*, 63–77. PMID:16604152

Kurtis, J. D., Higashi, A., Wu, H. W., Gundogan, F., McDonald, E. A., Sharma, S., PondTor, S., Jarilla, B., Sagliba, M. J., Gonzal, A., Olveda, R., Acosta, L., & Friedman, J. F. (2011). Maternal schistosomiasis japonica is associated with maternal, placental and fetal inflammation. *Infection and Immunity*, *79*(3), 1254–1261. doi:10.1128/IAI.01072-10 PMID:21149589

Kwok-Keung, L., Pao-Yun, C., Yen-Mei, L., Yu-Pei, L., Cheng, D., Won-Hsiung, L., & Mao-Hsiung, Y. (2013). The role of heat shock protein 70 in the protective effect of YC-1 on heat stroke rats. *European Journal of Pharmacology*, *699*(1-3), 67–73. doi:10.1016/j.ejphar.2012.11.044 PMID:23219797

Lawrence, J. L., Payton, R. R., Godkin, J. D., Saxton, A. M., Schrick, F. N., & Edwards, J. L. (2004). Retinol improves development of bovine oocytes compromised by heat stress during maturation. *Journal of Dairy Science*, *87*(8), 2449–2454. doi:10.3168/jds.S0022-0302(04)73368-8 PMID:15328267

Li, X., Ma, W., Wong, G., Ma, S., Li, S., Bi, Y., & Gao, G. F. (2018, August 2). A new threat to human reproduction system posed by Zika virus (ZIKV): From clinical investigations to experimental studies. *Virus Research*, *254*, 10–14. doi:10.1016/j.virusres.2017.09.005 PMID:28893652

Liang, Z., Lin, Y., Ma, Y., Zhang, L., Zhang, X., Li, L., Zhang, S., Cheng, Y., Zhou, X., Lin, H., Miao, H., & Zhao, Q. (2016). The association between ambient temperature and preterm birth in Shenzhen, China: a distributed lag non-linear time series analysis. *Environmental Health: A Global Access Science Source, 15*(1), 84.

Lin, S., Lin, Z., Ou, Y., Soim, A., Shrestha, S., Lu, Y., Sheridan, S., Luben, T. J., Fitzgerald, E., Bell, E., Shaw, G. M., Reefhuis, J., Langlois, P. H., Romitti, P., Feldkamp, M. L., Malik, S., Pantea, C., Nayak, S., Hwang, S. A., & Browne, M.National Birth Defects Prevention Study. (2018, September). Maternal ambient heat exposure during early pregnancy in summer and spring and congenital heart defects - A large US population-based, case-control study. *Environment International*, *118*, 211–221. doi:10.1016/j.envint.2018.04.043 PMID:29886237

Linzey, D. (Ed.). (2011). *Report: Saltwater Intrusion and Climate Change*. Prince Edward Island Department of Environment, Labour and Justice. http://www.gov.pe.ca/photos/original/cle_WA1.pdf

Liu, J. M., Chang, Y. H., Ho, T. W., Chang, F. W., Pang, S. T., Hsu, R. J., & Lin, P. H. (2017). Patients with Epididymo-Orchitis and Meteorological Impact in Taiwan: A Nationwide Population-Based Study. The Canadian journal of infectious diseases & medical microbiology =. *The Canadian Journal of Infectious Diseases & Medical Microbiology*, *2017*, 1506857. doi:10.1155/2017/1506857 PMID:28316630

Lu, J., Yang, Z., Ma, X., Ma, M., & Zhang, Z. (2020). The role of meteorological factors on mumps incidence among children in Guangzhou, Southern China. *PLoS One*, *15*(4), e0232273. doi:10.1371/journal.pone.0232273 PMID:32348370

Lyronis, I. D., Ploumis, N., Vlahakis, I., & Charissis, G. (2009). Acute scrotum-etiology, clinical presentation and seasonal variation. *Indian Journal of Pediatrics*, *76*(4), 407–410. doi:10.100712098-009-0008-2 PMID:19205631

Ma, W., Li, S., Ma, S., Jia, L., Zhang, F., Zhang, Y., Zhang, J., Wong, G., Zhang, S., Lu, X., Liu, M., Yan, J., Li, W., Qin, C., Han, D., Qin, C., Wang, N., Li, X., & Gao, G. F. (2016, December 1). Zika Virus Causes Testis Damage and Leads to Male Infertility in Mice. *Cell*, *167*(6), 1511–1524. doi:10.1016/j.cell.2016.11.016 PMID:27884405

Mabogunje, O. A. (1986). Testicular torsion and low relative humidity in a tropical country. *British Medical Journal*, *292*(6517), 363–364. doi:10.1136/bmj.292.6517.363 PMID:3080171

Makhseed, M., Musini, V. M., Ahmed, M. A., & Monem, R. A. (1999). Influence of seasonal variation on pregnancy-induced hypertension and/or preeclampsia. *Australian and New Zealand Journal of Obstetrics and Gynaecology*, *39*(2), 196–199. doi:10.1111/j.1479-828X.1999.tb03372.x PMID:10755779

Marai, I. F. M., Habeeb, A. A. M., & Gad, A. E. (2002). Rabbits' productive, reproductive and physiological performance traits as affected by heat stress: A review. *Livestock Science*, *78*(2), 71–90. doi:10.1016/S0301-6226(02)00091-X

Mathew, S., Mathur, D., Chang, A. B., McDonald, E., Singh, G. R., Nur, D., & Gerritsen, R. (2017). Examining the Effects of Ambient Temperature on Pre-Term Birth in Central Australia. *International Journal of Environmental Research and Public Health*, *14*(2), 147. doi:10.3390/ijerph14020147 PMID:28165406

Matte, T. D., Lane, K., & Ito, K. (2016). Excess Mortality Attributable to Extreme Heat in New York City, 1997-2013. *Health Security*, *14*(2), 64–70. doi:10.1089/hs.2015.0059 PMID:27081885

Mboera, L. E. G., Mayala, B. K., Kweka, E. J., & Mazigo, H. D. (2011). Impact of climate change on human health and health systems in Tanzania: A review. *Tanzania Journal of Health Research*, *13*, 407–426.

Mbonye, A. K., Neema, S., & Magnussen, P. (2006). Preventing malaria in pregnancy: A study of perceptions and policy implications in Mukono district, Uganda. *Health Policy and Planning*, *21*(1), 17–26. doi:10.1093/heapol/czj002 PMID:16317032

McCreesh, N., & Booth, M. (2013, November). Challenges in predicting the effects of climate change on Schistosoma mansoni and Schistosoma haematobium transmission potential. *Trends in Parasitology*, *29*(11), 548–555. doi:10.1016/j.pt.2013.08.007 PMID:24064438

McMichael. (2013). Globalization, climate change, and human health. *The New England Journal of Medicine*, *368*(14), 1335–1343.

Meerburg, B. G., & Kijlstra, A. (2009, July). Changing climate-changing pathogens: Toxoplasma gondii in North-Western Europe. *Parasitology Research*, *105*(1), 17–24. doi:10.100700436-009-1447-4 PMID:19418068

Mir, F., Fontaine, E., Reyes-Gomez, E., Carlus, M., & Fontbonne, A. (2012). Subclinical leishmaniasis associated with infertility and chronic prostatitis in a dog. *The Journal of Small Animal Practice*, *53*(7), 419–422. doi:10.1111/j.1748-5827.2012.01224.x PMID:22690941

Molina, O., & Saldarriaga, V. (2017). The perils of climate change: In utero exposure to temperature variability and birth outcomes in the Andean region. *Economics and Human Biology*, *24*, 111–124. doi:10.1016/j.ehb.2016.11.009 PMID:27984771

Moretti, M. E., Bar-Oz, B., Fried, S., & Koren, G. (2005). Maternal hyperthermia and the risk for neural tube defects in offspring systematic review and meta-analysis. *Epidemiology (Cambridge, Mass.)*, *16*(2), 216–219. doi:10.1097/01.ede.0000152903.55579.15 PMID:15703536

Mutheneni, S. R., Morse, A. P., Caminade, C., & Upadhyayula, S. M. (2017). Dengue burden in India: Recent trends and importance of climatic parameters. *Emerging Microbes & Infections*, *6*(8), e70. doi:10.1038/emi.2017.57 PMID:28790459

Narayansingh, R. M., Senchyna, M., Vijayan, M. M., & Carlson, J. C. (2004). Expression of prostaglandin G/H synthase (PGHS) and heat shock protein-70 (HSP-70) in the corpus luteum (CL) of prostaglandin F2 alpha-treated immature superovulated rats. *Canadian Journal of Physiology and Pharmacology*, *82*(6), 363–371. doi:10.1139/y04-032 PMID:15381959

Ngoula, F., Lontio, F. A., Tchoffo, H., Manfo Tsague, F. P., Djeunang, R. M., Vemo, B. N., Moffo, F., & Djuissi Motchewo, N. (2020). Heat Induces Oxidative Stress: Reproductive Organ Weights and Serum Metabolite Profile, Testes Structure, and Function Impairment in Male Cavy (Cavia porcellus). *Frontiers in Veterinary Science*, *7*, 37. doi:10.3389/fvets.2020.00037 PMID:32175332

Nicolino, M., & Forest, M. G. (2001). La puberté. In C. Thibault & M. C. Levasseur (Eds.), *La reproduction chez les mammifères domestiques et chez l'homme* (pp. 655–679). INRA Editions.

Oliveria Melo, A. S., Malinger, G., Ximenes, R., Szejnfeld, P. O., Alver Sampaio, S., & ... . (2016). Zika virus intrauterine infection causes fetal brain abnormality and microcephaly: Tip of the iceberg? *Ultrasound in Obstetrics & Gynecology*, *47*(1), 6–7. doi:10.1002/uog.15831 PMID:26731034

Panter-Brick, C., Lotstein, D. S., & Ellison, P. T. (1993). Seasonality of reproductive function and weight loss in rural Nepali women. *Human Reproduction (Oxford, England)*, *8*(5), 684–690. doi:10.1093/oxfordjournals.humrep.a138120 PMID:8314959

Parker, S. E., Mai, C. T., Canfield, M. A., Rickard, R., Wang, Y., Meyer, R. E., Anderson, P., Mason, C. A., Collins, J. S., Kirby, R. S., & Correa, A. (2010). Updated national birth prevalence estimates for selected birth defects in the United States, 2004–2006. *Birth Defects Research. Part A, Clinical and Molecular Teratology*, *88*(12), 1008–1016. doi:10.1002/bdra.20735 PMID:20878909

Pedersen, U. B., Stendel, M., Midzi, N., Mduluza, T., Soko, W., Stensgaard, A. S., Vennervald, B. J., Mukaratirwa, S., & Kristensen, T. K. (2014). Modelling climate change impact on the spatial distribution of fresh water snails hosting trematodes in Zimbabwe. *Parasites & Vectors*, *7*(1), 536. doi:10.118613071-014-0536-0 PMID:25498001

Pei, L., Zhu, H., Ye, R., Wu, J., Liu, J., Ren, A., Li, Z., & Zheng, X. (2015). Interaction between the SLC19A1 gene and maternal first trimester fever on offspring neural tube defects. *Birth Defects Research. Part A, Clinical and Molecular Teratology*, *103*(1), 3–11. doi:10.1002/bdra.23257 PMID:24917213

Petersen, L. R., Jamieson, D. J., Powers, A. M., & Honein, M. A. (2016). Zika virus. *The New England Journal of Medicine*, *374*(16), 1552–1563. doi:10.1056/NEJMra1602113 PMID:27028561

Pouliot, S. H., Xiong, X., Harville, E., Paz-Soldan, V., Tomashek, K. M., Breart, G., & Buekens, P. (2010). Maternal dengue and pregnancy outcomes: A systematic review. *Obstetrical & Gynecological Survey*, *65*(2), 107–118. PMID:20100360

Poursafa, P., Keikha, M., & Kelishadi, R. (2015). Systematic review on adverse birth outcomes of climate change. *Journal of Research in Medical Sciences*, *20*, 397–402. PMID:26109998

Putney, D. J., Mullins, S., Thatcher, W. W., Drost, M., & Gross, T. S. (1989). Embryonic development in superovulated dairy cattle exposed to elevated ambient temperatures between the onset of estrus and insemination. *Animal Reproduction Science*, *19*(1-2), 37–51. doi:10.1016/0378-4320(89)90045-6

Reiling, S. J., & Dixon, B. R. (2019). Toxoplasma gondii: How an Amazonian parasite became an Inuit health issue. *Canada Communicable Disease Report*, *45*(7-8), 183–190. doi:10.14745/ccdr.v45i78a03 PMID:31355827

Reiter P. (2001). Climate change and mosquito-borne disease. *Environ Health Perspect., 109*(Suppl 1), 141-61.

Rijken, M. J., McGready, R., Boel, M. E., Poespoprodjo, R., Singh, N., Syafruddin, D., Rogerson, S., & Nosten, F. (2012). Malaria in pregnancy in the Asia-Pacific region. *The Lancet. Infectious Diseases, 12*(1), 75–88. doi:10.1016/S1473-3099(11)70315-2 PMID:22192132

Rozenboim, I., Tako, E., Gal-Garber, O., Proudman, J. A., & Uni, Z. (2007). The effect of heat stress on ovarian function of laying hens. *Poultry Science, 86*(8), 1760–1765. doi:10.1093/ps/86.8.1760 PMID:17626822

Rylander, C., Odland, J. Ø., & Sandanger, T. M. (2013). Climate change and the potential effects on maternal and pregnancy outcomes: An assessment of the most vulnerable—the mother, fetus, and newborn child. *Global Health Action, 6*(1), 19538. doi:10.3402/gha.v6i0.19538 PMID:23481091

Sales, K., Vasudeva, R., Dickinson, M. E., Godwin, J. L., Lumley, A. J., Michalczyk, Ł., Hebberecht, L., Thomas, P., Franco, A., & Gage, M. (2018). Experimental heatwaves compromise sperm function and cause transgenerational damage in a model insect. *Nature Communications, 9*(1), 4771. doi:10.103841467-018-07273-z PMID:30425248

Salvetti, N. R., Baravalle, C., Mira, G. A., Gimeno, E. J., Dallard, B. E., Rey, F., & Ortega, H. H. (2009). Heat shock protein 70 and sex steroid receptors in the follicular structures of induced ovarian cysts. Reproduction in domestic animals. *Zuchthygiene, 44*(5), 805–814. doi:10.1111/j.1439-0531.2008.01086.x PMID:18992127

Sandström, C. A., Buma, A. G., Hoye, B. J., Prop, J., van der Jeugd, H., Voslamber, B., Madsen, J., & Loonen, M. J. (2013, May 1). Latitudinal variability in the seroprevalence of antibodies against Toxoplasma gondii in non-migrant and Arctic migratory geese. *Veterinary Parasitology, 194*(1), 9–15. doi:10.1016/j.vetpar.2012.12.027 PMID:23339847

Sarno, M., Sacramento, G. A., Khouri, R., do Rosário, M. S., Costa, F., Archanjo, G., Santos, L. A., Nery, N., Vasilakis, N., Ko, A. I., & de Almeida, A. R. P. (2016). Zika Virus Infection and Stillbirths: A Case of Hydrops Fetalis, Hydranencephaly and Fetal Demise. *PLoS Neglected Tropical Diseases, 10*(2), e0004517. doi:10.1371/journal.pntd.0004517 PMID:26914330

Schifano, P., Asta, F., Dadvand, P., Davoli, M., Basagana, X., & Michelozzi, P. (2016). Heat and air pollution exposure as triggers of delivery: A survival analysis of population-based pregnancy cohorts in Rome and Barcelona. *Environment International, 88*, 153–159. doi:10.1016/j.envint.2015.12.013 PMID:26760712

Schifano, P., Lallo, A., Asta, F., De Sario, M., Davoli, M., & Michelozzi, P. (2013). Effect of ambient temperature and air pollutants on the risk of preterm birth, Rome 2001-2010. *Environment International, 61*, 77–87. doi:10.1016/j.envint.2013.09.005 PMID:24103349

Setchell, B. P. (1998). The Parkes Lecture. Heat and the testis. *Journal of Reproduction and Fertility, 114*(2), 179–194. doi:10.1530/jrf.0.1140179 PMID:10070346

Shashar, S., Kloog, I., Erez, O., Shtein, A., Yitshak-Sade, M., Sarov, B., & Novack, L. (2020). Temperature and preeclampsia: Epidemiological evidence that perturbation in maternal heat homeostasis affects pregnancy outcome. *PLoS One*, *15*(5), e0232877. doi:10.1371/journal.pone.0232877 PMID:32421729

Sheffield, P. E., & Landrigan, P. J. (2011). Global climate change and children's health: Threats and strategies for prevention. National Institute of Environmental Health Sciences. *Environmental Health Perspectives*, *119*(3), 291–298. doi:10.1289/ehp.1002233 PMID:20947468

Shoghag, L., Hudson, S., Applewhite, L., Feldman, J., & Minkoff, H. L. (1997). An association between the heat-humidity index and preterm labor and delivery: A preliminary analysis. *American Journal of Public Health*, *87*(7), 1205–1207. doi:10.2105/AJPH.87.7.1205 PMID:9240114

Sirotkin, A. V. (2010). Effect of two types of stress (heat shock/high temperature and malnutrition/serum deprivation) on porcine ovarian cell functions and their response to hormones. *The Journal of Experimental Biology*, *213*(Pt 12), 2125–2130. doi:10.1242/jeb.040626 PMID:20511527

Sirotkin, A. V., Mlyncek, M., Kotwica, J., Makarevich, A. V., Florkovicová, I., & Hetényi, L. (2005). Leptin directly controls secretory activity of human ovarian granulosa cells: Possible inter-relationship with the IGF/IGFBP system. *Hormone Research*, *64*(4), 198–202. doi:10.1159/000089009 PMID:16244494

Slutsker, L., Heymann, D. L., & Breman, J. G. (1996). The effect of malaria and malaria prevention in pregnancy on offspring birthweight, prematurity, and intrauterine growth retardation in rural Malawi. *The American Journal of Tropical Medicine and Hygiene*, *55*(1, Suppl), 33–41. PMID:8702035

Solomon, S., Qin, D., Manning, M., Chen, Z., Marquis, M., Averyt, K. B., & Tignor, M. (2007). Summary for policymakers. Contribution of Working Group 1 to the Fourth Assessment Report of the Intergovernmental Panel on Climate Change. Intergovernmental Panel on Climate Change (IPCC). In *Climate Change 2007: The Physical Science Basis*. Cambridge University Press. https://www.ipcc.ch/site/assets/uploads/2018/02/ar4-wg1-spm-1.pdf

Sorensen, C., Saunik, S., Sehgal, M., Tewary, A., Govindan, M., Lemery, J., & Balbus, J. (2018, October 17). Climate Change and Women's Health: Impacts and Opportunities in India. *Geohealth.*, *2*(10), 283–297. doi:10.1029/2018GH000163 PMID:32159002

Srinivasan, A. K., Freyle, J., Gitlin, J. S., & Palmer, L. S. (2007). Climatic conditions and the risk of testicular torsion in adolescent males. *The Journal of Urology*, *178*(6), 2585–2588. doi:10.1016/j.juro.2007.08.049 PMID:17945301

Stensgaard, A. S., Vounatsou, P., Sengupta, M. E., & Utzinger, J. (2019, February). Schistosomes, snails and climate change: Current trends and future expectations. *Acta Tropica*, *190*, 257–268. doi:10.1016/j.actatropica.2018.09.013 PMID:30261186

Strand, L. B., Barnett, A. G., & Tong, S. (2012). Maternal exposure to ambient temperature and the risks of preterm birth and stillbirth in Brisbane, Australia. *American Journal of Epidemiology*, *175*(2), 99–107. doi:10.1093/aje/kwr404 PMID:22167749

Strand, L. B., Tong, S., & Barnett, A. G. (2011). The influence of season and ambient temperature on birth outcomes: A review of the epidemiological literature. *Environmental Research*, *111*(3), 451–462. doi:10.1016/j.envres.2011.01.023 PMID:21333980

Thonneau, P., Bujan, L., Multigner, L., & Mieusset, R. (1998, August). Occupational heat exposure and male fertility: A review. *Human Reproduction (Oxford, England)*, *13*(8), 2122–2125. doi:10.1093/humrep/13.8.2122 PMID:9756281

Tusell, L., David, L., Bodin, I., Legarra, O., Rafel, M., López, B., & Piles, M. (2014). Using the product threshold model for estimating separately the effect of temperature on male and female fertility. *Journal of Animal Science*, *89*(12), 3983–3995. doi:10.2527/jas.2011-3924 PMID:21764834

Van Zutphen, A. R., Lin, S., Fletcher, B. A., & Hwang, S.-A. (2012). A population-based case–control study of extreme summer temperature and birth defects. National Institute of Environmental Health Sciences. *Environmental Health Perspectives*, *120*(10), 1443–1449. doi:10.1289/ehp.1104671 PMID:23031822

Verma, S. K., Calero-Bernal, R., Cerqueira-Cézar, C. K., Kwok, O. C., Dudley, M., Jiang, T., Su, C., Hill, D., & Dubey, J. P. (2016, May). Toxoplasmosis in geese and detection of two new atypical Toxoplasma gondii strains from naturally infected Canada geese (Branta canadensis). *Parasitology Research*, *115*(5), 1767–1772. doi:10.100700436-016-4914-8 PMID:26796021

Vicedo-Cabrera, A. M., Iñíguez, C., Barona, C., & Ballester, F. (2014). Exposure to elevated temperatures and risk of preterm birth in Valencia, Spain. *Environmental Research*, *134*, 210–217. doi:10.1016/j.envres.2014.07.021 PMID:25173054

Vicedo-Cabrera, A. M., Olsson, D., & Forsberg, B. (2015). Exposure to seasonal temperatures during the last month of gestation and the risk of preterm birth in Stockholm. *International Journal of Environmental Research and Public Health*, *12*(4), 3962–3978. doi:10.3390/ijerph120403962 PMID:25867199

Wang, J., Williams, G., Guo, Y., Pan, X., & Tong, S. (2013). Maternal exposure to heatwave and preterm birth in Brisbane, Australia. *BJOG*, *120*(13), 1631–1641. doi:10.1111/1471-0528.12397 PMID:24034563

Wang, X. X., Ying, P., Diao, F., Wang, Q., Ye, D., Jiang, C., Shen, N., Xu, N., Chen, W. B., Lai, S. S., Jiang, S., Miao, X. L., Feng, J., Tao, W. W., Zhao, N. W., Yao, B., Xu, Z. P., Sun, H. X., Li, J. M., ... Li, C. J. (2013, July 29). Altered protein prenylation in Sertoli cells is associated with adult infertility resulting from childhood mumps infection. *The Journal of Experimental Medicine*, *210*(8), 1559–1574. doi:10.1084/jem.20121806 PMID:23825187

Watson, R. T. (2001). A Contribution of Working Groups I, II, and III to the Third Assessment Report of the Intergovernmental Panel on Climate Change. Intergovernmental Panel on Climate Change (IPCC). In *Climate Change 2001: Synthesis Report*. Cambridge University Press. https://www.ipcc.ch/site/assets/uploads/2018/05/SYR_TAR_full_report.pdf

Wolf, J., & Armstrong, B. (2012). The association of season and temperature with adverse pregnancy outcome in two German states, a time-series analysis. *PLoS One*, *7*(7), e40228. doi:10.1371/journal.pone.0040228 PMID:22792247

Woodall, P. A., & Kramer, M. R. (2018, April). Schistosomiasis and Infertility in East Africa. *The American Journal of Tropical Medicine and Hygiene*, *98*(4), 1137–1144. doi:10.4269/ajtmh.17-0280 PMID:29313478

World Health Organization. (n.d.). *Schistosomiasis*. https://www.who.int/mediacentre/factsheets/fs115/en/)

Wu, X., Lu, Y., Zhou, S., Chen, L., & Xu, B. (2016). Review article: impact of climate change on human infectious diseases: empirical evidence and human adaptation. *Environment International*, *86*, 14–23. doi:10.1016/j.envint.2015.09.007 PMID:26479830

Yang, Q., Yang, Z., Ding, H., Zhang, X., Dong, Z., Hu, W., Liu, X., Wang, M., Hu, G., & Fu, C. (2014). The relationship between meteorological factors and mumps incidence in Guangzhou, China, 2005-2012. *Human Vaccines & Immunotherapeutics*, *10*(8), 2421–2432. doi:10.4161/hv.29286 PMID:25424950

Yang, Y. T., & Sarfaty, M. (2016, August 12). Zika virus: A call to action for physicians in the era of climate change. *Preventive Medicine Reports*, *4*, 444–446. doi:10.1016/j.pmedr.2016.07.011 PMID:27617189

Yangli, P., Yingjie, W., Jie, C., & Yinghe, Q. (2012). Effects of chronic heat stress on the reproductive capacity of male Rex rabbits. *Livestock Science*, *146*(1), 13–21. doi:10.1016/j.livsci.2012.02.015

Yu, X., Feric, Z., Cordero, J. F., Meeker, J. D., & Alshawabkeh, A. (2018). Potential influence of temperature and precipitation on preterm birth rate in Puerto Rico. *Scientific Reports*, *8*(1), 16106. doi:10.103841598-018-34179-z PMID:30382121

Zha, W. T., Li, W. T., Zhou, N., Zhu, J. J., Feng, R., Li, T., Du, Y. B., Liu, Y., Hong, X. Q., & Lv, Y. (2020). Effects of meteorological factors on the incidence of mumps and models for prediction, China. *BMC Infectious Diseases*, *20*(1), 468. doi:10.118612879-020-05180-7 PMID:32615923

Zhang, W., Du, Z., Huang, S., Chen, L., Tang, W., Zheng, H., Yang, B., & Hao, Y. (2017, September 1). The association between human perceived heat and early-stage syphilis and its variance: Results from a case-report system. *The Science of the Total Environment*, *593-594*, 593–594. doi:10.1016/j.scitotenv.2017.03.194 PMID:28364611

Ziskin, M. C., & Morrissey, J. (2011). Thermal thresholds for teratogenicity, reproduction, and development. *International Journal of Hyperthermia*, *27*(4), 374–387. doi:10.3109/02656736.2011.553769 PMID:21591900

Compilation of References

Aagaard-Hansen, J., Nombela, N., & Alvar, J. (2010). Population movement: A key factor in the epidemiology of neglected tropical diseases. *Tropical Medicine & International Health*, *15*(11), 1281–1288. doi:10.1111/j.1365-3156.2010.02629.x PMID:20976871

Aarab, N., Minier, C., Lemaire, S., Unruh, E., Hansen, P. D., Larsen, B. K., Andersen, O.-K., & Narbonne, J.-F. (2004). Biochemical and histological responses in mussel (Mytilus edulis) exposed to North Sea oil and a mixture of North Sea oil and alkylphenols. *Marine Environmental Research*, *58*(2-5), 437–441. doi:10.1016/j.marenvres.2004.03.121

Abdelatty, A. M., Iwaniuk, M. E., Potts, S. B., & Gad, A. (2018). Influence of maternal nutrition and heat stress on bovine oocyte and embryo development. *International Journal of Veterinary Science and Medicine*, *6*(sup1), S1–S5. doi:10.1016/j.ijvsm.2018.01.005

Abdin, Z., & Khatoon, A. (2013). Heat stress in poultry and the beneficial effects of ascorbic acid (vitamin c) supplementation during periods of heat stress. *World's Poultry Science Journal*, *69*(1), 135–152. doi:10.1017/S0043933913000123

Abilay, T. A., Johnson, H. D., & Madam, L. M. (1975). Influence of environmental heath on peripheral plasma progesterone and cortisol during the bovine estrus cycle. *Journal of Dairy Science*, *58*(12), 1836–1840. doi:10.3168/jds.S0022-0302(75)84795-3 PMID:1239464

Abo El Gheit, R. E., Oun, A. M., & Al Arshal, A. (2020). The effect of maternal exposure to textile industry-induced pollution on pregnancy and its outcome. *International Journal of Medical Arts*, *2*, 162–172.

Abou-Okeil, A., El-Shafie, A., & El Zawahry, M. M. (2010). Eco-friendly laccase–hydrogen peroxide/ultrasound-assisted bleaching of linen fabrics and its influence on dyeing efficiency. *Ultrasonics Sonochemistry*, *17*(2), 383–390. doi:10.1016/j.ultsonch.2009.08.007 PMID:19748814

Abuajamieh, M., Kvidera, S. K., Mayorga, E. J., Kaiser, A., Lei, S., Seibert, J. T., Horst, E. A., Sanz Fernandez, M. V., Ross, J. W., Selsby, J. T., Keating, A. F., Rhoads, R. P., & Baumgard, L. H. (2018). The effect of recovery from heat stress on circulating bioenergetics and inflammatory biomarkers. *Journal of Animal Science*, *96*, 4599–4610. doi:10.1093/jasky345 PMID:30476152

Adekiya, T. A., Aruleba, R. T., Oyinloye, B. E., Okosun, K. O., & Kappo, A. P. (2019, December 26). The Effect of Climate Change and the Snail-Schistosome Cycle in Transmission and Bio-Control of Schistosomiasis in Sub-Saharan Africa. *International Journal of Environmental Research and Public Health*, *17*(1), 181. doi:10.3390/ijerph17010181 PMID:31887979

AE-96. (n.d.). Available from: https://www.extension.purdue.edu/extmedia/AE/AE-96.html

Agarwal, A., Saleh, R. A., & Bedaiwy, M. A. (2003). Role of reactive oxygen species in the pathophysiology of human reproduction. *Fertility and Sterility*, *79*(4), 829–843. doi:10.1016/S0015-0282(02)04948-8 PMID:12749418

Ahmad, N., Schrick, F. N., Butcher, R. L., & Inskeep, E. K. (1995). Effect of persistent follicles on early embryonic losses in beef cows. *Biology of Reproduction*, *52*(5), 1129–1135. doi:10.1095/biolreprod52.5.1129 PMID:7626713

Ahmed, A., Tiwari, R., Mishra, G., Jena, B., Dar, M., & Bhat, A. (2015). Effect of Environmental Heat Stress on Reproduction Performance of Dairy Cows- A Review. *International Journal of Livestock Research*, *5*(4), 10. doi:10.5455/ijlr.20150421122704

Aitken, R. J. (1999). The Amoroso Lecture. The human spermatozoon-a cell in crisis? *Journal of Reproduction and Fertility*, *115*(1), 1–7. doi:10.1530/jrf.0.1150001 PMID:10341716

Aitken, R. J., Clarkson, J. S., & Fishel, S. (1989). Generation of reactiveoxygen species, lipid peroxidation, and human spermfunction. *Biology of Reproduction*, *41*(1), 183–197. doi:10.1095/biolreprod41.1.183 PMID:2553141

Aitken, R. J., Gordon, E., Harkiss, D., Twigg, J. P., Milne, P., Jennings, Z., & Irvine, D. S. (1998). Relative impact of oxidative stress on the functional competence and genomic integrity of human spermatozoa. *Biology of Reproduction*, *59*(5), 1037–1046. doi:10.1095/biolreprod59.5.1037 PMID:9780307

Akarslan, E., & Demiralay, H. (2015). Effects of textile materials harmful to human health. *Acta Physica Polonica A*, *128*(2B), B-407–B-408. doi:10.12693/APhysPolA.128.B-407

Akingbemi, B. T., Ge, R., Klinefelter, G. R., Zirkin, B. R., & Hardy, M. P. (2004). Phthalate-induced Leydig cell hyperplasia is associated with multiple endocrine disturbances. *Proceedings of the National Academy of Sciences of the United States of America*, *101*(3), 775–780. doi:10.1073/pnas.0305977101

Al. Katanani, Y. M., Paula Lopes, F. F., & Hansen, P. J. (2002). Effect of season and exposure to heat stress on oocyte competence in Holstein cows. *Journal of Dairy Science*, *85*(2), 390–396. doi:10.3168/jds.S0022-0302(02)74086-1 PMID:11913699

Alavi, S. M. H., & Cosson, J. (2005). Sperm motility in fishes. I. Effects of temperature and pH: A review. *Cell Biology International*, *29*(2), 101–110. doi:10.1016/j.cellbi.2004.11.021 PMID:15774306

Alavi, S. M. H., Cosson, J., Karami, M., & Abdoulhay, H. (2004). Chemical composition and osmolality of seminal plasma ofAcipenserpersicus, their physiological relationship with spermmotility. *Aquaculture Research*, *35*(13), 1238–1243. doi:10.1111/j.1365-2109.2004.01132.x

Alexander, G., Hales, J. R., Stevens, D., & Donnelly, J. B. (1987). Effects of acute and prolonged exposure to heat on regional blood flows in pregnant sheep. *Journal of Developmental Physiology*, *9*, 1–15. PMID:3559062

Ali, H. M., Qater, A. N. A., & Selman, M. O. (2020). Relation of Some Heavy Metals with Male Infertility. *Medico-Legal Update*, *20*(1), 1440–1445.

Allen, J., Anderson, S., Collier, R., & Smith, J. (2013). *Managing heat stress and its impact on cow behavior.* Paper presented at the 28th Annual Southwest Nutrition and Management Conference.

Alonso-Fauste, I., Andres, M., Iturralde, M., Lampreave, F., Gallart, J., & Alava, M. A. (2012). Proteomic characterization by 2-DE in bovine serum and whey from healthy and mastitis affected farm animals. *Journal of Proteomics*, *75*(10), 3015–3030. doi:10.1016/j.jprot.2011.11.035

Alvarez, J. D., Hansen, A., Ord, T., Bebas, P., Chappell, P. E., Giebultowicz, J. M., Williams, C., Moss, S., & Sehgal, A. (2008). The circadian clock protein BMAL1 is necessary for fertility and proper testosterone production in mice. *Journal of Biological Rhythms*, *23*(1), 26–36. doi:10.1177/0748730407311254 PMID:18258755

Alvarez, J. G., Touchstone, J. C., Blasco, L., & Storey, B. (1987). Spontaneous lipid peroxidation and production of hydrogen peroxide and superoxide in human spermatozoa: Superoxide dismutase as a major enzyme protectant against oxygen toxicity. *Journal of Andrology*, *8*(5), 338–348. doi:10.1002/j.1939-4640.1987.tb00973.x PMID:2822642

Alves, J. R. A., Andrade, T. A. A., Assis, D. M., Gurjao, T. A., Melo, L. R. B., & Souza, B. B. (2017). Productive and reproductive performance, behavior and physiology of cattle under heat stress conditions. *Journal of Animal Behaviour and Biometeorology*, *5*(3), 91–96. doi:10.31893/2318-1265jabb.v5n3p91-96

Alves, M. B., Andrade, A. F., Arruda, R. P., Batissaco, L., Florez-Rodriguez, S. A., Oliveira, B. M., Torres, M. A., Lanconi, R., Ravagnani, G. M., Prado Filho, R. R., Vellone, V. S., Losano, J. D., Franci, C. R., Nichi, M., & Celeghini, E. C. (2016). Recovery of normal testicular temperature after scrotal heat stress in ramsassessed by infrared thermography and its effects on seminal characteristics and testosterone bloodserum concentration. *Theriogenology*, *86*(3), 795–805. e792. doi:10.1016/j.theriogenology.2016.02.034 PMID:27045627

Amaral, F. G., Castrucci, A. M., Cipolla-Neto, J., Poletini, M. O., Mendez, N., Richter, H. G., & Sellix, M. T. (2014). Environmental control of biological rhythms: Effects on development, fertility and metabolism. *Journal of Neuroendocrinology*, *26*(9), 603–612. doi:10.1111/jne.12144 PMID:24617798

Amegah, A. K., Rezza, G., & Jaakkola, J. J. K. (2016). Review article: temperature-related morbidity and mortality in sub-Saharan Africa: a systematic review of the empirical evidence. *Environment International*, *91*, 133–149. doi:10.1016/j.envint.2016.02.027 PMID:26949867

Amelar, R. D., Dubin, L., & Schoenfeld, C. (1980). Sperm motility. *Fertility and Sterility*, *134*, 197–215. PMID:6250914

Amin, T., Afrin, M., Haque, Z., & Islam, R. (2016). Effects of textile dye waste water on reproductive system of mice and their progeny. *International Journal of Scientific and Engineering Research*, *7*, 565–570.

Amireault, P., & Dubé, F. (2005). Serotonin and Its Antidepressant-Sensitive Transport in Mouse Cumulus-Oocyte Complexes and Early Embryos1. *Biology of Reproduction*, *73*(2), 358–365. doi:10.1095/biolreprod.104.039313 PMID:15858217

Amundson, J. L., Mader, T. L., Rasby, R. J., & Hu, Q. S. (2006). Environmental effects on pregnancy rate in beef cattle. *Journal of Animal Science*, *84*(12), 3415–3420. doi:10.2527/jas.2005-611

Andrady, A. L. (2011). Microplastics in the marine environment. *Marine Pollution Bulletin*, *62*(8), 1596–1605. doi:10.1016/j.marpolbul.2011.05.030

Andrady, A. L., & Neal, M. A. (2009). Applications and societal benefits of plastics. *Philosophical Transactions of the Royal Society of London. Series B, Biological Sciences*, *364*(1526), 1977–1984. doi:10.1098/rstb.2008.0304

Anisimov, V. N. A., Vinogradova, I. V., Panchenko, A. G., Popovich, I. A., & Zabezhinski, M. (2013). Light-at-Night-Induced Circadian Disruption, Cancer and Aging. *Current Aging Science*, *5*, 170–177. doi:10.2174/1874609811205030002 PMID:23237593

Anisimov, V. N., Popovich, I. G., & Zabezhinski, M. A. (1997). Melatonin and colon carcinogenesis: I. Inhibitory effect of melatonin on development of intestinal tumors induced by 1,2-dimethylhydrazine in rats. *Carcinogenesis*, *18*(8), 1549–1553. doi:10.1093/carcin/18.8.1549 PMID:9276629

Anjali, M. C., & Dhananjaya, B. C. (2019). Effect of Climate Change on Soil Chemical and Biological Properties-A Review. *International Journal of Current Microbiology and Applied Sciences*, *8*(2), 1502–1512. doi:10.20546/ijcmas.2019.802.174

Anjaneyulu, Y., Chary, N. S., & Raj, D. S. S. (2005). Decolorization of industrial effluents - available methods and emerging technologies: A review. *Reviews in Environmental Science and Biotechnology*, *4*(4), 245–273. doi:10.100711157-005-1246-z

Aplin, J. D., Spansewick, C., Behzad, F., Kimber, S. J., & Vićovac, L. (1996). Molecular aspects of implantation: Integrins β5, β3 and αv are apically distributed in endometrial epithelium. *MHR: Basic Science of Reproductive Medicine*, *2*(7), 527–534. doi:10.1093/molehr/2.7.527 PMID:9239663

Appell, R. A., Evans, P. R., & Blandy, J. P. (1977). The effect oftemperature on the motility and viability of sperm. *British Journal of Urology*, *49*(7), 751–756. doi:10.1111/j.1464-410X.1977.tb04566.x PMID:23196

Arbuckle, T.E., Savitz, D.A., Mery, L.S., & Curtis, K.M. (1999). Exposure to phenoxyherbicides and the risk of spontaneous abortion. *Epidemiology, 10*, 752–760.

Arbuckle, T. E., Lin, Z., & Mery, L. S. (2001). An exploratory analysis of the effect of pesticide exposure on the risk of spontaneous abortion in an Ontario farm population. *Environmental Health Perspectives*, *109*(8), 851–857. doi:10.1289/ehp.01109851 PMID:11564623

Arbuckle, T. E., & Sever, L. E. (1998). Pesticide exposures and fetal death: A review of the epidemiologic literature. *Critical Reviews in Toxicology*, *28*(3), 229–270. doi:10.1080/10408449891344218 PMID:9631282

Arechiga, C. F., & Hansen, P. J. (1998). Response of preimplantation murine embryos to heat shock as modified by developmental stage and glutathione status. *In Vitro Cellular & Developmental Biology. Animal*, *34*(8), 655–659. doi:10.100711626-996-0016-8 PMID:9769152

Arévalo-Gardini, E., Canto, M., Alegre, J., Loli, O., Julca, A., & Baligar, V. (2015). Changes in Soil Physical and Chemical Properties in Long Term Improved Natural and Traditional Agroforestry Management Systems of Cacao Genotypes in Peruvian Amazon. *PLoS One*, *10*(7), 1–18. doi:10.1371/journal.pone.0132147 PMID:26181053

Arici, A., Head, J. R., MacDonald, P. C., & Casey, M. L. (1993). Regulation of interleukin-8 gene expression in human endometrial cells in culture. *Molecular and Cellular Endocrinology*, *94*(2), 195–204. doi:10.1016/0303-7207(93)90168-J PMID:8224523

Arici, A., MacDonald, P. C., & Casey, M. L. (1995). Regulation of monocyte chemotactic protein-1 gene expression in human endometrial cells in cultures. *Molecular and Cellular Endocrinology*, *107*(2), 189–197. doi:10.1016/0303-7207(94)03442-V PMID:7768330

Arroyo, V., Díaz, J., Ortiz, C., Carmona, R., Sáez, M., & Linares, C. (2016). Short term effect of air pollution, noise and heat waves on preterm births in Madrid (Spain). *Environmental Research*, *145*, 162–168. doi:10.1016/j.envres.2015.11.034 PMID:26706568

Arslan, I., Balcioglu, A., & Tuhkanen, T. (1999). Oxidative treatment of simulated dye-house effluent by UV and near-UV light assisted fenton's reagent. *Chemosphere*, *39*(15), 2767–2783. doi:10.1016/S0045-6535(99)00211-8

Asabere, S. B., Zeppenfeld, T., Nketia, K. A., & Sauer, D. (2018). Urbanization Leads to Increases in pH, Carbonate, and Soil Organic Matter Stocks of Arable Soils of Kumasi, Ghana (West Africa). *Frontiers in Environmental Science*, *6*, 119–136. doi:10.3389/fenvs.2018.00119

Asamoah, B., Kjellstrom, T., & Östergren, P. (2018). Is ambient heat exposure levels associated with miscarriage or stillbirths in hot regions? A cross-sectional study using survey data from the Ghana Maternal Health Survey 2007. *International Journal of Biometeorology*, *62*(3), 319–330. doi:10.100700484-017-1402-5 PMID:28748383

Ashraf, S., Shah, S. M., Saini, N., Dhanda, S., Kumar, A., Goud, T. S., & Upadhyay, R. (2014). Developmental competence and expression pattern of bubaline (Bubalusbubalis) oocytes subjected to elevated temperatures during meiotic maturation in vitro. *Journal of Assisted Reproduction and Genetics*, *31*(10), 1349–1360. doi:10.100710815-014-0275-3

Aspholm, R., Lindbohm, M. L., Paakkulainen, H., Taskinen, H., Nurminen, T., & Tiitinen, A. (1999). Spontaneous abortions among Finnish flight attendants. *Journal of Occupational and Environmental Medicine*, *41*(6), 486–491. doi:10.1097/00043764-199906000-00015 PMID:10390700

Assis, V. P., Ribeiro, V. M., Rachid, M. A., Castro, A. C. S., & Valle, G. R. (2010). Dogs with Leishmania chagasi infection have semen abnormalities that partially revert during 150 days of allopurinol and amphotericin B therapy. *Animal Reproduction Science*, *117*(1-2), 183–186. doi:10.1016/j.anireprosci.2009.03.003 PMID:19345025

Auger, N., Naimi, A. I., Smargiassi, A., Lo, E., & Kosatsky, T. (2014). Extreme heat and risk of early delivery among preterm and term pregnancies. *Epidemiology (Cambridge, Mass.)*, *25*(3), 344–350. doi:10.1097/EDE.0000000000000074 PMID:24595396

Augugliaro, V., Prevol, A. B., Loddo, V., Marci, G., Palmisano, L., Pramauro, E., & ... , (2000). Photodegradation kinetics of aniline, 4-ethylaniline, and 4-chloroaniline in aqueous suspension of polycrystalline titanium dioxide. *Research on Chemical Intermediates*, *26*(5), 413–426. doi:10.1163/156856700X00426

Awal, R., Safeeq, M., Abbas, F., Fares, S., Deb, S. K., Ahmad, A., & Fares, A. (2019). Soil Physical Properties Spatial Variability under Long-Term No-Tillage Corn. *Agronomy (Basel)*, *9*(11), 750–767. doi:10.3390/agronomy9110750

Ayadi, I., Souissi, Y., Jlassi, I., Peixoto, F., & Mnif, W. (2016). Chemical Synonyms, molecular structure and toxicological risk assessment of synthetic textile dyes: A critical review. *Journal of Developing Drugs*, *5*, 151.

Ayrout, M., Simon, V., Bernard, V., Binart, N., Cohen-Tannoudji, J., Lombès, M., & Chauvin, S. (2017). A novel non genomic glucocorticoid signaling mediated by a membrane palmitoylated glucocorticoid receptor cross talks with GnRH in gonadotrope cells. *Scientific Reports*, *7*(1), 1–13. doi:10.103841598-017-01777-2 PMID:28484221

Baban, A., Yediler, A., & Ciliz, N. K. (2010). Integrated water management and CP implementation for wool and textile blend processes. *CLEAN Soil Air Water*, *38*(1), 84–90. doi:10.1002/clen.200900102

Babazadeh, T., Nadrian, H., Mosaferi, M., & Allahverdipour, H. (2018). Identifying challenges and barriers to participating in the source separation of waste program in Tabriz, northwest of Iran: A qualitative study from the citizens' perspective. *Resources*, *7*(3), 53. doi:10.3390/resources7030053

Babel, S., & Kurniawan, T. A. (2003). Low-cost adsorbents for heavy metals uptake from contaminated water: A review. *Journal of Hazardous Materials*, *97*(1-3), 219–243. doi:10.1016/S0304-3894(02)00263-7 PMID:12573840

Badroo, I. A., Khanday, A. H., Ali, Z. K., & Badroo, S. A. (2020). Environmental Phthalate Exposure in Relation to Reproductive Outcomes and Other Health Endpoints in Humans. In *Handbook of Research on Environmental and Human Health Impacts of Plastic Pollution* (pp. 288–313). IGI Global. doi:10.4018/978-1-5225-9452-9.ch015

Bafana, A., Devi, S. S., & Chakrabarti, T. (2011). Azo dyes: Past, present and the future. *Environmental Reviews*, *19*(NA), 350–370. doi:10.1139/a11-018

Baird, C., & Cann, M. (2012). Environmental chemistry. 5th ed. Toxic heavy metals. W. H. Freeman and Company.

Balboula, A. Z., Yamanaka, K., Sakatani, M., Kawahara, M., Hegab, A. O., Zaabel, S. M., & Takahashi, M. (2013). Cathepsin B activity has a crucial role in the developmental competence of bovine COCs exposed to heat-shock during in vitro maturation. *Reproduction (Cambridge, England)*, *146*, 407–417. doi:10.1530/REP-13-0179 PMID:23898216

Bamberger, A. M., Minas, V., Kalantaridou, S. N., Radde, J., Sadeghian, H., Löning, T., & Schulte, H. M. (2006). Corticotropin-releasing hormone modulates human trophoblast invasion through carcinoembryonic antigen-related cell adhesion molecule-1 regulation. *American Journal of Pathology*, *168*(1), 141–150. doi:10.2353/ajpath.2006.050167 PMID:16400017

Banks, S., King, S. A., Irvine, D. S., & Saunders, P. T. (2005). Impact of a mild scrotal heat stress on DNAintegrity in murine spermatozoa. *Reproduction (Cambridge, England)*, *129*(4), 505–514. doi:10.1530/rep.1.00531 PMID:15798026

Bantje, H. F. W. (1988). Female stress and birth seasonality in tanzania. *Journal of Biosocial Science*, *20*(2), 195–202. doi:10.1017/S0021932000017429 PMID:3384835

Bapayeva, G., Issayeva, R., Zhumadilova, A., Nurkasimova, R., Kulbayeva, S., & Tleuzhan, R. (2016). Organochlorine pesticides and female puberty in South Kazakhstan. *Reproductive Toxicology (Elmsford, N.Y.)*, *65*, 67–75. doi:10.1016/j.reprotox.2016.06.017

Barani, H., & Montazer, M. (2008). A review on applications of liposomes in textile processing. *Journal of Liposome Research*, *18*(3), 249–262. doi:10.1080/08982100802354665 PMID:18770074

Baranski, B. (1993). Effects of the workplace on fertility and related reproductiveoutcomes. *Environmental Health Perspectives*, *101*(Suppl 2), 81–90. PMID:8243410

Barathi, S., Karthik, C., Nadanasabapathi, S., & Padikasan, I. A. (2020). Biodegradation of textile dye Reactive Blue 160 by *Bacillus firmus* (Bacillaceae: Bacillales) and non-target toxicity screening of their degraded products. *Toxicology Reports*, *7*, 16–22. doi:10.1016/j.toxrep.2019.11.017 PMID:31871898

Barbieri, R. L. (2014). The endocrinology of the menstrual cycle. In *Human Fertility* (pp. 145–169). Humana Press. doi:10.1007/978-1-4939-0659-8_7

Barclay, J. L., Shostak, A., Leliavski, A., Tsang, A. H., Jöhren, O., Müller-Fielitz, H., Landgraf, D., Naujokat, N., van der Horst, G. T. J., & Oster, H. (2013). High-fat diet induced hyperinsulinemia and tissue-specific insulin resistance in Cry-deficient mice. *American Journal of Physiology. Endocrinology and Metabolism*, *304*(10), E1053–E1063. doi:10.1152/ajpendo.00512.2012 PMID:23531614

Barker, D. J. (2007). The origins of the developmental origins theory. *Journal of Internal Medicine*, *261*(5), 412–417. doi:10.1111/j.1365-2796.2007.01809.x PMID:17444880

Barnes, J. W., Tischkau, S. A., Barnes, J. A., Mitchell, J. W., Burgoon, P. W., Hickok, J. R., & Gillette, M. U. (2003). Requirement of mammalian timeless for circadian rhythmicity. *Science*, *302*(5644), 439–442. doi:10.1126cience.1086593 PMID:14564007

Barreca, A., Deschenes, O., & Guldi, M. (2018). Maybe Next Month? Temperature Shocks and Dynamic Adjustments in Birth Rates. *Demography*, *55*(4), 1269–1293. doi:10.100713524-018-0690-7 PMID:29968058

Bartocci, A., Pollard, J. W., & Stanley, E. R. (1986). Regulation of colony-stimulating factor 1 during pregnancy. *The Journal of Experimental Medicine*, *164*(3), 956–961. doi:10.1084/jem.164.3.956 PMID:3489064

Basu, R., Malig, B., & Ostro, B. (2010). High ambient temperature and the risk of preterm delivery. *American Journal of Epidemiology*, *172*(10), 1108–1117. doi:10.1093/aje/kwq170 PMID:20889619

Basu, R., Sarovar, V., & Malig, B. J. (2016). Association Between High Ambient Temperature and Risk of Stillbirth in California. *American Journal of Epidemiology*, *183*(10), 894–901. doi:10.1093/aje/kwv295 PMID:27037268

Baumgard, L., Rhoads, R., Rhoads, M., Gabler, N., Ross, J., & Keating, A. (2012). Impact of Climate Change on Livestock Production. *Environmental Stress and Amelioration in Livestock Production*, 413-468.

Baumgard, L. H., & Rhoads, R. P. Jr. (2013). Effects of heat stress on postabsorptive metabolism and energetics. *Annual Review of Animal Biosciences*, *1*(1), 311–337. doi:10.1146/annurev-animal-031412-103644 PMID:25387022

Beam, S. W., & Butler, W. (1997). Energy balance and ovarian follicle development prior to the first ovulation postpartum in dairy cows receiving three levels of dietary fat. *Biology of Reproduction*, *56*(1), 133–142. doi:10.1095/biolreprod56.1.133

Bedaiwy, M. A., Falcone, T., Goldberg, J., Attaran, M., Nelson, D., & Agarwal, A. (2002). Prediction of endometriosis with serum and peritoneal fluid markers: A prospective controlled trial. *Fertility and Sterility*, *17*, 426–431. PMID:11821289

Bedford, J. M. (1983). Significance of the need for sperm capacitation before fertilization in Eutherian mammals. *Biology of Reproduction*, *28*(1), 108–120. doi:10.1095/biolreprod28.1.108 PMID:6338941

Bedford, J. M. (1991). Effects of elevated temperature on the epididymis and testis: Experimental studies. *Advances in Experimental Medicine and Biology*, *286*, 19–32. doi:10.1007/978-1-4684-5913-5_3 PMID:2042501

Bedrosian, T. A., Vaughn, C. A., Galan, A., Daye, G., Weil, Z. M., & Nelson, R. J. (2013). Nocturnal light exposure impairs affective responses in a wavelength-dependent manner. *The Journal of Neuroscience: The Official Journal of the Society for Neuroscience*, *33*(32), 13081–13087. doi:10.1523/JNEUROSCI.5734-12.2013 PMID:23926261

Beere, H. M. (2004). 'The stress of dying': The role of heat shock proteins in the regulation of apoptosis. *Journal of Cell Science*, *117*(13), 2641–2651. doi:10.1242/jcs.01284 PMID:15169835

Beesley, S., Lee, J., & Olcese, J. (2015). Circadian clock regulation of melatonin MTNR1B receptor expression in human myometrial smooth muscle cells. *Molecular Human Reproduction*, *21*(8), 662–671. doi:10.1093/molehr/gav023 PMID:25939854

Behram Kandemir, Y., Aydin, C., & Gorgisen, G. (2017). The effects of melatonin on oxidative stressand prevention of primordial follicle loss via activation of mTOR pathway in the rat ovary. *Cellular and Molecular Biology*, *63*(2), 100–106. doi:10.14715/cmb/2017.63.2.16 PMID:28364790

Behrman, H. R., & Armstrong, D. T. (1969). Cholesterol esterase stimulation by luteinizing hormone in luteinized rat ovaries. *Endocrinology*, *85*(3), 474–480. doi:10.1210/endo-85-3-474

Bekkar, B., Pacheco, S., Basu, R., & DeNicola, N. (2020). Association of Air Pollution and Heat Exposure With Preterm Birth, Low Birth Weight, and Stillbirth in the US: A Systematic Review. *JAMA Netw Open.*, *3*(6).

Belhadj-Silmen, I., Najar, T., Ghram, A., & Abdrrabba, M. (2015). Heat stress effects on livestock: Molecular, cellular and metabolic aspects, a review. *Journal of Animal Physiology and Animal Nutrition*, *100*(3), 401–412. doi:10.1111/jpn.12379 PMID:26250521

Bell, A. W. (1987). Consequences of severe heat stress for fetal development. In J. R. S. Hales & D. A. B. Richards (Eds.), *Heat Stress, Physical Exertion and Environment* (pp. 313–333). Excerpta Medica.

Bell, C. E., Calder, M. D., & Watson, A. J. (2008). Genomic RNA profiling and the programme controlling preimplantation mammalian development. *Molecular Human Reproduction*, *14*(12), 691–701. doi:10.1093/molehr/gan063 PMID:19043080

Belova, N. V. (1981). Ecological and physiological characteristics of sperm ofpond Cyprinid fishes. [In Russian]. *VoprIchtiol*, *21*, 525–536.

Beltran, A. J., Wu, J., & Laurent, O. (2013). Associations of meteorology with adverse pregnancy outcomes: A systematic review of preeclampsia, preterm birth and birth weight. *International Journal of Environmental Research and Public Health*, *11*(1), 91–172. doi:10.3390/ijerph110100091 PMID:24362545

Benigni, R., & Passerini, L. (2002). Carcinogenicity of the aromatic amines from structure activity relationships to mechanisms of action and risk assessment. *Mutation Research*, *511*(3), 191–206. doi:10.1016/S1383-5742(02)00008-X PMID:12088717

Benmrad, M., & Stevenson, J. (1986). Gonadotropin-Releasing Hormone and Prostaglandin F2α for Postpartum Dairy Cows: Estrous, Ovulation, and Fertility Traits. *Journal of Dairy Science*, *69*(3), 800–811. doi:10.3168/jds.S0022-0302(86)80469-6

Benoff, S., Jacob, A., & Hurley, I. R. (2000). Male infertility and environmental exposure to lead and cadmium. *Human Reproduction Update*, *6*(2), 107–121. doi:10.1093/humupd/6.2.107 PMID:10782569

Benyei, B., Gaspard, A., & Barros, C. W. C. (2001). Changes in embryo production results and ovarian recrudescence during the acclimation to the semiarid tropics of embryo donor Holstein–Frisian cows raised in a temperate climate. *Animal Reproduction Science*, *68*(1-2), 57–68. doi:10.1016/S0378-4320(01)00144-0

Berg, D., Van Leeuwen, J., Beaumont, S., Berg, M., & Pfeffer, P. (2010). Embryo loss in cattle between Days 7 and 16 of pregnancy. *Theriogenology*, *73*(2), 250–260. doi:10.1016/j.theriogenology.2009.09.005

Berisha, B., & Schams, D. (2005). Ovarian function in ruminants. *Domestic Animal Endocrinology*, *29*(2), 305–317. doi:10.1016/j.domaniend.2005.02.035 PMID:15998502

Berman, A., Folman, Y., Kaim, M., Mamen, M., Herz, Z., Wolfenson, D., Arieli, A., & Graber, Y. (1985). Upper critical temperatures and forced ventilation effects for high yielding dairy cows in a subtropical climate. *Journal of Dairy Science*, *68*(6), 1488–1495. doi:10.3168/jds.S0022-0302(85)80987-5 PMID:4019887

Berman, A., & Horovitz, T. (2012). Radiant heat loss, an unexploited path for heat stress reduction in shaded cattle. *Journal of Dairy Science*, *95*(6), 3021–3031. doi:10.3168/jds.2011-4844

Bernabucci, U. (2019). Climate change: Impact on livestock and how we can adopt. *Animal Frontiers*, *9*(1), 3–5. doi:10.1093/af/vfy039 PMID:32002232

Berner, E. K. & Berner, R. A. (2012). Global environment: water, air, and geochemical cycles. Princeton University Press.

Bethea, C. L., Centeno, M. L., & Cameron, J. L. (2008). Neurobiology of stress-induced reproductive dysfunction in female macaques. *Molecular Neurobiology*, *38*(3), 199–230. doi:10.100712035-008-8042-z

Betteridge, K. (1986). Increasing productivity in farm animals. *Reproduction in Mammals, 5*, 1-47.

Bhakat, M., Mohanty, T. K., & Gupta, A. K. (2014). Effect of season on semen quality of crossbred (Karan Fries) bulls. *Advances in Animal and Veterinary Science*, *2*(11), 632–637. doi:10.14737/journal.aavs/2014/2.11.632.637

Bhat, P. N., Kumar, R., & Raheja, K. L. (1983). Breeding behavior of murrah buffalo. *Proceedings of the Fourth International Congress of SABRAO*, 91-97.

Bhatti, P., Cushing-Haugen, K. L., Wicklund, K. G., Doherty, J. A., & Rossing, M. A. (2013). Nightshift work and risk of ovarian cancer. *Occupational and Environmental Medicine*, *70*(4), 231–237. doi:10.1136/oemed-2012-101146 PMID:23343856

Bianca, W. (1964). Animal Housing in Hot Countries. *International Journal of Biometeorology*, *8*(2), 99–101. doi 10.1007/BF02187573

Bigelow, P. L., Jarrell, J., Young, M. R., Keefe, T. J., & Love, E. J. (1998). Associationof semen quality and occupational factors: Comparison of casecontrolanalysis and analysis of continuous variables. *Fertility and Sterility*, *69*(1), 11–18. doi:10.1016/S0015-0282(97)00437-8 PMID:9457925

Biggers, B., Geisert, R., Wetteman, R., & Buchanan, D. (1987). Effect of Heat Stress on Early Embryonic Development in the Beef Cow. *Journal of Animal Science*, *64*(5), 1512–1518. doi:10.2527/jas1987.6451512x

Billard, R., & Cosson, M. P. (1988). Sperm motility in Rainbow trout, Parasalmogairdneri; effects of pH and temperature. In Reproduction in fish basic and applied aspects in endocrinology and genetics (pp. 161-167). Paris, France: Academic Press.

Billard, R., Cosson, J., Crim, L. W., & Suquet, M. (1995a). Spermphysiology and quality. In Brood stockmanagement andegg and larval quality (pp. 25-52). Blackwell Science.

Billard, R., Cosson, M. P., & Christen, R. (1987). Some recent data on the biology on trout spermatozoa. In *Proceedings of the 3rd international symposium on reproductive physiology of fish* (pp. 187-190). St John's, Newfoundland, Canada: Academic Press.

Billard, R. (1978). Changes in structure and fertilizing ability of marine andfreshwater fish spermatozoa diluted in media of various salinities. *Aquaculture (Amsterdam, Netherlands)*, *14*(3), 187–198. doi:10.1016/0044-8486(78)90094-7

Billard, R. (1986). Spermatogenesis and spermatology of some teleost fishspecies. *Reproduction, Nutrition, Developpement*, *2*(4), 877–920. doi:10.1051/rnd:19860601

Billard, R. (2000). Biology and control of reproduction of sturgeon in fish farm. *Iranian Journal of Fisheries Science*, *2*, 1–20.

Billard, R., Cosson, J., Perchec, G., & Linhart, O. (1995b). Biology of sperm and artificial reproduction in carp. *Aquaculture (Amsterdam, Netherlands)*, *124*(1-4), 95–112. doi:10.1016/0044-8486(94)00231-C

Billard, R., & Cosson, M. P. (1989). Measurement of sperm motility in trout and carp. In N. De Pauw, E. Jaspers, H. Ackefors, & N. Wilkins (Eds.), *Aquaculture, a biotechnology in progress* (pp. 499–503). Bredene.

Billard, R., & Cosson, M. P. (1992). Some problems related to the assessment ofsperm motility in freshwater fish. *The Journal of Experimental Zoology*, *261*(2), 122–131. doi:10.1002/jez.1402610203

Bindari, Y. R., Shrestha, S., Shrestha, N., & Gaire, T. N. (2013). Effects of nutrition on reproduction-A review. *Advances in Applied Science Research*, *4*(1), 421–429.

Birch, E. L., Meleis, A., & Wachter, S. (2012). The urban water transition: Why we must address the new reality of urbanization, women, Water, and sanitation in sustainable development. *wH2O: The Journal of Gender and Water, 1*. Available at: https://repository.upenn.edu/wh2ojournal/vol1/iss1/1

Birck, A., Christensen, P., Labouriau, R., Pedersen, J., & Borchersen, S. (2010). In vitro induction of the acrosome reaction in bull sperm and the relationship to field fertility using low-dose inseminations. *Theriogenology*, *73*(9), 1180–119. doi:10.1016/j.theriogenology.2009.10.010

Birdsall, M. A., Hopkisson, J. F., Grant, K. E., Barlow, D. H., & Mardon, H. J. (1996). Endocrinology and paracrinology: Expression of heparin-binding epidermal growth factor messenger RNA in the human endometrium. *MHR: Basic Science of Reproductive Medicine*, *2*(1), 31–34. doi:10.1093/molehr/2.1.31 PMID:9238654

Bishaw, B., Henry, N., Jeremias, M., Abdu, A., Jonathan, M., & Gemedo, D. et al. (2013). *Farmers' Strategies for Adapting to and Mitigating Climate Variability and Change through Agroforestry in Ethiopia and Kenya*. Academic Press.

Bisschops, I. A. E., & Spanjers, H. (2003). Literature review on textile wastewater characterization. *Environmental Technology*, *24*(11), 1399–1411. doi:10.1080/09593330309385684 PMID:14733393

Bjørklund, G., Chirumbolo, S., Dadar, M., Pivina, L., Lindh, U., Butnariu, M., & Aaseth, J. (2019). Mercury exposure and its effects on fertility and pregnancy outcome. *Basic & Clinical Pharmacology & Toxicology*, *125*(4), 317–327. doi:10.1111/bcpt.13264

Blask, D. E., Brainard, G. C., Dauchy, R. T., Hanifin, J. P., Davidson, L. K., Krause, J. A., Sauer, L. A., Rivera-Bermudez, M. A., Dubocovich, M. L., Jasser, S. A., Lynch, D. T., Rollag, M. D., & Zalatan, F. (2005). Melatonin-depleted blood from premenopausal women exposed to light at night stimulates growth of human breast cancer xenografts in nude rats. *Cancer Research*, *65*(23), 11174–11184. doi:10.1158/0008-5472.CAN-05-1945 PMID:16322268

Bleiberg, F. M., Brun, T. A., Goihman, S., & Gouba, E. (1980). Duration of activities and energy expenditure of female farmers in dry and rainy seasons in Upper-Volta. *British Journal of Nutrition*, *43*(1), 71–82. doi:10.1079/BJN19800065 PMID:7189407

Boadi, D., Wittenberg, K., Scott, S., Burton, D., Buckley, K., Small, J., & Ominski, K. H. (2004). Effect of low and high forage diet on enteric and manure pack greenhouse gas emissions from a feedlot. *Canadian Journal of Animal Science*, *84*(3), 445–453. doi:10.4141/A03-079

Boden, M. J., Varcoe, T. J., Voultsios, A., & Kennaway, D. J. (2010). Reproductive biology of female Bmal1 null mice. *Reproduction (Cambridge, England)*, *139*(6), 1077–1090. doi:10.1530/REP-09-0523 PMID:20200203

Boekelheide, K., Darney, S. P., Daston, G. P., David, R. M., Luderer, U., Olshan, A. F., Sanderson, W. T., Willhite, C. C., & Woskie, S. (2004). NTP-CERHR Expert Panel Report on the reproductive and developmental toxicity of 2-bromopropane. *Reproductive Toxicology (Elmsford, N.Y.)*, *18*(2), 189–217. doi:10.1016/j.reprotox.2003.10.003

Bohmanova, J., Misztal, I., & Cole, J. B. (2007). Temperature humidity indices as indicators of milk production losses due to heat stress. *Journal of Dairy Science*, *83*(4), 2120–2125. doi:10.3168/jds.2006-513 PMID:17369235

Boivin, D. B., & James, F. O. (2002). Circadian adaptation to night-shift work by judicious light and darkness exposure. *Journal of Biological Rhythms*, *17*(6), 556–567. doi:10.1177/0748730402238238 PMID:12465889

Boivin, D. B., James, F. O., Wu, A., Cho-Park, P. F., Xiong, H., & Sun, Z. S. (2003). Circadian clock genes oscillate in human peripheral blood mononuclear cells. *Blood*, *102*(12), 4143–4145. doi:10.1182/blood-2003-03-0779 PMID:12893774

Bolocan, E. (2013). Effects of heat stress on sexual behavior in heifers. *Lucrari Stiintifice. Zootehnie si Biotehnologii*, *42*, 141–148.

Bonde, J. P. (1999). Environmental fertility research at the turn of the century. *Scandinavian Journal of Work, Environment & Health*, *25*(6), 529–536. doi:10.5271jweh.476 PMID:10884149

Bonde, J. P., & Kolstad, H. (1997). Fertility of Danish battery workers exposedto lead. *International Journal of Epidemiology*, *26*(6), 1281–1288. doi:10.1093/ije/26.6.1281 PMID:9447408

Bonfante, A., Terribile, F., & Bouma, J. (2019). Refining physical aspects of soil quality and soil health when exploring the effects of soil degradation and climate change on biomass production: An Italian case study. *Soil (Göttingen)*, *5*(1), 1–14. doi:10.5194oil-5-1-2019

Bongaerts, E., Bové, H., Ameloot, M., & Nawrot, T. (2019). Ambient Black Carbon Particles Reach the Fetal Side of Human Placenta. *Environmental Epidemiology*, *3*, 34–35. doi:10.1097/01.EE9.0000606004.04305.f0

Boni, R. (2019). Heat stress, a serious threat to reproductive function in animals and humans. *Molecular Reproduction and Development*, *86*(10), 1307–1323. doi:10.1002/mrd.23123 PMID:30767310

Boomsma, C. M., Keay, S. D., & Macklon, N. S. (2012). Peri-implantation glucocorticoid administration for assisted reproductive technology cycles. *Cochrane Database of Systematic Reviews*, (6), CD005996. doi:10.1002/14651858.CD005996.pub3 PMID:22696356

Bornick, H., Eppinger, P., Grischek, T., & Worch, E. (2001). Simulation of biological degradation of aromatic amines in river bed sediments. *Water Research*, *35*(3), 619–624. doi:10.1016/S0043-1354(00)00314-6 PMID:11228957

Börnick, H., & Schmidt, T. C. (2006). Amines. In T. Reemtsma & M. Jekel (Eds.), *Organic pollutants in the water cycle - properties, occurrence, analysis and environmental relevance of polar compounds* (pp. 181–208). Wiley-VCH. doi:10.1002/352760877X.ch8

Borniger, J. C., Maurya, S. K., Periasamy, M., & Nelson, R. J. (2014). Acute dim light at night increases body mass, alters metabolism, and shifts core body temperature circadian rhythms. *Chronobiology International*, *31*(8), 917–925. doi:10.3109/07420528.2014.926911 PMID:24933325

Bracci, M., Manzella, N., Copertaro, A., Staffolani, S., Strafella, E., Barbaresi, M., Copertaro, B., Rapisarda, V., Valentino, M., & Santarelli, L. (2014). Rotating-shift nurses after a day off: Peripheral clock gene expression, urinary melatonin, and serum 17-β-estradiol levels. *Scandinavian Journal of Work, Environment & Health*, *40*(3), 295–304. doi:10.5271jweh.3414 PMID:24402410

Brann, D. W., & Mahesh, V. B. (1991). Role of corticosteroids in female reproduction. *The FASEB Journal*, *5*(12), 2691–2698. doi:10.1096/fasebj.5.12.1655548 PMID:1655548

Brantly, S. (2013). *Mitigating heat stress in cattle*. Working Trees Info Sheet. USDA National Agroforestry Service. Retrieved from: http://nac.unl.edu/Working_Trees/infosheets.htm

Brasil, P., Pereira, J. P., Gabaglia, C. R., Damasceno, L., Wakimoto, M., & ... . (2016). Zika Virus Infection in Pregnant Women in Rio de Janeiro-Preliminary Report. *The New England Journal of Medicine*, *375*(24), 2321–2334. doi:10.1056/NEJMoa1602412 PMID:26943629

Breen, J. G., Claggett, T. W., Kimmel, G. L., & Kimmel, C. A. (1999). Heat shock during rat embryo development in vitro results in decreased mitosis and abundant cell death. *Reproductive Toxicology (Elmsford, N.Y.)*, *13*(1), 31–39. doi:10.1016/S0890-6238(98)00056-2 PMID:10080297

Brent, R. L. (2001). The cause and prevention of human birth defects: What have we learned in the past 50 years? *Congenital Anomalies*, *41*(1), 3–21. doi:10.1111/j.1741-4520.2001.tb00869.x

Bretveld, R., Brouwers, M., Ebisch, I., & Roeleveld, N. (2007). Influence of pesticides on male fertility. *Scandinavian Journal of Work, Environment & Health*, *33*(1), 13–28. doi:10.5271jweh.1060

Brevik, E. C. (2013). The Potential Impact of Climate Change on Soil Properties and Processes and Corresponding Influence on Food Security. *Agriculture*, *3*(3), 398–417. doi:10.3390/agriculture3030398

Bridges, G., Day, M., & Geary, T. (2013). Triennial Reproduction Symposium: deficiencies in the uterine environment and failure to support embryonic development. *Journal of Animal Science, 91*(7), 3002-3013.

Bridges, P., Brusie, M., & Fortune, J. (2005). Elevated temperature (heat stress) in vitro reduces androstenedione and estradiol and increases progesterone secretion by follicular cells from bovine dominant follicles. *Domestic Animal Endocrinology*, *29*(3), 508–522. doi:10.1016/j.domaniend.2005.02.017

Bridle, J. R., & Vines, T. H. (2007). Limits to evolution at range margins: When andwhy does adaptation fail? *Trends in Ecology & Evolution*, *22*(3), 140–147. doi:10.1016/j.tree.2006.11.002 PMID:17113679

Brito, L., Silva, A., Barbosa, R., & Kastelic, J. (2004). Testicular thermoregulation in Bos indicus, crossbred and Bostaurus bulls: Relationship with scrotal, testicular vascular cone and testicular morphology and effects on semen quality and sperm production. *Theriogenology*, *61*(3), 511–528. doi:10.1016/S0093-691X(03)00231-0 PMID:14662148

Britt, J. H. (1982). Phenotypic relationships between reproductive performance and milk yield in dairy cattle. In *Proceedings of the National Invitational Dairy Cattle Reprodion Workshop.*

Brodsky, J. L., & Chiosis, G. (2006). Hsp70 molecular chaperones: Emerging roles in human disease and identification of small molecule modulators. *Current Topics in Medicinal Chemistry*, *6*(11), 1215–1225. doi:10.2174/156802606777811997 PMID:16842158

Bronson, F. H. (2004). Are humans seasonally photoperiodic? *Journal of Biological Rhythms*, *19*(3), 180–192. doi:10.1177/0748730404264658 PMID:15155003

Bruckner, T. A., Modin, B., & Vågerö, D. (2014). Cold ambient temperature in utero and birth outcomes in Uppsala, Sweden, 1915-1929. *Annals of Epidemiology*, *24*(2), 116–121. doi:10.1016/j.annepidem.2013.11.005 PMID:24332864

Brzezinski, A., Seibel, M. M., Lynch, H. J., Deng, M. H., Wurtman, R. J. (1987). Melatonin in human preovulatory follicular fluid. *Journal of Clinical Endocrinology and Metabolism, 64*(4), 865–867.

Bucklin, R. A., Turner, L. W., Beede, D. K., Bray, D. R., & Hemken, R. W. (1991). Methods to relieve heat stress for dairy cows in hot, humid climates. *Applied Engineering in Agriculture*, *7*(2), 241–247. doi:10.13031/2013.26218

Bugge, A., Feng, D., Everett, L. J., Briggs, E. R., Mullican, S. E., Wang, F., Jager, J., & Lazar, M. A. (2012). Rev-erbα and Rev-erbβ coordinately protect the circadian clock and normal metabolic function. *Genes & Development*, *26*(7), 657–667. doi:10.1101/gad.186858.112 PMID:22474260

Bujan, L., Daudin, M., Charlet, J. P., Thonneau, P., & Mieusset, R. (2000). Increase in scrotal temperaturein car drivers. *Human Reproduction (Oxford, England)*, *15*(6), 1355–1357. doi:10.1093/humrep/15.6.1355 PMID:10831568

Bulmer, J. N., Morrison, L., Longfellow, M., Ritson, A., & Pace, D. (1991). Granulated lymphocytes in human endometrium: Histochemical and immunohistochemical studies. *Human Reproduction (Oxford, England)*, *6*(6), 791–798. doi:10.1093/oxfordjournals.humrep.a137430 PMID:1757516

Bünemann, E. K., Bongiorno, G., Bai, Z., Creamer, R. E., De Deyn, G., de Goede, R., & Brussaard, L. (2018). Soil quality – A critical review. *Soil Biology & Biochemistry*, *120*, 105–125. doi:10.1016/j.soilbio.2018.01.030

Bunker, A., Wildenhain, J., Vandenbergh, A., Henschke, N., Rocklöv, J., Hajat, S., & Sauerborn, R. (2016). Research paper: Effects of air temperature on climate-sensitive mortality and morbidity outcomes in the elderly; a systematic review and meta-analysis of epidemiological evidence. *EBioMedicine*, *6*, 258–268. doi:10.1016/j.ebiom.2016.02.034 PMID:27211569

Burden, E. B. (2005). The Pollution in Newborns. A Benchmark Investigation of Industriacl Chemicals, Pollutants and Pesticides in Umbilical Cord Blood. Environmental Working Group (EWG).

Burfening, P. J., Elliott, D. S., Eisen, E., & Ulberg, L. C. (1970). Survival of embryosresulting from spermatozoa produced by mice exposed to elevated ambienttemperature. *Journal of Animal Science*, *30*(4), 578–582. doi:10.2527/jas1970.304578x

Burfening, P. J., & Ulberg, L. C. (1968). Embryonic survival subsequent to culture of rabbit spermatozoa at 38 and 40 C. *Journal of Reproduction and Fertility*, *15*(1), 87–92. doi:10.1530/jrf.0.0150087 PMID:5643479

Burkholder, W. J. (2000). Use of body condition scores in clinical assessment of the provision of optimal nutrition. *Journal of the American Veterinary Medical Association*, *217*(5), 650–654. doi:10.2460/javma.2000.217.650

Butrimovitz, G. P., Sharlip, I., & Lo, R. (1983). Extremely low seminal lead concentrations and male fertility. *Clinica Chimica Acta*, *135*(2), 229–231. doi:10.1016/0009-8981(83)90139-0

Caballero-Campo, P., Domínguez, F., Coloma, J., Meseguer, M., Remohí, J., Pellicer, A., & Simón, C. (2002). Hormonal and embryonic regulation of chemokines IL-8, MCP-1 and RANTES in the human endometrium during the window of implantation. *Molecular Human Reproduction*, *8*(4), 375–384. doi:10.1093/molehr/8.4.375 PMID:11912286

Caballero-Ortega, H., Uribe-Salas, F. J., Conde-Glez, C. J., Cedillo-Pelaez, C., Vargas-Villavicencio, J. A., Luna-Pastén, H., Cañedo-Solares, I., Ortiz-Alegría, L. B., & Correa, D. (2012). Seroprevalence and national distribution of human toxoplasmosis in Mexico: Analysis of the 2000 and 2006 National Health Surveys. *Transactions of the Royal Society of Tropical Medicine and Hygiene*, *106*(11), 653–659. doi:10.1016/j.trstmh.2012.08.004 PMID:22998951

Cai, H., Ren, Y., Li, X. X., Yang, J. L., Zhang, C. P., Chen, M., Fan, C. H., Hu, X. Q., Hu, Z. Y., Gao, F., & Liu, Y. X. (2011). Scrotal heat stress causes a transient alteration in tight junctions and induction of TGF-betaexpression. *International Journal of Andrology*, *34*(4pt1), 352–362. doi:10.1111/j.1365-2605.2010.01089.x PMID:20633196

Callegari, C., Everett, S., Ross, M., & Brasel, J. A. (1987). Anogenital ratio: Measure of fetal virilization in premature and full-term newborn infants. *The Journal of Pediatrics*, *111*(2), 240–243. doi:10.1016/S0022-3476(87)80075-6

Cameron, R. D., & Blackshaw, A. W. (1980). The effect of elevatedambient temperature on spermatogenesis in the boar. *Journal of Reproduction and Fertility*, *59*(1), 173–179. doi:10.1530/jrf.0.0590173 PMID:7401033

Canle, L. M., Santaballa, J. A., & Vulliet, E. (2005). On the mechanism of TiO_2-photocatalyzed degradation of aniline derivatives. *Journal of Photochemistry and Photobiology A Chemistry*, *175*(2-3), 192–200. doi:10.1016/j.jphotochem.2005.05.001

Cann, K. F., Thomas, D. R., Salmon, R. L., Wyn-Jones, A. P., & Kay, D. (2013). Extreme water-related weather events and waterborne disease. *Epidemiology and Infection*, *141*(4), 671–686. doi:10.1017/S0950268812001653 PMID:22877498

Cariati, F., D'Uonno, N., Borrillo, F., Iervolino, S., Galdiero, G., & Tomaiuolo, R. (2019). Bisphenol a: An emerging threat to male fertility. *Reproductive Biology and Endocrinology*, *17*(1), 6. doi:10.118612958-018-0447-6

Carlson, A. E., Quill, T. A., Westenbroek, R. E., Schuh, S. M., Hille, B., & Babcock, D. F. (2005). Identical phenotypes of CatSper1and CatSper2 null sperm. *The Journal of Biological Chemistry*, *280*(37), 32238–32244. doi:10.1074/jbc.M501430200 PMID:16036917

Carmen, Z., & Daniela, S. (2012). Textile organic dyes - characteristics, polluting effects and separation/elimination procedures from industrial effluents: A critical overview. In T. Puzyn (Ed.), *Organic pollutants ten years after the Stockholm convention - Environmental and analytical update*. doi:10.5772/32373

Carneiro, P. A., Osugi, M. E., Sene, J. J., Anderson, M. A., & Zanoni, M. V. B. (2004). Evaluation of color removal and degradation of a reactive textile azo dye on nanoporous TiO thin-film electrodes. *Electrochimica Acta*, *49*(22-23), 3807–3820. doi:10.1016/j.electacta.2003.12.057

Carson, D. D., Rohde, L. H., & Surveyor, G. (1994). Cell surface glycoconjugates as modulators of embryo attachment to uterine epithelial cells. *The International Journal of Biochemistry*, *26*(10-11), 1269–1277. doi:10.1016/0020-711X(94)90096-5 PMID:7851630

Carson, R. (1962). Silent Spring. Fawcett Press, New York, USA.Chapin RE, Stevens JT, Hughes CL, Kele WR, Hess RA and Daston GP(1996) Endocrine modulation of reproduction. *Fundamental and Applied Toxicology*, *29*, 1–17.

Carter, B. D., Diver, W. R., Hildebrand, J. S., Patel, A. V, Gapstur, S. M. (2014). Circadian Disruption and Fatal Ovarian Cancer. *American Journal of Preventive Medicine, 46*(3), S34–S41.

Casey, G., Shayegh, S., Moreno-Cruz, J., Bunzl, M., Galor, O., & Caldeira, K. (2019). The impact of climate change on fertility. *Environmental Research Letters*, *14*(5), 054007. doi:10.1088/1748-9326/ab0843

Catalano, R. D., Critchley, H. O., Heikinheimo, O., Baird, D. T., Hapangama, D., Sherwin, J. R. J. R., Charnock-Jones, D. S., Smith, S. K., & Sharkey, A. M. (2007). Mifepristone induced progesterone withdrawal reveals novel regulatory pathways in human endometrium. *Molecular Human Reproduction*, *13*(9), 641–654. doi:10.1093/molehr/gam021 PMID:17584828

Cawdell, J., Upfold, J., Edwards, M., & Smith, M. (1992). Neural tube and other developmental anomalies in the guinea pig following maternal hyperthermia during early neural tube development. *Teratogen Carcinogen Mutagen, 12*, 1-9.

Ceelen, M., Van Weissenbruch, M. M., Prein, J., Smit, J. J., Vermeiden, J. P., Spreeuwenberg, M., van Leeuwen, F. E., & Delemarre-van De Waal, H. A. (2009). Growth during infancy and early childhood in relation to blood pressure and body fat measures at age 8–18 years of IVF children and spontaneously conceived controls born to subfertile parents. *Human Reproduction (Oxford, England)*, *24*(11), 2788–2795. doi:10.1093/humrep/dep273 PMID:19648588

Ceelen, M., van Weissenbruch, M. M., Roos, J. C., Vermeiden, J. P., van Leeuwen, F. E., & Delemarre-van de Waal, H. A. (2007). Body composition in children and adolescents born after in vitro fertilization or spontaneous conception. *The Journal of Clinical Endocrinology and Metabolism*, *92*(9), 3417–3423. doi:10.1210/jc.2006-2896 PMID:17595253

Ceelen, M., van Weissenbruch, M. M., Vermeiden, J. P., van Leeuwen, F. E., & Delemarre-van de Waal, H. A. (2008). Cardiometabolic differences in children born after in vitro fertilization: Follow-up study. *The Journal of Clinical Endocrinology and Metabolism*, *93*(5), 1682–1688. doi:10.1210/jc.2007-2432 PMID:18285409

Centers for Disease Control and Prevention. (2005). Third National Report on Human Exposure to Environmental Chemicals. National Center for Environmental Health, Division of Laboratory Sciences, Atlanta, GA.

Cetica, P. D., Pintos, L. N., Dalvit, G. C., & Beconi, M. T. (2001). Antioxidant enzyme activity and oxidative stress in bovine oocyte in vitro maturation. *IUBMB Life*, *51*(1), 57–64. doi:10.1080/15216540152035073 PMID:11419698

Chandler, M. A., & Sohl, L. E. (2000). Climate forcings and the initiation of low latitude ice sheets during the Neoproterozoic Varanger glacial interval. *Journal of Geophysical Research, D, Atmospheres*, *105*(D16), 20737–20756. doi:10.1029/2000JD900221

Chang, W., Tran, H., Park, D., Zhang, R., & Ahn, D. (2009). Ammonium nitrogen removal characteristics of zeolite media in a Biological Aerated Filter (BAF) for the treatment of textile wastewater. *Journal of Industrial and Engineering Chemistry*, *15*(4), 524–528. doi:10.1016/j.jiec.2009.01.009

Chapin, R. E., Stevens, J. T., Hughes, C. L., Kele, W. R., Hess, R. A., & Daston, G. P. (1996). Endocrine modulation of reproduction. *Fundamental and Applied Toxicology*, *29*(1), 1–17. doi:10.1006/faat.1996.0001 PMID:8838635

Charoensook, R., Gatphayak, K., Sharifi, A., Chaisongkram, C., Brenig, B., & Knorr, C. (2011). Polymorphisms in the bovine HSP90AB1 gene are associated with heat tolerance in Thai indigenous cattle. *Tropical Animal Health and Production*, *44*(4), 921–928. doi:10.100711250-011-9989-8

Chatterjee, A., & Harper, M. J. K. (1970). Interruption of implantation and gestation in rats by reserpine, chlorpromazine and ACTH: Possible mode of action. *Endocrinology*, *87*(5), 966–969. doi:10.1210/endo-87-5-966 PMID:4320640

Chatterjee, D., Kuots, R., Ao, M., Saha, S., Ray, S. K., & Ngachan, S. V. (2019). Does rise in temperature adversely affect soil fertility, carbon fractions, microbial biomass and enzyme activities under different land uses? *Current Science*, *116*(12), 2044–2054. doi:10.18520/cs/v116/i12/2044-2054

Chauhan, D., & Ghosh, N. (2014). Impact of Climate Change on Livestock Production: A Review. *Journal of Animal Research*, *4*(2), 223. doi:10.5958/2277-940X.2014.00009.6

Chauhan, S. S., Celi, P., Leury, B., Liu, F., & Dunshea, F. R. (2015). Exhaled breath condensate hydrogen peroxide concentration, a novel biomarker for assessment of oxidative stress in sheep during heat stress. *Animal Production Science*, *56*(7), 1105–1112. doi:10.1071/AN14070

Chawla, M. K., Gutierrez, G. M., Young, W. S. III, McMullen, N. T., & Rance, N. E. (1997). Localization of neurons expressing substance P and neurokinin B gene transcripts in the human hypothalamus and basal forebrain. *The Journal of Comparative Neurology*, *384*(3), 429–442. doi:10.1002/(SICI)1096-9861(19970804)384:3<429::AID-CNE8>3.0.CO;2-5 PMID:9254037

Chen, I., Hill, J. K., Ohlemüller, R., Roy, D. B., & Thomas, C. D. (2011). Rapid rangeshifts of species associated with high levels of climate warming. *Science*, *333*(6045), 1024–1026. doi:10.1126cience.1206432 PMID:21852500

Chen, Q., Peng, H., Zhang, Y., Lei, L., Cao, Y., & Duan, E. (2009). Embryo implantation: A time for recalling and forwarding. *Chinese Science Bulletin*, *54*(22), 4083–4093. doi:10.100711434-009-0631-3

Chen, Y., Breen, K., & Pepling, M. E. (2009). Estrogen can signal through multiple pathways to regulate oocyte cyst breakdown and primordial follicle assembly in the neonatal mouse ovary. *The Journal of Endocrinology*, *202*(3), 407–417. doi:10.1677/JOE-09-0109

Chequer, F. M. D., Lizier, T. M., Felicio, R., Zanoni, M. V. B., Debonsi, H. M., Lopes, N. P., & ... , (2011). Analyses of the genotoxic and mutagenic potential of the products formed after the biotransformation of the azo dye Disperse Red 1. *Toxicology In Vitro*, *25*(8), 2054–2063. doi:10.1016/j.tiv.2011.05.033 PMID:21907275

Chernovetz, M. E., Justesen, D. R., & Oke, A. F. (1977). A teratological study of the rat: Microwave and infrared radiations compared. *Radio Science*, *12*(6S), 191–197. doi:10.1029/RS012i06Sp00191

Chevrier, C., Dananché, B., Bahuau, M., Nelva, A., Herman, C., Francannet, C., Robert-Gnansia, E., & Cordier, S. (2006). Occupational exposure to organic solvent mixtures during pregnancy and the risk of non-syndromic oral clefts. *Occupational and Environmental Medicine*, *63*(9), 617–623. doi:10.1136/oem.2005.024067

Chia, M. A., & Musa, R. I. (2014). Effect of indigo dye effluent on the growth, biomass production and phenotypic plasticity of Scenedesmus quadricauda (Chlorococcales). *Journal of the Brazilian Academy of Science*, *36*(1), 419–428. doi:10.1590/0001-3765201420130225 PMID:24676177

Chini, B., & Fanelli, F. (2000). Molecular basis of ligand binding and receptor activation in the oxytocin and vasopressin receptor family. *Experimental Physiology*, *85*(s1), 59s–66s. doi:10.1111/j.1469-445X.2000.tb00008.x PMID:10795907

Chiu, B., Chen, C.-S., Keller, J. J., Lin, C.-C., & Lin, H.-C. (2012). Seasonality of testicular torsion: A 10-year nationwide population based study. *The Journal of Urology*, *187*(5), 1781–1785. doi:10.1016/j.juro.2011.12.104 PMID:22425116

Cho, H., Zhao, X., Hatori, M., Yu, R. T., Barish, G. D., Lam, M. T., Chong, L. W., Ditacchio, L., Atkins, A. R., Glass, C. K., Liddle, C., Auwerx, J., Downes, M., Panda, S., & Evans, R. M. (2012). Regulation of Circadian Behavior and Metabolism by Rev-erbα and Rev-erbβ. *Nature*, *485*(7396), 123–127. doi:10.1038/nature11048 PMID:22460952

Choi, B. H. (1991). Effects of methylmercury on the developing brain. In *Advances in Mercury Toxicology* (pp. 315–337). Springer. doi:10.1007/978-1-4757-9071-9_20

Chou, P. Y., Wu, M. H., Pan, H. A., Hung, K. H., & Chang, F. M. (2011). Use of an oxytocin antagonist in in vitro fertilization–embryo transfer for women with repeated implantation failure: A retrospective study. *Taiwanese Journal of Obstetrics & Gynecology*, *50*(2), 136–140. doi:10.1016/j.tjog.2011.04.003 PMID:21791296

Christians, E., Michel, E., Adenot, P., Mezger, V., Rallu, M., Morange, M., & Renard, J. P. (1997). Evidence for the involvement of mouse heat shock factor 1 in the atypical expression of the HSP70.1 heat shock gene during mouse zygotic genome activation. *Molecular and Cellular Biology*, *17*(2), 778–788. doi:10.1128/MCB.17.2.778 PMID:9001232

Christiansen, T., & Voigt, T. (2008). *Impact of Europe's Changing Climate—2008 Indicator-Based Assessment*. https://ec.europa.eu/jrc/sites/jrcsh/files/jrc_reference_report_2008_09_climate_change.pdf

Christidis, N., Jones, G. S., & Stott, P. A. (2015). Dramatically increasing chance ofextremely hot summers since the 2003 European heatwave. *Nature Climate Change*, *5*(1), 3–7. doi:10.1038/nclimate2468

Christie, R. (2001). *Colour chemistry*. The Royal Society of Chemistry.

Chung, F. F., Yao, C. C., & Wan, G. H. (2005). The associations between menstrual function and lifestyle/working conditions among nurses in Taiwan. *Journal of Occupational Health*, *47*(2), 149–156. doi:10.1539/joh.47.149 PMID:15824480

Chung, J. J., Shim, S. H., Everley, R. A., Gygi, S. P., Zhuang, X., & Clapham, D. E. (2014). Structurally distinct Ca(2+) signalling domains of sperm flagella orchestrate tyrosine phosphorylation and motility. *Cell*, *157*(4), 808–822. doi:10.1016/j.cell.2014.02.056 PMID:24813608

Chung, K. T. (2015). Occurrence, uses and carcinogenicity of arylamines. *Frontiers in Bioscience*, *7*(2), 367–393. doi:10.2741/e737 PMID:25553383

Chung, K. T., & Cerniglia, C. E. (1992). Mutagenicity of azo dyes: Structure-activity relationships. *Mutation Research*, *277*(3), 201–220. doi:10.1016/0165-1110(92)90044-A PMID:1381050

Chung, K. T., Stevens, S. E. J., & Cerniglia, C. E. (1992). The reduction of azo dyes by the intestinal microflora. *Critical Reviews in Microbiology*, *18*(3), 175–197. doi:10.3109/10408419209114557 PMID:1554423

Chung, M. K., Lee, H., Mizuno, A., Suzuki, M., & Caterina, M. J. (2004). TRPV3 and TRPV4 mediate warmth-evoked currents in primary mouse keratinocytes. *The Journal of Biological Chemistry*, *279*(20), 21569–21575. doi:10.1074/jbc.M401872200 PMID:15004014

Chu, W., Choy, W. K., & So, T. Y. (2007). The effect of solution pH and peroxide in the TiO_2-induced photocatalysis of chlorinated aniline. *Journal of Hazardous Materials*, *141*(1), 86–91. doi:10.1016/j.jhazmat.2006.06.093 PMID:16916576

Ciechanowska, M., Łapot, M., Paruszewska, E., Radawiec, W., & Przekop, F. (2018). The influence of dopaminergic system inhibition on biosynthesis of gonadotrophin-releasing hormone (GnRH) and GnRH receptor in anoestrous sheep; hierarchical role of kisspeptin and RFamide-related peptide-3 (RFRP-3). *Reproduction, Fertility, and Development*, *30*(4), 672–680. doi:10.1071/RD16309 PMID:29019791

Cini, G., Coronnello, M., Mini, E., & Neri, B. (1998). Melatonin's growth-inhibitory effect on hepatoma AH 130 in the rat. *Cancer Letters*, *125*(1-2), 51–59. Advance online publication. doi:10.1016/S0304-3835(97)00480-1 PMID:9566696

Classen, A. T., Sundqvist, M. K., Henning, J. A., Newman, G. S., Moore, J. A. M., Cregger, M. A., Moorhead, L. C., & Patterson, C. M. (2015). Direct and indirect effects of climate change on soil microbial and soil microbial-plant interactions: What lies ahead? *Ecosphere*, *6*(8), 1–24. doi:10.1890/ES15-00217.1

Claxton, L., Hughes, T., & Chung, K. (2001). Using base-specific Salmonella tester strains to characterize the types of mutation induced by benzidine and benzidine congeners after reductive metabolism. *Food and Chemical Toxicology*, *39*(12), 1253–1261. doi:10.1016/S0278-6915(01)00072-2 PMID:11696399

Cocco, P. (2002). On the rumors about the silent spring. Review of thescientific evidence linking occupational and environmental pesticideexposure to endocrine disruption health effects. *Cadernos de Saude Publica*, *18*(2), 39–402. doi:10.1590/S0102-311X2002000200003 PMID:11923880

Cohen, J., Fehilly, C. B., & Waiters, D. E. (1985). Prolonged storage ofhuman spermatozoa at room temperature or in arefrigerator. *Fertility and Sterility*, *44*(2), 254–262. doi:10.1016/S0015-0282(16)48747-9 PMID:4018281

Cohen, M., Lippman, M., & Chabner, B. (1978). Role of pineal gland in aetiology and treatment of breast cancer. *Lancet*, *2*, 1001–1002. PMID:81980

Colborn, T. (1991) Nontraditional evaluation of risk from fish contaminants. In *Proceedings of a Symposium on Issues in Seafood Safety*. National Academy of Sciences, Institute of Medicine, Food and Nutrition Board.

Colborn, T. (1995). Pesticides—how research has succeeded and failed to translate science into policy: Endocrinological effects on wildlife. *Environmental Health Perspectives*, *103*(Suppl 6), 81–85. PMID:8549498

Colborn, T., Dumanoski, D., & Meyers, J. P. (1996). *Our Stolen Future*. Dutton.

Colborn, T., Vom Saal, F. S., & Soto, A. M. (1993). Developmental effects ofendocrine disrupting chemicals in wildlife and humans. *Environmental Health Perspectives*, *101*(5), 378–384. doi:10.1289/ehp.93101378 PMID:8080506

Colenbrander, B., Feitsma, H., & Grooten, H. J. (1993). Optimizingsemen production for artificial insemination in swine. *Journal of Reproduction and Fertility. Supplement*, *48*, 207–215. PMID:8145205

Colicchia, M., Campagnolo, L., Baldini, E., Ulisse, S., Valensise, H., & Moretti, C. (2014). Molecular basis of thyrotropin and thyroid hormone action during implantation and early development. *Human Reproduction Update*, *20*(6), 884–904. doi:10.1093/humupd/dmu028 PMID:24943836

Collier, R. J., Dahl, G. E., & VanBaale, M. J. (2006). Major advances associated with environmental effects on dairy cattle. *Journal of Dairy Science*, *89*(4), 1244–1253. doi:10.3168/jds.S0022-0302(06)72193-2 PMID:16537957

Collier, R. J., Doelger, S. G., Head, H. H., Thatcher, W. W., & Wilcox, C. J. (1982). Effects of heat stress during pregnancy on maternal hormone concentrations, calf birth weight and postpartum milk yield of Holstein cow. *Journal of Animal Science*, *54*(2), 309–319. doi:10.2527/jas1982.542309x PMID:7076593

Colon, I., Caro, D., Bourdony, C. J., & Rosario, O. (2000). Identification of phthalate esters in the serum of young Puerto Rican girls with premature breast development. *Environmental Health Perspectives*, *108*(9), 895.

Conaway, C. (1971). Ecological adaptation and mammalian reproduction. *Biology of Reproduction*, *4*(3), 239–247. doi:10.1093/biolreprod/4.3.239

Conneely, O. M., Mulac-Jericevic, B., DeMayo, F., Lydon, J. P., & O'Malley, B. W. (2002). Reproductive functions of progesterone receptors. *Recent Progress in Hormone Research*, *57*(1), 339–355. doi:10.1210/rp.57.1.339 PMID:12017551

Cooper, R. L., Goldman, J. M., & Stoker, T. E. (1999). Neuroendocrine and reproductive effects of contemporary-use pesticides. *Toxicology and Industrial Health*, *15*, 26–36.

Cooper, R. L., Goldman, J. M., & Stoker, T. E. (1999). Neuroendocrine and reproductiveeffects of contemporary-use pesticides. *Toxicology and Industrial Health*, *15*(1), 26–36. doi:10.1191/074823399678846727 PMID:10188189

Cordier, S. (2008). Evidence for a role of paternal exposures in developmental toxicity. *Basic & Clinical Pharmacology & Toxicology*, *102*(2), 176–181. doi:10.1111/j.1742-7843.2007.00162.x

Correa, A., Gray, R. H., Cohen, R., Rothman, N., Shah, F., Seacat, H., & Corn, M. (1996). Ethylene glycol ethers and risks of spontaneous abortion and subfertility. *American Journal of Epidemiology*, *143*(7), 707–717. doi:10.1093/oxfordjournals.aje.a008804 PMID:8651233

Cos, S., Fernández, R., Güézmes, A., & Sánchez-Barcelo, E. J. (1998). Influence of melatonin on invasive and metastatic properties of MCF-7 human breast cancer cells. *Cancer Research*, *58*, 4383–4390. PMID:9766668

Cos, S., Martínez-Campa, C., Mediavilla, M. D., & Sánchez-Barceló, E. J. (2005). Melatonin modulates aromatase activity in MCF-7 human breast cancer cells. *Journal of Pineal Research*, *38*(2), 136–142. doi:10.1111/j.1600-079X.2004.00186.x PMID:15683469

Cosson, J. (2004). The ionic and osmotic factors controlling motility of fishspermatozoa. *Aquaculture International, 12*(1), 69–85. doi:10.1023/B:AQUI.0000017189.44263.bc

Cosson, J., Billard, R., Cibert, C., Dreanno, C., & Suquet, M. (1999). Ionic factorsregulating the motility of fish sperm. In C. Gagnon (Ed.), *The male gamete: From basic to clinical applications* (pp. 161–186). Cache Rive Press.

Cosson, J., Billard, R., Gibert, C., Dreanno, C., Linhart, O., & Suquet, M. (1997). Movements of fish sperm flagella studied by high speed videomicroscopycoupled to computer assisted image analysis. *Polskie Archiwum Hydrobiologii, 10*, 518–527.

Coss, R. A., Dewey, W. C., & Bamburg, J. R. (1982). Effects of hyperthermia on dividing Chinese hamster ovary cells and microtubules in vitro. *Cancer Research, 42*, 1059–1071. PMID:7199378

Costello, A., Abbas, M., Allen, A., Ball, S., Bell, S., Bellamy, R., Friel, S., Groce, N., Johnson, A., Kett, M., Lee, M., Levy, C., Maslin, M., McCoy, D., McGuire, B., Montgomery, H., Napier, D., Pagel, C., Patel, J., ... Patterson, C. (2009). Managing the health effects of climate change: Lancet and University College London Institute for Global Health Commission. *Lancet, 373*(9676), 1693–1733. doi:10.1016/S0140-6736(09)60935-1 PMID:19447250

Council, N. R. (1981). *Effect of environment on nutrient requirements of domestic animals.* National Academies Press.

Courtney, L. P. (1984). *The temporal relationship of the ovary and ovarian morphology for the onset and duratin of behavioralestrus in prostaglandin Fa,,-treated ewes.* Montana State University-Bozeman, College of Agriculture.

Couto, S. R. (2009). Dye removal by immobilised fungi. *Biotechnology Advances, 27*(3), 227–235. doi:10.1016/j.biotechadv.2008.12.001 PMID:19211032

Couzinet, B., Strat, N. L., Ulmann, A., Baulieu, E. E., & Schaison, G. (1986). Termination of early pregnancy by the progesterone antagonist RU 486 (mifepristone). *The New England Journal of Medicine, 315*(25), 1565–1570. doi:10.1056/NEJM198612183152501 PMID:3785321

Cox, B., Vicedo-Cabrera, A. M., Gasparrini, A., Roels, H. A., Martens, E., Vangronsveld, J., Forsberg, B., & Nawrot, T. S. (2016). Ambient temperature as a trigger of preterm delivery in a temperate climate. *Journal of Epidemiology and Community Health, 70*(12), 1191–1199. doi:10.1136/jech-2015-206384 PMID:27261529

Craig, Z. R., Wang, W., & Flaws, J. A. (2011). Endocrine-disrupting chemicals in ovarian function: Effects on steroidogenesis, metabolism and nuclear receptor signaling. *Reproduction (Cambridge, England), 142*(5), 633–646. doi:10.1530/REP-11-0136

Creighton, K. W. (2004). *Heifer development systems for March-born heifers and improving pregnancy rates in June-calving cows.* The University of Nebraska- Lincoln.

Cremonese, C., Piccoli, C., Pasqualotto, F., Clapauch, R., Koifman, R. J., Koifman, S., & Freire, C. (2017). Occupational exposure to pesticides, reproductive hormone levels and sperm quality in young Brazilian men. *Reproductive Toxicology (Elmsford, N.Y.), 67*, 174–185. doi:10.1016/j.reprotox.2017.01.001

Crew, F. A. E. (1927). XX.—On Fertility in the Domestic Fowl. *Proceedings of the Royal Society of Edinburgh, 46*, 230–238. doi:10.1017/S0370164600022021

Crinnion, W. J. (2009). Maternal levels of xenobiotics that affect fetal development and childhood health. *Alternative Medicine Review, 14*(3). PMID:19803547

Cross, J. C., Werb, Z., & Fisher, S. J. (1994). Implantation and the placenta: Key pieces of the development puzzle. *Science, 266*(5190), 1508–1518. doi:10.1126cience.7985020 PMID:7985020

Cunningham, F. G., Leveno, K. J., Bloom, S. L., Hauth, J. C., Rouse, D. J., & Spong, C. Y. (2010). Overview of obstetrics. In F. G. Cunningham, K. J. Leveno, S. L. Bloom, J. C. Hauth, D. J. Rouse, & C. Y. Spong (Eds.), *Williams obstetrics* (23rd ed.). McGraw-Hill.

Curtis Hewitt, S., Goulding, E. H., Eddy, E. M., & Korach, K. S. (2002). Studies using the estrogen receptor alpha knock-out uterus demonstrate that implantation but not decidualization-associated signaling is estrogen dependent. *Biology of Reproduction*, *67*(4), 1268–1277. doi:10.1095/biolreprod67.4.1268 PMID:12297545

Curtis, K. M., Savitz, D. A., Weinberg, C. R., & Arbuckle, T. E. (1999). The effect ofpesticide exposure on time to pregnancy. *Epidemiology (Cambridge, Mass.)*, *10*(2), 112–117. doi:10.1097/00001648-199903000-00005 PMID:10069244

Curtis, V., & Lambert, W. V. (1929). A study of fertility in poultry. *Poultry Science*, *8*(3), 142–150. doi:10.3382/ps.0080142

Daader, A. H., & Seleem, T. S. T. (1999). Recent trends in rabbit production. *Proceedings of 1st International Conference on Indigenous Versus Acclimatized Rabbits,* 23-50.

Dadvand, P., Basagaña, X., Sartini, C., Figueras, F., Vrijheid, M., de Nazelle, A., Sunyer, J., & Nieuwenhuijsen, M. J. (2011). Climate extremes and the length of gestation. National Institute of Environmental Health Sciences. *Environmental Health Perspectives*, *119*(10), 1449–1453. doi:10.1289/ehp.1003241 PMID:21659038

Dadvand, P., Nazelle, A., Triguero-Mas, M., Schembari, A., Cirach, M., Amoly, E., Figueras, F., Basagana, X., & Ostro, B. (2012). Nieuwenhuijsen, M. Surrounding greenness and exposure to air pollution during pregnancy: An analysis of personal monitoring data. *Environmental Health Perspectives*, *120*(9), 1286–1290. doi:10.1289/ehp.1104609 PMID:22647671

Dadvand, P., Ostro, B., Figueras, F., Foraster, M., Basagaña, X., Valentín, A., Martinez, D., Beelen, R., Cirach, M., Hoek, G., Jerrett, M., Brunekreef, B., & Nieuwenhuijsen, M. J. (2014). Residential proximity to major roads and term low birth weight: The roles of air pollution, heat, noise, and road-adjacent trees. *Epidemiology (Cambridge, Mass.)*, *25*(4), 518–525. doi:10.1097/EDE.0000000000000107 PMID:24787556

Dahl, G. E., Tao, S., & Monteiro, A. P. A. (2016). Effects of late-gestation heat stress on immunity and performance of calves. *Journal of Dairy Science*, *99*(4), 3193–3198. doi:10.3168/jds.2015-9990 PMID:26805989

Dahl, J. E., Sunbdy, J., Hensten-Pettersen, A., & Jacobsen, N. (1999). Dental workplace exposure and effect on fertility. *Scandinavian Journal of Work, Environment & Health*, *25*(3), 285–290. doi:10.5271jweh.436 PMID:10450781

Damgaard, I. N., Main, K. M., Toppari, J., & Skakkebaek, N. E. (2002). Impact ofexposure to endocrine disrupters in utero and in childhood and adultreproduction. *Best Practice & Research. Clinical Endocrinology & Metabolism*, *16*(2), 289–309. doi:10.1053/beem.2002.0205 PMID:12064894

Damstra, T., Barlow, S., Bergman, A., Kavlock, R., & Van der Kraak, G. (2002). *Global assessment on the state-of-the-science of endocrine disruptors. WHO publication no. WHO/PCS/EDC/02.2*. World Health Organization.

Damstra, T., Barlow, S., Bergman, A., Kavlock, R., & Van Der Kraak, G. (Eds.). (2004). *Global assessment of the state of the science of endocrine disruptors*. International Programme on Chemical Safety.

Danforth, D. N. Jr, Tamarkin, L., Mulvihill, J. J., Bagley, C. S., & Lippman, M. E. (1985). Plasma melatonin and the hormone-dependency of human breast cancer. *Journal of Clinical Oncology*, *3*(7), 941–948. doi:10.1200/JCO.1985.3.7.941 PMID:4020407

Dash, S., Chakravarty, A. K., Singh, A., Upadhyay, A., Singh, M., & Yousuf, S. (2016). Effect of heat stress on reproductive performances of dairy cattle and buffaloes. *Veterinary World*, *9*(3), 235–244. doi:10.14202/vetworld.2016.235-244

Das, R., Sailo, L., Verma, N., Bharti, P., Saikia, J., Imtiwati, & Kumar, R. (2016). Impact of heat stress on health and performance of dairy animals: A review. *Veterinary World*, *9*(3), 260–268. doi:10.14202/vetworld.2016.260-268 PMID:27057109

Das, S. (2017). Impact of Climate Change on Livestock, Various Adaptive and Mitigative Measures for Sustainable Livestock Production. *Approaches in Poultry, Dairy & Veterinary Sciences*, *1*(4).

Das, S. (2017). Impact of Climate Change on Livestock, Various Adaptive and Mitigative Measures for Sustainable Livestock Production. *Approaches in Poultry. Dairy and Veterinary Sciences*, *1*(4), 33–42.

Das, S. K., Chakraborty, I., Paria, B. C., Wang, X. N., Plowman, G., & Dey, S. K. (1995). Amphiregulin is an implantation-specific and progesterone-regulated gene in the mouse uterus. *Molecular Endocrinology (Baltimore, Md.)*, *9*(6), 691–705. PMID:8592515

Das, S. K., Karunakaran, M., Barbuddhe, S. B., & Singh, N. P. (2015). Effect of Orientation, Ventilation, Floor Space Allowance and Cooling Arrangement on Milk Yield and Microclimate of Dairy Shed in Goa. *Journal of Animal Research*, *5*(2), 231–235. doi:10.5958/2277-940X.2015.00040.6

Das, S. K., Wang, X. N., Paria, B. C., Damm, D., Abraham, J. A., Klagsbrun, M., Andrews, G. K., & Dey, S. K. (1994). Heparin-binding EGF-like growth factor gene is induced in the mouse uterus temporally by the blastocyst solely at the site of its apposition: A possible ligand for interaction with blastocyst EGF-receptor in implantation. *Development*, *120*(5), 1071–1083. PMID:8026321

Das, S., & Dash, H. R. (2014). Microbial bioremediation: A potential tool for restoration of contaminated areas. In D. S. Elsevier (Ed.), *Microbial Biodegradation and Bioremediation* (pp. 1–21). doi:10.1016/B978-0-12-800021-2.00001-7

Das, S., & Dash, H. R. (2017). *Handbook of metal-microbe interactions and bioremediation*. CRC Press. doi:10.1201/9781315153353

Daston, G. P., Gooch, J. W., Shuey, D. L., Nikiforov, A. I., Fico, T. A., & Gorsuch, J. W. (1997). Environmental estrogens and reproductive health: A discussion ofthe human, environmental data. *Reproductive Toxicology (Elmsford, N.Y.)*, *11*(4), 465–481. doi:10.1016/S0890-6238(97)00014-2 PMID:9241667

Davidson, J. F., Whyte, B., Bissinger, P. H., & Schiestl, R. H. (1996). Oxidative stress is involved in heat-induced cell death in Saccharomyces cerevisiae. *Proceedings of the National Academy of Sciences of the United States of America*, *93*(10), 5116–5121. doi:10.1073/pnas.93.10.5116 PMID:8643537

Davis, B. J., Maronpot, R. R., & Heindel, J. J. (1994). Di-(2-ethylhexyl) phthalate suppresses estradiol and ovulation in cycling rats. *Toxicology and Applied Pharmacology*, *128*(2), 216–223. doi:10.1006/taap.1994.1200

Davis, B. J., Weaver, R., Gaines, L. J., & Heindel, J. J. (1994). Mono-(2-ethylhexyl) phthalate suppresses estradiol production independent of FSH-cAMP stimulation in rat granulosa cells. *Toxicology and Applied Pharmacology*, *128*(2), 224–228. doi:10.1006/taap.1994.1201

Davis, J. B., Gray, J., Gunthorpe, M. J., Hatcher, J. P., Davey, P. T., Overend, P., Harries, M. H., Latcham, J., Clapham, C., Atkinson, K., Hughes, S. A., Rance, K., Grau, E., Harper, A. J., Pugh, P. L., Rogers, D. C., Bingham, S., Randall, A., & Sheardown, S. A. (2000). Vanilloid receptor-1 is essential for inflammatory thermal hyperalgesia. *Nature*, *405*(6783), 183–187. doi:10.1038/35012076 PMID:10821274

De Alba, & Riera, S. (1966). Sexual maturity and spermatogenesis under heat stress in the bovine. *Animal Production*, *8*, 137–144.

De Angelis, C., Galdiero, M., Pivonello, C., Salzano, C., Gianfrilli, D., Piscitelli, P., Lenzi, A., Colao, A., & Pivonello, R. (2017). The environment and male reproduction: The effect of cadmium exposure on reproductive function and its implication in fertility. *Reproductive Toxicology (Elmsford, N.Y.)*, *73*, 105–127. doi:10.1016/j.reprotox.2017.07.021

De Bacquer, D., Van Risseghem, M., Clays, E., Kittel, F., De Backer, G., & Braeckman, L. (2009). Rotating shift work and the metabolic syndrome: A prospective study. *International Journal of Epidemiology*, *38*(3), 848–854. doi:10.1093/ije/dyn360 PMID:19129266

De Oliveira, G. A. R., de Lapuente, J., Teixido, E., Porredon, C., Borras, M., & de Oliveira, D. P. (2015). Textile dyes induce toxicity on zebrafish early life stages. *Environmental Toxicology and Chemistry*, *35*(2), 429–434. doi:10.1002/etc.3202 PMID:26267709

De Rensis, F., & Scaramuzzi, R. J. (2003). Heat stress and seasonal effects on reproduction in the dairy cow-a review. *Theriogenology*, *60*(6), 1139–1151. doi:10.1016/S0093-691X(03)00126-2

de Sherbinin, A., Carr, D., Cassels, S., & Jiang, L. (2007). Population and Environment. *Annual Review of Environment and Resources*, *32*(1), 345–373. doi:10.1146/annurev.energy.32.041306.100243 PMID:20011237

De Silva, H. J., Samarawickrema, N. A., & Wickremasinghe, A. R. (2006). Toxicity due to organophosphorus compounds: What about chronic exposure? *Transactions of the Royal Society of Tropical Medicine and Hygiene*, *100*(9), 803–806. doi:10.1016/j.trstmh.2006.05.001

Deb, R., Sajjanar, B., Singh, U., Kumar, S., Singh, R., Sengar, G., & Sharma, A. (2014). Effect of heat stress on the expression profile of Hsp90 among Sahiwal (Bos indicus) and Frieswal (Bos indicus × Bos taurus) breed of cattle. *Gene*, *536*(2), 435–440. doi:10.1016/j.gene.2013.11.086

DeCelis, R., Pedron-Nuevo, N., & Feria-Velasco, A. (1996). Toxicology ofmale reproduction in animals and humans. *Archives of Andrology*, *37*(3), 201–218. doi:10.3109/01485019608988523 PMID:8939299

DeCock, J., Heederik, D., Tielemans, E., te Velde, E., & Van Kooij, R. (1995). Offspring sex ratio as an indicator of reproductive hazards associated with pesticides. *Occupational and Environmental Medicine*, *52*, 429–430. PMID:7627322

Demott, R. P., & Suarez, S. S. (1992). Hyperactivated sperm progress in the mouse oviduct. *Biology of Reproduction*, *46*(5), 779–785. doi:10.1095/biolreprod46.5.779 PMID:1591334

Denver, R. (2009). Stress hormones mediate environment-genotype interactions during amphibian development. *General and Comparative Endocrinology*, *164*(1), 20–31. doi:10.1016/j.ygcen.2009.04.016

Desai, N., Sabanegh, E. Jr, Kim, T., & Agarwal, A. (2010). Free radical theory of aging: Implications in male infertility. *Urology*, *75*(1), 14–19. doi:10.1016/j.urology.2009.05.025 PMID:19616285

DeShazer, J. A., Hahn, G. L., & Xin, H. (2009). Basic principles of the thermal environment and livestock energetics. In Livestock energetics and thermal environmental management. American Society of Agricultural and Biological Engineers (ASABE).

DeShazer, J. A., Hahn, G. L., & Xin, H. (2009). Basic principals of the thermal environment and livestock energetics. In J. A. DeShazer (Ed.), *Livestock energetics and thermal environment management*. ASABE.

Devi, P., Debbarma, S., Kumar, G., & Thakur, P. (2019). Effect of heat stress on reproduction in farm animals and its mitigation: A review. *Journal of Entomology and Zoology Studies*, *7*(2), 342–345.

Dey, S. K. (2010). How we are born. *The Journal of Clinical Investigation*, *120*(4), 952–955. doi:10.1172/JCI42380 PMID:20364092

Dhimal, M., Dahal, S., Dhimal, M.L., Mishra, S.R., Karki, K.B., Aryal, K.K., Haque, U., Kabir, M.I., Guin, P., Butt, A.M., Harapan, H., Liu, Q.Y., Chu, C., Montag, D., Groneberg, D.A., Pandey, B.D., Kuch, U., & Müller, R. (2018). Threats of Zika virus transmission for Asia and its Hindu-Kush Himalayan region. *Infect Dis Poverty, 15*(7), 40.

Diamanti-Kandarakis, E., Bourguignon, J.-P., Giudice, L. C., Hauser, R., Prins, G. S., Soto, A. M., Zoeller, R. T., & Gore, A. C. (2009). Endocrine-disrupting chemicals: An Endocrine Society scientific statement. *Endocrine Reviews, 30*(4), 293–342. doi:10.1210/er.2009-0002

Díaz-Gimeno, P., Ruiz-Alonso, M., Blesa, D., Bosch, N., Martínez-Conejero, J. A., Alamá, P., Garrido, N., Pellicer, A., & Simón, C. (2013). The accuracy and reproducibility of the endometrial receptivity array is superior to histology as a diagnostic method for endometrial receptivity. *Fertility and Sterility, 99*(2), 508–517. doi:10.1016/j.fertnstert.2012.09.046 PMID:23102856

Dobson, H., & Smith, R. F. (2000). What is stress and how does itaffects fertility. *Animal Reproduction Science, 60*, 743–752. doi:10.1016/S0378-4320(00)00080-4 PMID:10844239

Dolatshad, H., Campbell, E. A., O'Hara, L., Maywood, E. S., Hastings, M. H., & Johnson, M. H. (2006). Developmental and reproductive performance in circadian mutant mice. *Human Reproduction (Oxford, England), 21*(1), 68–79. doi:10.1093/humrep/dei313 PMID:16210390

Donaghay, M., & Lessey, B. A. (2007). Uterine receptivity: Alterations associated with benign gynecological disease. *Seminars in Reproductive Medicine, 25*(6), 461–475. doi:10.1055-2007-991044 PMID:17960531

Dong, N., Liu, Z., Luo, M., Fang, C., & Lin, H. (2019). The effects of anthropogenic land use changes on climate in China driven by global socioeconomic and emission scenarios. *Earth's Future, 7*(7), 784–804. doi:10.1029/2018EF000932

Dougherty, K. A., Emilson, L. B., Cockett, A. T., & Urry, R. L. (1975). Acomparison of subjective measurements of humansperm motility and viability with two live-dead stainingtechniques. *Fertility and Sterility, 26*, 700–703. doi:10.1016/S0015-0282(16)41238-0 PMID:50237

Drabkina, B. M. (1961). Effect of different water salinities on the survival rate ofspermatozoa, eggs and larvae of sturgeon. *Doklady Akademii Nauk SSSR, 138*, 492–495.

Dreanno, C., Suquet, M., Desbruyeres, E., Cosson, J., Le Delliou, H., & Billard, R. (1998). Effect of urine on semen quality in Turbot, *Psettamaxima. Aquaculture (Amsterdam, Netherlands), 169*(3-4), 247–262. doi:10.1016/S0044-8486(98)00262-2

Dreier, J. W., Andersen, A.-M. N., & Berg-Beckhoff, G. (2014). Systematic review and meta-analyses: Fever in pregnancy and health impacts in the offspring. *Pediatrics, 133*(3), e674–e688. doi:10.1542/peds.2013-3205 PMID:24567014

Dreiling, C. E., Carman, F. S. III, & Brown, D. E. (1991). Maternal endocrine and fetal metabolic responses to heat stress. *Journal of Dairy Science, 74*(1), 312–327. doi:10.3168/jds.S0022-0302(91)78175-7 PMID:2030175

Dris, R., Gasperi, J., Rocher, V., Saad, M., Renault, N., & Tassin, B. (2015). Microplastic contamination in an urban area: A case study in Greater Paris. *Environmental Chemistry, 12*(5), 592–599. doi:10.1071/EN14167

Du Ruisseau, P., Tache, Y., Brazeau, P., & Collu, R. (1979). Effects of chronic immobilization stress on pituitary hormone secretion, on hypothalamic factor levels, and on pituitary responsiveness to LHRH and TRH in female rats. *Neuroendocrinology, 29*(2), 90–99. doi:10.1159/000122910 PMID:116141

Duncan, K. (2006). Global climate change, air pollution, and women's health. *WIT Trans Ecol and the Envir., 99*, 633–643. doi:10.2495/RAV060611

Dunlap, K., Brown, J., Keith, A., & Satterfield, M. (2015). Factors controlling nutrient availability to the developing fetus in ruminants. *Journal of Animal Science and Biotechnology, 6*(1), 16. doi:10.118640104-015-0012-5

Durairajanayagam, D., Agarwal, A., & Ong, C. (2015). Causes, effects and molecular mechanisms oftesticular heat stress. *Reproductive Biomedicine Online*, *30*(1), 14–27. doi:10.1016/j.rbmo.2014.09.018 PMID:25456164

Dutt, R. H. (1964). Detrimental effects of high ambient temperature on fertility and early embryo survival in sheep. *International Journal of Biometeorology*, *8*(1), 47–56. doi:10.1007/BF02186927 PMID:5888798

Duty, S. M., Ackerman, R. M., Calafat, A. M., & Hauser, R. (2005). Personal care product use predicts urinary concentrations of some phthalate monoesters. *Environmental Health Perspectives*, *113*(11), 1530–1535. doi:10.1289/ehp.8083

Duty, S. M., Calafat, A. M., Silva, M. J., Ryan, L., & Hauser, R. (2005). Phthalate exposure and reproductive hormones in adult men. *Human Reproduction (Oxford, England)*, *20*(3), 604–610. doi:10.1093/humrep/deh656

Duty, S. M., Silva, M. J., Barr, D. B., Brock, J. W., Ryan, L., Chen, Z., Herrick, R. F., Christiani, D. C., & Hauser, R. (2003). Phthalate exposure and human semen parameters. *Epidemiology (Cambridge, Mass.)*, *14*(3), 269–277. doi:10.1097/01.EDE.0000059950.11836.16

Dwevedi, S. (2013). Effect of textile dyes on Spirulina platensis. *Journal of Chemical and Pharmaceutical Research*, *5*, 66–80.

Ealy, A. D., Drost, M., & Hansen, P. J. (1993). Developmental changes in embryonic resistance to adverse effects of maternal heat stress in cows. *Journal of Dairy Science*, *76*(10), 2899–2905. doi:10.3168/jds.S0022-0302(93)77629-8 PMID:8227617

Ealy, A. D., Howell, J. L., Monterroso, V. H., Arechiga, C. F., & Hansen, P. J. (1995). Developmental changes in sensitivity of bovine embryos to heat shock and use of antioxidants as thermoprotectants. *Journal of Animal Science*, *73*(5), 1401–1407. doi:10.2527/1995.7351401x

Ecker, D. J., Stein, P., Xu, Z., Williams, C. J., Kopf, G. S., Bilker, W. B., Abel, T., & Schultz, R. M. (2004). Stein P, Xu Z, Williams CJ, Kopf GS, Bilker WB, Abel T, Schultz RM. Long-term effects of culture of preimplantation mouse embryos on behavior. *Proceedings of the National Academy of Sciences of the United States of America*, *101*(6), 1595–1600. doi:10.1073/pnas.0306846101 PMID:14747652

Edwards, J. L., & Hansen, P. J. (1996). Elevated temperature increases heat shock protein 70 synthesis in bovine two-cell embryos and compromises function of maturing oocytes. *Biology of Reproduction*, *55*(2), 340–346. doi:10.1095/biolreprod55.2.341 PMID:8828838

Edwards, J. L., & Hansen, P. J. (1997). Differential responses of bovine oocytes and pre- implantation embryos to heat shock. *Molecular Reproduction and Development*, *46*(2), 138–145. doi:10.1002/(SICI)1098-2795(199702)46:2<138::AID-MRD4>3.0.CO;2-R

Edwards, J. L., King, W. A., Kawarsky, S. J., & Ealy, A. D. (2001). Responsiveness of early embryos to environmental insults: Potential protective roles of HSP70 and glutathione. *Theriogenology*, *55*(1), 209–223. doi:10.1016/S0093-691X(00)00455-6 PMID:11198083

Edwards, J. L., Saxton, A. M., Lawrence, J. L., Payton, R. R., & Dunlap, J. R. (2005). Exposure to a physiologically relevant elevated temperature hastens in vitro maturation in bovine oocytes. *Journal of Dairy Science*, *88*(12), 4326–4333. doi:10.3168/jds.S0022-0302(05)73119-2 PMID:16291624

Edwards, M. J. (1971). The experimental production of arthrogryposis multiplex congenita in guinea pigs by maternal hyperthermia during gestation. *The Journal of Pathology*, *104*(4), 221–229. doi:10.1002/path.1711040403 PMID:5135658

Edwards, M. J. (2006). Review: Hyperthermia and fever during pregnancy. *Birth Defects Research. Part A, Clinical and Molecular Teratology*, *76*(7), 507–516. doi:10.1002/bdra.20277 PMID:16933304

Edwards, M. J., Saunders, R. D., & Shiota, K. (2003). Effects of heat on embryos and foetuses. *International Journal of Hyperthermia, 19*(3), 295–324. doi:10.1080/0265673021000039628 PMID:12745973

Edwards, M. J., Walsh, D. A., & Li, Z. (1999). Hyperthermia, teratogenesis and the heat shock response in mammalian embryos in culture. *The International Journal of Developmental Biology, 41*, 345–358. PMID:9184344

Effler, E., Isaacson, M., Arntzen, L., Heenan, R., Canter, P., Barrett, T., Lee, L., Mambo, C., Levine, W., Zaidi, A., & Griffin, P. M. (2001). Factors contributing to the emergence of Escherichia Coli O157 in Africa. *Emerging Infectious Diseases, 7*(5), 812–819. doi:10.3201/eid0705.017507 PMID:11747693

Egbunike, G. N., & Dede, T. I. (1979). The influence of short-term exposure to tropical sunlight on boar seminal characteristics. *Biology of Reproduction, 20*, 235–241.

Einarsson, S., Brandt, Y., Lundeheim, N., & Madej, A. (2008). Stress and its influence on reproduction in pigs: A review. *Acta Veterinaria Scandinavica, 50*(1), 48. doi:10.1186/1751-0147-50-48

Elhakim, A. F. (2016). Estimation of soil permeability. *Alexandria Engineering Journal, 55*(3), 2631–2638. doi:10.1016/j.aej.2016.07.034

Elias, S. G., van Noord, P. A., Peeters, P. H., den Tonkelaar, I., Kaaks, R., & Grobbee, D. E. (2007). Menstruation during and after caloric restriction: The 1944–1945 Dutch famine. *Fertility and Sterility, 88*(4), 1101–1107. doi:10.1016/j.fertnstert.2006.12.043 PMID:17544416

El-Raey, M., Geshi, M., Somfai, T., Kaneda, M., Hirako, M., Abdel-Ghaffar, A. E., Sosa, G. A., El-Roos, M. E. A. A., & Nagai, T. (2011). Evidence of melatonin synthesis in the cumulus oocyte complexes and its role in enhancing oocyte maturation in vitro in cattle. *Molecular Reproduction and Development, 78*(4), 250–262. doi:10.1002/mrd.21295 PMID:21381146

Elster, J., & Mann, H. (1952). Weitereuntersuchungenuber die physiologiederbefruchtung und die zuordnug der gametenbeifischen. *Arch HydrobiolSuppl, 20*, 267–276.

El-Tarabany, M. S., El-Bayomi, K., & Abdelhamid, T. (2015). Semen characteristics of purebred and crossbred male rabbits. *PLoS One, 10*(5), e0128435-e0128435. doi:10.1371/journal.pone.0128435

El-Tarabany, M. S., & El-Tarabany, A. A. (2015). Impact of thermal stress on the efficiency of ovulation synchronization protocols in Holstein cows. *Animal Reproduction Science, 160*, 138–145. doi:10.1016/j.anireprosci.2015.08.002 PMID:26297623

Ema, M., Okuda, H., Gamo, M., Honda, K. A. (2017). Review of reproductive and developmental toxicity of silver nanoparticles in laboratory animals. *Reproductive Toxicology, 67*, 149-64.

Enders, A. C. (1991). Structural responses of the primate endometrium to implantation. *Placenta, 12*(4), 309–325. doi:10.1016/0143-4004(91)90340-L PMID:1946242

Enders, A. C., Hendrickx, A. G., & Schlafke, S. (1983). Implantation in the rhesus monkey: Initial penetration of endometrium. *The American Journal of Anatomy, 167*(3), 275–298. doi:10.1002/aja.1001670302 PMID:6881070

EPA. (1996). *Textile Mills Effluent Guidelines*. https://www.epa.gov/eg/textile-mills-effluent-guidelines

Erkkola, R. U., Pirhonen, J. P., & Kivijarvi, A. K. (1992). Flow velocity waveforms in uterine and umbi- lical arteries during submaximal bicycle exercise in normal pregnancy. *Obstetrician & Gynaecologist, 79*, 611–615. PMID:1553187

Escarcha, J., Lassa, J., & Zander, K. (2018). Livestock under Climate Change: A Systematic Review of Impacts and Adaptation. *Climate (Basel), 6*(3), 54. doi:10.3390/cli6030054

Eskew, A., Reschke, L., Broughton, D. E., Schulte, M., & Jungheim, E. (2017). Chronodisruption and decreased antral follicle count. *Fertility and Sterility*, *108*(3), e323. doi:10.1016/j.fertnstert.2017.07.955

Eswaramoorthi, S., Dhanapal, K., & Chauhan, D. (2008). *Advanced in textile waste water treatment: The case for UV-ozonation and membrane bioreactor for common effluent treatment plants in Tirupur, Tamil Nadu, India*. Environment with People's Involvement and Coordination in India.

Eugster, A., & Vingerhoets, A. J. J. M. (1999). Psychological aspects of in vitro fertilization: A review. *Social Science & Medicine*, *48*(5), 575–589. doi:10.1016/S0277-9536(98)00386-4 PMID:10080360

Evans, N. P., North, T., Dye, S., & Sweeney, T. (2004). Differential effects of the endocrine-disrupting compounds bisphenol-A and octylphenol on gonadotropin secretion, in prepubertal ewe lambs. *Domestic Animal Endocrinology*, *26*(1), 61–73. doi:10.1016/j.domaniend.2003.09.005 PMID:14732453

EWA. (2005). *Efficient use of water in the textile finishing industry. Official Publication of European Water Association*. EWA.

Falany, J. L., & Falany, C. N. (1996). Regulation of estrogen sulfotransferase in human endometrial adenocarcinoma cells by progesterone. *Endocrinology*, *137*(4), 1395–1401. doi:10.1210/endo.137.4.8625916 PMID:8625916

FAO. (2007). *Adaptation to climate change in agriculture, forestry, and fisheries: perspective, framework and priorities.* http://www.fao.org/3/a-au030e.pdf

FAO. (2009). *The state of food and agriculture: Livestock in the balance.* http://www.fao.org/3/a-i0680e.pdf

Fatma, B. A., Nozha, C. F., Ines, D., Hamadi, A., Basma, H., & Leila, A. K. (2009). Sperm quality improvement after date seed oil in vitro supplementation in spontaneous and induced oxidative stress. *Asian Journal of Andrology*, *11*(3), 393–398. doi:10.1038/aja.2008.6 PMID:19151733

Fazleabas, A. T., Bell, S. C., Fleming, S., Sun, J., & Lessey, B. A. (1997). Distribution of integrins and the extracellular matrix proteins in the baboon endometrium during the menstrual cycle and early pregnancy. *Biology of Reproduction*, *56*(2), 348–356. doi:10.1095/biolreprod56.2.348 PMID:9116133

Fear, J. M., & Hansen, P. J. (2011). Developmental changes in expression of genes involved in regulation of apoptosis in the bovine pre-implantation embryo. *Biology of Reproduction*, *84*(1), 43–51. doi:10.1095/biolreprod.110.086249

Feichtinger, W. (1991). Environmental factors and fertility. *Human Reproduction (Oxford, England)*, *6*, 1170–1175. doi:10.1093/oxfordjournals.humrep.a137505 PMID:1806580

Feinberg, R. F., & Kliman, H. J. (1997). *MAG (mouse ascites Golgi) mucin in the endometrium: a potential marker of endometrial receptivity to implantation. Endometrium and Endometriosis*. Blackwell Science.

Fenkes, M., Fitzpatrick, J. L., Ozolina, K., Shiels, H. A., & Nudds, R. L. (2017). Sperm in hot water: Direct and indirect thermal challenges interact to impact on brown trout sperm quality. *The Journal of Experimental Biology*, *220*(14), 2513–2520. doi:10.1242/jeb.156018 PMID:28455440

Fenster, L., Schaefer, C., Mathur, A., Hiatt, R. A., Pieper, C., Hubbard, A. E., von Behren, J., & Swan, S. H. (1995). Psychologic stress in the workplaceand spontaneous abortion. *American Journal of Epidemiology*, *142*(11), 1176–1183. doi:10.1093/oxfordjournals.aje.a117576 PMID:7485064

Fernandes, F. H., Bustos-Obregon, E., & Salvadori, D. M. F. (2015). Disperse Red 1 (textile dye) induces cytotoxic and genotoxic effects in mouse germ cells. *Reproductive Toxicology (Elmsford, N.Y.)*, *53*, 75–81. doi:10.1016/j.reprotox.2015.04.002 PMID:25883024

Ferraz, E. R. A., Umbuzeiro, G. A., de-Almeida, G., Caloto-Oliveira, A., Chequer, F. M. D., Zanoni, M. V. B., Dorta, D. J., & Oliveira, D. P. (2011). Differential toxicity of Disperse Red 1 and Disperse Red 13 in the ames test, HepG2 cytotoxicity assay, and daphnia acute toxicity test. *Environmental Toxicology*, *26*(5), 489–497. doi:10.1002/tox.20576 PMID:20549607

Ferrero, F., & Periolatto, M. (2012). Ultrasound for low temperature dyeing of wool with acid dye. *Ultrasonics Sonochemistry*, *19*(3), 601–606. doi:10.1016/j.ultsonch.2011.10.006 PMID:22055328

Ferronato, N., & Torretta, V. (2019). Waste Mismanagement in Developing Countries: A Review of Global Issues. *International Journal of Environmental Research and Public Health*, *16*(6), 1060. doi:10.3390/ijerph16061060 PMID:30909625

Fiandanese, N., Borromeo, V., Berrini, A., Fischer, B., Schaedlich, K., Schmidt, J.-S., Secchi, C., & Pocar, P. (2016). Maternal exposure to a mixture of di (2-ethylhexyl) phthalate (DEHP) and polychlorinated biphenyls (PCBs) causes reproductive dysfunction in adult male mouse offspring. *Reproductive Toxicology (Elmsford, N.Y.)*, *65*, 123–132. doi:10.1016/j.reprotox.2016.07.004

Figa-Talamanca, I., Cini, C., Varricchio, G. C., Dondero, F., Gandini, L., Lenzi, A., Lombardo, F., Angelucci, L., DiGrezia, R., & Patacchioli, F. R. (1996). Effects of prolonged autovehicle driving on malereproduction function: A study among taxi drivers. *American Journal of Industrial Medicine*, *30*, 750–758. doi:10.1002/(SICI)1097-0274(199612)30:6<750::AID-AJIM12>3.0.CO;2-1 PMID:8914722

Figueiro, M. G., Radetsky, L., Plitnick, B., & Rea, M. S. (2017). Glucose tolerance in mice exposed to light-dark stimulus patterns mirroring dayshift and rotating shift schedules. *Scientific Reports*, *7*(1), 40661. doi:10.1038rep40661 PMID:28079162

Finklea, H. O. (1998). Semiconductor electrodes. Elsevier.USA, p-199. ISBN: 0-444-42926-3.

Firmino, P. I. M., Silva, M. E. R., Cervantes, F. J., & Santos, A. B. (2010). Colour removal of dyes from synthetic and real textile wastewaters in one- and two-stage anaerobic systems. *Bioresource Technology*, *101*(20), 7773–7779. doi:10.1016/j.biortech.2010.05.050 PMID:20542688

Fisch, H., Hyun, G., & Golden, R. (2000). The possible effects of environmentalestrogen disrupters on reproductive health. *Current Urology Reports*, *1*(4), 253–261. doi:10.100711934-000-0004-6 PMID:12084301

Fisher, J.S. (2004) Environmental anti-androgens and male reproductive health: Focus on Phthalates and testicular dysgenesis syndrome. *Reproduction, 127*, 305–315.

Fisher, J. S. (2004). Environmental anti-androgens and male reproductive health: Focus on phthalates and testicular dysgenesis syndrome. *Reproduction (Cambridge, England)*, *127*(3), 305–315. doi:10.1530/rep.1.00025 PMID:15016950

Fisher, J. S., Macpherson, S., Marchetti, N., & Sharpe, R. M. (2003). Human 'testicular dysgenesis syndrome': A possible model using in-utero exposure of the rat to dibutyl phthalate. *Human Reproduction (Oxford, England)*, *18*(7), 1383–1394. doi:10.1093/humrep/deg273

Fleming, C., Nice, L., Hughes, A. O., & Hull, M. G. R. (1994). Apparent lack ofseasonal variation in implantation rates and in-vitro fertilization. *Human Reproduction (Oxford, England)*, *9*, 2164–2166. doi:10.1093/oxfordjournals.humrep.a138411 PMID:7868692

Flores, E. R., Perez, F., & Torre, M. (1997). Scale-up of Bacillus thuringiensis fermentation based on oxygen transfer. *Journal of Fermentation and Bioengineering*, *83*(6), 561–564. doi:10.1016/S0922-338X(97)81137-3

Foley, G. L., & Schlafer, D. H. (1994). Bacterial endotoxemia and reproductive effects in ruminants. *The Veterinary Clinics of North America. Food Animal Practice*, *10*(3), 491–501. doi:10.1016/S0749-0720(15)30534-X PMID:7728632

Fonken, L. K., Aubrecht, T. G., Meléndez-Fernández, O. H., Weil, Z. M., & Nelson, R. J. (2013). Dim light at night disrupts molecular circadian rhythms and increases body weight. *Journal of Biological Rhythms*, *28*(4), 262–271. doi:10.1177/0748730413493862 PMID:23929553

Fonken, L. K., & Nelson, R. J. (2014). The effects of light at night on circadian clocks and metabolism. *Endocrine Reviews*, *35*(4), 648–670. doi:10.1210/er.2013-1051 PMID:24673196

Fonken, L. K., Workman, J. L., Walton, J. C., Weil, Z. M., Morris, J. S., Haim, A., & Nelson, R. J. (2010). Light at night increases body mass by shifting the time of food intake. *Proceedings of the National Academy of Sciences of the United States of America*, *107*(43), 18664–18669. doi:10.1073/pnas.1008734107 PMID:20937863

Fonseca, F. A., Britt, J. H., McDaniel, B. T., Wilk, J. C., & Rakes, A. H. (1983). Reproductive traits of Holsteins and Jerseys. Effects of age, milk yield, and clinical abnormalities on involution of cervix and uterus, ovulation, estrous cycles, detection ofestrus, conception rate, and days open. *Journal of Dairy Science*, *66*(5), 1128–1147. doi:10.3168/jds.S0022-0302(83)81910-9 PMID:6683729

Foote, R. (1978). Factors influencing the quantity and quality of semen harvested from bulls, rams, boars and stallions. *Journal of Animal Science, 47*(suppl_II), 1-11.

Fornazari, F., da Silva, R. C., Richini-Pereira, V. B., Beserra, H. E., Luvizotto, M. C., & Langoni, H. (2012). Comparison of conventional PCR, quantitative PCR, bacteriological culture and the Warthin Starry technique to detect Leptospira spp. in kidney and liver samples from naturally infected sheep from Brazil. *Journal of Microbiological Methods*, *90*(3), 321–326. doi:10.1016/j.mimet.2012.06.005 PMID:22713608

Foster, P. M. (2006). Disruption of reproductive development in male rat offspring following in utero exposure to phthalate esters. *International Journal of Andrology*, *29*(1), 140–147. doi:10.1111/j.1365-2605.2005.00563.x

Foster, P. M. D., Blacburn, D. M., Moore, R. B., & Llovd, S. C. (1986). Testicular toxicity of 2-methyoxyacetaldehyde, a possible metabolite of ethylene glycol monomethyl ether, in the rat. *Toxicology Letters*, *32*(1-2), 73–80. doi:10.1016/0378-4274(86)90051-2 PMID:3738932

Foster, P. M. D., Mylchreest, E., Gaido, K. W., & Sar, M. (2001). Effects of phthalate esters on the developing reproductive tract of male rats. *APMIS*, *109*(S103), S272–S277. doi:10.1111/j.1600-0463.2001.tb05776.x

Foster, P. M. D., Mylchreest, E., Gaido, K. W., & Sar, M. (2001). Effects of phthalateesters on the developing reproductive tract of male rats. *Human Reproduction Update*, *7*(3), 231–235. doi:10.1093/humupd/7.3.231 PMID:11392369

Foster, W. G., & Holloway, A. C. (2003). Do environmental contaminantsadversely affect human reproductive physiology? *Journal of Obstetrics and Gynaecology Canada*, *25*(1), 33–44. doi:10.1016/S1701-2163(16)31080-5 PMID:12548323

Franceschini, M. D., Custer, C. M., Custer, T. W., Reed, J. M., & Romero, L. M. (2008). Corticosterone stress response in tree swallows nesting near polychlorinated biphenyl-and dioxin-contaminated rivers. *Environmental Toxicology and Chemistry: An International Journal*, *27*(11), 2326–2331. doi:10.1897/07-602.1 PMID:18476749

Francisco, I., Pugnaire, Morillo, J. A., Peñuelas, J., Reich, P. B., & Bardgett, R. D. (2019). Climate change effects on plant-soil feedbacks and consequences for biodiversity and functioning of terrestrial ecosystems. *Science Advances*, *5*(11), 1–12. PMID:31807715

Franks, S. (1995). Polycystic ovary syndrome. *The New England Journal of Medicine*, *333*(13), 853–861. doi:10.1056/NEJM199509283331307 PMID:7651477

Franks, S., & Hardy, K. (2010). Aberrant follicle development and anovulation in polycystic ovary syndrome. *Annales d'Endocrinologie*, *71*(3), 228–230. doi:10.1016/j.ando.2010.02.007 PMID:20362969

Fraser, F. C., & Skelton, J. (1978). Possible teratogenicity of maternal fever. *Lancet*, *2*(8090), 634–634. doi:10.1016/S0140-6736(78)92859-3 PMID:80560

Frazer, L. (2009). Climate change: Will warmer soil be as fertile? *Environmental Health Perspectives*, *117*(2), 58–61. doi:10.1289/ehp.117-a59 PMID:19270772

Freeman, A. E. (1984). Secondary traits: Sire evaluation and the reproductive complex. *Journal of Dairy Science*, *67*(2), 449–458. doi:10.3168/jds.S0022-0302(84)81324-7 PMID:6371077

Freund, M., & Peterson, R. N. (1976). Semen evaluation and fertility. In E. S. Hafez (Ed.), Human Semen and Fertility-Regulation in Men (p. 34454). Academic Press.

Frumkin, H., Hess, J., Luber, G., Malilay, J., & McGeehin, M. (2008). Climate change: The public health response. *American Journal of Public Health*, *98*(3), 435–445. doi:10.2105/AJPH.2007.119362 PMID:18235058

Fukuda, M. N., Sato, T., Nakayama, J., Klier, G., Mikami, M., Aoki, D., & Nozawa, S. (1995). Trophinin and tastin, a novel cell adhesion molecule complex with potential involvement in embryo implantation. *Genes & Development*, *9*(10), 1199–1210. doi:10.1101/gad.9.10.1199 PMID:7758945

Fukuda, M., Fukuda, K., Shimizu, T., Nobunaga, M., Mamsen, L. S., & Yding Andersen, C. (2014). Climate change is associated with male:female ratios of fetal deaths and newborn infants in Japan. *Fertility and Sterility*, *102*(5), 1364–1370. doi:10.1016/j.fertnstert.2014.07.1213 PMID:25226855

Fukushima, M., Tatsumi, K., & Morimoto, K. (2010). The fate of aniline after a photo-fenton reaction in an aqueous system containing iron (III), humic acid, and hydrogen peroxide. *Environmental Science & Technology*, 34.

Fuortes, L., Clark, M. K., Kirchner, H. L., & Smith, E. M. (1997). Associationbetween female infertility and agricultural work history. *Am J IndustrMed*, *31*(4), 445–451. doi:10.1002/(SICI)1097-0274(199704)31:4<445::AID-AJIM11>3.0.CO;2-# PMID:9093660

Galan, H. L., Anthony, R. V., Rigano, S., Parker, T. A., de Vrijer, B., Ferrazzi, E., & Regnault, T. R. H. (2005). Fetal hypertension and abnormal Doppler velocimetry in an ovine model of intrauterine growth restriction. *American Journal of Obstetrics and Gynecology*, *192*(1), 272–279. doi:10.1016/j.ajog.2004.05.088 PMID:15672036

Gallinelli, A., Roncaglia, R., Matteo, M. L., Ciaccio, I., Volpe, A., & Facchinetti, F. (2001). Immunological changes and stress are associated with differentimplantation rates in patients undergoing in vitro fertilization-embryotransfer. *Fertility and Sterility*, *76*(1), 85–91. doi:10.1016/S0015-0282(01)01826-X PMID:11438324

Gallis, J. L., Fedrigo, E., Jatteau, P., Bonpunt, E., & Billard, R. (1991). Siberiansturgeon spermatozoa: Effects of dilution, pH, osmotic pressure,sodium and potassium ions on motility. In Acipenser (pp. 143-51). Bordeaux, France: Academic Press.

Gamble, K. L., Resuehr, D., & Johnson, C. H. (2013). Shift work and circadian dysregulation of reproduction. *Frontiers in Endocrinology*, *4*, 1–9. doi:10.3389/fendo.2013.00092 PMID:23966978

Gangwar, P. C., Branton, C., & Evans, D. L. (1965). Reproductive and physiological response of Holstein heifers to controlled and natural climatic conditions. *Journal of Dairy Science*, *48*(2), 222–227. doi:10.3168/jds.S0022-0302(65)88200-5 PMID:14277425

Gao, C., Han, H. B., Tian, X. Z., Tan, D. X., Wang, L., Zhou, G., Zhu, S.-E., & Liu, G.-S. (2012). Melatonin promotes embryonic development and reduces reactive oxygen species in vitrified mouse 2-cell embryos. *Journal of Pineal Research*, *52*(3), 305–311. doi:10.1111/j.1600-079X.2011.00944.x PMID:22225541

Garaulet, M., Ordovás, J. M., & Madrid, J. A. (2010). The chronobiology, etiology and pathophysiology of obesity. *International Journal of Obesity*, *34*(12), 1667–1683. doi:10.1038/ijo.2010.118 PMID:20567242

Garcia-Ispierto, I., Lopez-Gatius, F., Santolaria, P., Yaniz, J. L., Nogareda, C., Lopez-Bejar, M., & Rensis, F. (2005). Relationship between heat stress during the peri-implantation period and early fetal loss in dairy cattle. *Theriogenology, 2006*(65), 799–807. PMID:16085298

Garrigue, M. (2017). *Effet du stress thermique sur les paramètres séminologiques de taureaux de centre d'insémination* (Thèse d'exercice, Médecine vétérinaire). Ecole Nationale Vétérinaire de Toulouse, Toulouse, France.

Gaughan, J. B., Mader, T. L., Holt, S. M., Sullivan, M. L., & Hahn, G. L. (2010). Assessing the heat tolerance of 17 beef cattle genotypes. *International Journal of Biometeorology, 54*(6), 617–627. doi:10.100700484-009-0233-4 PMID:19458966

Gauthier, D., Coulaud, G., & Varo, H. (1986). The influence of season and shade on estrous behaviour, timing of preovulatory LH surge and the pattern of progesterone secretion in FFPN and Creole heifers in a tropical climate. *Reproduction, Nutrition, Developpement, 26*(3), 767–775. doi:10.1051/rnd:19860502 PMID:3749595

Geisert, R., Morgan, G., Short, E., & Zavy, M. (1992). Endocrine events associated with endometrial function and conceptus development in cattle. *Reproduction, Fertility, and Development, 4*(3), 301–305. doi:10.1071/RD9920301

Gelybó, G., Tóth, E., Farkas, C., Horel, Á., Kása, I., & Bakacsi, Z. (2018). Potential impacts of climate change on soil properties. *Agrokémia és Talajtan, 67*(1), 121–141. doi:10.1556/0088.2018.67.1.9

Gendelman, M., Aroyo, A., Yavin, S., & Roth, Z. (2010). Seasonal effects on gene expression, cleavage timing, and developmental competence of bovine preimplantation embryos. *Reproduction (Cambridge, England), 140*(1), 73–82. doi:10.1530/REP-10-0055

Gendelman, M., & Roth, Z. (2012). In vivo vs. in vitro models for studying the effects of elevated temperature on the GV-stage oocyte, subsequent developmental competence and gene expression. *Animal Reproduction Science, 134*(3-4), 125–134. doi:10.1016/j.anireprosci.2012.07.009 PMID:22898494

Gerber, W.L., De la Pena, V.E., & Mobley, W.C. (1988). Infertility, chemical lexposure, and farming in Iowa: Absence of an association. *Urology, 31*, 46–50.

Gerber, P. J., Steinfeld, H., Henderson, B., Mottet, A., Opio, C., Dijkman, J., Falcucci, A., & Tempio, G. (2013). *Tackling Climate Change through Livestock: A Global Assessment of Emissions and Mitigation Opportunities*. FAO.

Germain, M. A., Webster, W. S., & Edwards, M. J. (1985). Hyperthermia as a teratogen: Parameters determining hyperthermia-induced defects in the rat. *Teratology, 31*(2), 265–272. doi:10.1002/tera.1420310212 PMID:3992495

Ghasemi, N., Babaei, H., Azizallahi, S., & Kheradmand, A. (2009). Effect of long-term administration ofzinc after scrotal heating on mice spermatozoa and subsequent offspring quality. *Andrologia, 41*(4), 222–228. doi:10.1111/j.1439-0272.2009.00920.x PMID:19601933

Ghosh, D., De, P., & Sengupta, J. (1994). Luteal phase ovarian oestrogen is not essential for implantation and maintenance of pregnancy from surrogate embryo transfer in the rhesus monkey. *Human Reproduction (Oxford, England), 9*(4), 629–637. doi:10.1093/oxfordjournals.humrep.a138561 PMID:8046014

Gilad, E., Meidan, R., Berman, A., Graber, Y., & Wolfenson, D. (1993). Effect of heat stress on tonic and GnRH-induced gonadotrophin secretion in relation to concentration of oestradiol in plasma of cyclic cows. *Journal of Reproduction and Fertility, 99*(2), 315–321. doi:10.1530/jrf.0.0990315 PMID:8107012

Gilbert-Barness, E. (2010). Review: Teratogenic causes of malformations. *Annals of Clinical and Laboratory Science, 40*, 99–114. PMID:20421621

Ginzburg, A. S. (1968). Fertilization of fishes and the problem of polyspermy. New York: Academy of Science USSR.

Giudice, L. C. (1999). Potential biochemical markers of uterine receptivity. *Human Reproduction (Oxford, England), 14*(suppl 2), 3–16. doi:10.1093/humrep/14.suppl_2.3 PMID:10690796

Glasser, S. R., Mulholland, J., Mani, S. K., & Julian, J. (1991). Blastocyst-endometrial relationships: Reciprocal interactions between uterine epithelial and stromal cells and blastocysts. *Trophoblast Research, 5*, 229–229.

Glaze, W. H., Kang, J. W., & Chapin, D. H. (1987). The chemistry of water treatment processes involving ozone, hydrogen peroxide and ultraviolet radiation. *Ozone Science and Engineering, 9*, 335–342.

Gogoi, P. P., Vinoj, V., Swain, D., Roberts, G., Dash, J., & Tripathy, S. (2019). Land use and land cover change effect on surface temperature over Eastern India. *Scientific Reports, 9*(1), 8859–8869. doi:10.103841598-019-45213-z PMID:31222135

Goldhaber, M. K., Polen, M. R., & Hiatt, R. A. (1988). The risk of miscarriage andbirth defects among women who use visual display terminals duringpregnancy. *American Journal of Industrial Medicine, 13*(6), 695–706. doi:10.1002/ajim.4700130608 PMID:3389364

Goldstein, C. A. & Smith, Y. R. (2016). Sleep, Circadian Rhythms, and Fertility. *Current Sleep Medicine Reports.*

Gómez-Acebo, I., Dierssen-Sotos, T., Papantoniou, K., García-Unzueta, M. T., Santos-Benito, M. F., & Llorca, J. (2015). Association between exposure to rotating night shift versus day shift using levels of 6-sulfatoxymelatonin and cortisol and other sex hormones in women. *Chronobiology International, 32*(1), 128–135. doi:10.3109/07420528.2014.958494 PMID:25216206

Gordon, M. (1975). Cyclic changes in the fine structure of the epithelial cells of human endometrium. *International Review of Cytology, 42*, 127–172. doi:10.1016/S0074-7696(08)60980-8 PMID:172466

Gore, A. C., Attardi, B., & DeFranco, D. B. (2006). Glucocorticoid repression of the reproductive axis: Effects on GnRH and gonadotropin subunit mRNA levels. *Molecular and Cellular Endocrinology, 256*(1-2), 40–48. doi:10.1016/j.mce.2006.06.002 PMID:16839661

Graham, C. H., Lysiak, J. J., McCrae, K. R., & Lala, P. K. (1992). Localization of transforming growth factor-β at the human fetal-maternal interface: Role in trophoblast growth and differentiation. *Biology of Reproduction, 46*(4), 561–572. doi:10.1095/biolreprod46.4.561 PMID:1374270

Graham, C. H., McCrae, K. R., & Lala, P. K. (1993). Molecular mechanisms controlling trophoblast invasion of the uterus. *Placenta, 14*, 237–250. doi:10.1016/S0143-4004(05)80297-8

Graham, J. D., Yeates, C., Balleine, R. L., Harvey, S. S., Milliken, J. S., Bilous, A. M., & Clarke, C. L. (1996). Progesterone receptor A and B protein expression in human breast cancer. *The Journal of Steroid Biochemistry and Molecular Biology, 56*(1-6), 93–98. doi:10.1016/0960-0760(95)00226-X PMID:8603052

Graham, J. M. Jr. (2005). Marshall J. Edwards: Discoverer of maternal hyperthermia as a human teratogen. *Birth Defects Research. Part A, Clinical and Molecular Teratology, 73*(11), 857–864. doi:10.1002/bdra.20185 PMID:16265640

Graham, J. M. Jr, Edwards, M. J., & Edwards, M. J. (1998). Teratogen update: Gestational effects of maternal hyperthermia due to febrile illness and result patterns of defects in human. *Teratology, 58*(5), 209–221. doi:10.1002/(SICI)1096-9926(199811)58:5<209::AID-TERA8>3.0.CO;2-Q PMID:9839360

Graves, K., Seibert, J., Keating, A., Baumgard, L., & Ross, J. (2018). Characterizing the acute heat stress response in gilts: II.Assessing repeatability and association with fertility. *Journal of Animal Science, 96*(6), 10. doi:10.1093/jaskx037 PMID:29788126

Gray, L. E. Jr, & Ostby, J. S. (1993). The effect of prenatal administration of azo dyes on testicular development in the mouse: A structure activity profile of dyes derived from benzidine, dimethyl benzidine or dimethoxy benzidine. *Fundamental and Applied Toxicology*, *20*(2), 177–183. doi:10.1006/faat.1993.1024 PMID:8449389

Gray, L. E. Jr, Ostby, J. S., Kavlock, R. J., & Marshall, R. (1992). Gonadal effects of fetal exposure to the azo dye Congo red in mice: Infertility in female but not male offspring. *Fundamental and Applied Toxicology*, *19*(3), 411–422. doi:10.1016/0272-0590(92)90180-P PMID:1459372

Gray, L. E., Ostby, J., Furr, J., Wolf, C. J., Lambright, C., Parks, L., Veeramachaneni, D. N., Wilson, V., Price, M., & Hotchkiss, A. (2001). Effects of environmentalantiandrogens on reproductive development in experimental animals. *Human Reproduction Update*, *7*(3), 248–264. doi:10.1093/humupd/7.3.248 PMID:11392371

Gray, T. J. B., & Beamand, J. A. (1984). Effect of some phthalate esters and other testicular toxins on primary cultures of testicular cells. *Food and Chemical Toxicology*, *22*(2), 123–131. doi:10.1016/0278-6915(84)90092-9

Greer, E. (1983). Heat stress and reproduction in pigs: Its role in seasonal infertility. *Recent Advances in Animal Nutrition in Australia*, *7*, 216–223.

Grimaldi, B., Bellet, M. M., Katada, S., Astarita, G., Hirayama, J., Amin, R. H., Granneman, J. G., Piomelli, D., Leff, T., & Sassone-Corsi, P. (2010). PER2 controls lipid metabolism by direct regulation of PPARγ. *Cell Metabolism*, *12*(5), 509–520. doi:10.1016/j.cmet.2010.10.005 PMID:21035761

Groll, J. M., Usadi, R. S., Lessey, B. A., Lininger, R., Young, S. L., & Fritz, M. A. (2009). Effects of variations in serum estradiol concentrations on secretory endometrial development and function in experimentally induced cycles in normal women. *Fertility and Sterility*, *92*(6), 2058–2061. doi:10.1016/j.fertnstert.2009.06.018 PMID:19608171

Grossi, G., Goglio, P., Vitali, A., & Williams, A. (2018). Livestock and climate change: Impact of livestock on climate and mitigation strategies. *Animal Frontiers*, *9*(1), 69–76. doi:10.1093/af/vfy034

Guaratini, C. C. I., & Zanoni, M. V. B. (2000). Textile dyes. *Quimica Nova*, *23*(1), 71–78. doi:10.1590/S0100-40422000000100013

Guerine, P., Mouatassim, S., & Menezo, Y. (2001). Oxidative stress and protection against reactive oxygen species in the pre-implantation embryo and its surroundings. *Human Reproduction Update*, *7*(d), 175–189. doi:10.1093/humupd/7.2.175 PMID:11284661

Guillette, L. J. Jr, & Gunderson, M. P. (2001). Alterations in development ofreproductive and endocrine system of wildlife populations exposed toendocrine-disrupting contaminants. *Reproduction (Cambridge, England)*, *122*, 857–864. doi:10.1530/rep.0.1220857 PMID:11732981

Gupta, R. S., Sharma, R., Chaudhary, R., Yadav, R. K., & Khan, T. I. (2003). Effect of textile waste water on the spermatogenesis of male albino rats. *Journal of Applied Toxicology*, *23*(3), 171–175. doi:10.1002/jat.862 PMID:12794938

Gurpide, E., Gusberg, S. B., & Tseng, L. (1976). Estradiol binding and metabolism in human endometrial hyperplasia and adenocarcinoma. *Journal of Steroid Biochemistry*, *7*(11-12), 891–896. doi:10.1016/0022-4731(76)90007-8 PMID:1025366

Gwatkin, R. (2012). *Fertilization mechanisms in man and mammals*. Springer Science and Business Media.

Gwazdauskas, F. C., Thatcher, W. W., Kiddy, C. A., Pape, M. J., & Wilcox, C. J. (1981). Hormonal pattern during heat stress following PGF2alpha-tham salt induced luteal regression in heifers. *Theriogenology*, *16*(3), 271–285. doi:10.1016/0093-691X(81)90012-1 PMID:16725640

Habbema, J. D. F., Collins, J., Leridon, H., Evers, J. L., Lunenfeld, B., & teVelde, E. R. (2004). Towards less confusing terminology in reproductive medicine: A proposal. *Human Reproduction (Oxford, England)*, *19*(7), 1497–1501. doi:10.1093/humrep/deh303

Habeeb, A., Gad, A., & Atta, M. (2018). Temperature-Humidity Indices as Indicators to Heat Stress of Climatic Conditions with Relation to Production and Reproduction of Farm Animals. *International Journal of Biotechnology and Recent Advances*, *1*(1), 35–50. doi:10.18689/ijbr-1000107

Ha, E., Cho, S.-I., Chen, D., Chen, C., Ryan, L., Smith, T. J., Xu, X., & Christiani, D. C. (2002). Parental exposure to organic solvents and reduced birth weight. *Archives of Environmental Health: An International Journal*, *57*(3), 207–214. doi:10.1080/00039890209602938

Halden, R. U. (2010). Plastics and public health. *Annual Review of Public Health*, *31*(1), 179–194. doi:10.1146/annurev.publhealth.012809.103714

Hale, B. J., Hager, C. L., Seibert, J. T., Selsby, J. T., Baumgard, L. H., Keating, A. F., & Ross, J. W. (2017). Heat stress induces autophagy in pig ovaries during follicular development. *Biology of Reproduction*, *97*(3), 426–437. doi:10.1093/biolre/iox097 PMID:29025092

Hallberg, I., Kjellgren, J., Persson, S., Örn, S., & Sjunnesson, Y. (2019). Perfluorononanoic acid (PFNA) alters lipid accumulation in bovine blastocysts after oocyte exposure during in vitro maturation. *Reproductive Toxicology (Elmsford, N.Y.)*, *84*, 1–8. doi:10.1016/j.reprotox.2018.11.005

Hamada, T., Antle, M. C., & Silver, R. (2004). Temporal and spatial expression patterns of canonical clock genes and clock-controlled genes in the suprachiasmatic nucleus. *The European Journal of Neuroscience*, *19*(7), 1741–1748. doi:10.1111/j.1460-9568.2004.03275.x PMID:15078548

Hamerezaee, M., Dehghan, S. F., Golbabaei, F., Fathi, A., Barzegar, L., & Heidarnejad, N. (2018, June). Assessment of Semen Quality among Workers Exposed to Heat Stress: A Cross-Sectional Study in a Steel Industry. *Safety and Health at Work*, *9*(2), 232–235. doi:10.1016/j.shaw.2017.07.003 PMID:29928539

Hamidov, A., Helming, K., Bellocchi, G., Bojar, W., Dalgaard, T., Ghaley, B. B., Hoffmann, C., Holman, I., Holzkämper, A., Krzeminska, D., Kvaernø, S. H., Lehtonen, H., Niedrist, G., Øygarden, L., Reidsma, P., Roggero, P. P., Rusu, T., Santos, C., Seddaiu, G., ... Schönhart, M. (2018). Impacts of climate change adaptation options on soil functions: A review of European case-studies. *Land Degradation & Development*, *29*(8), 2378–2389. doi:10.1002/ldr.3006 PMID:30393451

Hamilton, G. S., Lysiak, J. J., Han, V. K. M., & Lala, P. K. (1998). Autocrine-paracrine regulation of human trophoblast invasiveness by insulin-like growth factor (IGF)-II and IGF-binding protein (IGFBP)-1. *Experimental Cell Research*, *244*(1), 147–156. doi:10.1006/excr.1998.4195 PMID:9770358

Hammarberg, K., Astbury, J., & Baker, H. W. G. (2001). Women's experience ofIVF: A follow-up study. *Human Reproduction (Oxford, England)*, *16*(2), 374–383. doi:10.1093/humrep/16.2.374 PMID:11157838

Hansen, J., Sato, M., Ruedy, R., Lo, K., Lea, D. W., & Medina-Elizade, M. (2006). Global temperature change. *Proceedings of the National Academy of Sciences of the United States of America*, *103*(39), 14288–14293. doi:10.1073/pnas.0606291103 PMID:17001018

Hansen, P. (2018). Cellular and molecular basis of therapies to ameliorate effects of heat stress on embryonic development in cattle. *Animal Reproduction (AR)*, *10*(3), 322–333.

Hansen, P. J. (1999). Possible roles for heat shock protein 70 and glutathione in protection of the mammalian preimplantation embryo from heat shock. *Annual Review of Biomedical Science*, *1*, 5–29.

Hansen, P. J. (2005). Managing the heat-stressed cow to improve reproduction. *Proceedings of the 7th Western Dairy Management Conference*, 9-11.

Hansen, P. J. (2007). Exploitation of genetic and physiological determinants of embryonic resistance to elevated temperature to improve embryonic survival in dairy cattle during heat stress. *Theriogenology*, *68S*, S242–S249. doi:10.1016/j.theriogenology.2007.04.008

Hansen, P. J. (2007). To be or not to bedeterminants of embryonic survival following heat shock. *Theriogenology*, *68*, 40–S48. doi:10.1016/j.theriogenology.2007.03.013

Hansen, P. J. (2009). Effects of heat stress on mammalian reproduction. Phil Trans Royal Soc London. Series B. *Biological Science*, *364*(1534), 3341–3350. doi:10.1098/rstb.2009.0131

Hansen, P. J. (2013). Cellular and molecular basis of therapies to ameliorate effects of heat stress on embryonic development in cattle. *Animal Reproduction*, *10*, 322–333.

Hansen, P. J., & Arechiga, C. F. (1999). Strategies for managing reproduction in the heat-stressed dairy cow. *Journal of Animal Science*, *77*(2), 36–50.

Hansen, S., Spears, J., Lloyd, K., & Whisnant, C. (2006). Growth, reproductive performance, and manganese status of heifers fed varying concentrations of manganese. *Journal of Animal Science*, *84*(12), 3375–3380. doi:10.2527/jas.2005-667

Han, V. K., Bassett, N., Walton, J., & Challis, J. R. (1996). The expression of insulin-like growth factor (IGF) and IGF-binding protein (IGFBP) genes in the human placenta and membranes: Evidence for IGF-IGFBP interactions at the feto-maternal interface. *The Journal of Clinical Endocrinology and Metabolism*, *81*(7), 2680–2693. PMID:8675597

Hardin, B. D. (1983). Reproductive toxicity of the glycol ethers. *Toxicology*, *27*(2), 91–102. doi:10.1016/0300-483X(83)90014-8 PMID:6351353

Harrison, P. T., Holmes, P., & Humfrey, C. D. (1997). Reproductive health inhumans and wildlife: Are adverse trends associated with environmentalchemical exposure? *The Science of the Total Environment*, *205*(2-3), 97–106. doi:10.1016/S0048-9697(97)00212-X PMID:9372623

Hartig, T., & Catalano, R. (2013). Cold summer weather, constrained restoration, and very low birth weight in Sweden. *Health & Place*, *22*, 68–74. doi:10.1016/j.healthplace.2013.03.008 PMID:23603428

Hart, R. J. (2016). Physiological aspects of female fertility: Role of the environment, modern lifestyle, and genetics. *Physiological Reviews*, *96*(3), 873–909. doi:10.1152/physrev.00023.2015

Ha, S., Liu, D., Zhu, Y., Soo Kim, S., Sherman, S., Grantz, K. L., & Mendola, P. (2017). Ambient Temperature and Stillbirth: A Multi-Center Retrospective Cohort Study. *Environmental Health Perspectives*, *125*(6), 067011. doi:10.1289/EHP945 PMID:28650842

Hassaan, M. A., & El Nemr, A. (2017). Advanced oxidation processes for textile wastewater treatment. *International Journal of Photochemistry and Photobiology*, *2*, 85–93.

Hassaan, M. A., El Nemr, A., & Madkour, F. F. (2016). Application of ozonation and UV-assisted ozonation for decolorization of direct yellow 50 in seawater. *Pharmaceutical Chemistry Journal*, *3*, 131–138.

Hassaan, M., El Nemr, A., & Madkour, F. F. (2017). Advanced oxidation processes of mordant violet 40 dye in freshwater and seawater. *Egyptian Journal of Aquatic Research*, *43*(1), 1–9. doi:10.1016/j.ejar.2016.09.004

Hauser, R., Gaskins, A. J., Souter, I., Smith, K. W., Dodge, L. E., Ehrlich, S., Meeker, J. D., Calafat, A. M., & Williams, P. L. (2016). Urinary phthalate metabolite concentrations and reproductive outcomes among women undergoing in vitro fertilization: Results from the EARTH study. *Environmental Health Perspectives, 124*(6), 831–839. doi:10.1289/ehp.1509760

Hawkes, S. J. (1997). What is a "heavy metal?". *Journal of Chemical Education, 74*(11), 1374. doi:10.1021/ed074p1374

Heba, F., Hozen Hodallah, H. A., Shalaby, S. I. A., & Essawy, G. E. S. (2016). Seasonal Heat Stress Effect on Cholesterol, Estradiol and Progesterone during Follicular Development in Egyptian Buffalo. *Wayamba Journal of Animal Science.* http://www.wayambajournal.com/documents/1304195534.pdf

He, J. R., Liu, Y., Xia, X. Y., Ma, W. J., Lin, H. L., Kan, H. D., Lu, J. H., Feng, Q., Mo, W. J., Wang, P., Xia, H. M., Qiu, X., & Muglia, L. J. (2016). Ambient Temperature and the Risk of Preterm Birth in Guangzhou, China (2001-2011). *Environmental Health Perspectives, 124*(7), 1100–1106. doi:10.1289/ehp.1509778 PMID:26672059

Hejmej, A., Kotula-Balak, M., & Bilińska, B. (2011). *Antiandrogenic and estrogenic compounds: effect on development and function of male reproductive system.* Steroids–Clinical Aspect.

Hemminki, K., Franssila, E., & Vainio, H. (1980). Spontaneous abortions among female chemical workers in Finland. *International Archives of Occupational and Environmental Health, 45*(2), 123–126. doi:10.1007/BF01274131 PMID:7364488

Hendricks, K. E. M., Martins, L., & Hansen, P. J. (2009). Consequences for the bovine embryo of being derived from a spermatozoon subjected to post-ejaculatory aging and heat shock: Development to the blastocyst stage and sex ratio. *The Journal of Reproduction and Development, 55*(1), 69–74. doi:10.1262/jrd.20097 PMID:18957823

Hendrickx, A. G., Stone, G. W., Henrickson, R. V., & Matayoshi, K. (1979). Teratogenic effects of hyperthermia in the bonnet monkey (macaca radiata). *Teratology, 19*(2), 177–182. doi:10.1002/tera.1420190208 PMID:112705

Hennig, J., Kieferdorf, P., Moritz, C., Huwe, S., & Netter, P. (1998). Changes in cortisol secretion during shiftwork: Implications for tolerance to shift work? *Ergonomics, 41*(5), 610–621. doi:10.1080/001401398186784 PMID:9613222

Henryson, K., Sundberg, C., Kätterer, T., & Hansson, P. A. (2018). Accounting for long-term soil fertility effects when assessing the climate impact of crop cultivation. *Agricultural Systems, 164*, 185–192. doi:10.1016/j.agsy.2018.03.001

Hess, R. A. (1998). Effects of environmental toxicants on the efferent ducts, epididymis and fertility. *Journal of Reproduction and Fertility. Supplement*, 247–259.

Hicks, S. P., & D'Amato, C. P. (1966). Effects of ionizing radiation on mammalian development. In D. H. M. Woollam (Ed.), *Advances in Teratology* (pp. 196–243). Logos Press.

Hillers, J. K., Senger, P. L., Darlington, R. L., & Fleming, W. N. (1984). Effects of production, season, age of cow, days dry and days in milk on conception to first service in large commercial dairy herds. *Journal of Dairy Science, 67*(4), 861–867. doi:10.3168/jds.S0022-0302(84)81378-8 PMID:6539346

Hillhouse, E. W., & Grammatopoulos, D. K. (2006). The molecular mechanisms underlying the regulation of the biological activity of corticotropin-releasing hormone receptors: Implications for physiology and pathophysiology. *Endocrine Reviews, 27*(3), 260–286. doi:10.1210/er.2005-0034 PMID:16484629

Hinoue, A., Fushiki, S., Nishimura, Y., & Shiota, K. (2001). In utero exposure to brief hyperthermia interferes with the production and migration of neocortical neurons and induces apoptotic cell death in the fetal mouse brain. *Dev Brain Res, 132*, 59–67.

Hinoue, A., Fushiki, S., Nishimura, Y., & Shiota, K. (2001). In utero exposure to brief hyperthermia interferes with the production and migration of neocortical neurons and induces apoptotic cell death in the fetal mouse brain. *Brain Research. Developmental Brain Research*, *132*(1), 59–67. doi:10.1016/S0165-3806(01)00295-4 PMID:11744107

Hirao, Y., & Yanagimachi, R. (1978). Temperature dependence of sperm-egg fusion and post-fusion events in hamster fertilization. *The Journal of Experimental Zoology*, *205*(3), 433–437. doi:10.1002/jez.1402050312 PMID:702084

Hjollund, N. H. I., Bonde, J. P. E., Henriksen, T. B., Giwercman, A., & Olsen, J. (2004). Reproductive effects of male psychologic stress. *Epidemiology (Cambridge, Mass.)*, *15*(1), 21–27. doi:10.1097/01.ede.0000100289.82156.8b PMID:14712143

Hjollund, N. H., Bonde, J. P., Ernst, E., Lindenberg, S., Andersen, A. N., & Olsen, J. (2004b). Pesticide exposure in male farmers and survival of in vitro fertilizedpregnancies. *Human Reproduction (Oxford, England)*, *19*(6), 1331–1337. doi:10.1093/humrep/deh203 PMID:15117907

Hjollund, N. H., Bonde, J. P., & Hansen, K. S. (1995). Male-mediated risk of spontaneousabortion with reference to stainless steel welding. *Scandinavian Journal of Work, Environment & Health*, *21*(4), 272–276. doi:10.5271jweh.37 PMID:8553001

Hjollund, N. H., Skotte, J. H., Kolstad, H. A., & Bonde, J. P. (1999). Extremely lowfrequency magnetic fields and fertility: A follow up study of couplesplanning first pregnancies. The Danish First Pregnancy Planner StudyTeam. *Occupational and Environmental Medicine*, *56*, 253–255. doi:10.1136/oem.56.4.253 PMID:10450242

Ho, H. C., Granish, K. A., & Suarez, S. S. (2002). Hyperactivated motility of bull sperm is triggered at the axoneme by $Ca^{2+}$and not cAMP. *Developmental Biology*, *250*(1), 208–217. doi:10.1006/dbio.2002.0797 PMID:12297107

Holt, W. (2000). Basic aspects of frozen storage of semen. *Animal Reproduction Science*, *62*(1-3), 3–22. doi:10.1016/S0378-4320(00)00152-4

Homer, C. S., Hanna, E., & McMichael, A. J. (2009). Climate change threatens the achievement of the millennium development goal for maternal health. *Midwifery*, *25*(6), 606–612. doi:10.1016/j.midw.2009.09.003 PMID:19836867

Hooper, L. M., Payton, R. R., Rispoli, L. A., Saxton, A. M., & Edwards, J. L. (2015). Impact of heat stress on germinal vesicle breakdown and lipolytic changes during in vitro maturation of bovine oocytes. *The Journal of Reproduction and Development*, *61*(5), 459–464. doi:10.1262/jrd.2014-168 PMID:26120041

Hoque, M. N., Talukder, A. K., Akter, M., & Shamsuddin, M. (2014). Evaluation of ovsynch protocols for timed artificial insemination in water buffaloes in Bangladesh. *Turkish Journal of Veterinary and Animal Sciences*, *38*, 418–424. doi:10.3906/vet-1302-35

Hoshino, H., Abe, T., Watanabe, H., Katsuoka, Y., & Kawamura, N. (1993). Correlation between atmospheric temperature and testicular torsion. *Acta Urologica Japonica.*, *39*(11), 1031–1033. PMID:8266870

Hossain, F., Ali, O., D'souza, U. J., & Naing, D. K. S. (2010). Effects of pesticide use on semen quality among farmers in rural areas of Sabah, Malaysia. *Reproductive Toxicology*, *56*, 118-140.

Hossain, S. (2020). Salinity and Miscarriage: Is There a Link? Impact of Climate Change in Coastal Areas of Bangladesh - A Systematic Review. *European Journal of Environment and Public Health*, *4*(1), em0036.

Houghton, J. T. (2001). *Climate Change 2001: The Scientific Basis. Contribution of Working Group I to the third assessment report of the Intergovernmental Panel on Climate Change.* Cambridge University Press.

Hovatta, O., Venäläinen, E. R., Kuusimäki, L., Heikkilä, J., Hirvi, T., & Reima, I. (1998). Aluminium, lead and cadmium concentrations in seminal plasma and spermatozoa, and semen quality in Finnish men. *Human Reproduction (Oxford, England)*, *13*(1), 115–119. doi:10.1093/humrep/13.1.115

HSE. (2016). *Dyes and chemicals in textile finishing: An introduction. Dyeing and Finishing Information Sheet No 1-Health and Safety Executive (HSE) information sheet*. HSE.

Huang, F., Wang, H., Zou, Y., Liu, Q., Cao, J., & Yin, T. (2013). Effect of GnRH-II on the ESC proliferation, apoptosis and VEGF secretion in patients with endometriosis in vitro. *International Journal of Clinical and Experimental Pathology*, *6*(11), 2487. PMID:24228111

Huang, P. C., Kuo, P. L., Guo, Y. L., Liao, P. C., & Lee, C. C. (2007). Associations between urinary phthalate monoesters and thyroid hormones in pregnant women. *Human Reproduction (Oxford, England)*, *22*(10), 2715–2722. doi:10.1093/humrep/dem205

Hughes, E. G., Beecroft, M. L., Wilkie, V., Burville, L., Claman, P., Tummon, I., Greenblatt, E., Fluker, M., & Thorpe, K. (2004). A multicentre randomizedcontrolled trial of expectant management versus IVF in women with fallopiantube patency. *Human Reproduction (Oxford, England)*, *19*(5), 1105–1109. doi:10.1093/humrep/deh209 PMID:15044399

Hurley, L. L., McDiarmid, C. S., Friesen, C. R., Griffith, S. C., & Rowe, M. (2018). Experimental heatwaves negatively impact sperm quality in the zebra finch. *Proceedings. Biological Sciences*, *285*(1871), 2017–2547. doi:10.1098/rspb.2017.2547 PMID:29343605

Hussain, M. S., & Javadi, A. A. (2016). Assessing impacts of sea level rise on seawater intrusion in a coastal aquifer with sloped shoreline boundary. *Journal of Hydro-environment Research*, *11*, 29–41. doi:10.1016/j.jher.2016.01.003

Husse, J., Eichele, G., & Oster, H. (2015). Synchronization of the mammalian circadian timing system: Light can control peripheral clocks independently of the SCN clock: alternate routes of entrainment optimize the alignment of the body's circadian clock network with external time. *BioEssays*, *37*(10), 1119–1128. doi:10.1002/bies.201500026 PMID:26252253

Hu, W., Li, Y., Han, W., Xue, L., Zhang, W., Ma, W., & Bi, P. (2018, April 1). Meteorological factors and the incidence of mumps in Fujian Province, China, 2005-2013: Non-linear effects. *The Science of the Total Environment*, *619-620*, 1286–1298. doi:10.1016/j.scitotenv.2017.11.108 PMID:29734606

Huynh, T. T., Aarnink, A. J., Verstegen, M. W., Gerrits, W. J., Heetkamp, M. J., Kemp, B., & Canh, T. T. (2005). Effects of increasing temperatures on physiological changes in pigs at different relative humidities. *Journal of Animal Science*, *83*(6), 1385–1396. doi:10.2527/2005.8361385x PMID:15890816

Hvidtjørn, D., Schieve, L., Schendel, D., Jacobsson, B., Sværke, C., & Thorsen, P. (2009). Cerebral palsy, autism spectrum disorders, and developmental delay in children born after assisted conception: A systematic review and meta-analysis. *Archives of Pediatrics & Adolescent Medicine*, *163*(1), 72–83. doi:10.1001/archpediatrics.2008.507 PMID:19124707

Hyatt, O. M., Lemke, B., & Kjellstrom, T. (2010). Regional maps of occupational heat exposure: Past, present, and potential future. *Global Health Action*, *3*(1), 5715. doi:10.3402/gha.v3i0.5715 PMID:21165172

Ibtisham, F., Nawab, A., Li, G., Xiao, M., An, L., & Naseer, G. (2018). Effect of nutrition on reproductive efficiency of dairy animals. *Medycyna Weterynaryjna*, *74*(1), 356–361. doi:10.21521/mw.6025

ILO. (2015). *Textiles, clothing, leather and footwear sector*. http://www.ilo.org/global/industriesand-sectors/textiles-clothing-leather-footwear/lang--ja/index.htm

Inchul, C., & Amarnath, D. (2015). Effects of prolonged exposure of mouse embryos to elevated temperature on embryonic developmental competence. *Reproductive Biomedicine*, *31*(2), 171–179. doi:10.1016/j.rbmo.2015.04.017

Indu, S., Sejian, V., & Naqvi, S. (2014). Impact of simulated heat stress on growth, physiological adaptability, blood metabolites and endocrine responses in Malpura ewes under semiarid tropical environment. *Animal Production Science*, *55*(6), 766. doi:10.1071/AN14085

Infante-Rivard, C., Siemiatycki, J., Lakhani, R., & Nadon, L. (2005). Maternal exposure to occupational solvents and childhood leukemia. *Environmental Health Perspectives*, *113*(6), 787–792. doi:10.1289/ehp.7707

Ingermann, R. L., Holcomb, M., Robinson, M. L., & Cloud, J. G. (2002). Carbondioxide and pH affect sperm motility of white sturgeon (*Acipensertransmontanus*). *The Journal of Experimental Biology*, *205*, 2885–2890. PMID:12177152

Ingraham, R., Kappel, L., Morgan, E., & Srikandakumar, A. (1987). Correction of subnormal fertility with copper and magnesium supplementation. *Journal of Dairy Science*, *70*(1), 167–180. doi:10.3168/jds.S0022-0302(87)79991-3

Inhorn, M. C., & Patrizio, P. (2015). Infertility around the globe: New thinking on gender, reproductive technologies and global movements in the 21st century. *Human Reproduction Update*, *21*(4), 411–426. doi:10.1093/humupd/dmv016

Inskeep, E., & Dailey, R. (2005). Embryonic death in cattle. *Veterinary Clinics: Food Animal Practice*, *21*, 437–461. PMID:15955439

International Labour Office. (1998). *General Hazards: Heat and Cold website*. https://www.iloencyclopaedia.org/part-vi-16255/heat-and-cold

IPCC. (2007a). Summary for Policymakers. In Climate change 2007: Impacts, adaptation and vulnerability. Contribution of Working Group II to the Fourth Assessment Report of the Intergovernmental Panel on Climate Change. Cambridge University Press.

IPCC. (2013). Summary for policymakers. In Climate change 2013: the physical science basis. Contribution of Working Group I to the Fifth Assessment Report of the Intergovernmental Panel on Climate Change. Cambridge University Press.

Irving, J. A., & Lala, P. K. (1995). Functional role of cell surface integrins on human trophoblast cell migration: Regulation by TGF-β, IGF-II, and IGFBP-1. *Experimental Cell Research*, *217*(2), 419–427. doi:10.1006/excr.1995.1105 PMID:7535237

Irwin, J. C., & Giudice, L. C. (1998). IGFBP-1 binds to the α5β1 integrin in human cytotrophoblasts and inhibits invasion into decidualized endometrium stromal cells in vitro. *Growth Hormone & IGF Research*, *8*, 21–23. doi:10.1016/S1096-6374(98)80318-3 PMID:10990441

Isaak, D. J., Wollrab, S., Horan, D., & Chandler, G. (2011). Climate changeeffects on stream and river temperatures across the northwest U.S. from1980–2009 and implications for salmonid fishes. *Climatic Change*, *113*(2), 499–524. doi:10.100710584-011-0326-z

Ishaya, K. S., Maracus, D. N. K., & Issac, J. (2011). The influence of waste water on soil chemical properties on irrigated fields in Kaduna South Township, North Central Nigeria. *Journal of Sustainable Development in Africa*, *13*, 91–101.

Ishijima, S., Mohri, H., Overstreet, J. W., & Yudin, A. I. (2006). Hyperactivation of monkey spermatozoa is triggered by Ca^{2+} and completed by cAMP. *Molecular Reproduction and Development*, *73*(9), 1129–1139. doi:10.1002/mrd.20420 PMID:16804884

Issoufou, A. A., Soumana, I., Maman, G., Konate, S., & Mahamane, A. (2020). Dynamic relationship of traditional soil restoration practices and climate change adaptation in semi-arid Niger. *Heliyon*, *6*(1), 1–7. doi:10.1016/j.heliyon.2020.e03265 PMID:32042980

Iwasaki, S., Nakazawa, K., Sakai, J., Kometani, K., Iwashita, M., Yoshimura, Y., & Maruyama, T. (2005). Melatonin as a local regulator of human placental function. *Journal of Pineal Research*, *39*(3), 261–265. doi:10.1111/j.1600-079X.2005.00244.x PMID:16150106

James, F. O., Boivin, D. B., Charbonneau, S., Belanger, V., & Cermakian, N. (2007). Expression of clock genes in human peripheral blood mono nuclear cells throughout the sleep/wake and circadian cycles. *Chronobiology International*, *24*(6), 1009–1034. doi:10.1080/07420520701800736 PMID:18075796

James, F. O., Cermakian, N., & Boivin, D. B. (2007). Circadian rhythms of melatonin, cortisol and clock gene expression during simulated night shiftwork. *Sleep*, *30*(11), 1427–1436. doi:10.1093leep/30.11.1427 PMID:18041477

Jana, K., Samanta, P. K., & De, D. K. (2010). Nicotine diminishes testicular gametogenesis, steroidogenesis and steroidogenic acute regulatory protein expression in adult albino rats: Possible influence on pituitary gonadotropins and alteration of testicular antioxidant status. *Toxicological Sciences*, *116*(2), 647–159. doi:10.1093/toxsci/kfq149 PMID:20498001

Jang, D. H., & Hoffman, R. S. (2011). Heavy metal chelation in neurotoxic exposures. *Neurologic Clinics*, *29*(3), 607–622. doi:10.1016/j.ncl.2011.05.002

Jarfelt, K., Dalgaard, M., Hass, U., Borch, J., Jacobsen, H., & Ladefoged, O. (2005). Antiandrogenic effects in male rats perinatally exposed to a mixture of di (2-ethylhexyl) phthalate and di (2-ethylhexyl) adipate. *Reproductive Toxicology (Elmsford, N.Y.)*, *19*(4), 505–515. doi:10.1016/j.reprotox.2004.11.005

Jarrell, J.F., Labelle, R., Goeree, R., Milner, R., & Collins, J. (1993a). In vitro fertilizationand embryo transfer: A randomized controlled trial. *J Curr Clin Trials, 73.*

Jensen, T. K., Bonde, J. P., & Joffe, M. (2006). The influence of occupational exposure on male reproductive function. *Occupational Medicine*, *56*(8), 544–553. doi:10.1093/occmed/kql116

Jezierska, B., & Witeska, M. (1999). The effect of time and temperature on motilityof spermatozoa of common and grass carp. *Electronic Journal of the Polish Agricultural University*, *2*, 1–8.

Jia, Y., Lu, Y., Wu, K., Lin, Q., Shen, W., Zhu, M., Huang, S., & Chen, J. (2013). Does night work increase the risk of breast cancer? A systematic review and meta-analysis of epidemiological studies. *Cancer Epidemiology*, *37*(3), 197–206. doi:10.1016/j.canep.2013.01.005 PMID:23403128

Jin, J., Jin, N., Zheng, H., Ro, S., Tafolla, D., Sanders, K. M., & Yan, W. (2007). Catsper3 and Catsper4 are essential for spermhyperactivated motility and male fertility in the mouse. *Biology of Reproduction*, *77*(1), 37–44. doi:10.1095/biolreprod.107.060186 PMID:17344468

Jockenhovel, F., Grawe, A., & Nieschlag, E. (1990). A portable digital data recorder for long-termmonitoring of scrotal temperatures. *Fertility and Sterility*, *54*, 694–700. doi:10.1016/S0015-0282(16)53832-1 PMID:2209892

Joffe, M. (2003). Infertility and environmental pollutants. *British Medical Bulletin*, *68*(1), 47–70. doi:10.1093/bmb/ldg025 PMID:14757709

Johnson, H. D. (1987). Bioclimates and livestock. In Bioclimatology and adaptation of livestock. Elsevier.

Johnson, J. S. (2018). Heat stress: Impact on livestock well-being and productivity and mitigation strategies to alleviate the negative effects. *Animal Production Science*, *58*(8), 1404. doi:10.1071/AN17725

Jonson, K. M., Lyle, J. G., Edwards, M. J., & Penny, R. H. C. (1976). Effect of prenatal heat stress on brain growth and serial discrimination reversal learning in the guinea pig. *Brain Research Bulletin*, *1*(1), 135–150. doi:10.1016/0361-9230(76)90056-3 PMID:974792

Josephy, P. D., & Novak, M. (2013). Reactive electrophilic metabolites of aromatic amine and amide carcinogens. *Frontiers in Bioscience*, *S5*(1), 341–359. doi:10.2741/S376 PMID:23277055

Joshi, V. J., & Santani, D. D. (2012). Physicochemical characterization and heavy metal concentration in effluent of textile industry. *Universal Journal of Environmental Research and Technology*, *2*, 93–96.

Ju, J. C., Jiang, S., Tseng, J. K., Parks, J. E., & Yang, X. (2005). Heat shock reduces developmental competence and alters spindle configuration of bovine oocytes. *Theriogenology*, *64*(8), 1677–1689. doi:10.1016/j.theriogenology.2005.03.025 PMID:15951010

Ju, J. C., Parks, J. E., & Yang, X. (1999). Thermotolerance of IVM-derived bovine oocytes and embryos after short-term heat shock. *Molecular Reproduction and Development*, *53*(3), 336–340. doi:10.1002/(SICI)1098-2795(199907)53:3<336::AID-MRD9>3.0.CO;2-M PMID:10369394

Julvez, J., & Grandjean, P. (2009). Neurodevelopmental toxicity risks due to occupational exposure to industrial chemicals during pregnancy. *Industrial Health*, *47*(5), 459–468. doi:10.2486/indhealth.47.459

Jung, A., & Schuppe, H. C. (2007). Influence of genital heat stress on semen quality in humans. *Andrologia*, *39*(6), 203–215. doi:10.1111/j.1439-0272.2007.00794.x PMID:18076419

Kacew, S. (1994). Current issues in lactation: Advantages, environment, silicone. Biomedical and environmental sciences. *BES*, *7*(4), 307–319.

Kadokawa, H., Sakatani, M., & Hansen, P. (2012). Perspectives on improvement of reproduction incattle during heat stress in a future Japan. *Animal Science Journal*, *83*(6), 439–445. doi:10.1111/j.1740-0929.2012.01011.x PMID:22694326

Kalantaridou, S. N., Zoumakis, E., Makrigiannakis, A., Lavasidis, L. G., Vrekoussis, T., & Chrousos, G. P. (2010). Corticotropin-releasing hormone, stress and human reproduction: An update. *Journal of Reproductive Immunology*, *85*(1), 33–39. doi:10.1016/j.jri.2010.02.005 PMID:20412987

Kamrin, M. A., Carney, E. W., Chou, K., Cummings, A., Dostal, L. A., Harris, C., Dostal, L. A., Harris, C., Henck, J. W., & Loch-Caruso, R. (1994). Femalereproductive and developmental toxicology: Overview and currentapproaches. *Toxicology Letters*, *74*(2), 99–119. doi:10.1016/0378-4274(94)90089-2 PMID:7940600

Kanai, Y., Yagyu, N., & Shimizu, T. (2001). Hypogonadism in heat stressed goats: Poor responsiveness of the ovary to the pulsatile LH stimulation induced by hourly injections of a small dose of GnRH. *The Journal of Reproduction and Development*, *41*(2), 133–139. doi:10.1262/jrd.41.133

Kanter, M., Aktas, C., & Erboga, M. (2013). Heat stress decreases testicular germ cell proliferation andincreases apoptosis in short term: An immune histochemical and ultrastructural study. *Toxicology and Industrial Health*, *29*(2), 99–113. doi:10.1177/0748233711425082 PMID:22082826

Karmakar, R., Das, I., Dutta, D., & Rakshit, A. (2016). Potential Effects of Climate Change on Soil Properties: A Review. *Science International*, *4*(2), 51–73. doi:10.17311ciintl.2016.51.73

Karthikeyan, S., Jambulingam, M., Sivakumar, P., Shekhar, A. P., & Krithika, J. (2006). Impact of textile effluents on fresh water fish *Mastacembelus arimatus*. *Journal of Chemistry*, *3*, 303–306.

Kasai, Y. (2011). Molecular technologies for analysis of petroleum bioremediation. In A. Koukkou (Ed.), *Microbial Bioremediation of non-metals: Current Research* (pp. 233–252). Caister Acedemic Press.

Kastelic, J. P. (2013). Male involvement in fertility and factors affecting semen quality in bulls. *Animal Frontiers*, *3*(4), 20–25. doi:10.2527/af.2013-0029

Kay, D., & McDonald, A. (1983, September). Predicting coliform concentrations in upland impoundments: Design and calibration of a multivariate model. *Applied and Environmental Microbiology*, *46*(3), 611–618. doi:10.1128/AEM.46.3.611-618.1983 PMID:6639016

Kdasi, A., Idris, A., Saed, K., & Guan, C. (2004). Treatment of textile wastewater by advanced oxidation processes: A review. *Global Nest International Journal*, *6*, 222–230.

Keiler, M., Knight, J., & Harrison, S. (2010). Climate change and geomorphological hazards in the eastern European Alps. Philosophical transactions. *Series A, Mathematical, Physical, and Engineering Sciences, 368*(1919), 2461–2479.

Kendall, P., Nielsen, P., Webster, J., Verkerk, G., Littlejohn, R., & Matthews, L. (2006). The effects of providing shade to lactating dairy cows in a temperate climate. *Livestock Science*, *103*(1-2), 148–157. doi:10.1016/j.livsci.2006.02.004

Kennaway, D. J., Boden, M. J., & Varcoe, T. J. (2012). Circadian rhythms and fertility. *Molecular and Cellular Endocrinology*, *349*(1), 56–61. doi:10.1016/j.mce.2011.08.013 PMID:21872642

Kennaway, D. J., Varcoe, T. J., & Mau, V. J. (2003). Rhythmic expression of clock and clock-controlled genes in the rat oviduct. *Molecular Human Reproduction*, *9*(9), 503–507. doi:10.1093/molehr/gag067 PMID:12900508

Kent, S. T., McClure, L. A., Zaitchik, B. F., Smith, T. T., & Gohlke, J. M. (2014). Heat waves and health outcomes in Alabama (USA): The importance of heat wave definition. *Environmental Health Perspectives*, *S122*(2), 151–158. doi:10.1289/ehp.1307262 PMID:24273236

Kestioglu, K., Yonar, T., & Azbar, N. (2005). Feasibility of physico-chemical treatment and advanced oxidation processes (AOPs) as a means of pretreatment of olive mill effluent (OME). *Process Biochemistry*, *40*(7), 2409–2416. doi:10.1016/j.procbio.2004.09.015

Kettner, N. M., Mayo, S. A., Hua, J., Lee, C., Moore, D. D., & Fu, L. (2015). Circadian Dysfunction Induces Leptin Resistance in Mice. *Cell Metabolism*, *22*(3), 448–459. doi:10.1016/j.cmet.2015.06.005 PMID:26166747

Khai, N. M., Tuan, P. T., Vinh, C. N., & Oborn, I. (2008). Effects of using wastewater as nutrient sources on soil chemical properties in periurban agricultural systems. VNU Journal of Science: Earth and Sciences (Paris), *24*, 87–95.

Khairy, M., Dhillon, R. K., Chu, J., Rajkhowa, M., & Coomarasamy, A. (2016). The effect of peri-implantation administration of uterine relaxing agents in assisted reproduction treatment cycles: A systematic review and meta-analysis. *Reproductive Biomedicine Online*, *32*(4), 362–376. doi:10.1016/j.rbmo.2016.01.004 PMID:26936145

Khan, A., & Malik, A. (2013). Environmental and health effects of textile industry wasterwater. In A. Malik, E. Grohman, & R. Akhtar (Eds.), *Environmental deterioration and human health: Natural and anthropogenic determinants* (pp. 55–71). Springer, Netherlands.

Khan, F. A., Das, G. K., Pande, M., Sarkar, M., Mahapatra, R. K., & Shankar, U. (2012). Alterations in follicular fluid estradiol, progesterone and insulin concentrations during ovarian acyclicity in water buffalo (Bubalus bubalis). *Animal Reproduction Science*, *130*(1-2), 27–32. doi:10.1016/j.anireprosci.2011.12.020

Khan, I. (2016). *Evaluation of different stress indicators associated with repproductivecyclicity of crossbred dairy cows*. The University of Agriculture.

Khattak,, S., Moghtader, G., McMartin, K., Barrera, M., Kennedy, D., & Koren, G. (1999). Pregnancy outcome following gestational exposure to organicsolvents: A prospective controlled study. *Journal of the American Medical Association*, *281*, 1106–1109. doi:10.1001/jama.281.12.1106 PMID:10188661

Khodaei-Motlagh, M., Shahneh, A. Z., Masoumi, R., & Derensis, F. (2011). Alterations in reproductive hormones during heat stress in dairy cattle. *African Journal of Biotechnology*, *10*(29), 5552–5558.

Kimler, B. F. (1998). Prenatal irradiation: A major concern for the developing brain. *International Journal of Radiation Biology, 73*(4), 423–434. doi:10.1080/095530098142266 PMID:9587081

Kim, S. K., Han, E. J., Kim, S. M., Lee, J. R., Jee, B. C., Suh, C. S., & Kim, S. H. (2016). Efficacy of oxytocin antagonist infusion in improving in vitro fertilization outcomes on the day of embryo transfer: A meta-analysis. *Clinical and Experimental Reproductive Medicine, 43*(4), 233. doi:10.5653/cerm.2016.43.4.233 PMID:28090463

King, A., Gardner, L., & Loke, Y. W. (1996). Evaluation of oestrogen and progesterone receptor expression in uterine mucosal lymphocytes. *Human Reproduction (Oxford, England), 11*(5), 1079–1082. doi:10.1093/oxfordjournals.humrep.a019300 PMID:8671394

Kinsey-Jones, J. S., Li, X. F., Knox, A. M. I., Wilkinson, E. S., Zhu, X. L., Chaudhary, A. A., Milligan, S. R., Lightman, S. L., O'byrne, K. T., & O'byrne, K. T. (2009). Down-regulation of hypothalamic kisspeptin and its receptor, Kiss1r, mRNA expression is associated with stress-induced suppression of luteinising hormone secretion in the female rat. *Journal of Neuroendocrinology, 21*(1), 20–29. doi:10.1111/j.1365-2826.2008.01807.x PMID:19094090

Kirk-Othmer. (2004). *Encyclopedia of chemical technology* (5th ed., Vol. 7). Wiley-Interscience.

Kjellstrom, T. (2009). Climate change, direct heat exposure, health and well-being in low and middle-income countries. *Global Health Action*, 2. PMID:20027264

Klaine, S. J., & Lewis, M. A. (1995). Algae and plant toxicity testing. In D. J. Hoffman, B. A. Rattner, G. A. Burton Jr, & J. Cairns Jr., (Eds.), *Handbook of ecotoxicology* (p. 371). CRC Press.

Klemola, K., Pearson, J., Wright, A. V., Liesivuori, J., & Lindstrom-Seppa, P. (2007). Evaluating the toxicity of reactive dyes and dyed fabrics with the hepa-1 cytotoxicity test. *AUTEX Research Journal, 7*, 224–230.

Klik, A., & Eitzinger, J. (2010). Impact of climate change on soil erosion and the efficiency of soil conservation practices in Austria. *The Journal of Agricultural Science, 148*(05), 529–541. doi:10.1017/S0021859610000158

Kline, J., Stein, Z., Susser, M., & Warburton, D. (1985). Fever during pregnancy and spontaneous abortion. *American Journal of Epidemiology, 121*(6), 832–842. doi:10.1093/oxfordjournals.aje.a114054 PMID:4014176

Klonoff-Cohen, H., Chu, E., Natarajan, L., & Sieber, W. (2001a). A prospectivestudy of stress among women undergoing in vitro fertilization or gameteintrafallopian transfer. *Fertility and Sterility, 76*(4), 675–687. doi:10.1016/S0015-0282(01)02008-8 PMID:11591398

Kloog, I., Novack, L., Erez, O., Just, A. C., & Raz, R. (2018). Associations between ambient air temperature, low birth weight and small for gestational age in term neonates in southern Israel. *Environmental Health: A Global Access Science Source, 17*(1), 76.

Kloog, I., Melly, S. J., Coull, B. A., Nordio, F., & Schwartz, J. D. (2015). Using Satellite-Based Spatiotemporal Resolved Air Temperature Exposure to Study the Association between Ambient Air Temperature and Birth Outcomes in Massachusetts. *Environmental Health Perspectives, 123*(10), 1053–1058. doi:10.1289/ehp.1308075 PMID:25850104

Knoester, M., Helmerhorst, F. M., van der Westerlaken, L. A., Walther, F. J., & Veen, S. (2007). Matched follow-up study of 5–8-year-old ICSI singletons: Child behaviour, parenting stress and child (health-related) quality of life. *Human Reproduction (Oxford, England), 22*(12), 3098–3107. doi:10.1093/humrep/dem261 PMID:17905745

Knöfler, M., Haider, S., Saleh, L., Pollheimer, J., Gamage, T. K., & James, J. (2019). Human placenta and trophoblast development: Key molecular mechanisms and model systems. *Cellular and Molecular Life Sciences, 76*(18), 1–18. doi:10.100700018-019-03104-6

Kobayashi, R., Terakawa, J., Omatsu, T., Hengjan, Y., Mizutani, T., Ohmori, Y., & Hondo, E. (2017). The window of implantation is closed by estrogen via insulin-like growth factor 1 pathway. *Journal of Reproduction & Infertility*, *18*(2), 231. PMID:28868248

Kobayashi, Y., Itoh, M. T., Kondo, H., Okuma, Y., Sato, S., Kanishi, Y., Hamada, N., Kiguchi, K., & Ishizuka, B. (2003). Melatonin binding sites in estrogen receptor-positive cells derived from human endometrial cancer. *Journal of Pineal Research*, *35*(2), 71–74. doi:10.1046/j.1439-0434.2002.00691_2.x-i1 PMID:12887647

Ko, C. H., & Takahashi, J. S. (2006). Molecular components of the mammalian circadian clock. *Human Molecular Genetics*, *15*(suppl_2), R271–R277. doi:10.1093/hmg/ddl207 PMID:16987893

Kodaman, P. H., & Taylor, H. S. (2004). Hormonal regulation of implantation. *Obstetrics and Gynecology Clinics*, *31*(4), 745–766. PMID:15550333

Kolbe, I., & Oster, H. (2019). Chronodisruption, metabolic homeostasis,and the regulation of inflammation in adipose tissues. *The Yale Journal of Biology and Medicine*, *92*, 317–325. PMID:31249492

Konkel, L. (2019). Taking the Heat: Potential Fetal Health Effects of Hot Temperatures. *Environmental Health Perspectives*, *127*(10), 102002. doi:10.1289/EHP6221 PMID:31652107

Konstantinou, I. K., & Albanis, T. A. (2004). TiO2-assisted photocatalytic degradation of azo dyes in aqueous solution: Kinetic and mechanistic investigations - A review. *Applied Catalysis B: Environmental*, *49*(1), 1–14. doi:10.1016/j.apcatb.2003.11.010

Kopp, R., Martínez, I. O., Legradi, J., & Legler, J. (2017). Exposure to endocrine disrupting chemicals perturbs lipid metabolism and circadian rhythms. *Journal of Environmental Sciences (China)*, *62*, 133–137. doi:10.1016/j.jes.2017.10.013

Kourtis, A. P., Read, J. S., & Jamieson, D. J. (2014). Pregnancy and infection. *The New England Journal of Medicine*, *370*(23), 2211–2218. doi:10.1056/NEJMra1213566 PMID:24897084

Kozin, L. F., & Hansen, S. C. (2013). *Mercury handbook: chemistry, applications and environmental impact*. Royal Society of Chemistry.

Kraus, W. L., & Katzenellenbogen, B. S. (1993). Regulation of progesterone receptor gene expression and growth in the rat uterus: Modulation of estrogen actions by progesterone and sex steroid hormone antagonists. *Endocrinology*, *132*(6), 2371–2379. doi:10.1210/endo.132.6.8504742 PMID:8504742

Kregel, K. C. (2002). Heat shock proteins: Modifying factors in physiological stress response and acquired thermotolerance. *Journal of Applied Physiology (Bethesda, Md.)*, *92*(5), 2177–2186. doi:10.1152/japplphysiol.01267.2001 PMID:11960972

Kriegsfeld, L., Gibson, E., Williams, W. III, Zhao, S., Mason, A., Bentley, G., & Tsutsui, K. (2010). The Roles of RFamide-Related Peptide-3 in Mammalian Reproduc tive Function and Behaviour. *Journal of Neuroendocrinology*, *22*(7), 692–700. doi:10.1111/j.1365-2826.2010.02031.x

Kríninger, C. E. III, Stephens, S. H., & Hansen, P. J. (2002). Developmental changes in inhibitory effects of arsenic and heat shock on growth of preimplantation bovine embryos. *Molecular Reproduction and Development*, *63*(3), 335–340. doi:10.1002/mrd.90017 PMID:12237949

Krishnan, G., Bagath, M., Pragna, P., Vidya, M. K., Aleena, J., Archana, P. R., Sejian, V., & Bhatta, R. (2017). *Mitigation of the Heat Stress Impact in Livestock Reproduction in Theriogenology*. InTech Publishers.

Krishnan, G., Bagath, M., Pragna, P., Vidya, M., Aleena, J., & Archana, P. (2017). Mitigation of the Heat Stress Impact in Livestock Reproduction. *Theriogenology*, 63–87.

Krogenaes, A. K., Taubøll, E., Stien, A., Oskam, I. C., Lyche, J. L., Dahl, E., Thomassen, R. F., Sweeney, T., & Ropstad, E. (2008). Valproate affects reproductive endocrine function, testis diameter and some semen variables in non-epileptic adolescent goat bucks. *Theriogenology*, *70*(1), 15–26. doi:10.1016/j.theriogenology.2008.01.029 PMID:18394693

Kruse, S., Bridges, G., Funnell, B., Bird, S., Lake, S., Arias, R., Amundson, O., Larimore, E., Keisler, D., & Perry, G. (2017). Influence of post-insemination nutrition on embryonic development in beef heifers. *Theriogenology*, *90*, 185–190. doi:10.1016/j.theriogenology.2016.11.021

Kuehn, L., & McCormick, S. (2017). Heat Exposure and Maternal Health in the Face of Climate Change. *Int J Environ Res Public Health, 14*(8), 853.

Kulp, S. A., & Strauss, B. H. (2019). New elevation data triple estimates of global vulnerability to sea-level rise and coastal flooding. *Nature Communications*, *10*(1), 4844–4856. doi:10.103841467-019-12808-z PMID:31664024

Kumar, D., De, K., Sejian, V., & Naqvi, S. (2017). Impact of Climate Change on Sheep Reproduction. *Sheep Production Adapting to Climate Change*, 71–93.

Kumar, B., Singh, G., & Meur, S. (2010). Effects of Addition of Electrolyte and Ascorbic Acid in Feed during Heat Stress in Buffaloes. *Asian-Australasian Journal of Animal Sciences*, *23*(7), 880–888. doi:10.5713/ajas.2010.90053

Kumar, R., & Das, A. J. (2014). Climate Change and its Impact on Land Degradation: Imperative Need to Focus. *Journal of Climatology and Weather Forecasting*, *2*(1), 1–3. doi:10.4172/2332-2594.1000108

Kumar, S., Pandey, A. K., Razzaque, W., & Dwivedi, D. (2011). Importance of micro minerals in reproductive performance of livestock. *Veterinary World*, *4*(5), 230. doi:10.5455/vetworld.2011.230-233

Kunavongkrita, A., Suriyasomboonb, A., Lundeheimc, N., & Hearda, T. (2005). Management and sperm production of boars under differing environmental conditions. *Theriogenology*, *63*(2), 657–667. doi:10.1016/j.theriogenology.2004.09.039 PMID:15626423

Kurita, T., Young, P., Brody, J. R., Lydon, J. P., O'Malley, B. W., & Cunha, G. R. (1998). Stromal progesterone receptors mediate the inhibitory effects of progesterone on estrogen-induced uterine epithelial cell deoxyribonucleic acid synthesis. *Endocrinology*, *139*(11), 4708–4713. doi:10.1210/endo.139.11.6317 PMID:9794483

Kuroki, T., Ikeda, S., Okada, T., Maoka, T., Kitamura, A., Sugimoto, M., & Kume, S. (2013). Astaxanthin ameliorates heat stress-induced impairment of blastocyst development in vitro:– astaxanthin colocalization with and action on mitochondria. *Journal of Assisted Reproduction and Genetics*, *30*(5), 623–631. doi:10.100710815-013-9987-z PMID:23536152

Kurowicka, B., Gajewska, A., & Franczak, A. (2006). Effect of early thermal experience on pituitary-gonadal axis in female rats. *Reproductive Biology*, *6*, 63–77. PMID:16604152

Kurtis, J. D., Higashi, A., Wu, H. W., Gundogan, F., McDonald, E. A., Sharma, S., PondTor, S., Jarilla, B., Sagliba, M. J., Gonzal, A., Olveda, R., Acosta, L., & Friedman, J. F. (2011). Maternal schistosomiasis japonica is associated with maternal, placental and fetal inflammation. *Infection and Immunity*, *79*(3), 1254–1261. doi:10.1128/IAI.01072-10 PMID:21149589

Kusanagi, H., Mishima, K., Satoh, K., Echizenya, M., Katoh, T., & Shimizu, T. (2004). Similar profiles in human period1 gene expression in peripheral mononuclear and polymor-pho nuclear cells. *Neuroscience Letters*, *365*(2), 124–127. doi:10.1016/j.neulet.2004.04.065 PMID:15245792

Kwok-Keung, L., Pao-Yun, C., Yen-Mei, L., Yu-Pei, L., Cheng, D., Won-Hsiung, L., & Mao-Hsiung, Y. (2013). The role of heat shock protein 70 in the protective effect of YC-1 on heat stroke rats. *European Journal of Pharmacology*, *699*(1-3), 67–73. doi:10.1016/j.ejphar.2012.11.044 PMID:23219797

Kwong, W. Y., Wild, A. E., Roberts, P., Willis, A. C., & Fleming, T. P. (2000). Maternal undernutrition during the preimplantation period of rat development causes blastocyst abnormalities and programming of postnatal hypertension. *Development*, *127*(19), 4195–4202. PMID:10976051

Kyei-Mensah, C., Kyerematen, R., & Adu-Acheampong, S. (2019). Impact of Rainfall Variability on Crop Production within the Worobong Ecological Area of Fanteakwa District. Ghana, *Advances in Agriculture*. *Article ID*, *7930127*, 1–7.

Labyak, S., Lava, S., Turek, F., & Zee, P. (2002). Effects of shift work on sleep and menstrual function in nurses. *Health Care for Women International*, *23*(6-7), 703–714. doi:10.1080/07399330290107449 PMID:12418990

Lacetera, N. (2019). Impact of climate change on animal health and welfare. *Animal Frontiers*, *9*(1), 26–31. doi:10.1093/af/vfy030

Lähdetie, J. (1995). Occupation-and exposure-related studies on human sperm. *Journal of Occupational and Environmental Medicine*, *37*(8), 922–930. doi:10.1097/00043764-199508000-00006

Lahnsteiner, F., Berger, B., Weismann, T., & Patzner, R. A. (1996). Motility ofspermatozoa of *Alburnusalburnus* (Cyprinidae) and its relationshipto seminal plasma composition and sperm metabolism. *Fish Physiology and Biochemistry*, *15*(2), 167–179. doi:10.1007/BF01875596 PMID:24194090

Lahnsteiner, F., & Mansour, N. (2012). The effect of temperature on spermmotility and enzymatic activity in brown trout *Salmotrutta, burbotLotalota* andgrayling *Thymallusthymallus*. *Journal of Fish Biology*, *81*(1), 197–209. doi:10.1111/j.1095-8649.2012.03323.x PMID:22747813

Lam, D. A., & Miron, J. A. (1994). Global patterns of seasonal variation inhuman fertility. *Annals of the New York Academy of Sciences*, *709*(1), 9–28. doi:10.1111/j.1749-6632.1994.tb30385.x PMID:8154738

Lanzafame, F. M., La Vignera, S., Vicari, E., & Calogero, A. E. (2009). Oxidative stress and medical antioxidant treatment in male infertility. *Reproductive Biomedicine Online*, *19*(5), 638–659. doi:10.1016/j.rbmo.2009.09.014 PMID:20021713

Larsen, S. B., Joffe, M., & Bonde, J. P.ASCLEPIOS Study Group. (1998). Time to pregnancy and exposure topesticides in Danish farmers. *Occupational and Environmental Medicine*, *55*(4), 278–283. doi:10.1136/oem.55.4.278 PMID:9624283

Larsson, K., & Einarsson, S. (1984). Seminal changes in boarsafter heat stress. *Acta Veterinaria Scandinavica*, *25*, 57–66. PMID:6464925

Lartey, R. T.Robert T. Lartey. (2006). Dynamics of Soil Flora and Fauna in Biological Control of Soil Inhabiting Plant Pathogens. *Plant Pathology Journal*, *5*(2), 125–142. doi:10.3923/ppj.2006.125.142

Lary, J. M., Conover, D. L., Johnson, P. H., & Burg, J. R. (1983). Teratogenicity of 27.12-MHz radiation in rats is related to duration of hyperthermic exposure. *Bioelectromagnetics*, *4*(3), 249–255. doi:10.1002/bem.2250040305 PMID:6626274

Lassen, T. H., Frederiksen, H., Jensen, T. K., Petersen, J. H., Joensen, U. N., Main, K. M., Skakkebaek, N. E., Juul, A., Jørgensen, N., & Andersson, A.-M. (2014). Urinary bisphenol A levels in young men: Association with reproductive hormones and semen quality. *Environmental Health Perspectives*, *122*(5), 478–484. doi:10.1289/ehp.1307309

Latini, G., Gallo, F., & De Felice, C. (2004). Birth characteristics and hepatoblastoma risk in young children. Cancer. *Interdisciplinary International Journal of the American Cancer Society*, *101*(1), 210–210.

Lawrence, A. B. (2008). Applied animal behaviour science: Past, present and future prospects. *Applied Animal Behaviour Science*, *115*(1-2), 1–24. doi:10.1016/j.applanim.2008.06.003

Lawrence, J. L., Payton, R. R., Godkin, J. D., Saxton, A. M., Schrick, F. N., & Edwards, J. L. (2004). Retinol improves development of bovine oocytes compromised by heat stress during maturation. *Journal of Dairy Science*, *87*(8), 2449–2454. doi:10.3168/jds.S0022-0302(04)73368-8 PMID:15328267

Lecha, L. B. (2007). *Fourth Assessment Report*. Cambridge University Press.

Lee, N. (1993). Environmental stress effect on bovine reproduction. *The Veterinary Clinics of North America*, *9*, 263–273. PMID:8348371

Lee, S., Min, J. Y., & Min, K. B. (2020). Female Infertility Associated with Blood Lead and Cadmium Levels. *International Journal of Environmental Research and Public Health*, *17*(5), 1794. doi:10.3390/ijerph17051794

Leibfried-Rutledge, M. L., Dominko, T., & Critser, E. S. (1997). Tissue maturation in vivo and in vitro: gamete and early embryo ontogeny. In Reproductive tissue banking (pp. 23-138). Elsevier.

Lellis, B., Fávaro-Polonio, C. Z., Pamphile, J. A., & Polonio, J. C. (2019). Effects of textile dyes on health and the environment and bioremediation potential of living organisms. *Biotechnology Research and Innovation*, *3*(2), 275–290. doi:10.1016/j.biori.2019.09.001

Lenz, R. W., Ball, G. D., Leibfried, M. L., Ax, R. L., & First, N. L. (1983). In vitro maturation and fertilization of bovine oocytes are temperature-dependent processes. *Biology of Reproduction*, *29*(1), 173–179. doi:10.1095/biolreprod29.1.173 PMID:6615963

Leonhardt, H., Gull, B., Kishimoto, K., Kataoka, M., Nilsson, L., Janson, P. O., Stener-Victorin, E., & Hellström, M. (2012). Uterine morphology and peristalsis in women with polycystic ovary syndrome. *Acta Radiologica*, *53*(10), 1195–1201. doi:10.1258/ar.2012.120384 PMID:23081959

Le, P. V. V., Kumar, P., Ruiz, M. O., Mbogo, C., & Muturi, E. J. (2019). Predicting the direct and indirect impacts of climate change on malaria in coastal Kenya. *PLoS One*, *14*(2), 1–18. doi:10.1371/journal.pone.0211258 PMID:30726279

Lepock, J. R. (1982). Involvement of membranes in cellular responses to hyperthermia. *Radiation Research*, *92*(3), 433–438. doi:10.2307/3575914 PMID:7178412

Leroy, J., Vanholder, T., Mateusen, B., Christophe, A., Opsomer, G., de Kruif, A., & Van Soom, A. (2005). Non-esterified fatty acids in follicular fluid of dairy cows and their effect on developmental capacity of bovine oocytes in vitro. *Reproduction (Cambridge, England)*, *130*(4), 485–495. doi:10.1530/rep.1.00735

Lesoin, L., Crampon, C., Boutin, O., & Badens, E. (2011). Preparation of liposomes using the supercritical anti-solvent (SAS) process and comparison with a conventional method. *The Journal of Supercritical Fluids*, *57*(2), 162–174. doi:10.1016/j.supflu.2011.01.006

Lessey, B. A. (2011). Assessment of endometrial receptivity. *Fertility and Sterility*, *96*(3), 522–529. doi:10.1016/j.fertnstert.2011.07.1095 PMID:21880273

Lessey, B. A., & Castelbaum, A. J. (2002). Integrins and implantation in the human. *Reviews in Endocrine & Metabolic Disorders*, *3*(2), 107–117. doi:10.1023/A:1015450727580 PMID:12007287

Lessey, B. A., Gui, Y., Apparao, K. B. C., Young, S. L., & Mulholland, J. (2002). Regulated expression of heparin-binding EGF-like growth factor (HB-EGF) in the human endometrium: A potential paracrine role during implantation. *Molecular Reproduction and Development*, *62*(4), 446–455. doi:10.1002/mrd.10129 PMID:12112577

Lessey, B. A., Killam, A. P., Metzger, D. A., Haney, A. F., Greene, G. L., & McCARTY, K. S. Jr. (1988). Immunohistochemical analysis of human uterine estrogen and progesterone receptors throughout the menstrual cycle. *The Journal of Clinical Endocrinology and Metabolism*, *67*(2), 334–340. doi:10.1210/jcem-67-2-334 PMID:2455728

Lessey, B. A., & Young, S. L. (2014). Homeostasis imbalance in the endometrium of women with implantation defects: The role of estrogen and progesterone. *Seminars in Reproductive Medicine, 32*(5), 365–375. doi:10.1055-0034-1376355 PMID:24959818

Levine, W. G. (1991). Metabolism of azo dyes: Implication for detoxication and activation. *Drug Metabolism Reviews, 23*(3-4), 253–309. doi:10.3109/03602539109029761 PMID:1935573

Liang, Z., Lin, Y., Ma, Y., Zhang, L., Zhang, X., Li, L., Zhang, S., Cheng, Y., Zhou, X., Lin, H., Miao, H., & Zhao, Q. (2016). The association between ambient temperature and preterm birth in Shenzhen, China: a distributed lag non-linear time series analysis. *Environmental Health: A Global Access Science Source, 15*(1), 84.

Liang, X. H., Deng, W. B., Li, M., Zhao, Z. A., Wang, T. S., Feng, X. H., & Yang, Z. M. (2014). Egr1 protein acts downstream of estrogen-leukemia inhibitory factor (LIF)-STAT3 pathway and plays a role during implantation through targeting Wnt4. *The Journal of Biological Chemistry, 289*(34), 23534–23545. doi:10.1074/jbc.M114.588897 PMID:25012664

Li, D. X., Romans, S., De Souza, M. J., Murray, B., & Einstein, G. (2015). Actigraphic and self-reported sleep quality in women: Associations with ovarian hormones and mood. *Sleep Medicine, 16*(10), 1217–1224. Advance online publication. doi:10.1016/j.sleep.2015.06.009 PMID:26429749

Lie, J. A. S., Andersen, A., & Kjærheim, K. (2007). Cancer risk among 43 000 Norwegian nurses. *Scandinavian Journal of Work, Environment & Health, 33*(1), 66–73. doi:10.5271jweh.1066 PMID:17353967

Lie, J. A. S., Kjuus, H., Zienolddiny, S., Haugen, A., Stevens, R. G., & Kjærheim, K. (2011). Night work and breast cancer risk among norwegian nurses: Assessment by different exposure metrics. *American Journal of Epidemiology, 173*(11), 1272–1279. doi:10.1093/aje/kwr014 PMID:21454824

Limesand, S., Regnault, T.R., & Hay, W.W. (2004). Characterization of of glucose transporter 8 (GLUT8) in the ovine placenta of normal and growth restricted fetuses. *Placenta, 25*,70 – 77.

Lim, H. J., & Wang, H. (2010). Uterine disorders and pregnancy complications: Insights from mouse models. *The Journal of Clinical Investigation, 120*(4), 1004–1015. doi:10.1172/JCI41210 PMID:20364098

Lim, J. M., Liou, S. S., & Hansel, W. (1996). Intracytoplasmic glutathione concentration and the role of β-mercaptoethanol in preimplantation development of bovine embryos. *Theriogenology, 46*(3), 429–439. doi:10.1016/0093-691X(96)00165-3 PMID:16727911

Lindbohm, M. L. (1995). Effects of parental exposure to solvents on pregnancy outcome. *Journal of Occupational and Environmental Medicine, 37*(8), 908–914. doi:10.1097/00043764-199508000-00004

Lindbohm, M. L. (1999). Women's reproductive health: Some recent developmentsin occupational epidemiology. *American Journal of Industrial Medicine, 36*(1), 18–24. doi:10.1002/(SICI)1097-0274(199907)36:1<18::AID-AJIM3>3.0.CO;2-N PMID:10361582

Lindbohm, M. L., Hemminki, K., Bonhomme, M. G., Anttila, A., Rantala, K., Heikkila, P., & Rosenberg, M. J. (1991). Effects of paternal occupationalexposure on spontaneous abortions. *American Journal of Public Health, 81*(8), 1029–1033. doi:10.2105/AJPH.81.8.1029 PMID:1853994

Lindbohm, M. L., Hietanen, M., Kyyronen, P., Sallmen, M., von Nandelstadh, P., Taskinen, H., Pekkarinen, M., Ylikoski, M., & Hemminki, K. (1992). Magneticfields of video display terminals and spontaneous abortion. *American Journal of Epidemiology, 136*(9), 1041–1051. doi:10.1093/oxfordjournals.aje.a116569 PMID:1462964

Lindbohm, M. L., Taskinen, H., Sallmen, M., & Hemminki, K. (1990). Spontaneousabortions among women exposed to organic solvents. *American Journal of Industrial Medicine, 17*(4), 449–463. doi:10.1002/ajim.4700170404 PMID:2327413

Lindhard, A., Bentin-Ley, U., Ravn, V., Islin, H., Hviid, T., Rex, S., & Sørensen, S. (2002). Biochemical evaluation of endometrial function at the time of implantation. *Fertility and Sterility*, *78*(2), 221–233. doi:10.1016/S0015-0282(02)03240-5 PMID:12137855

Lin, H., Decuypere, E., & Buyse, J. (2006). Acute heat stress induces oxidative stress in broiler chickens. *Comparative Biochemistry and Physiology. Part A, Molecular & Integrative Physiology*, *144*(1), 11–17. doi:10.1016/j.cbpa.2006.01.032

Linhart, O., Cosson, J., Mims, S. D., Shelton, W. L., & Rodina, M. (2002). Effects ofions on the motility of fresh and demembranated paddlefish(*Polyodonspathula*) spermatozoa. *Reproduction (Cambridge, England)*, *124*, 713–719. doi:10.1530/rep.0.1240713 PMID:12417010

Linhart, O., Mims, S. D., & Shelton, W. L. (1995). Motility of spermatozoa fromShovelnose sturgeon, *Scaphirhynchus-platorynchus*, and Paddlefish,*Polyodonspathula*. *Journal of Fish Biology*, *47*, 902–909. doi:10.1111/j.1095-8649.1995.tb06011.x

Linhart, O., Slechta, V., & Slavik, T. (1991). Fish sperm composition andbiochemistry. *Bulletin of the Institute of Zoology, Academia Sinica*, *16*, 285–311.

Lin, S. H., & Chen, M. L. (1997). Purification of textile wastewater effluents by a combined Fenton process and the ion exchange. *Desalination*, *109*(2), 121–130. doi:10.1016/S0011-9164(97)00058-1

Lin, S., Lin, Z., Ou, Y., Soim, A., Shrestha, S., Lu, Y., Sheridan, S., Luben, T. J., Fitzgerald, E., Bell, E., Shaw, G. M., Reefhuis, J., Langlois, P. H., Romitti, P., Feldkamp, M. L., Malik, S., Pantea, C., Nayak, S., Hwang, S. A., & Browne, M.National Birth Defects Prevention Study. (2018, September). Maternal ambient heat exposure during early pregnancy in summer and spring and congenital heart defects - A large US population-based, case-control study. *Environment International*, *118*, 211–221. doi:10.1016/j.envint.2018.04.043 PMID:29886237

Linzey, D. (Ed.). (2011). *Report: Saltwater Intrusion and Climate Change*. Prince Edward Island Department of Environment, Labour and Justice. http://www.gov.pe.ca/photos/original/cle_WA1.pdf

Lipscomb, J. A., Fenster, L., Wrensch, M., Shusterman, D., & Swan, S. (1991). Pregnancy outcomes in women potentially exposed to occupational solventsand women working in the electronics industry. *Journal of Occupational Medicine.*, *33*, 597–604. PMID:1870012

Li, R., Cheng, S., & Wang, Z. (2015). Circadian clock gene plays a key role on ovarian cycle and spontaneous abortion. *Cellular Physiology and Biochemistry*, *37*(3), 911–920. doi:10.1159/000430218 PMID:26390085

Lishko, P. V., Kirichok, Y., Ren, D., Navarro, B., Chung, J. J., & Clapham, D. E. (2012). The control of male fertility byspermatozoan ion channels. *Annual Review of Physiology*, *74*(1), 453–475. doi:10.1146/annurev-physiol-020911-153258 PMID:22017176

Liu, F., Gentles, A., & Theodorakis, C. W. (2007). Arsenate and perchlorate toxicity, growth effects, and thyroid histopathology in hypothyroid zebrafish Danio rerio. *Chemosphere*, *71*(7), 1369–1379. doi:10.1016/j.chemosphere.2007.11.036 PMID:18164748

Liu, F., & Li, R. (2019). *The Role of Neuroendocrine in Embryo Implantation*. Embryology - Theory and Practice, Bin Wu and Huai L. Feng, IntechOpen. doi:10.5772/intechopen.87863

Liu, J. M., Chang, Y. H., Ho, T. W., Chang, F. W., Pang, S. T., Hsu, R. J., & Lin, P. H. (2017). Patients with Epididymo-Orchitis and Meteorological Impact in Taiwan: A Nationwide Population-Based Study. The Canadian journal of infectious diseases & medical microbiology =. *The Canadian Journal of Infectious Diseases & Medical Microbiology*, *2017*, 1506857. doi:10.1155/2017/1506857 PMID:28316630

Liu, X., Nie, J., & Guo, S. W. (2011). Elevated immunoreactivity to tissue factor and its association with dysmenorrhea severity and the amount of menses in adenomyosis. *Human Reproduction (Oxford, England)*, *26*(2), 337–345. doi:10.1093/humrep/deq311 PMID:21115504

Li, X., Ma, W., Wong, G., Ma, S., Li, S., Bi, Y., & Gao, G. F. (2018, August 2). A new threat to human reproduction system posed by Zika virus (ZIKV): From clinical investigations to experimental studies. *Virus Research*, *254*, 10–14. doi:10.1016/j.virusres.2017.09.005 PMID:28893652

Li, Y., Cheng, S., Li, L., Zhao, Y., Shen, W., & Sun, X. (2019). Light-exposure at night impairs mouse ovary development via cell apoptosis and DNA damage. *Bioscience Reports*, *39*(5), BSR20181464. doi:10.1042/BSR20181464 PMID:30962269

Li, Z., Huo, W., Li, Z., Wang, B., Zhang, J., & Ren, A. (2016). Association between titanium and silver concentrations in maternal hair and risk of neural tube defects in offspring: A case-control study in north China. *Reproductive Toxicology (Elmsford, N.Y.)*, *66*, 115–121. doi:10.1016/j.reprotox.2016.10.006

Llobet, J. M., Falco, G., Casas, C., Teixido, A., & Domingo, J. L. (2003). Concentrations of arsenic, cadmium, mercury, and lead in common foods and estimated daily intake by children, adolescents, adults, and seniors of Catalonia, Spain. *Journal of Agricultural and Food Chemistry*, *51*(3), 838–842. doi:10.1021/jf020734q

Lopez-Carrillo, L., Hernandez-Ramirez, R. U., Calafat, A. M., Torres-Sanchez, L., Galvan Portillo, M., Needham, L. L., & Cebrian, M. E. (2010). Ex- posure to phthalates and breast cancer risk in Northen Mexico. *Environmental Health Perspectives*, *118*(4), 539–544. doi:10.1289/ehp.0901091

López-Gatius, F. (2003). Is fertility declining in dairy cattle? A retrospective study in northeastern Spain. *Theriogenology*, *60*(1), 89–99. doi:10.1016/S0093-691X(02)01359-6 PMID:12620583

Lord-Fontaine, S., & Averill-Bates, D. A. (2002). Heat shock inactivates cellular antioxidant defenses against hydrogen peroxide: Protection by glucose. *Free Radical Biology & Medicine*, *32*(8), 752–765. doi:10.1016/S0891-5849(02)00769-4 PMID:11937301

Loren, P., Sánchez, R., Arias, M. E., Felmer, R., Risopatrón, J., & Cheuquemán, C. (2017). Melatonin scavenger properties against oxidative and nitrosative stress: Impact on gamete handling and in vitro embryo production in humans and other mammals. *International Journal of Molecular Sciences*, *18*(6), 1119. doi:10.3390/ijms18061119 PMID:28613231

Lori, M., Symnaczik, S., Mäder, P., De Deyn, G., & Gattinger, A. (2017). Organic farming enhances soil microbial abundance and activity—A meta-analysis and meta-regression. *PLoS One*, *12*(7), 1–14. doi:10.1371/journal.pone.0180442 PMID:28700609

Lotthammer, K. (1991). Influence of nutrition on reproductive performance of the milking/gestating cow in the tropics. *Feeding dairy cows in the tropics*, 36-47.

Lovekamp-Swan, T., & Davis, B. J. (2003). Mechanisms of phthalate ester toxicity in the female reproductive system. *Environmental Health Perspectives*, *111*(2), 139–145. doi:10.1289/ehp.5658

Loven, D. P. (1988). A role for reduced oxygen species in heat induced cell killing and the induction of thermotolerance. *Medical Hypotheses*, *26*(1), 39–50. doi:10.1016/0306-9877(88)90111-9 PMID:2840559

Luboshits, G., Shina, S., Kaplan, O., Engelberg, S., Nass, D., Lifshitz-Mercer, B., Chaitchik, S., Keydar, I., & Ben-Baruch, A. (1999). Elevated expression of the CC chemokine regulated on activation, normal T cell expressed and secreted (RANTES) in advanced breast carcinoma. *Cancer Research*, *59*, 4681–4687. PMID:10493525

Lucas, B., Fields, C., & Hofmann, M. C. (2009). Signaling pathways in spermatogonial stem cells and their disruption by toxicants. *Birth Defects Research. Part C, Embryo Today*, *87*(1), 35–42. doi:10.1002/bdrc.20145

Lucio, A. C., Alves, B. G., Alves, K. A., Martins, M. C., Braga, L. S., Miglio, L., Alves, B. G., Silva, T. H., Jacomini, J. O., & Beletti, M. E. (2016). Selected sperm traits are simultaneously altered after scrotal heatstress and play specific roles in in vitro fertilization and embryonic development. *Theriogenology*, *86*(4), 924–933. doi:10.1016/j.theriogenology.2016.03.015 PMID:27087533

Lucy, M. C. (2001). Reproductive loss in high-producing dairy cattle: Where will it end? *Journal of Dairy Science*, *84*(6), 1277–1293. doi:10.3168/jds.S0022-0302(01)70158-0 PMID:11417685

Ludwig, H., & Spornitz, U. M. (1991). Microarchitecture of the human endometrium by scanning electron microscopy: Menstrual desquamation and remodeling. *Annals of the New York Academy of Sciences*, *622*(1), 28–46. doi:10.1111/j.1749-6632.1991.tb37848.x PMID:2064187

Lue, Y. H., Lasley, B. L., Laughlin, L. S., Swerdloff, R. S., Hikim, A. P., Leung, A., Overstreet, J. W., & Wang, C. (2002). Mild testicular hyperthermia induces profound transitional spermatogenic suppression throughincreased germ cell apoptosis in adult cynomolgus monkeys (Macacafascicularis). *Journal of Andrology*, *23*, 799–805. PMID:12399525

Lu, J., Yang, Z., Ma, X., Ma, M., & Zhang, Z. (2020). The role of meteorological factors on mumps incidence among children in Guangzhou, Southern China. *PLoS One*, *15*(4), e0232273. doi:10.1371/journal.pone.0232273 PMID:32348370

Lundsberg, L. S., Bracken, M. B., & Belanger, K. (1995). Occupationally relatedmagnetic field exposure and male subfertility. *Fertility and Sterility*, *63*(2), 384–391. doi:10.1016/S0015-0282(16)57373-7 PMID:7843447

Luo, M. J., Han, J. B., Li, W., Sui, H. S., & Tan, J. H. (2005). Serum and follicular fluid steroid levels as related to follicular development and granulosa cell apoptosis during the estrous cycle of goats. *Small Ruminant Research*, *57*(1), 57–65. doi:10.1016/j.smallrumres.2004.06.015

Lutfalla, P. S. S., Riley, W. J., Torn, M. S., Schmidt, M. W. I., & Soussana, J.-F. (2017). The changing faces of soil organic matter research. *European Journal of Soil Science*, *69*(1), 23–30.

Lyche, J. L., Oskam, I. C., Skaare, J. U., Reksen, O., Sweeney, T., Dahl, E., & Ropstad, E. (2004). Effects of gestational and lactational exposure to low doses of PCBs 126 and 153 on anterior pituitary and gonadal hormones and on puberty in female goats. *Reproductive Toxicology (Elmsford, N.Y.)*, *19*(1), 87–95. doi:10.1016/j.reprotox.2004.05.005 PMID:15336716

Lynch, J. J., Hinch, G. N., & Adams, D. B. (1992). *The behavior of sheep:biological principles and implications for production*. C.A.B. International.

Lyronis, I. D., Ploumis, N., Vlahakis, I., & Charissis, G. (2009). Acute scrotum-etiology, clinical presentation and seasonal variation. *Indian Journal of Pediatrics*, *76*(4), 407–410. doi:10.100712098-009-0008-2 PMID:19205631

Mabogunje, O. A. (1986). Testicular torsion and low relative humidity in a tropical country. *British Medical Journal*, *292*(6517), 363–364. doi:10.1136/bmj.292.6517.363 PMID:3080171

MacCracken, M. C., Moore, F., & Topping, J. C. (Eds.). (2008). *Sudden and disruptive climate change: exploring the real risks and how we can avoid them*. Earthscan.

Maganhin, C. C., Fuchs, L. F. P., Simões, R. S., Oliveira-Filho, R. M., De Jesus Simões, M., Baracat, E. C., & Soares, J. M. Jr. (2013). Effects of melatonin on ovarian follicles. *European Journal of Obstetrics, Gynecology, and Reproductive Biology*, *166*(2), 178–184. doi:10.1016/j.ejogrb.2012.10.006 PMID:23102587

Maggi, R., Cariboni, A. M., Marelli, M. M., Moretti, R. M., Andre, V., Marzagalli, M., & Limonta, P. (2016). GnRH and GnRH receptors in the pathophysiology of the human female reproductive system. *Human Reproduction Update*, *22*(3), 358–381. doi:10.1093/humupd/dmv059 PMID:26715597

Makarevich, A., Olexikova, L., Chrenek, P., Kubovicova, E., Freharova, K., & Pivko, J. (2007). The effect of hyperthermia in vitro on vitality of rabbit preimplantation embryos. *Physiological Research*, *56*, 789–796. PMID:17087597

Makena, P. S., & Chung, K. T. (2007). Evidence of oxidative genotoxicity induced by 4-aminobiphenyl, benzidine, and benzidine congeners. *Environmental and Molecular Mutagenesis*, *48*, 404–413. doi:10.1002/em.20288 PMID:17370336

Makey, C. M., McClean, M. D., Braverman, L. E., Pearce, E. N., Sjödin, A., Weinberg, J., & Webster, T. F. (2016). Polybrominated diphenyl ether exposure and reproductive hormones in North American men. *Reproductive Toxicology (Elmsford, N.Y.)*, *62*, 46–52. doi:10.1016/j.reprotox.2016.04.009

Makhseed, M., Musini, V. M., Ahmed, M. A., & Monem, R. A. (1999). Influence of seasonal variation on pregnancy-induced hypertension and/or preeclampsia. *Australian and New Zealand Journal of Obstetrics and Gynaecology*, *39*(2), 196–199. doi:10.1111/j.1479-828X.1999.tb03372.x PMID:10755779

Makrigiannakis, A., Zoumakis, E., Kalantaridou, S., Coutifaris, C., Margioris, A. N., Coukos, G., Rice, K. C., Gravanis, A., & Chrousos, G. P. (2001). Corticotropin-releasing hormone promotes blastocyst implantation and early maternal tolerance. *Nature Immunology*, *2*(11), 1018–1024. doi:10.1038/ni719 PMID:11590404

Ma, L., Benson, G. V., Lim, H., Dey, S. K., & Maas, R. L. (1998). Abdominal B (AbdB) HoxaGenes: Regulation in adult Uterus by estrogen and progesterone and repression in müllerian duct by the synthetic estrogen diethylstilbestrol (DES). *Developmental Biology*, *197*(2), 141–154. doi:10.1006/dbio.1998.8907 PMID:9630742

Malayer, J., & Hansen, P. (1990). Differences between Brahman and Holstein cows in heat-shock induced alterations of protein synthesis and secretion by oviducts and uterine endometrium. *Journal of Animal Science*, *68*(1), 266–280. doi:10.2527/1990.681266x

Mandal, D. K., Nagpaul, P. K., & Gupta, A. K. (2000). Seasonal variation in seminal attributes and sexual behaviour of Murrah buffalo bulls. *Indian Journal of Dairy Science*, *53*(4), 278–283.

Manimaran, D., Sulthana, A. S., & Elangovan, N. (2018). Reactive black 5 induced developmental defects via potentiating apoptotic cell death in Zebrafish (*Danio rerio*) embryos. *Pharmacy and Pharmacology International Journal*, *6*, 449–451.

Manipalviratn, S., DeCherney, A., & Segars, J. (2009). Imprinting disorders and assisted reproductive technology. *Fertility and Sterility*, *91*(2), 305–315. doi:10.1016/j.fertnstert.2009.01.002 PMID:19201275

Manuel, R., Machado, A., & Serralheiro, R. P. (2017). Soil Salinity: Effect on Vegetable Crop Growth. Management Practices to Prevent and Mitigate Soil Salinizatio. In Soil Science: An Introduction, (pp.353-384). Indian Society of Soil Science.

Marai, F. M. I., & Haeeb, A. A. M. (2010). Buffaloes' reproductive and productive traits as affected by heat stress. *Tropical and Subtropical Agroecosystems*, *12*(2), 193–217.

Marai, I. F. M., Habeeb, A. A. M., & Gad, A. E. (2002). Rabbits' productive, reproductive and physiological performance traits as affected by heat stress: A review. *Livestock Science*, *78*(2), 71–90. doi:10.1016/S0301-6226(02)00091-X

Marcheva, B., Ramsey, K. M., Buhr, E. D., Kobayashi, Y., Su, H., Ko, C. H., Ivanova, G., Omura, C., Mo, S., Vitaterna, M. H., Lopez, J. P., Philipson, L. H., Bradfield, C. A., Crosby, S. D., Jebailey, L., Wang, X., Takahashi, J. S., & Bass, J. (2010). Disruption of the clock components CLOCK and BMAL1 leads to hypoinsulinaemia and diabetes. *Nature*, *466*(7306), 627–631. doi:10.1038/nature09253 PMID:20562852

Margalioth, E. J., Ben-Chetrit, A., Gal, M., & Eldar-Geva, T. (2006). Investigation and treatment of repeated implantation failure following IVF-ET. *Human Reproduction (Oxford, England)*, *21*(12), 3036–3043. doi:10.1093/humrep/del305 PMID:16905766

Mariasegaram, R., Chase, C. C., Jr Chaparro, J. X., Olson, T. A., Brenneman, R. A., & Niedz, R. P. (2007). The slick air coat locus maps to chromosome 20 in Senepol-derived cattle. *Animal Genetics*, *38*(1), 54–59. doi:10.1111/j.1365-2052.2007.01560.x

Marimuthua, S., Antonisamya, A. J., Malayandia, S., Rajendrana, K., Tsaib, P. C., & Pugazhendhi, A. (2020). Silver nanoparticles in dye effluent treatment: A review on synthesis, treatment methods, mechanisms, photocatalytic degradation, toxic effects and mitigation of toxicity. *Journal of Photochemistry and Photobiology. B, Biology*, *205*, 111823. doi:10.1016/j.jphotobiol.2020.111823 PMID:32120184

Maroni, M., Fanetti, A. C., & Metruccio, F. (2006). Risk assessment and management of occupational exposure to pesticides in agriculture. *La Medicina del Lavoro*, *97*(2), 430–437.

Martínez-Sales, M., García-Ximénez, F., & Espinós, F. J. (2015). Zebrafish as a possible bioindicator of organic pollutants with effects on reproduction in drinking waters. *Journal of Environmental Sciences (China)*, *33*, 254–260. doi:10.1016/j.jes.2014.11.012

Martinho, V. J. P. D. (2020). Exploring the Topics of Soil Pollution and Agricultural Economics: Highlighting Good Practices. *Agriculture*, *10*(1), 24–43.

Martin, K. L., Barlow, D. H., & Sargent, I. L. (1998). Heparin-binding epidermal growth factor significantly improves human blastocyst development and hatching in serum-free medium. *Human Reproduction (Oxford, England)*, *13*(6), 1645–1652. doi:10.1093/humrep/13.6.1645 PMID:9688406

Maske, P., Dighe, V., & Vanage, G. (2018). n-butylparaben exposure during perinatal period impairs fertility of the F1 generation female rats. *Chemosphere*, *213*, 114–123. doi:10.1016/j.chemosphere.2018.08.130

Mason, H. J. (1990). Occupational cadmium exposure and testicular endocrine function. *Human and Experimental Toxicology*, *9*(2), 91–94. doi:10.1177/096032719000900205

Mathew, S., Mathur, D., Chang, A. B., McDonald, E., Singh, G. R., Nur, D., & Gerritsen, R. (2017). Examining the Effects of Ambient Temperature on Pre-Term Birth in Central Australia. *International Journal of Environmental Research and Public Health*, *14*(2), 147. doi:10.3390/ijerph14020147 PMID:28165406

Mathur, N., Bhatnagar, P., & Sharma, P. (2012). Review of the mutagenicity of textile dye products. *Universal Journal of Environmental Research and Technology*, *2*, 1–18.

Matsuzaki, S., Fukaya, T., Suzuki, T., Murakami, T., Sasano, H., & Yajima, A. (1999). Oestrogen receptor α and β mRNA expression in human endometrium throughout the menstrual cycle. *Molecular Human Reproduction*, *5*(6), 559–564. doi:10.1093/molehr/5.6.559 PMID:10341004

Matsuzuka, T., Sakamoto, N., Ozawa, M., Ushitani, A., Hirabayashi, M., & Kanai, Y. (2005b). Alleviation of maternal hyperthermia-induced early embryonic death by administration of melatonin to mice. *Journal of Pineal Research*, *39*(3), 217–222. doi:10.1111/j.1600-079X.2005.00260.x PMID:16150100

Matte, T. D., Lane, K., & Ito, K. (2016). Excess Mortality Attributable to Extreme Heat in New York City, 1997-2013. *Health Security*, *14*(2), 64–70. doi:10.1089/hs.2015.0059 PMID:27081885

Ma, W. G., Song, H., Das, S. K., Paria, B. C., & Dey, S. K. (2003). Estrogen is a critical determinant that specifies the duration of the window of uterine receptivity for implantation. *Proceedings of the National Academy of Sciences of the United States of America*, *100*(5), 2963–2968. doi:10.1073/pnas.0530162100 PMID:12601161

Ma, W., Li, S., Ma, S., Jia, L., Zhang, F., Zhang, Y., Zhang, J., Wong, G., Zhang, S., Lu, X., Liu, M., Yan, J., Li, W., Qin, C., Han, D., Qin, C., Wang, N., Li, X., & Gao, G. F. (2016, December 1). Zika Virus Causes Testis Damage and Leads to Male Infertility in Mice. *Cell*, *167*(6), 1511–1524. doi:10.1016/j.cell.2016.11.016 PMID:27884405

Mboera, L. E. G., Mayala, B. K., Kweka, E. J., & Mazigo, H. D. (2011). Impact of climate change on human health and health systems in Tanzania: A review. *Tanzania Journal of Health Research*, *13*, 407–426.

Mbonye, A. K., Neema, S., & Magnussen, P. (2006). Preventing malaria in pregnancy: A study of perceptions and policy implications in Mukono district, Uganda. *Health Policy and Planning*, *21*(1), 17–26. doi:10.1093/heapol/czj002 PMID:16317032

McCreesh, N., & Booth, M. (2013, November). Challenges in predicting the effects of climate change on Schistosoma mansoni and Schistosoma haematobium transmission potential. *Trends in Parasitology*, *29*(11), 548–555. doi:10.1016/j.pt.2013.08.007 PMID:24064438

McDonald, A. D. (1958). Maternal health and congenital defect. *The New England Journal of Medicine*, *258*(16), 767–773. doi:10.1056/NEJM195804172581601 PMID:13541657

McDonald, A. D. (1961). Maternal health in early pregnancy and cogeital defect. Final report on a prospective inquiry. *British Journal of Preventive & Social Medicine*, *15*, 154–166.

McElgunn, B. (1998). Reproductive and developmental hazards in the workplace. *Clinical Excellence for Nurse Practitioners*, *2*, 140–145. PMID:12675082

McFadden, E., Jones, M. E., Schoemaker, M. J., Ashworth, A., & Swerdlow, A. J. (2014). The relationship between obesity and exposure to light at night: Cross-sectional analyses of over 100,000 women in the breakthrough generations study. *American Journal of Epidemiology*, *180*(3), 245–250. doi:10.1093/aje/kwu117 PMID:24875371

McGinnis, W., & Krumlauf, R. (1992). Homeobox genes and axial patterning. *Cell*, *68*(2), 283–302. doi:10.1016/0092-8674(92)90471-N PMID:1346368

McGregor, A. J., & Mason, H. J. (1991). The effects of occupational exposure to cadmium, lead and mercury vapour on male reproductive endocrine function. Heavy Metals in the Environment. Edinburgh. *CEP Consultants*, *1*, 375–378.

McKee, R. H., Butala, J. H., David, R. M., & Gans, G. (2004). NTP center for the evaluation of risks to human reproduction reports on phthalates: Addressing the data gaps. *Reproductive Toxicology (Elmsford, N.Y.)*, *18*(1), 1–22. doi:10.1016/j.reprotox.2003.09.002

McKeon, G. M., Stone, G. S., Syktus, J. I., Carter, J. O., Flood, N. R., Ahrens, D. G., Bruget, D. N., Chilcott, C. R., Cobon, D. H., Cowley, R. A., Crimp, S. J., Fraser, G. W., Howden, S. M., Johnston, P. W., Ryan, J. G., Stokes, C. J., & Day, K. A. (2009). Climate change impacts on northern Australian rangeland livestock carrying capacity: A review of issues. *The Rangeland Journal*, *31*(1), 1–29. doi:10.1071/RJ08068

McLeod, A. (2011). *World livestock 2011-livestock in food security*. Food and Agriculture Organization of the United Nations (FAO). http://www.fao.org/docrep/014/i2373e/i2372e00.htm

McMartin, K. I., Chu, M., Kopecky, E., Einarson, T. R., & Koren, G. (1998). Pregnancyoutcome following maternal organic solvent exposure: A meta-analysisof epidemiologic studies. *American Journal of Industrial Medicine*, *34*(3), 288–292. doi:10.1002/(SICI)1097-0274(199809)34:3<288::AID-AJIM12>3.0.CO;2-Q PMID:9698999

McMichael. (2013). Globalization, climate change, and human health. *The New England Journal of Medicine*, *368*(14), 1335–1343.

McMurray, R. G., & Katz, V. L. (1990). Thermoregulation in pregnancy: Implications for exercise. *Sports Medicine (Auckland, N.Z.)*, *10*(3), 146–158. doi:10.2165/00007256-199010030-00002 PMID:2237032

Meehl, G. A., & Tebaldi, C. (2004). More intense, more frequent, and longer lastingheat waves in the 21st century. *Science*, *305*(5686), 994–997. doi:10.1126cience.1098704 PMID:15310900

Meeker, J. D., Calafat, A. M., & Hauser, R. (2007). Di (2-ethylhexyl) phthalate metabolites may alter thyroid hormone levels in men. *Environmental Health Perspectives*, *115*(7), 1029–1034. doi:10.1289/ehp.9852

Meeker, J. D., Rossano, M. G., Protas, B., Diamond, M. P., Puscheck, E., Daly, D., Paneth, N., & Wirth, J. J. (2008). Cadmium, lead, and other metals in relation to semen quality: Human evidence for molybdenum as a male reproductive toxicant. *Environmental Health Perspectives*, *116*(11), 1473–1479. doi:10.1289/ehp.11490

Meeker, J. D., Yang, T., Ye, X., Calafat, A. M., & Hauser, R. (2011). Urinary concentrations of parabens and serum hormone levels, semen quality parameters, and sperm DNA damage. *Environmental Health Perspectives*, *119*(2), 252–257. doi:10.1289/ehp.1002238

Meerburg, B. G., & Kijlstra, A. (2009, July). Changing climate-changing pathogens: Toxoplasma gondii in North-Western Europe. *Parasitology Research*, *105*(1), 17–24. doi:10.100700436-009-1447-4 PMID:19418068

Megdal, S. P., Kroenke, C. H., Laden, F., Pukkala, E., & Schernhammer, E. S. (2005). Night work and breast cancer risk: A systematic review and meta-analysis. *European Journal of Cancer*, *41*(13), 2023–2032. doi:10.1016/j.ejca.2005.05.010 PMID:16084719

Mehlis, M., & Bakker, T. C. M. (2014). The influence of ambient water temperature on sperm performance and fertilization success in three-spined sticklebacks(*Gasterosteusaculeatus*). *Evolutionary Ecology*, *28*(4), 655–667. doi:10.100710682-014-9707-x

Mehrotra, V. P., & Redford, C. R. (2005). *Apparatus and method for detection of estrus and non-pregnancy*. Google Patents.

Mehta, R., & Yadav, K. (2013). Soil contamination due to textile effluent-case study on the printing cluster of Jaipur. *J Text Assoc*, *73*, 367–370.

Melendez, P. (2006). Nutritional management of the transition period to optimize fertility in dairy cattle. *Proceedings 3rd Florida and Georgia Dairy Road Show Conference*.

Menashe, I., Figueroa, J. D., Garcia-Closas, M., Chatterjee, N., Malats, N., Picornell, A., & et al., (2012). Large-scale pathway-based analysis of bladder cancer genome-wide association data from five studies of European background. *PLoS One*, *7*(1), e29396. doi:10.1371/journal.pone.0029396 PMID:22238607

Mendiola, J., Jørgensen, N., Andersson, A.-M., Calafat, A. M., Ye, X., Redmon, J. B., Drobnis, E. Z., Wang, C., Sparks, A., Thurston, S. W., Liu, F., & Swan, S. H. (2010). Are environmental levels of bisphenol A associated with reproductive function in fertile men? *Environmental Health Perspectives*, *118*(9), 1286–1291. doi:10.1289/ehp.1002037

Mendola, P., Messer, L. C., & Rappazzo, K. (2008). Science linking environmental contaminant exposures with fertility and reproductive health impacts in the adult female. *Fertility and Sterility*, *89*(2), 81–94. doi:10.1016/j.fertnstert.2007.12.036

Menegassi, S., Barcellos, J., Dias, E., Koetz, C. Jr, Pereira, G., Peripolli, V., McManus, C., Canozzi, M. E. A., & Lopes, F. G. (2015). Scrotal infrared digital thermography as a predictor of seasonal effects on sperm traits in Braford bulls. *International Journal of Biometeorology*, *19*(3), 357–364. doi:10.100700484-014-0847-z PMID:24848445

Meng, L., Rijntjes, E., Swarts, H., Bunschoten, A., van der Stelt, I., Keijer, J., & Teerds, K. (2016). Dietary-induced chronic hypothyroidism negatively affects rat follicular development and ovulation rate and is associated with oxidative stress. *Biology of Reproduction*, *94*(4), 90, 91–11. doi:10.1095/biolreprod.115.136515

Michael, A. E., Pester, L. A., Curtis, P., Shaw, R. W., Edwards, C. R. W., & Cooke, B. A. (1993). Direct inhibition of ovarian steroidogenesis by Cortisol and the modulatory role of 11β-hydroxysteroid dehydrogenase. *Horumon To Rinsho*, *38*(6), 641–644. doi:10.1111/j.1365-2265.1993.tb02147.x PMID:8334750

Michel, C. L, Chastel, O. & Bonnet, X. (2011). Ambient temperature and pregnancy influence cortisol levels in female guinea pigs and entail long-term effects on the stress response of their offspring. *General and Comparative Endocrinology, 171*, 275-282.

Mieusset, R., Bujan, L., Mondinat, C., Monsat, A., Pontonnier, F., & Grandjean, H. (1987). Association of scrotal hyperthermia with impaired spermatogenesis ininfertile men. *Fertility and Sterility*, *48*(6), 1006–1011. doi:10.1016/S0015-0282(16)59600-9 PMID:3678498

Milad, M. P., Klock, S. C., Moses, S., & Chatterton, R. (1998). Stress and anxietydo not result in pregnancy stage. *Human Reproduction (Oxford, England)*, *13*(8), 296–300. doi:10.1093/humrep/13.8.2296 PMID:9756314

Miles, H. L., Hofman, P. L., Peek, J., Harris, M., Wilson, D., Robinson, E. M., Gluckman, P. D., & Cutfield, W. S. (2007). In vitro fertilization improves childhood growth and metabolism. *The Journal of Clinical Endocrinology and Metabolism*, *92*(9), 3441–3445. doi:10.1210/jc.2006-2465 PMID:17566097

Millar, R. P. (2003). GnRH II and type II GnRH receptors. *Trends in Endocrinology and Metabolism*, *14*(1), 35–43. doi:10.1016/S1043-2760(02)00016-4 PMID:12475610

Millar, R., Lowe, S., Conklin, D., Pawson, A., Maudsley, S., Troskie, B., Ott, T., Millar, M., Lincoln, G., Sellar, R., Faurholm, B., Scobie, G., Kuestner, R., Terasawa, E., & Katz, A. (2001). A novel mammalian receptor for the evolutionarily conserved type II GnRH. *Proceedings of the National Academy of Sciences of the United States of America*, *98*(17), 9636–9641. doi:10.1073/pnas.141048498 PMID:11493674

Miller, B. H., Olson, S. L., Levine, J. E., Turek, F. W., Horton, T. H., & Takahashi, J. S. (2006). Vasopressin regulation of the proestrous luteinizing hormone surge in wild-type and Clock mutant mice. *Biology of Reproduction*, *75*(5), 778–784. doi:10.1095/biolreprod.106.052845 PMID:16870944

Miller, B. H., Olson, S. L., Turek, F. W., Levine, J. E., Horton, T. H., & Takahashi, J. S. (2004). Circadian clock mutation disrupts estrous cyclicity and maintenance of pregnancy. *Current Biology*, *14*(15), 1367–1373. doi:10.1016/j.cub.2004.07.055 PMID:15296754

Miller, M. R., Mannowetz, N., Iavarone, A. T., Safavi, R., Gracheva, E. O., Smith, J. F., Hill, R. Z., Bautista, D. M., Kirichok, Y., & Lishko, P. V. (2016). Unconventional endocannabinoidsignalling governs sperm activation via the sex hormoneprogesterone. *Science*, *352*(6285), 555–559. doi:10.1126cience.aad6887 PMID:26989199

Miller, M. R., Mansell, S. A., Meyers, S. A., & Lishko, P. V. (2015). Flagellar ion channels of sperm: Similarities and differencesbetween species. *Cell Calcium*, *58*(1), 105–113. doi:10.1016/j.ceca.2014.10.009 PMID:25465894

Miller, M. W. (1993). Migration of cortical neurons is altered by gestational exposure to ethanol. *Alcoholism, Clinical and Experimental Research*, *17*(2), 304–314. doi:10.1111/j.1530-0277.1993.tb00768.x PMID:8488973

Mims, S. D. (1991). Evaluation of activator solutions, motility duration andshort-term storage of paddlefish spermatozoa. *Journal of the World Aquaculture Society*, *22*(4), 224–229. doi:10.1111/j.1749-7345.1991.tb00738.x

Mimura, N. (2013). Sea-level rise caused by climate change and its implications for society. *Proceedings of the Japan Academy. Series B, Physical and Biological Sciences*, *89*(7), 281–301. doi:10.2183/pjab.89.281 PMID:23883609

Minkel, J., Moreta, M., Muto, J., Htaik, O., Jones, C., Basner, M., & Dinges, D. (2014). Sleep deprivation potentiates HPA axis stress reactivity in healthy adults. *Health Psychology*, *33*(11), 1430–1434. doi:10.1037/a0034219 PMID:24818608

Mir, F., Fontaine, E., Reyes-Gomez, E., Carlus, M., & Fontbonne, A. (2012). Subclinical leishmaniasis associated with infertility and chronic prostatitis in a dog. *The Journal of Small Animal Practice*, *53*(7), 419–422. doi:10.1111/j.1748-5827.2012.01224.x PMID:22690941

Mirghani, M. E. S., Kabbashi, N. A., Qudsieh, I. Y., & Elfaki, F. A. (2008). A new method for the determination of toxic dye using FTIR spectroscopy. *IIUM Engineering Journal*, *9*, 27–38. doi:10.31436/iiumej.v9i2.98

Mirkes, P. E., Cornel, L. M., Park, H. W., & Cunningham, L. M. (1997). Induction of thermotolerance in early postimplantation rat embryos is associated with increased resistance to hyperther- mia-induced apoptosis. *Teratology*, *56*(3), 210–219. doi:10.1002/(SICI)1096-9926(199709)56:3<210::AID-TERA4>3.0.CO;2-4 PMID:9358608

Mirkes, P. E., & Dogget, B. (1992). Accumulation of heat shock protein 72 (hsp72) in postimplantation rat embryos after exposure to various periods of hyperthermia (40–438C) in vitro: Evidence that heat shock protein 72 is a biomarker of heat-induced embryotoxicity. *Teratology*, *46*(3), 301–309. doi:10.1002/tera.1420460315 PMID:1523587

Mitchell, M., Schulz, S. L., Armstrong, D. T., & Lane, M. (2009). Metabolic and mitochondrial dysfunction in early mouse embryos following maternal dietary protein intervention. *Biology of Reproduction*, *80*(4), 622–630. doi:10.1095/biolreprod.108.072595 PMID:19129514

Mojiri, A. (2011). Effects of municipal waste water on physical and chemical properties of saline soil. *Journal of Biological and Environmental Sciences*, *5*, 71–76.

Mokhlesi, B., Scoccia, B., Mazzone, T., & Sam, S. (2012). Risk of obstructive sleep apnea in obese and nonobese women with polycystic ovary syndrome and healthy reproductively normal women. *Fertility and Sterility*, *97*(3), 786–791. doi:10.1016/j.fertnstert.2011.12.024 PMID:22264851

Mok-Lin, E., Ehrlich, S., Williams, P. L., Petrozza, J., Wright, D. L., Calafat, A. M., Ye, X., & Hauser, R. (2010). Urinary bisphenol A concentrations and ovarian response among women undergoing IVF. *International Journal of Andrology*, *33*(2), 385–393. doi:10.1111/j.1365-2605.2009.01014.x

Molina, O., & Saldarriaga, V. (2017). The perils of climate change: In utero exposure to temperature variability and birth outcomes in the Andean region. *Economics and Human Biology*, *24*, 111–124. doi:10.1016/j.ehb.2016.11.009 PMID:27984771

Monograph, I. A. R. C. (1998). Aromatic amines, in: Some aromatic amines, hydrazine and related substances, N-nitroso compounds and miscellaneous alkylating agents. IARC Monographs on the evaluation of the carcinogenic risk of chemicals to humans, 4.

Monograph, I. A. R. C. (2010). Some aromatic amines, organic dyes and related exposures. *IARC Monographs on the Evaluation of Carcinogenic Risks to Humans*, *99*. PMID:21528837

Moore, R. W., Rudy, T. A., Lin, T. M., Ko, K., & Peterson, R. E. (2001). Abnormalities of sexual development in male rats with in utero and lactational exposure to the antiandrogenic plasticizer Di (2-ethylhexyl) phthalate. *Environmental Health Perspectives*, *109*(3), 229–237. doi:10.1289/ehp.01109229

Moore, S. B., & Ausley, L. W. (2004). Systems thinking and green chemistry in the textile industry: Concepts, technologies and benefits. *Journal of Cleaner Production*, *12*(6), 585–601. doi:10.1016/S0959-6526(03)00058-1

Moosvi, S., Kher, X., & Madamwar, D. (2007). Isolation, characterization and decolorization of textile dyes by a mixed bacterial consortium JW2. *Dyes and Pigments*, *74*(3), 723–729. doi:10.1016/j.dyepig.2006.05.005

Moretti, M. E., Bar-Oz, B., Fried, S., & Koren, G. (2005). Maternal hyperthermia and the risk for neural tube defects in offspring systematic review and meta-analysis. *Epidemiology (Cambridge, Mass.)*, *16*(2), 216–219. doi:10.1097/01.ede.0000152903.55579.15 PMID:15703536

Morgan, M. A. (1992). Different patterns of myometrial activity and 24-h rhythms in myometrial contractility in the gravid baboon during the second half of pregnancy. *Biology of Reproduction*, *46*(6), 1158–1164. doi:10.1095/biolreprod46.6.1158 PMID:1391314

Morishima, H. O., Glaser, B., Biermann, W. H., & James, L. S. (1975). Increased uterine activity and fetal deterioration during maternal hyperthermia. *American Journal of Obstetrics and Gynecology*, *121*(4), 531–538. doi:10.1016/0002-9378(75)90087-3 PMID:807107

Mortensen, J. T. (1988). Risk for reduced sperm quality among metal workerswith special reference to welders. *Scandinavian Journal of Work, Environment & Health*, *14*(1), 27–30. doi:10.5271jweh.1954 PMID:3353693

Moshkdanian, G., Moghani-Ghoroghi, F., Pasbakhsh, P., Nematollahi-Mahani, S. N., Najafi, A., & Kashani, I. R. (2017). Melatonin upregulates ErbB1 and ErbB4, two primary implantation receptors, in pre-implantation mouse embryos. *Iranian Journal of Basic Medical Sciences.*, *20*, 655–661. PMID:28868121

Moss, E. J., Thomas, L. V., Cook, M. W., Walters, D. G., Foster, P. M. D., Creasv, D. M., & ... (1985). The role of metabolism in 2-methoxyethanol-induced testicular toxicity. *Toxicology and Applied Pharmacology*, *79*(3), 480–489. doi:10.1016/0041-008X(85)90145-0 PMID:4035689

Moussa, M., Shu, J., Zhang, X., & Zeng, F. (2015). Maternal control of oocyte quality in cattle "a review". *Animal Reproduction Science*, *155*, 11–27. doi:10.1016/j.anireprosci.2015.01.011

Mucio-Ramírez, S., Sánchez-Islas, E., Sánchez-Jaramillo, E., Currás-Collazo, M., Juárez-González, V. R., Álvarez-González, M. Y., Orser, L. E., Hou, B., Pellicer, F., Kodavanti, P. R. S., & León-Olea, M. (2017). Perinatal exposure to organohalogen pollutants decreases vasopressin content and its mRNA expression in magnocellular neuroendocrine cells activated by osmotic stress in adult rats. *Toxicology and Applied Pharmacology*, *329*, 173–189. doi:10.1016/j.taap.2017.05.039 PMID:28579251

Muehleis, P. M., & Long, S. Y. (1976). The effects of altering the pHof seminal fluid on the sex ratio of rabbit offspring. *Fertility and Sterility*, *27*(12), 1438–1445. doi:10.1016/S0015-0282(16)42261-2 PMID:12027

Mundt, N., Spehr, M., & Lishko, P. V. (2018). TRPV4 is the temperature-sensitive ion channel of human sperm. *eLife*, *7*, e35853. doi:10.7554/eLife.35853 PMID:29963982

Munro, S. S. (1938). Fowl sperm immobilization by a temperature-media interaction and its biological significance. *Quarterly Journal of Experimental Physiology: Translation and Integration*, *27*(3), 281–291. doi:10.1113/expphysiol.1938.sp000753

Murata, M., & Kawanish, S. (2011). Mechanisms of oxidative DNA damage induced by carcinogenic arylamines. *Frontiers in Bioscience*, *16*(1), 1132–1143. doi:10.2741/3739 PMID:21196222

Murphy, A. A., Palinski, W., Rankin, S., Morales, A. J., & Parthasarathy, S. (1998). Macrophage scavenger receptor(s) and oxidatively modified proteins in endometriosis. *Fertility and Sterility*, *69*(6), 1085–1091. doi:10.1016/S0015-0282(98)00088-0 PMID:9627297

Musumeci, G., Castorina, S., Castrogiovanni, P., Loreto, C., Leonardi, R., Aiello, F. C., & Imbesi, R. (2015). A journey through the pituitary gland: Development, structure and function, with emphasis on embryo-foetal and later development. *Acta Histochemica*, *117*(4-5), 355–366. doi:10.1016/j.acthis.2015.02.008 PMID:25858531

Muter, J., Lucas, E. S., Chan, Y. W., Brighton, P. J., Moore, J. D., Lacey, L., Quenby, S., Lam, E. W. F., & Brosens, J. J. (2015). The clock protein period 2 synchronizes mitotic expansion and decidual transformation of human endometrial stromal cells. *The FASEB Journal*, *29*(4), 1603–1614. doi:10.1096/fj.14-267195 PMID:25573754

Mutheneni, S. R., Morse, A. P., Caminade, C., & Upadhyayula, S. M. (2017). Dengue burden in India: Recent trends and importance of climatic parameters. *Emerging Microbes & Infections*, *6*(8), e70. doi:10.1038/emi.2017.57 PMID:28790459

Myers, J. P., Dumanoski, D., & Colborn, T. (2006). Our Stolen Future: A Decade Later. *San Francisco Medicine*, *79*, 32–33.

Myers, J. P., Guillette, L. J., Palanza, P., Parmigiani, S., Swan, S. H., & Vom Saal, F. S. (2003). The emerging science of endocrine disruption. *Intl Sem NuclearWar Plan Emerg*, *28*, 1–13.

Nabenishi, H., Ohta, H., Nishimoto, T., Morita, T., Ashizawa, K., & Tsuzuki, Y. (2012). The effects of cysteine addition during in vitro maturation on the developmental competence, ROS, GSH and apoptosis level of bovine oocytes exposed to heat stress. *Zygote (Cambridge, England)*, *20*(3), 249–259. doi:10.1017/S0967199411000220 PMID:21729376

Nagy, A., Gertsenstein, M., Vintersten, K., & Behringer, R. (2003). *Manipulating the mouse embryo: a laboratory manual* (Vol. 10). Cold Spring Harbor Laboratory Press.

Nagy, A., Gertsenstein, M., Vintersten, K., & Behringer, R. (2006). *A Laboratory Manual Manipulating the Mouse Embryo* (3rd ed.). Chemical Industry Press.

Nakamura, Y., Tamura, H., Takayama, H., & Kato, H. (2003). Increased endogenous level of melatonin in preovulatory human follicles does not directly influence progesterone production. *Fertility and Sterility*, *80*(4), 1012–1016. doi:10.1016/S0015-0282(03)01008-2 PMID:14556825

Nandi, S., Chauhan, M. S., & Palta, P. (1998). Influence of cumulus cells and sperm concentration on cleavage rate and subsequent embryonic development of buffalo (Bubalusbubalis) oocytes matured and fertilized in vitro. *Theriogenology*, *50*(8), 1251–1262. doi:10.1016/S0093-691X(98)00224-6

Narayansingh, R. M., Senchyna, M., Vijayan, M. M., & Carlson, J. C. (2004). Expression of prostaglandin G/H synthase (PGHS) and heat shock protein-70 (HSP-70) in the corpus luteum (CL) of prostaglandin F2 alpha-treated immature superovulated rats. *Canadian Journal of Physiology and Pharmacology*, *82*(6), 363–371. doi:10.1139/y04-032 PMID:15381959

Nardo, L. G., Nikas, G., Makrigiannakis, A., Sinatra, F., & Nardo, F. (2003). Synchronous expression of pinopodes and alpha v beta 3 and alpha 4 beta 1 integrins in the endometrial surface epithelium of normally menstruating women during the implantation window. *The Journal of Reproductive Medicine*, *48*(5), 355–361. PMID:12815909

Nardone, A., Ronchi, B., Lacetera, N., Ranieri, M. S., & Bernabucci, U. (2010). Effects of climate changes on animal production and sustainability of livestock systems. *Livestock Science*, *130*(1-3), 57–69. doi:10.1016/j.livsci.2010.02.011

Naruse, K., Lash, G. E., Innes, B. A., Otun, H. A., Searle, R. F., Robson, S. C., & Bulmer, J. N. (2009). Localization of matrix metalloproteinase (MMP)-2, MMP-9 and tissue inhibitors for MMPs (TIMPs) in uterine natural killer cells in early human pregnancy. *Human Reproduction (Oxford, England)*, *24*(3), 553–561. doi:10.1093/humrep/den408 PMID:19088110

Navarro, E., Funtikova, A. N., Fíto, M., & Schröder, H. (2017). Prenatal nutrition and the risk of adult obesity: Long-term effects of nutrition on epigenetic mechanisms regulating gene expression. *The Journal of Nutritional Biochemistry*, *39*, 1–14. doi:10.1016/j.jnutbio.2016.03.012

Negro-Vilar, A. (1993). Stress and other environmental factors affecting fertilityin men and women: Overview. *Environmental Health Perspectives*, *101*(Suppl 2), 59–64. PMID:8243408

Ngoula, F., Lontio, F. A., Tchoffo, H., Manfo Tsague, F. P., Djeunang, R. M., Vemo, B. N., Moffo, F., & Djuissi Motchewo, N. (2020). Heat Induces Oxidative Stress: Reproductive Organ Weights and Serum Metabolite Profile, Testes Structure, and Function Impairment in Male Cavy (Cavia porcellus). *Frontiers in Veterinary Science*, *7*, 37. doi:10.3389/fvets.2020.00037 PMID:32175332

Ng, T. P., Goh, H. H., Ng, Y. L., Ong, H. Y., Ong, C. N., Chia, K. S., Chia, S. E., & Jeyaratnam, J. (1991). Male endocrine functions in workers with moderate exposure to lead. *Occupational and Environmental Medicine*, *48*(7), 485–491. doi:10.1136/oem.48.7.485

Nichi, M., Bols, P., Zuche, R., Barnabe, V., Goovaerts, I., Barnabe, R., & Cortada, C. (2006). Seasonal variation in semen quality in Bos indicus and Bos taurus bulls raised under tropical conditions. *Theriogenology*, *66*(4), 822–828. doi:10.1016/j.theriogenology.2006.01.056 PMID:16529802

Nicolaides, C. (1934). Fertility studies in poultry. *Poultry Science*, *13*(3), 178–183. doi:10.3382/ps.0130178

Nicolino, M., & Forest, M. G. (2001). La puberté. In C. Thibault & M. C. Levasseur (Eds.), *La reproduction chez les mammifères domestiques et chez l'homme* (pp. 655–679). INRA Editions.

Nicolopoulou-Stamati, P., & Pitsos, M. A. (2001). The impact of endocrinedisrupters on female reproductive system. *Human Reproduction Update*, *7*(3), 323–330. doi:10.1093/humupd/7.3.323 PMID:11392379

Nikas, G. (1999). Cell-surface morphological events relevant to human implantation. *Human Reproduction (Oxford, England)*, *14*(suppl 2), 37–44. doi:10.1093/humrep/14.suppl_2.37 PMID:10690799

Nikas, G., & Aghajanova, L. (2002). Endometrial pinopodes: Some more understanding on human implantation. *Reproductive Biomedicine Online*, *4*, 18–23. doi:10.1016/S1472-6483(12)60111-4 PMID:12470560

Nilsen, N. O. (1985). Vascular abnormalities due to hyperthermia in chick embryos. *Teratology*, *30*(2), 237–251. doi:10.1002/tera.1420300211 PMID:6495224

Nisenblat, V., & Norman, R. J. (2009). Androgens and polycystic ovary syndrome. *Current Opinion in Endocrinology, Diabetes, and Obesity*, *16*(3), 224–231. doi:10.1097/MED.0b013e32832afd4d PMID:19390322

No, A. C. O. (2013). Exposure to toxic environmental agents. *Fertility and Sterility*, *100*(4), 931–934. doi:10.1016/j.fertnstert.2013.08.043

Nobel, R. L., Jobst, S. M., Dransfield, M. B. G., Pandolfi, S. M., & Balley, T. L. (1997). Use of radio frequency data communication system, Heat Watch, to describe behavioural estrus in dairy cattle. *Journal of Dairy Science*, 179.

Nordengren, J., Pilka, R., Noskova, V., Ehinger, A., Domanski, H., Andersson, C., Hoyer-Hansen, G., Hansson, S. R., & Casslen, B. (2004). Differential localization and expression of urokinase plasminogen activator (uPA), its receptor (uPAR), and its inhibitor (PAI-1) mRNA and protein in endometrial tissue during the menstrual cycle. *Molecular Human Reproduction*, *10*(9), 655–663. doi:10.1093/molehr/gah081 PMID:15243126

Norwitz, E. R., Schust, D. J., & Fisher, S. J. (2001). Implantation and the survival of early pregnancy. *The New England Journal of Medicine*, *345*(19), 1400–1408. doi:10.1056/NEJMra000763 PMID:11794174

Novaraa, A., & Pisciottaa, A. (2018). The impact of soil erosion on soil fertility and vine vigor. A multidisciplinary approach based on field, laboratory and remote sensing approaches. *The Science of the Total Environment*, *622–623*, 474–480. doi:10.1016/j.scitotenv.2017.11.272 PMID:29223076

NTP-CERHR expert panel report on the reproductive and developmental toxicity of ethylene glycol. (2004a). *Reprod Toxicol, 18*, 457–532.

NTP-CERHR expert panel report on the reproductive and developmental toxicityof ethylene glycol. (2004b). *Reprod Toxicol, 18*, 533–579.

Nunes, L., Meireles, C., Pinto Gomes, C., de Almeida Ribeiro, N. (2019). Global Climate Change Outlook. *Springer Briefs in Environmental Science*, 9-17.

Nyika, J. (2020). Climate change situation in Kenya and measures towards adaptive management in the water sector. *International Journal of Environmental Sustainability and Green Technologies, 11*(2), 34–47. doi:10.4018/IJESGT.2020070103

O'Neill, C., Hawkes, F. R., Hawkes, D. L., Lourenço, N. D., Pinheiro, H. M., & Delée, W. (1999). Colour in textile effluents - sources, measurement, discharge consents and simulation: A review. *Journal of Chemical Technology and Biotechnology (Oxford, Oxfordshire), 74*(11), 1009–1018. doi:10.1002/(SICI)1097-4660(199911)74:11<1009::AID-JCTB153>3.0.CO;2-N

Oak, M. K., Chantler, E. N., Vaughan Williams, C. A., & Elstein, M. (1985). Sperm survival studies in peritoneal fluid from infertile women with endometriosis and unexplained infertility. *Clinical Reproduction and Fertility, 3*, 297–303. PMID:3841649

OECD. (2017). The future of global value chains: Business as usual or a "new normal"? OECD Publishing, No. 41.

Ohnishi, S., Murata, M., Degawa, M., & Kawanishi, S. (2001). Oxidative DNA damage induced by an N-hydroxy metabolite of carcinogenic 4-aminoaminoazobenzene. *Japanese Journal of Cancer Research, 92*(1), 23–29. doi:10.1111/j.1349-7006.2001.tb01043.x PMID:11173540

Okabe, M. (2013). The cell biology of mammalian fertilization. *Development, 140*(22), 4471–4479. doi:10.1242/dev.090613 PMID:24194470

Oliva, A., Spira, A., & Multigner, L. (2001). Contribution of environmental factorsto the risk of male infertility. *Human Reproduction (Oxford, England), 16*(8), 1768–1776. doi:10.1093/humrep/16.8.1768 PMID:11473980

Oliveira, G. A. R., Ferraz, E. R. A., Chequer, F. M. D., Grando, M. D., Angeli, J. P. F., Tsuboy, M. S., Marcarini, J. C., Mantovani, M. S., Osugi, M. E., Lizier, T. M., Zanoni, M. V. B., & Oliveira, D. P. (2010). Chlorination treatment of aqueous samples reduces, but does not eliminate, the mutagenic effect of the azo dyes Disperse Red 1, Disperse Red 13 and Disperse Orange 1. *Mutation Research, 703*(2), 200–208. doi:10.1016/j.mrgentox.2010.09.001 PMID:20851777

Oliveria Melo, A. S., Malinger, G., Ximenes, R., Szejnfeld, P. O., Alver Sampaio, S., & … . (2016). Zika virus intrauterine infection causes fetal brain abnormality and microcephaly: Tip of the iceberg? *Ultrasound in Obstetrics & Gynecology, 47*(1), 6–7. doi:10.1002/uog.15831 PMID:26731034

Olivius, C., Friden, B., Borg, G., & Bergh, C. (2004). Why do couples discontinuein vitro fertilization treatment? A cohort study. *Fertility and Sterility, 81*(2), 258–261. doi:10.1016/j.fertnstert.2003.06.029 PMID:14967352

Oltenacu, P. A., Britt, J. H., Braun, R. K., & Mellenberger, R. W. (1983). Relationships among type of parturition, type of discharge from genital tract, involution of cervix, and subsequent reproductive performance in Holstein cows. *Journal of Dairy Science, 66*(3), 612–619. doi:10.3168/jds.S0022-0302(83)81832-3 PMID:6841757

Ombelet, W., Cooke, I., Dyer, S., Serour, G., & Devroey, P. (2008). Infertility and the provision of infertility medical services in developing countries. *Human Reproduction Update, 14*(6), 605–621. doi:10.1093/humupd/dmn042

Onwuka, B., & Mang, B. (2018). Effects of soil temperature on some soil properties and plant growth. *Advances in Plants & Agriculture Research, 8*(1), 34–37. doi:10.15406/apar.2018.08.00288

Opperhuizen, A. L., Stenvers, D. J., Jansen, R. D., Foppen, E., Fliers, E., & Kalsbeek, A. (2017). Light at night acutely impairs glucose tolerance in a time-, intensity- and wavelength-dependent manner in rats. *Diabetologia*, *60*(7), 1333–1343. doi:10.100700125-017-4262-y PMID:28374068

Oskam, I. C., Lyche, J. L., Krogenæs, A., Thomassen, R., Skaare, J. U., Wiger, R., Dahl, E., Sweeney, T., Stien, A., Ropstad, E., & Ropstad, E. (2005). Effects of long-term maternal exposure to low doses of PCB126 and PCB153 on the reproductive system and related hormones of young male goats. *Reproduction (Cambridge, England)*, *130*(5), 731–742. doi:10.1530/rep.1.00690 PMID:16264102

Otak, E. M., & Schull, W. J. (1998). Review: radiation-related brain damage and growth retardation among the prenatally exposed bomb survivors. *International Journal of Radiation Biology*, *74*, 159–71.

Otake, M., & Schull, W. J. (1998). Review: Radiation-related brain damage and growth retardation among the prenatally exposed bomb survivors. *International Journal of Radiation Biology*, *74*(2), 159–171. doi:10.1080/095530098141555 PMID:9712546

Oudiz, D., & Zenick, H. (1986). In vivo and in vitro evaluations of spermatoxicity induced by 2-ethoxyethanol treatment. *Toxicology and Applied Pharmacology*, *84*(3), 576–583. doi:10.1016/0041-008X(86)90263-2 PMID:3726877

Ozawa, M., Hirabayashi, M., & Kanai, Y. (2002). Developmental competence and oxidative state of mouse zygotes heat-stressed maternally or in vitro. *Reproduction (Cambridge, England)*, *124*, 683–689. doi:10.1530/rep.0.1240683 PMID:12417007

Ozawa, M., Tabayashi, D., Latief, T. A., Shimizu, T., & Kanal, Y. (2005). Alterations in follicular dynamics and Steroidogenic abilities induced by heat stress during follicular recruitment in goats. *Reproduction (Cambridge, England)*, *2005*(129), 621–630. doi:10.1530/rep.1.00456 PMID:15855625

Padmanabhan, V., & Veiga-Lopez, A. (2011). Developmental origin of reproductive and metabolic dysfunctions: Androgenic versus estrogenic reprogramming. *Seminars in Reproductive Medicine*, *29*(03), 173–186. doi:10.1055-0031-1275519 PMID:21710394

Pang, Y. W., An, L., Wang, P., Yu, Y., Yin, Q. D., Wang, X. H., Xin-Zhang, Qian-Zhang, Yang, M.-L., Min-Guo, Wu, Z.-H., & Tian, J.-H. (2013). Treatment of porcine donor cells and reconstructed embryos with the antioxidant melatonin enhances cloning efficiency. *Journal of Pineal Research*, *54*(4), 389–397. doi:10.1111/jpi.12024 PMID:24325731

Panter-Brick, C., Lotstein, D. S., & Ellison, P. T. (1993). Seasonality of reproductive function and weight loss in rural Nepali women. *Human Reproduction (Oxford, England)*, *8*(5), 684–690. doi:10.1093/oxfordjournals.humrep.a138120 PMID:8314959

Pant, N., Upadhyay, G., Pandey, S., Mathur, N., Saxena, D. K., & Srivastava, S. P. (2003). Lead and cadmium concentration in the seminal plasma of men in the general population: Correlation with sperm quality. *Reproductive Toxicology (Elmsford, N.Y.)*, *17*(4), 447–450. doi:10.1016/S0890-6238(03)00036-4

Paras, M. K., Ekwall, D., Pal, R., Curteza, A., Chen, Y., & Wang, L. (2018). An exploratory study of Swedish charities to develop a model for the reuse-based clothing value chain. *Sustainibility*, *10*(4), 1176. doi:10.3390u10041176

Pareek, N. (2017). Climate change impact on soils: Adaptation and mitigation. *Shengtai Huanjing Xuebao*, *2*(3), 136–139. doi:10.15406/mojes.2017.02.00026

Parker, S. E., Mai, C. T., Canfield, M. A., Rickard, R., Wang, Y., Meyer, R. E., Anderson, P., Mason, C. A., Collins, J. S., Kirby, R. S., & Correa, A. (2010). Updated national birth prevalence estimates for selected birth defects in the United States, 2004–2006. *Birth Defects Research. Part A, Clinical and Molecular Teratology*, *88*(12), 1008–1016. doi:10.1002/bdra.20735 PMID:20878909

Parmesan, C., & Yohe, G. (2003). A globally coherent fingerprint of climate changeimpacts across natural systems. *Nature*, *421*(6918), 37–42. doi:10.1038/nature01286 PMID:12511946

Parry, M. (2009). Climate change is a development issue, and only sustainable development can confront the challenge. *Climate and Development*, *1*(1), 5–9. doi:10.3763/cdev.2009.0012

Pati, A., & Lamnganbi, M. (2018). Impct of climate change on soil health: A review. *International Journal of Chemical Studies*, *6*(3), 2399–2404.

Patterson, D., Perry, R., Kiracofe, G., Bellows, R., Staigmiller, R., & Corah, L. (1992). Management considerations in heifer development and puberty. *Journal of Animal Science*, *70*(12), 4018–4035. doi:10.2527/1992.70124018x

Patz, J. A., Graczyk, T. K., Geller, N., & Vitto, Y. A. (2000). Effects of environmental change on emerging parasitic diseases. *International Journal for Parasitology*, *30*(12-13), 1395–1405. doi:10.1016/S0020-7519(00)00141-7

Paula-Lopes, F. F., & Hansen, P. J. (2002). Heat-shock induced apoptosis in preimplantation bovine embryos is a developmentally-regulated phenomenon. *Biology of Reproduction*, *66*(4), 1169–1177. doi:10.1093/biolreprod/66.4.1169 PMID:11906938

Paula-Lopes, F. F., Lima, R. S., Satrapa, R. A., & Barros, C. M. (2013). Physiology and Endocrinology Symposium: Influence of cattle genotype (Bos indicus vs. Bos taurus) on oocyte and preimplantation embryo resistance to increased temperature. *Journal of Animal Science*, *91*(3), 1143–1153. doi:10.2527/jas.2012-5802 PMID:23296831

Paul, C., Teng, S., & Saunders, P. T. (2009). A single, mild, transientscrotal heat stress causes hypoxia and oxidative stress in mousetestes, which induces germ cell death. *Biology of Reproduction*, *80*(5), 913–919. doi:10.1095/biolreprod.108.071779 PMID:19144962

Payton, R. R., Romar, R., Coy, P., Saxton, A. M., Lawrence, J. L., & Edwards, J. L. (2004). Susceptibility of bovine germinal vesicle-stage oocytes from antral follicles to direct effects of heat stress in vitro. *Biology of Reproduction*, *71*(4), 1303–1308. doi:10.1095/biolreprod.104.029892 PMID:15201201

Pedersen, U. B., Stendel, M., Midzi, N., Mduluza, T., Soko, W., Stensgaard, A. S., Vennervald, B. J., Mukaratirwa, S., & Kristensen, T. K. (2014). Modelling climate change impact on the spatial distribution of fresh water snails hosting trematodes in Zimbabwe. *Parasites & Vectors*, *7*(1), 536. doi:10.118613071-014-0536-0 PMID:25498001

Pei, Zhu, H., Ye, R., Wu, J., Liu, J., Ren, A., Li, Z., & Zheng, X. (2015). Interaction between the SLC19A1 gene and maternal first trimester fever on offspring neural tube defects. *Birth Defects Research. Part A, Clinical and Molecular Teratology*, *103*(1), 3–11. doi:10.1002/bdra.23257 PMID:24917213

Penn, C. J., & Camberato, J. J. (2019). A Critical Review on Soil Chemical Processes that Control How Soil pH Affects Phosphorus Availability to Plants. *Agriculture*, *9*(6), 120–138. doi:10.3390/agriculture9060120

Pennington, J. A., Albright, J. L., Diekman, M. A., & Callahan, C. J. (1985). Sexual activity of Holstein cows: Seasonal effects. *Journal of Dairy Science*, *68*(11), 3023–3030. doi:10.3168/jds.S0022-0302(85)81197-8 PMID:4078128

Pereira, R. S. (2004). Identificação e caracterização das fontes de poluição em sistemas hídricos. *Revista Eletrônica de Recursos Hídricos*, *1*, 20–36.

Pereira, W. S., & Freire, R. S. (2005). Ferro zero: Uma nova abordagem para o tratamento de águas contaminadas com compostos orgânicos poluentes. *Quimica Nova*, *28*(1), 130–136. doi:10.1590/S0100-40422005000100022

Perez-Cerezales, S., Ramos-Ibeas, P., Lopez-Cardona, A., Pericuesta, E., Fernandez-Gonzalez, R., Pintado, B., & Gutierrez-Adan, A. (2017). Elimination of methylation marks at lysines 4 and 9 of histone 3(H3K4 and H3K9) of spermatozoa alters offspring phenotype. *Reproduction, Fertility, and Development*, *29*(4), 740–746. doi:10.1071/RD15349 PMID:26678259

Perez-Crespo, M., Pintado, B., & Gutierrez-Adan, A. (2008). Scrotal heat stress effects on sperm viability,sperm DNA integrity, and the offspring sex ratio in mice. *Molecular Reproduction and Development, 75*(1), 40–47. doi:10.1002/mrd.20759 PMID:17474098

Perez-Crespo, M., Ramirez, M. A., Fernandez-Gonzalez, R., & Gutierrez-Adan, A. (2005). Differential sensitivity of male and female mouse embryos to oxidative induced heat-stress is mediated by glucose-6-phosphate dehydrogenase gene expression. *Molecular Reproduction and Development, 72*(4), 502–510. doi:10.1002/mrd.20366 PMID:16149081

Perry, M. J. (2008). Effects of environmental and occupational pesticide exposure on human sperm: A systematic review. *Human Reproduction Update, 14*(3), 233–242. doi:10.1093/humupd/dmm039

Perry, M. J., Ouyang, F., Korrick, S. A., Venners, S. A., Chen, C., Xu, X., Lasley, B. L., & Wang, X. (2006). A prospective study of serum DDT and progesterone and estrogen levels across the menstrual cycle in nulliparous women of reproductive age. *American Journal of Epidemiology, 164*(11), 1056–1064. doi:10.1093/aje/kwj329

Perry, M. J., Venners, S. A., Chen, X., Liu, X., Tang, G., Xing, H., Barr, D. B., & Xu, X. (2011). Organophosphorous pesticide exposures and sperm quality. *Reproductive Toxicology (Elmsford, N.Y.), 31*(1), 75–79. doi:10.1016/j.reprotox.2010.08.006

Perry, S. F., & Gilmour, K. (1993). An evaluation of factors limiting carbon dioxide excretion by trout red blood cells in vitro. *The Journal of Experimental Biology, 180*(1), 39–54.

Petersen, L. R., Jamieson, D. J., Powers, A. M., & Honein, M. A. (2016). Zika virus. *The New England Journal of Medicine, 374*(16), 1552–1563. doi:10.1056/NEJMra1602113 PMID:27028561

Phillip, M., De, C. B., Pilpel, D., Karplus, M., & Sofer, S. (1996). Clitoral and penile sizes of full term newborns in two different ethnic groups. Journal of pediatric endocrinology & metabolism. *Journal of Pediatric Endocrinology & Metabolism, 9*(2), 175–179.

Phillips, R. W., Knapp, B. Jr, Hemstra, L. C., & Eaton, O. N. (1943). Seasonal variation in the semen of bulls. *American Journal of Veterinary Research, 4*, 115.

Phumsatitpong, C., & Moenter, S. M. (2018). Estradiol-dependent stimulation and suppression of gonadotropin-releasing hormone neuron firing activity by corticotropin-releasing hormone in female mice. *Endocrinology, 159*(1), 414–425. doi:10.1210/en.2017-00747 PMID:29069304

Piccirilli, D., Baldini, E., Massimiani, M., Camaioni, A., Salustri, A., Bernardini, R., Centanni, M., Ulisse, S., Moretti, C., & Campagnolo, L. (2018). Thyroid hormone regulates protease expression and activation of Notch signaling in implantation and embryo development. *The Journal of Endocrinology, 236*(1), 1–12. doi:10.1530/JOE-17-0436 PMID:28993437

Pierzynski, P. (2011). Oxytocin and vasopressin V1A receptors as new therapeutic targets in assisted reproduction. *Reproductive Biomedicine Online, 22*(1), 9–16. doi:10.1016/j.rbmo.2010.09.015 PMID:21130036

Pilorz, V., & Steinlechner, S. (2008). Low reproductive success in Per1 and Per2 mutant mouse females due to accelerated ageing? *Reproduction (Cambridge, England), 135*(4), 559–568. doi:10.1530/REP-07-0434 PMID:18367514

Pinto, M., de-Alencar-Araripe, M., & Alencar, A. (2001). Effects of scrotal insulation on testissize and semen criteria in Santa Ines hairy sheep raised in the State of Ceara. *Northeast of Brazil Rev. Brasileira Zootec, 30*(6), 1704–1711.

Pirhonen, J. P., Vähä-Eskeli, K., Seppänen, A., Vuorinen, J., & Erkkola, R. (1994). Does thermal stress decrease uterine blood ñow in hypertensive pregnancies. *American Journal of Perinatology, 11*(05), 313–316. doi:10.1055-2007-994542 PMID:7993506

Pleet, H., Graham, J. M., & Smith, D. W. (1981). Central nervous system and facial defects associated with maternal hyperthermia at 4 to 14 weeks gestation. *Pediatrics*, *67*, 785–789. PMID:7232042

Poehl, M., Bichler, K., Wicke, V., Dorner, V., & Feichtinger, W. (1999). Psychotherapeuticcounseling and pregnancy rates in *in vitro* fertilization. *Journal of Assisted Reproduction and Genetics*, *16*(6), 302–305. doi:10.1023/A:1020449831333 PMID:10394525

Polak, G., Koziol-Montewka, M., Gogacz, M., Blaszkowska, I., & Kotarski, J. (2001). Total antioxidant status of peritoneal fluid in infertile women. *European Journal of Obstetrics, Gynecology, and Reproductive Biology*, *94*(2), 261–263. doi:10.1016/S0301-2115(00)00352-3 PMID:11165736

Politch, J. A., & Herrenkohl, L. R. (1984). Effects of prenatal stress on reproduction in male and female mice. *Physiology & Behavior*, *32*(1), 95–99. doi:10.1016/0031-9384(84)90077-5 PMID:6718542

Pollack, A. Z., Schisterman, E. F., Goldman, L. R., Mumford, S. L., Perkins, N. J., Bloom, M. S., Rudra, C. B., Browne, R. W., & Wactawski-Wende, J. (2012). Relation of blood cadmium, lead, and mercury levels to biomarkers of lipid peroxidation in premenopausal women. *American Journal of Epidemiology*, *175*(7), 645–652. doi:10.1093/aje/kwr375

Pouliot, S. H., Xiong, X., Harville, E., Paz-Soldan, V., Tomashek, K. M., Breart, G., & Buekens, P. (2010). Maternal dengue and pregnancy outcomes: A systematic review. *Obstetrical & Gynecological Survey*, *65*(2), 107–118. PMID:20100360

Poursafa, P., Keikha, M., & Kelishadi, R. (2015). Systematic review on adverse birth outcomes of climate change. *Journal of Research in Medical Sciences*, *20*, 397–402. PMID:26109998

Poursafa, P., & Kelishadi, R. (2011). What health professionals should know about the health effects of air pollution and climate change on children and pregnant mothers. *Iranian Journal of Nursing and Midwifery Research*, *16*, 257–264. PMID:22224116

Power, G. G. (1989). Biology of temperature: The mammalian fetus. *Journal of Developmental Physiology*, *12*, 295–304. PMID:2701105

Pradhan, R., & Nakagoshi, N. (2008). Reproductive disorders in cattle due to nutritional status. *Journal of International Development and Cooperation, 14*(1), 45-66.

Pramauro, E., Prevot, A. B., Augugliaro, V., & Palmisano, L. (1995). Photocatalytic treatment of laboratory wastes containing aromatic amines. *Analyst (London)*, *120*(2), 237–242. doi:10.1039/an9952000237

Pratt, M. (2000). Disperse blue dyes 106 and 124 are common causes of textile dermatitis and should serve as screening allergens for this condition. *American Journal of Clinical Dermatology*, *11*(1), 30–41. doi:10.1016/S1046-199X(00)90030-7 PMID:10684387

Praveen, D., Andimuthu, R., & Palanivelu, K. (2016). The urgent call for land degradation vulnerability assessment for conserving land quality in the purview of climate change: Perspective from South Indian Coast. *AIMS Agriculture and Food*, *1*(3), 330–341. doi:10.3934/agrfood.2016.3.330

Prosser, C. L., Heath, J. E., & Prosser, C. L. (1991) Temperature. Comparative animal physiology environmental and metabolic animal physiology. New York: Wiley.

Psychoyos, A. (1986). Uterine Receptivity for Nidation. *Annals of the New York Academy of Sciences*, *476*(1), 36–42. doi:10.1111/j.1749-6632.1986.tb20920.x PMID:3541745

Püntener, A., & Pages, C. (2013). *European on certain azo dyes.* http://wwot.destcorp.com/documents/Azo_Dye_Information.pdf

Pursley, J., Silcox, R. W., & Wiltbank, M. C. (1998). Effect of time of artificial insemination on preg- nancy rates, calving rates, pregnancy loss and gender ratio after synchronization of ovulation in lactating dairy cows. *Journal of Dairy Science*, *81*(8), 2139–2144. doi:10.3168/jds.S0022-0302(98)75790-X PMID:9749378

Putney, D. J., Gross, T. S., & Thatcher, W. W. (1988b). Prostaglandin secretion by endometrium of pregnant and cyclic cattle at day 17 after estrus in response to in vitro heat stress. *Journal of Reproduction and Fertility*, *84*(2), 475–483. doi:10.1530/jrf.0.0840475

Putney, D., Drost, M., & Thatcher, W. (1988a). Embryonic development in super ovulated dairy cattle exposed to elevated ambient temperatures between Days 1 to 7 post insemination. *Theriogenology*, *30*(2), 195–209. doi:10.1016/0093-691X(88)90169-0

Putney, D., Mullins, S., Thatcher, W., Drost, M., & Gross, T. (1989). Embryonic development in superovulated dairy cattle exposed to elevated ambient temperatures between the onset of estrus and insemination. *Animal Reproduction Science*, *19*(1-2), 37–51. doi:10.1016/0378-4320(89)90045-6

Puvaneswari, N., Muthukrishnan, J., & Gunasekaran, P. (2006). Toxicity assessment and microbial degradation of azo dyes. *Indian Journal of Experimental Biology*, *44*, 618–626. PMID:16924831

Qambrani, N. A., Rahman, M. M., Won, S., Shim, S., & Ra, C. (2017). Biochar properties and eco-friendly applications for climate change mitigation, waste management, and wastewater treatment: A review. *Renewable & Sustainable Energy Reviews*, *79*, 255–273. doi:10.1016/j.rser.2017.05.057

Quinteiro-Filho, W., Rodrigues, M., Ribeiro, A., Ferraz-de-Paula, V., Pinheiro, M., Sá, L., Ferreira, A. J. P., & Palermo-Neto, J. (2012). Acute heat stress impairs performance parameters and induces mild intestinal enteritis in broiler chickens: Role of acute hypothalamic-pituitary-adrenal axis activation1. *Journal of Animal Science*, *90*(6), 1986–1994. doi:10.2527/jas.2011-3949

Rachelle, S., Brendan, M., & Cullena, R. (2018). Potential impacts of climate change on soil organic carbon and productivity in pastures of south eastern Australia. *Agricultural Systems*, *167*, 34–46. doi:10.1016/j.agsy.2018.08.010

Rachootin, P., & Olsen, J. (1983). The risk of infertility and delayed conceptionassociated with exposures in the Danish workplace. *Journal of Occupational Medicine.*, *25*, 394–402. PMID:6854429

Radwan, M., Wielgomas, B., Dziewirska, E., Radwan, P., Kałużny, P., Klimowska, A., Hanke, W., & Jurewicz, J. (2018). Urinary bisphenol A levels and male fertility. *American Journal of Men's Health*, *12*(6), 2144–2151. doi:10.1177/1557988318799163

Raftery, A. E., Zimmer, A., Frierson, D. M. W., Startz, R., & Liu, P. (2017). Less than 2°C warming by 2100 unlikely. *Nature Climate Change*, *7*(9), 637–645. doi:10.1038/nclimate3352 PMID:30079118

Rainforth, M. (2019). The effect of maternal nutritional restriction on fetal development and performance of offspring in beef cattle. *Alberta Academic Review*, *1*(2), 1–11. doi:10.29173/aar12

Raman, C. D., & Kanmani, S. (2016). Textile dye degradation using nano zero valent iron: A review. *Journal of Environmental Management*, *177*, 341–355. doi:10.1016/j.jenvman.2016.04.034 PMID:27115482

Ramsey, I. S., Moran, M. M., Chong, J. A., & Clapham, D. E. (2006). A voltage-gated proton-selective channel lacking thepore domain. *Nature*, *440*(7088), 1213–1216. doi:10.1038/nature04700 PMID:16554753

Rance, N. E., Young, W. S. III, & McMullen, N. T. (1994). Topography of neurons expressing luteinizing hormone-releasing hormone gene transcripts in the human hypothalamus and basal forebrain. *The Journal of Comparative Neurology*, *339*(4), 573–586. doi:10.1002/cne.903390408 PMID:8144747

Rao, M., Zhao, X. L., Yang, J., Hu, S. F., Lei, H., Xia, W., & Zhu, C. H. (2015). Effect of transient scrotalhyperthermia on sperm parameters, seminal plasma biochemical markers, and oxidative stress inmen. *Asian Journal of Andrology*, *17*(4), 668–675. doi:10.4103/1008-682X.146967 PMID:25652627

Rasooli, A., Jalali, M. T., Nouri, M., Mohammadian, B., & Barati, F. (2010). Effects of chronic heat stress on testicular structures, serum testosterone and cortisol concentrations in developing lambs. *Animal Reproduction Science*, *117*(1-2), 55–59. doi:10.1016/j.anireprosci.2009.03.012 PMID:19428197

Ratajczak, C. K., Boehle, K. L., & Muglia, L. J. (2009). Impaired steroidogenesis and implantation failure in Bmal1-/- mice. *Endocrinology*, *150*(4), 1879–1885. doi:10.1210/en.2008-1021 PMID:19056819

Raza, A., Razzaq, A., Mehmood, S., Zou, X., Zhang, X., Lv, Y., & Xu, J. (2019). Impact of Climate Change on Crops Adaptation and Strategies to Tackle Its Outcome: A Review. *Plants*, *8*(2), 34–63. doi:10.3390/plants8020034 PMID:30704089

Recio-Vega, R., Ocampo-Gómez, G., Borja-Aburto, V. H., Moran-Martínez, J., & Cebrian-Garcia, M. E. (2008). Organophosphorus pesticide exposure decreases sperm quality: Association between sperm parameters and urinary pesticide levels. *Journal of Applied Toxicology*, *28*(5), 674–680. doi:10.1002/jat.1321

Redmond, G. P. (2004). Thyroid dysfunction and women's reproductive health. *Thyroid, 14*(3), 5-15.

Reed, B. G., & Carr, B. R. (2018). The normal menstrual cycle and the control of ovulation. In Endotext. MDText. com, Inc.

Rehfeld, A., Andersson, A. M., & Skakkebæk, N. E. (2020). Bisphenol A Diglycidyl Ether (BADGE) and Bisphenol Analogs, but Not Bisphenol A (BPA), Activate the CatSper Ca^{2+} Channel in Human Sperm. *Frontiers in Endocrinology*, *11*, 324. doi:10.3389/fendo.2020.00324

Reiling, S. J., & Dixon, B. R. (2019). Toxoplasma gondii: How an Amazonian parasite became an Inuit health issue. *Canada Communicable Disease Report*, *45*(7-8), 183–190. doi:10.14745/ccdr.v45i78a03 PMID:31355827

Reiter P. (2001). Climate change and mosquito-borne disease. *Environ Health Perspect., 109*(Suppl 1), 141-61.

Reiter, R. J., Tamura, H., Tan, D. X., & Xu, X. Y. (2014). Melatonin and the circadian system: Contributions to successful female reproduction. In Fertility and Sterility (Vol. 102, Issue 2, pp. 321–328).

Reiter, R. J., Tan, D.-X., Manchester, L. C., Paredes, S. D., Mayo, J. C., & Sainz, R. M. (2009). Melatonin and Reproduction Revisited. *Biology of Reproduction, 81*(3), 445–456.

Renaudeau, D., & Noblet, J. (2001). Effects of exposure to high ambient temperature and dietary protein level on sow milk production and performance of piglets. *Journal of Animal Science*, *79*(6), 1540–1548. doi:10.2527/2001.7961540x PMID:11424692

Report, L. G. C. (1999). The risk of cancer caused by textiles and leather goods coloured with azo dyes. CSTEE plenary meeting, Brussels, Belgium.

Reppert, S. M., & Schwartz, W. J. (1983). Maternal coordination of the fetal biological clock in utero. *Science*, *220*(4600), 969–971. doi:10.1126cience.6844923 PMID:6844923

Reppert, S. M., & Schwartz, W. J. (1986). Maternal suprachiasmatic nuclei are necessary for maternal coordination of the developing circadian system. *The Journal of Neuroscience: The Official Journal of the Society for Neuroscience*, *6*(9), 2724–2729. doi:10.1523/JNEUROSCI.06-09-02724.1986 PMID:3746430

Rhind, S. M., Evans, N. P., Bellingham, M., Sharpe, R. M., Cotinot, C., Mandon-Pepin, B., Centanni, M., Ulisse, S., Moretti, C., Campagnolo, L., & Fischer, B. (2010). Effects of environmental pollutants on the reproduction and welfare of ruminants. *Animal*, *4*(7), 1227–1239. doi:10.1017/S1751731110000595 PMID:20582145

Rhynes, W. E., & Ewing, L. L. (1973). Testicular endocrine function in Hereford bulls exposed to high ambient temperature. *Endocrinology*, *92*(2), 509–515. doi:10.1210/endo-92-2-509 PMID:4682863

Ribeiro, A. R., & de Aragão Umbuzeiro, G. (2014). Effects of a textile azo dye on mortality, regeneration, and reproductive performance of the planarian, Girardiatigrina. *Environmental Sciences Europe*, *26*(1), 22. doi:10.118612302-014-0022-5

Rice, D., & Barone, S. Jr. (2000). Critical periods of vulnerability for the developing nervous system: Evidence from humans and animal models. *Environmental Health Perspectives*, *108*(suppl 3), 511–533. PMID:10852851

Richter, C. A., Birnbaum, L. S., Farabollini, F., Newbold, R. R., Rubin, B. S., Talsness, C. E., Vandenbergh, J. G., Walser-Kuntz, D. R., & vom Saal, F. S. (2007). In vivo effects of bisphenol A in laboratory rodent studies. *Reproductive Toxicology (Elmsford, N.Y.)*, *24*(2), 199–224. doi:10.1016/j.reprotox.2007.06.004

Rignell-Hydbom, A., Rylander, L., Giwercman, A., Jönsson, B. A. G., Lindh, C., Eleuteri, P., Rescia, M., Leter, G., Cordelli, E., Spano, M., & Hagmar, L. (2005). Exposure to PCBs and p, p¢-DDE and human sperm chromatin integrity. *Environmental Health Perspectives*, *113*(2), 175–179. doi:10.1289/ehp.7252

Rijken, M. J., McGready, R., Boel, M. E., Poespoprodjo, R., Singh, N., Syafruddin, D., Rogerson, S., & Nosten, F. (2012). Malaria in pregnancy in the Asia-Pacific region. *The Lancet. Infectious Diseases*, *12*(1), 75–88. doi:10.1016/S1473-3099(11)70315-2 PMID:22192132

Rim, K. T. (2017). Reproductive Toxic chemicals at work and efforts to protect workers' health: A literature review. *Safety and Health at Work*, *8*(2), 143–150. doi:10.1016/j.shaw.2017.04.003

Rispoli, L. A., Lawrence, J. L., Payton, R. R., Saxton, A. M., Schrock, G. E., Schrick, F. N., & Edwards, J. L. (2011). Disparate consequences of heat stress exposure during meiotic maturation: Embryo development after chemical activation vs fertilization of bovine oocytes. *Reproduction (Cambridge, England)*, *142*(6), 831–843. doi:10.1530/REP-11-0032

Rispoli, L. A., Payton, R. R., Gondro, C., Saxton, A. M., Nagle, K. A., Jenkins, B. W., Schrick, F. N., & Edwards, J. L. (2013). Heat stress effects on the cumulus cells surrounding the bovine oocyte during maturation: Altered matrix metallopeptidase 9 and progesterone production. *Reproduction (Cambridge, England)*, *146*(2), 193–207. doi:10.1530/REP-12-0487 PMID:23744615

Rivera, R. M., & Hansen, P. J. (2001). Development of cultural bovine embryos after exposure to high temperature in the physiological range. *Reproduction (Cambridge, England)*, *121*, 107–115. doi:10.1530/rep.0.1210107 PMID:11226033

Rizzo, A., Roscino, M. T., Minoia, G., Trisolini, C., Spedicato, M., Mutinati, M., Pantaleo, M., Jirillo, F., & Sciorsci, R. L. (2009). Serum levels of reactive oxygen species (ROS) in the bitch. *Immunopharmacology and Immunotoxicology*, *31*(2), 310–313. doi:10.1080/08923970802683954 PMID:19235534

Rizzo, P., Raffone, E., & Benedetto, V. (2010). Effect of the treatment with myo-inositol plus folic acid plus melatonin in comparison with a treatment with myo-inositol plus folic acid on oocyte quality and pregnancy outcome in IVF cycles. A prospective, clinical trial. *European Review for Medical and Pharmacological Sciences*, *14*, 555–561. PMID:20712264

Robinson, S. I., McLaughlin, Ó. B., Marteinsdóttir, B., & O'Gorman, E. J. (2018). Soil temperature effects on the structure and diversity of plant and invertebrate communities in a natural warming experiment. *Journal of Animal Ecology*, *87*(3), 634–646. doi:10.1111/1365-2656.12798 PMID:29368345

Rocha, A., Randel, R. D., Broussard, J. R., Lim, J. M., Blair, R. M., Roussel, J. D., Godke, R. A., & Hansel, W. (1998). High environmental temperature and humidity decrease oocyte quality in Bos taurus but not in Bos indicus cows. *Theriogenology*, *49*, 657–665. doi:10.1016/S0093-691X(98)00016-8 PMID:10732044

Rocha, R. M. P., Lima, L. F., Alves, A. M. C. V., Celestino, J. J. H., Matos, M. H. T., Lima-Verde, I. B., Bernuci, M. P., Lopes, C. A. P., Báo, S. N., Campello, C. C., Rodrigues, A. P. R., & Figueiredo, J. R. (2013). Interaction between melatonin and follicle-stimulating hormone promotes in vitro development of caprine preantral follicles. *Domestic Animal Endocrinology*, *44*(1), 1–9. doi:10.1016/j.domaniend.2012.07.001 PMID:22920266

Rochester, J. R. (2013). Bisphenol A and human health: A review of the literature. *Reproductive Toxicology (Elmsford, N.Y.)*, *42*, 132–155. doi:10.1016/j.reprotox.2013.08.008

Rockett, J. C., Mapp, F. L., Garges, J. B., Luft, J. C., Mori, C., & Dix, D. J. (2001). Effects of hyperthermia onspermatogenesis, apoptosis, gene expression, and fertility in adult male mice. *Biology of Reproduction*, *65*(1), 229–239. doi:10.1095/biolreprod65.1.229 PMID:11420244

Rodier, P. M., Aschner, M., & Sager, P. R. (1984). Mitotic arrest in the developing CNS after prenatal exposure to methylmercury. *Neurobehavioral Toxicology and Teratology*, *6*(5), 379–385. PMID:6514102

Rodrigues-Cunha, M. C., Mesquita, L. G., Bressan, F., del Collado, M., Balieiro, J. C. C., Schwarz, K. R. L., de Castro, F. C., Watanabe, O. Y., Watanabe, Y. F., de Alencar Coelho, L., & Leal, C. L. V. (2016). Effects of melatonin during IVM in defined medium on oocyte meiosis, oxidative stress, and subsequent embryo development. *Theriogenology*, *86*(7), 1685–1694. doi:10.1016/j.theriogenology.2016.05.026 PMID:27471183

Rodriguez-Martinez, H., & Barth, A. D. (2007). In vitro evaluation of sperm quality related to in vivo function and fertility. *Society of Reproduction and Fertility Supplement*, *64*(1), 39–54. doi:10.5661/RDR-VI-39 PMID:17494215

Rodriguez-Osorio, N., Kim, I. J., Wang, H., Kaya, A., & Memili, E. (2007). Melatonin increases cleavage rate of porcine preimplantation embryos in vitro. *Journal of Pineal Research*, *43*(3), 283–288. doi:10.1111/j.1600-079X.2007.00475.x PMID:17803526

Rohila, A. K., Duhan, A., Mann, D., & Kumar, A. (2017). Impact of agricultural practices on environment. Asian Journal of Microbiology. *Biotechnology and Environmental Sciences*, *19*(2), 381–384.

Rojansky, N., Brzezinski, A., & Schenker, J. G. (1992). Seasonality in humanreproduction. *Human Reproduction (Oxford, England)*, *7*(6), 735–745. doi:10.1093/oxfordjournals.humrep.a137729 PMID:1323571

Rojas-Downing, M. M., Nejadhashemi, A. P., Harrigan, T., & Woznicki, S. A. (2017). Climate change and livestock: Impacts, adaptation, and mitigation. *Climate Risk Management*, *16*, 145–163. doi:10.1016/j.crm.2017.02.001

Roman-Ponce, H., Thatcher, W. W., & Wilcox, C. J. (1981). Hormonal interrelationships and physiological responses of lactating dairy cows to shade management system in a tropical environment. *Theriogenology*, *16*(2), 139–154. doi:10.1016/0093-691X(81)90097-2 PMID:16725628

Ronchi, B., Stradaioli, G., Verini Supplizi, A., Bernabucci, U., Lacetera, N., Accorsi, P. A., Nardone, A., & Seren, E. (2001). Influence of heat stress and feed restriction on plasma progesterone, estradiol-17β LH, FSH, prolactin and cortisol in Holstein heifers. *Livestock Production Science*, *68*(2-3), 231–241. doi:10.1016/S0301-6226(00)00232-3

Ronis, M. J., Badger, T. M., Shema, S. J., Roberson, P. K., & Shaikh, F. (1998). Effects on pubertal growth and reproduction in rats exposed to lead perinatally or continuously throughout development. *Journal of Toxicology and Environmental Health*, *53*(4), 327–341. doi:10.1080/009841098159312 PMID:9490329

Ron, M., Bar-Anan, R., & Wiggans, G. R. (1984). Factors affecting conception rate of Israeli Holstein cattle. *Journal of Dairy Science*, *67*(4), 861. doi:10.3168/jds.S0022-0302(84)81377-6 PMID:6725731

Rönnberg, L., Kauppila, A., Leppäluoto, J., Martikainen, H., & Vakkuri, O. (1990). Circadian and seasonal variation in human preovulatory follicular fluid melatonin concentration. *The Journal of Clinical Endocrinology and Metabolism*, *71*(2), 492–496. doi:10.1210/jcem-71-2-493 PMID:2380343

Rosario, G. X., Modi, D. N., Sachdeva, G., Manjramkar, D. D., & Puri, C. P. (2005). Morphological events in the primate endometrium in the presence of a preimplantation embryo, detected by the serum preimplantation factor bioassay. *Human Reproduction (Oxford, England)*, *20*(1), 61–71. doi:10.1093/humrep/deh534 PMID:15576397

Rosenberg, M., Folman, Y., Herz, Z., Flamenbaum, I., Berman, A., & Kaim, M. (1982). Effect of climatic conditions on peripheral concentrations of LH, progesterone and estradiol-17b in high milk-yielding cows. *Journal of Reproduction and Fertility*, *66*(1), 139–146. doi:10.1530/jrf.0.0660139 PMID:7120177

Roth, Z. (2008). Heat stress, the follicle, and its enclosed oocyte: Mechanisms and potential strategies to improve fertility in dairy cows. *Reproduction in Domestic Animals*, *43*, 238–244. doi:10.1111/j.1439-0531.2008.01168.x

Roth, Z. (2015). Physiology and endocrinology symposium: Cellular and molecular mechanisms of heat stress related to bovine ovarian function. *Journal of Animal Science*, *93*(5), 2034–2044. doi:10.2527/jas.2014-8625 PMID:26020299

Roth, Z. (2017). Effect of heat stress on reproduction in dairy cows: Insights into the cellular and molecular responses of the oocyte. *Annual Review of Animal Biosciences*, *5*(1), 151–170. doi:10.1146/annurev-animal-022516-022849

Roth, Z., Arav, A., Bor, A., Zeron, Y., Braw-Tal, R., & Wolfenson, D. (2001). Improvement of quality of oocytes collected in the autumn by enhanced removal of impaired follicles from previously heat- stressed cows. *Reproduction (Cambridge, England)*, *122*, 737–744. doi:10.1530/rep.0.1220737 PMID:11690534

Roth, Z., Aroyo, A., Yavin, S., & Arav, A. (2008). The antioxidantepigallocatechin gallate (EGCG) moderates thedeleterious effects of maternal hyperthermia on follicle enclosedoocytes in mice. *Theriogenology*, *70*(6), 887–897. doi:10.1016/j.theriogenology.2008.05.053 PMID:18585774

Roth, Z., & Hansen, P. J. (2005). Disruption of nuclear maturation and rearrangement of cytoskeletal elements in bovine oocytes exposed to heat shock during maturation. *Reproduction (Cambridge, England)*, *129*(2), 235–244. doi:10.1530/rep.1.00394 PMID:15695618

Roth, Z., & Hansen, P. J. (2009). Involvement of apoptosis in disruption of developmental competence of bovine oocytes by heat shock during maturation. *Biology of Reproduction*, *71*(6), 1898–1906. doi:10.1095/biolreprod.104.031690 PMID:15306551

Roth, Z., Meidan, R., Braw-Tal, R., & Wolfenson, D. (2000). Immediate and delayed effects of heat stress on follicular development and its association with plasma FSH and inhibin concentration in cows. *Journal of Reproduction and Fertility*, *120*(1), 83–90. doi:10.1530/jrf.0.1200083 PMID:11006149

Roth, Z., Meidan, R., Shaham-Albalancy, A., Braw-Tal, R., & Wolfenson, D. (2001). Delayed effect of heat stress on steroid production in medium-sized and preovulatory bovine follicles. *Reproduction (Cambridge, England)*, *121*, 745–751. doi:10.1530/rep.0.1210745

Roth, Z., & Wolfenson, D. (2016). Comparing the effects of heat stress and mastitis on ovarian function in lactating cows: Basic and applied aspects. *Domestic Animal Endocrinology*, *56*, S218–S227. doi:10.1016/j.domaniend.2016.02.013

Rowan, E., Evans, C., Riley-Gilbert, M., Hyman, R., Kafalenos, R., Beucler, B., Rodehorst, B., Choate, A., & Schultz, P. (2013). Assessing the Sensitivity of Transportation Assets to Extreme Weather Events and Climate Change. *Transportation Research Record: Journal of the Transportation Research Board*, *2326*(1), 16–23. doi:10.3141/2326-03

Royal, M. D., Darwash, A. O., Flint, A. P. F., Webb, R., Woolliams, J. A., & Lamming, G. E. (2000). Declining fertility in dairy cattle: Changes in traditional and endocrine parameters of fertility. *Animal Science (Penicuik, Scotland)*, *70*(3), 487–501. doi:10.1017/S1357729800051845

Rozenboim, I., Tako, E., Gal-Garber, O., Proudman, J. A., & Uni, Z. (2007). The effect of heat stress on ovarian function of laying hens. *Poultry Science*, *86*(8), 1760–1765. doi:10.1093/ps/86.8.1760 PMID:17626822

Ruediger, F., Chacur, M., Alves, F., Oba, E., & Ramos, A. (2016). Digital infrared thermography of the scrotum, semen quality, serumtestosterone levels in Nellore bulls (Bostaurusindicus) and theircorrelation with climatic factors. *CiênciasA grárias Londrina*, *37*(1), 221-232.

Rusan, M., Hinnawi, S., & Rousan, L. (2007). Long term effect wastewater irrigation of forage crops on soil and plant quality parameters. *Desalination*, *215*(1-3), 143–152. doi:10.1016/j.desal.2006.10.032

Russart, K. L. G., & Nelson, R. J. (2018). Light at night as an environmental endocrine disruptor. *Physiology & Behavior*, *190*, 82–89. doi:10.1016/j.physbeh.2017.08.029 PMID:28870443

Rust, J., & Rust, T. (2013). Climate change and livestock production: A review with emphasis on Africa. *South African Journal of Animal Science*, *43*(3), 255. doi:10.4314ajas.v43i3.3

Ryan, D. P., Blakewood, E. G., Lynn, J. W., Munyakazi, L., & Godke, R. A. (1992). Effect of heat-stress on bovine embryo development in vitro. *Journal of Animal Science*, *70*(11), 3490–3497. doi:10.2527/1992.70113490x

Rylander, C., Oyvind Odland, J., & Manning Sandanger, T. (2013). Climate change and the potential effects on maternal and pregnancy outcomes: An assessment of the most vulnerable–the mother, fetus, and newborn child. *Global Health Action*, *6*(1), 19538. doi:10.3402/gha.v6i0.19538 PMID:23481091

Sáenz de Miera, C., Bothorel, B., Jaeger, C., Simonneaux, V., & Hazlerigg, D. (2017). Maternal photoperiod programs hypothalamic thyroid status via the fetal pituitary gland. *Proceedings of the National Academy of Sciences of the United States of America*, *114*(31), 8408–8413. doi:10.1073/pnas.1702943114 PMID:28716942

Saha, S., Singh, K. M., & Gupta, B. B. P. (2019). Melatonin synthesis and clock gene regulation in the pineal organ of teleost fish compared to mammals: Similarities and differences. *General and Comparative Endocrinology*, *279*, 27–34. doi:10.1016/j.ygcen.2018.07.010 PMID:30026020

Sakaguchi, K., Itoh, M. T., Takahashi, N., Tarumi, W., & Ishizuka, B. (2013). The rat oocyte synthesises melatonin. *Reproduction, Fertility, and Development*, *25*(4), 674–682. doi:10.1071/RD12091 PMID:22951050

Sakai, S., Hagihara, N., Kuse, M., Kimura, K., & Okuda, K. (2018). Heat stress affects prostaglandin synthesis in bovine endometrial cells. *The Journal of Reproduction and Development*, *64*(4), 64–2051. doi:10.1262/jrd.2018-051 PMID:29710018

Sakamoto, M., Ahmed, T., Begum, S., & Huq, H. (2019). Water Pollution and the Textile Industry in Bangladesh: Flawed Corporate Practices or Restrictive Opportunities? *Sustainability*, *11*(7), 1951–1964. doi:10.3390u11071951

Sakamoto, N., Ozawa, M., Yokotani-Tomita, K., Morimoto, A., Matsuzuka, T., Ijiri, D., Hirabayashi, M., Ushitani, A., & Kanai, Y. (2008). dl-α-Tocopherol acetate mitigates maternal hyperthermia-induced pre-implantation embryonic death accompanied by a reduction of physiological oxidative stress in mice. *Reproduction (Cambridge, England)*, *135*(4), 489–496. doi:10.1530/REP-07-0379 PMID:18367509

Sakatani, M., Alvarez, N. V., Takahashi, M., & Hansen, P. J. (2012). Consequences of physiological heat shock beginning at the zygote stage on embryonic development and expression of stress response genes in cattle. *Journal of Dairy Science*, *95*(6), 3080–3091. doi:10.3168/jds.2011-4986 PMID:22612944

Sakatani, M., Kobayashi, S. I., & Takahashi, M. (2004). Effects of heat shock on invitro development and intracellular oxidative state of bovine pre-implantation embryos. *Molecular Reproduction and Development*, *67*(1), 77–82. doi:10.1002/mrd.20014 PMID:14648877

Sakatani, M., Yamanaka, K., Balboula, A. Z., Takenouchi, N., & Takahashi, M. (2015). Heat stress during in vitro fertilization decreases fertilization success by disrupting anti-polyspermy systems of the oocytes. *Molecular Reproduction and Development*, *82*(1), 36–47. doi:10.1002/mrd.22441 PMID:25462627

Sakka, S. D., Malamitsi-Puchner, A., Loutradis, D., Chrousos, G. P., & Kanaka-Gantenbein, C. (2009). Euthyroid hyperthyrotropinemia in children born after in vitro fertilization. *The Journal of Clinical Endocrinology and Metabolism*, *94*(4), 1338–1341. doi:10.1210/jc.2008-1624 PMID:19190111

Salehi, A., Fallah, S., & Sourki, A. A. (2017). Organic and inorganic fertilizer effect on soil CO2 flux, microbial biomass, and growth of Nigella sativa. *International J. of Agrophysics*, *31*(1), 103–116. doi:10.1515/intag-2016-0032

Sales, K., Vasudeva, R., Dickinson, M. E., Godwin, J. L., Lumley, A. J., Michalczyk, Ł., Hebberecht, L., Thomas, P., Franco, A., & Gage, M. J. (2018). Experimental heatwaves compromise sperm function and cause transgenerational damage in a model insect. *Nature Communications*, *9*(1), 1–11. doi:10.103841467-018-07273-z PMID:30425248

Saletta, F., Matullo, G., Manuguerra, M., Bardelli, S. A., & Vineis, P. (2007). Exposure to the tobacco smoke constituent 4-aminobiphenyl induces chromosomal instability in human cancer cells. *Cancer Research*, *67*(15), 7088–7094. doi:10.1158/0008-5472.CAN-06-4420 PMID:17671175

Sallmen, M., Lindbohm, M. L., Anttila, A., Kyyronen, P., Taskinen, H., & Nykyri Eand Hemminki, K. (1998). Time to pregnancy among the wives of menexposed to organic solvents. *Occupational and Environmental Medicine*, *55*(1), 24–30. doi:10.1136/oem.55.1.24 PMID:9536159

Salvetti, N. R., Baravalle, C., Mira, G. A., Gimeno, E. J., Dallard, B. E., Rey, F., & Ortega, H. H. (2009). Heat shock protein 70 and sex steroid receptors in the follicular structures of induced ovarian cysts. Reproduction in domestic animals. *Zuchthygiene*, *44*(5), 805–814. doi:10.1111/j.1439-0531.2008.01086.x PMID:18992127

Samal, L. (2013). Heat stress in dairy cows - reproductive problems and control measures. *International Journal of Livestock Research*, *3*(3), 14–23.

Samal, L. (2013). Heat Stress in Dairy Cows - Reproductive Problems and Control Measures. *International Journal of Livestock Research*, *3*, 14–23.

Samuelsson, A. M., Matthews, P. A., Argenton, M., Christie, M. R., McConnell, J. M., Jansen, E. H., Piersma, A. H., Ozanne, S. E., Twinn, D. F., Remacle, C., Rowlerson, A., Poston, L., & Taylor, P. D. (2008). Diet-induced obesity in female mice leads to offspring hyperphagia, adiposity, hypertension, and insulin resistance: A novel murine model of developmental programming. *Hypertension*, *51*(2), 383–392. doi:10.1161/HYPERTENSIONAHA.107.101477 PMID:18086952

Sandin, G., & Peters, G. M. (2018). Environmental impact of textile reuse and recycling: A review. *Journal of Cleaner Production*, *184*, 353–365. doi:10.1016/j.jclepro.2018.02.266

Sandström, C. A., Buma, A. G., Hoye, B. J., Prop, J., van der Jeugd, H., Voslamber, B., Madsen, J., & Loonen, M. J. (2013, May 1). Latitudinal variability in the seroprevalence of antibodies against Toxoplasma gondii in non-migrant and Arctic migratory geese. *Veterinary Parasitology*, *194*(1), 9–15. doi:10.1016/j.vetpar.2012.12.027 PMID:23339847

Santangeli, S., Maradonna, F., Olivotto, I., Piccinetti, C. C., Gioacchini, G., & Carnevali, O. (2017). Efects of BPA on female reproductive function: Te involvement of epigenetic mechanism. *General and Comparative Endocrinology*, *245*, 122–126. doi:10.1016/j.ygcen.2016.08.010

Santos, J., Bilby, T., Thatcher, W., Staples, C., & Silvestre, F. (2008). Long chain fatty acids of diet as factors influencing reproduction in cattle. *Reproduction in Domestic Animals*, *43*, 23–30. doi:10.1111/j.1439-0531.2008.01139.x

Santos, J., Thatcher, W., Chebel, R., Cerri, R., & Galvao, K. (2004). The effect of embryonic death rates in cattle on the efficacy of estrus synchronization programs. *Animal Reproduction Science*, *82*, 513–535. doi:10.1016/j.anireprosci.2004.04.015

Sarno, M., Sacramento, G. A., Khouri, R., do Rosário, M. S., Costa, F., Archanjo, G., Santos, L. A., Nery, N., Vasilakis, N., Ko, A. I., & de Almeida, A. R. P. (2016). Zika Virus Infection and Stillbirths: A Case of Hydrops Fetalis, Hydranencephaly and Fetal Demise. *PLoS Neglected Tropical Diseases*, *10*(2), e0004517. doi:10.1371/journal.pntd.0004517 PMID:26914330

Sartori, R., Sartor-Bergfelt, R., Mertens, S. A., Guenther, J. N., Parrish, J. J., & Wiltbank, M. C. (2002). Fertilization and early embryonic development in heifers and lactating cows in summer and lactating and dry cows in winter. *Journal of Dairy Science*, *85*(11), 2803–2812. doi:10.3168/jds.S0022-0302(02)74367-1 PMID:12487447

Sasaki, M., Takagi, M., & Okamura, Y. (2006). A voltage sensor-domain protein is a voltage-gated proton channel. *Science*, *312*(5773), 589–592. doi:10.1126cience.1122352 PMID:16556803

Savabieasfahani, M., Kannan, K., Astapova, O., Evans, N. P., & Padmanabhan, V. (2006). Developmental programming: Differential effects of prenatal exposure to bisphenol-A or methoxychlor on reproductive function. *Endocrinology*, *147*(12), 5956–5966. doi:10.1210/en.2006-0805 PMID:16946013

Savsani, H., Padodara, R., Bhadaniya, A., Kalariya, V., Javia, B., Ghodasara, S., & Ribadiya, N. (2015). Impact of climate on feeding, production and reproduction of animals-A Review. *Agricultural Reviews (Karnal)*, *36*(1), 26–36. doi:10.5958/0976-0741.2015.00003.3

Sawyer, G., & Jitik Narayan, E. (2019). A Review on the Influence of Climate Change on Sheep Reproduction. In E. Narayan (Ed.), *Comparative Endocrinology of Animals*. doi:10.5772/intechopen.86799

Sayem, A. S. M., Giribabu, N., Muniandy, S., & Salleh, N. (2017). Effects of thyroxine on expression of proteins related to thyroid hormone functions (TR-α, TR-β, RXR and ERK1/2) in uterus during peri-implantation period. *Biomedicine and Pharmacotherapy*, *96*, 1016–1021. doi:10.1016/j.biopha.2017.11.128 PMID:29221723

Scenna, F., Edwards, J., Rohrbach, N., Hockett, M., Saxton, A., & Schrick, F. (2004). Detrimental effects of prostaglandin F2alpha on preimplantation bovine embryos. *Prostaglandins Other Lipid Meditation*, *73*(3-4), 215–226. doi:10.1016/j.prostaglandins.2004.02.001 PMID:15287153

Schernhammer, E. S., Laden, F., Speizer, F. E., Willett, W. C., Hunter, D. J., Kawachi, I., Fuchs, C. S., & Colditz, G. A. (2003). Night-shift work and risk of colorectal cancer in the Nurses' Health Study. *Journal of the National Cancer Institute*, *95*(11), 825–828. doi:10.1093/jnci/95.11.825 PMID:12783938

Schernhammer, E. S., & Schulmeister, K. (2004). Melatonin and cancer risk: Does light at night compromise physiologic cancer protection by lowering serum melatonin levels? *British Journal of Cancer*, *90*(5), 941–943. doi:10.1038j.bjc.6601626 PMID:14997186

Schifano, P., Asta, F., Dadvand, P., Davoli, M., Basagana, X., & Michelozzi, P. (2016). Heat and air pollution exposure as triggers of delivery: A survival analysis of population-based pregnancy cohorts in Rome and Barcelona. *Environment International*, *88*, 153–159. doi:10.1016/j.envint.2015.12.013 PMID:26760712

Schifano, P., Lallo, A., Asta, F., De Sario, M., Davoli, M., & Michelozzi, P. (2013). Effect of ambient temperature and air pollutants on the risk of preterm birth, Rome 2001-2010. *Environment International*, *61*, 77–87. doi:10.1016/j.envint.2013.09.005 PMID:24103349

Schillo, K. K., Alliston, C. W., & Malven, P. V. (1978). Plasma concentrations of luteinizing hormone and prolactin in the ovariectomized ewe during induced hyperthermia. *Biology of Reproduction*, *19*(2), 306–313. doi:10.1095/biolreprod19.2.306 PMID:719089

Schjoerrin, J. K., Cakmak, I., & White, P. J. (2019). Plant nutrition and soil fertility: Synergies for acquiring global green growth and sustainable development. *Plant and Soil*, *434*(1-2), 1–6. doi:10.100711104-018-03898-7

Schlenk, W., & Kahmann, H. (1938). The chemical composition of seminal fluidsand their physiological importance study with trout sperm. *Biochemical Zoology*, *295*, 283–301.

Schoonover, J. E., & Crim, J. F. (2015). An Introduction to Soil Concepts and the Role of Soils in Watershed Management,*Journal of Contemporary Water Research andamp. Education*, *154*(1), 21–47.

Schroder, H. J., & Power, G. G. (1997). Engine and radiator: Fetal and placental interactions for heat dissipation. *Experimental Physiology*, *82*(2), 403–414. doi:10.1113/expphysiol.1997.sp004035 PMID:9129954

Schuetz, A. W., & Dubin, N. H. (1981). Progesterone and prostaglandin secretion by ovulated rat cumulus cell-oocyte complexes. *Endocrinology*, *108*(2), 457–463. doi:10.1210/endo-108-2-457 PMID:7449734

Schuller, L. K., Burfeind, O., & Heuwieser, W. (2014). Impact of heat stress on conception rate of dairy cows in the moderate climate considering different temperature humidity index thresholds, periods relative to breeding, and heat load indices. *Theriogenology*, *81*(8), 1050–1057. doi:10.1016/j.theriogenology.2014.01.029

Sejian, V. (2012). Climate change, green house gas emission and sheepproduction. In Status Papers on Future Research in Sheep Production and Production Development, 50 Years Research Contributions 1962-2012. Central Sheep and Wool Research Institute.

Sejian, V., Samal, L., Haque, N., Bagath, M., Hyder, I., Maurya, V. P., Bhatta, R., Ravindra, J. P., Prasad, C. S., & Lal, R. (2015). Overview on adaptation, mitigation and amelioration strategies to improve livestock production under the changing climatic scenario. Sheep Production Adapting to Climate Change, 359-398.

Sejian, V. (2013). Climate Change: Impact on Production And Reproduction, Adaptation Mechanisms And Mitigation Strategies In Small Ruminants: A Review. *Indian Journal of Small Ruminants*, *19*(1), 1–22.

Sejian, V., Lakritz, J., Ezeji, T., & Lal, R. (2011). Forage and Flaxseed impact on enteric methane emission in dairy cows. *Research Journal of Veterinary Sciences*, *4*(1), 1–8. doi:10.3923/rjvs.2011.1.8

Sekar, P., Prasad, S. H., & Raman, M. D. (2009). Effect of textile dye industry effluent on the nutritive value of fresh water female crab *Spiralothelphusa hydrodroma*. *Journal of Applied Sciences Research*, *5*, 2041–2048.

Sellix, M. T. (2015). Circadian clock function in the mammalian ovary. *Journal of Biological Rhythms*, *30*(1), 7–19. doi:10.1177/0748730414554222 PMID:25367899

Sellix, M. T., & Menaker, M. (2011). Circadian clocks in mammalian reproductive physiology: Effects of the "other" biological clock on fertility. *Discovery Medicine*, *11*, 273–281. PMID:21524381

Sellix, M. T., Murphy, Z. C., & Menaker, M. (2013). Excess androgen during puberty disrupts circadian organization in female rats. *Endocrinology*, *154*(4), 1636–1647. doi:10.1210/en.2012-2066 PMID:23417420

Selvaraj, D., Leena, R., & Kamal, D. C. (2015). Toxicological and histopathological impacts of textile dyeing industry effluent on a selected teleost fish Poeciliareticulata. *Journal of Pharmacology and Toxicology*, *3*, 26–30.

Sen, N., Liu, X., & Craig, Z. R. (2015). Short term exposure to di-n-butyl phthalate (DBP) disrupts ovarian function in young CD-1 mice. *Reproductive Toxicology (Elmsford, N.Y.)*, *53*, 15–22. doi:10.1016/j.reprotox.2015.02.012

Setchell, B. (2006). The effects of heat on the testes of mammals. *Animal Reproduction*, *3*, 81–91.

Setchell, B. (2018). The effects of heat on the testes of mammals. *Animal Reproduction (AR)*, *3*(2), 81–91.

Setchell, B. P. (1998). The Parkes Lecture.Heat and the testis. *Journal of Reproduction and Fertility*, *114*(2), 179–194. doi:10.1530/jrf.0.1140179 PMID:10070346

Setyawan, H., Pratiwi, Q. C., Sjarifah, I., & Atmojo, T. B. (2017). Environmental heat stress enhances crystallization in urine. *IOP Conference Series: Earth and Environmental Science, 129*, 1-6.

Setyawan, H., Qodrijati, I., Widjanarti, M. P., Rinawati, S., Atmojo, T. B., & Fajariani, R. (2018). The impact of hot work climate on textile industry productivity. *IOP Conference Series: Earth and Environmental Science, 200*, 1-5. 10.1088/1755-1315/200/1/012053

Seuanez, H. N. (1980). Chromosomes and spermatozoa of the African great apes. *Journal of Reproduction and Fertility. Supplement*, *28*, 91–104. PMID:6934315

Shaikh MA (2009). Environmental issues related with textile sector. *Pakistan Textile Journal*, 36-40.

Sharara, F. I., Seifer, D. B., & Flaws, J. A. (1998). Environmental toxicants andfemale reproduction. *Fertility and Sterility*, *70*, 613–622. doi:10.1016/S0015-0282(98)00253-2 PMID:9797086

Sharpe, R. M. (2001). Hormones and testis development and the possibleadverse effects of environmental chemicals. *Toxicology Letters*, *120*(1-3), 221–232. doi:10.1016/S0378-4274(01)00298-3 PMID:11323180

Sharpe, R. M., & Irvine, D. S. (2004). How strong is the evidence of a link between environmental chemicals and adverse effects on human reproductive health? *BMJ (Clinical Research Ed.)*, *328*(7437), 447–451. doi:10.1136/bmj.328.7437.447

Shashar, S., Kloog, I., Erez, O., Shtein, A., Yitshak-Sade, M., Sarov, B., & Novack, L. (2020). Temperature and pre-eclampsia: Epidemiological evidence that perturbation in maternal heat homeostasis affects pregnancy outcome. *PLoS One*, *15*(5), e0232877. doi:10.1371/journal.pone.0232877 PMID:32421729

Sheffield, P. E., & Landrigan, P. J. (2011). Global climate change andchildren's health: Threats and strategies for prevention. *Environmental Health Perspectives*, *119*(3), 291–298. doi:10.1289/ehp.1002233 PMID:20947468

Shehab-El-Deen, M. A. M. M., Leroy, J. L. M. R., Fadel, M., Saleh, S., Maes, D., & Van Soom, A. (2010). Biochemical changes in the follicular fluid of the dominant follicle of high producing dairy cows exposed to heat stress early post partum. *Animal Reproduction Science*, *117*(3-4), 189–200. doi:10.1016/j.anireprosci.2009.04.013 PMID:19481380

Sheikh, A., Bhagat, R., Islam, S., Dar, R., Sheikh, S., Wani, J., & Dogra, P. (2013). Effect of climate change on reproduction and milk production performance of livestock: A review. *Journal of Pharmacognosy and Phytochemistry*, *6*(6), 2062–2064.

Sheth, K. N., & Marvania, M. (2017). A critical study on heat stress in textile industry. *International Journal for Innovative Research in Science and Technology*, *4*, 47–49.

Shi, D., & Vine, D. F. (2012). Animal models of polycystic ovary syndrome: A focused review of rodent models in relationship to clinical phenotypes and cardiometabolic risk. *Fertility and Sterility*, *98*(1), 185–193. doi:10.1016/j.fertnstert.2012.04.006 PMID:22607890

Shimba, S., Ogawa, T., Hitosugi, S., Ichihashi, Y., Nakadaira, Y., Kobayashi, M., Tezuka, M., Kosuge, Y., Ishige, K., Ito, Y., Komiyama, K., Okamatsu-Ogura, Y., Kimura, K., & Saito, M. (2011). Deficient of a clock gene, brain and muscle Arnt-like protein-1 (BMAL1), induces dyslipidemia and ectopic fat formation. *PLoS One*, *6*(9), e25231. doi:10.1371/journal.pone.0025231 PMID:21966465

Shimizu, T., Ohshima, I., Ozawa, M., Takahashi, S., Tajima, A., Shiota, M., Miyazaki, H., & Kanai, Y. (2005). Heat stress diminishes gonadotropin receptor expression and enhances susceptibility to apoptosis of rat granulose cells. *Reproduction*, *129*, 46-472.

Shimizu, T., Ohshima, I., Ozawa, M., Takahashi, S., Tajima, A., Shiota, M., Miyazaki, H., & Kanai, Y. (2005). Heat stress diminishes gonadotropin receptor expression and enhances susceptibility to apoptosis of rat granulose cells. *Reproduction (Cambridge, England)*, *129*(4), 463–472. doi:10.1530/rep.1.00502 PMID:15798021

Shiozawa, T., Suyama, K., Nakano, K., Nukaya, H., Sawanishi, H., Oguri, A., Wakabayashi, K., & Terao, Y. (1999). Mutagenic activity of 2 phenylbenzotriazole derivatives related to a mutagen, PBTA-1, in river water. *Mutation Research*, *442*(2), 105–111. doi:10.1016/S1383-5718(99)00069-8 PMID:10393279

Shoghag, L., Hudson, S., Applewhite, L., Feldman, J., & Minkoff, H. L. (1997). An association between the heat-humidity index and preterm labor and delivery: A preliminary analysis. *American Journal of Public Health*, *87*(7), 1205–1207. doi:10.2105/AJPH.87.7.1205 PMID:9240114

Short, R. E., Bellows, R. A., Staigmiller, R. B., Berardinelli, J. G., & Custer, E. E. (1990). Physiological mechanisms controlling anestrus and infertility in postpartum beef cattle. *Journal of Animal Science*, *68*(3), 799–816. doi:10.2527/1990.683799x

Shreeve, N., Cagampang, F., Sadek, K., Tolhurst, M., Houldey, A., Hill, C. M., Brook, N., MacKlon, N., & Cheong, Y. (2013). Poor sleep in PCOS; Is melatonin the culprit? *Human Reproduction (Oxford, England)*, *28*(5), 1348–1353. doi:10.1093/humrep/det013 PMID:23438443

Shrivastava, P., & Kumar, R. (2015). Soil salinity: A serious environmental issue and plant growth promoting bacteria as one of the tools for its alleviation. *Saudi Journal of Biological Sciences*, *22*(2), 123–131. doi:10.1016/j.sjbs.2014.12.001 PMID:25737642

Shwartz, G., Rhoads, M. L., VanBaale, M. J., Rhoads, R. P., & Baumgard, L. H. (2009). Effects of a supplemental yeast culture on heat-stressed lactating holstein cows. *Journal of Dairy Science*, *92*(3), 935–942. doi:10.3168/jds.2008-1496

Sidibé, Y., Foudi, S., Pascual, U., & Termansen, M. (2018). Adaptation to Climate Change in Rainfed Agriculture in the Global South: Soil Biodiversity as Natural Insurance. *Ecological Economics*, *146*, 588–596. doi:10.1016/j.ecolecon.2017.12.017

Sikka, S. C. (2004). Role of oxidative stress and antioxidants in andrology and assisted reproductive technology. *Journal of Andrology*, *25*(1), 5–18. doi:10.1002/j.1939-4640.2004.tb02751.x PMID:14662779

Silanikove, N. (2000). Effects of heat stress on the welfare of extensively managed domestic ruminants. *Livestock Production Science*, *67*(1-2), 1–18. doi:10.1016/S0301-6226(00)00162-7

Silva, C., Sartorelli, E., Castilho, A., Satrapa, R., Puelker, R., Razza, E., & Barros, C. M. (2013). Effects of heat stress on development, quality and survival of Bosindicus and Bostaurus embryos produced in vitro. *Theriogenology*, *79*(2), 351–357. doi:10.1016/j.theriogenology.2012.10.003

Simonneaux, V., & Bahougne, T. (2015). A multi-oscillatory circadian system times female reproduction. *Frontiers in Endocrinology (Lausanne)*, *6*, 157. doi:10.3389/fendo.2015.00157 PMID:26539161

Sinclair, K. D., Allegrucci, C., Singh, R., Gardner, D. S., Sebastian, S., Bispham, J., Thurston, A., Huntley, J. F., Rees, W. D., Maloney, C. A., Lea, R. G., Craigon, J., McEvoy, T. G., & Young, L. E. (2007). DNA methylation, insulin resistance, and blood pressure in offspring determined by maternal periconceptional B vitamin and methionine status. *Proceedings of the National Academy of Sciences of the United States of America*, *104*(49), 19351–19356. doi:10.1073/pnas.0707258104 PMID:18042717

Singh, A. K., Chandra, G., Ojha, B. K., & Rajoriya, J. S. (2019). Effect of heat stress on livestock reproduction. *Haryana Veterinarian*, *58*, 16–24.

Singh, K. B. (1969). Induction of polycystic ovarian disease in rats by continuous light. II. Observations on mating, pregnancy, and the postpartum period. *American Journal of Obstetrics and Gynecology*, *104*(7), 1004–1007. doi:10.1016/0002-9378(69)90692-9 PMID:5794828

Singh, M., Chaudhari, B. K., Singh, J. K., Singh, A. K., & Maurya, P. K. (2013). Effects of thermal load on buffalo reproductive performance during summer season. *The Journal of Biological Sciences*, *1*(1), 1–8.

Sinha Hikim, A. P., Lue, Y., Yamamoto, C., Vera, Y., Rodriguez, S., Yen, P. H., & … . (2003). Key apoptotic pathways for heat-induced programmed germ cell death in the testis. *Endocrinology*, *144*(7), 3167–3175. doi:10.1210/en.2003-0175

Sinha, R., Ranjan, A., Lone, S., Rahim, A., Devi, I., & Tiwari, S. (2017). The impact of climate change on livestock production and reproduction: Ameliorative management. *International Journal of Livestock Research*, *7*(6), 1–8. doi:10.5455/ijlr.20170417042102

Sirotkin, A. V. (2010). Effect of two types of stress (heat shock/high temperature and malnutrition/serum deprivation) on porcine ovarian cell functions and their response to hormones. *The Journal of Experimental Biology*, *213*(Pt 12), 2125–2130. doi:10.1242/jeb.040626 PMID:20511527

Sirotkin, A. V., & Kacaniova, M. (2010). The effect of high temperature on swine ovarian function in vitro. *Veterinarni Medicina*, *55*(8), 377–382. doi:10.17221/2953-VETMED

Sirotkin, A. V., Mlyncek, M., Kotwica, J., Makarevich, A. V., Florkovicová, I., & Hetényi, L. (2005). Leptin directly controls secretory activity of human ovarian granulosa cells: Possible inter-relationship with the IGF/IGFBP system. *Hormone Research*, *64*(4), 198–202. doi:10.1159/000089009 PMID:16244494

Skovmand, A., Jensen, A. C., Maurice, C., Marchetti, F., Lauvås, A. J., Koponen, I. K., Jensen, K. A., Goericke-Pesch, S., Vogel, U., & Hougaard, K. S. (2019). Effects of maternal inhalation of carbon black nanoparticles on reproductive and fertility parameters in a four-generation study of male mice. *Particle and Fibre Toxicology*, *16*(1), 13. doi:10.118612989-019-0295-3

Slaem, M. B., & Bouraoui, R. (2011). Heat stress in Tunisia: Effects on dairy cows and potential means of alleviating. SA. *Journal of Animal Science*, *39*, 256–259.

Slimen, I., Najar, T., Ghram, A., Dabbebi, H., & Ben Mrad, M. (2014). Reactive oxygen species, heat stress and oxidative-induced mitochondrial damage. *Internal Journal of Hyperthermia*, *30*(7), 513–523. doi:10.3109/02656736.2014.971446 PMID:25354680

Slutsker, L., Heymann, D. L., & Breman, J. G. (1996). The effect of malaria and malaria prevention in pregnancy on offspring birthweight, prematurity, and intrauterine growth retardation in rural Malawi. *The American Journal of Tropical Medicine and Hygiene*, *55*(1, Suppl), 33–41. PMID:8702035

Smith, J. F., Syritsyna, O., Fellous, M., Serres, C., Mannowetz, N., Kirichok, Y., & Lishko, P. V. (2013). Disruption of theprincipal, progesterone-activated sperm Ca^{2+} channel in a CatSper2-deficient infertile patient. *Proceedings of the National Academy of Sciences of the United States of America*, *110*(17), 6823–6828. doi:10.1073/pnas.1216588110 PMID:23530196

Smith, M. S. R., Upfold, J. B., Edwards, M. J., Shiota, K., & Cawdell-Smith, J. (1992). The induction of neural tube defects by maternal hyperthermia; a comparison of the guinea-pig and human. *Neuropathology and Applied Neurobiology*, *18*(1), 71–80. doi:10.1111/j.1365-2990.1992.tb00765.x PMID:1579201

Smith, P., Juengel, J., Maclean, P., Rand, C., & Stanton, J.-A. L. (2019). Gestational nutrition 2: Gene expression in sheep fetal ovaries exposed to gestational under nutrition. *Reproduction (Cambridge, England)*, *157*, 13–25. doi:10.1530/REP-18-0358

Soares, J. C., Santos, C. S., Carvalho, S. M. P., Pintado, M. M., & Vasconcelos, M. W. (2019). Preserving the nutritional quality of crop plants under a changing climate: Importance and strategies. *Plant and Soil*, *443*(1-2), 1–26. doi:10.100711104-019-04229-0

Soares, J. M. Jr, Masana, M. I., Erşahin, C., & Dubocovich, M. L. (2003). Functional melatonin receptors in rat ovaries at various stages of the estrous cycle. *The Journal of Pharmacology and Experimental Therapeutics*, *306*(2), 694–702. doi:10.1124/jpet.103.049916 PMID:12721330

Soares, J. M., Simões, M. J., Oshima, C. T. F., Mora, O. A., De Lima, G. R., & Baracat, E. C. (2003). Pinealectomy changes rat ovarian interstitial cell morphology and decreases progesterone receptor expression. *Gynecological Endocrinology*, *17*(2), 115–123. doi:10.1080/gye.17.2.115.123 PMID:12737672

Sodango, T. H., Li, X., Sha, J., & Bao, Z. (2018). Review of the Spatial Distribution, Source and Extent of Heavy Metal Pollution of Soil in China: Impacts and Mitigation Approaches. *Journal of Health and Pollution*, *8*(17), 53–70. doi:10.5696/2156-9614-8.17.53 PMID:30524849

Soetan, K., Olaiya, C., & Oyewole, O. (2010). The importance of mineral elements for humans, domestic animals and plants: A review. *African Journal of Food Science*, *4*(5), 200–222.

Solomon, S., Qin, D., Manning, M., Chen, Z., Marquis, M., Averyt, K. B., & Tignor, M. (2007). Summary for policy-makers. Contribution of Working Group 1 to the Fourth Assessment Report of the Intergovernmental Panel on Climate Change. Intergovernmental Panel on Climate Change (IPCC). In *Climate Change 2007: The Physical Science Basis* Cambridge University Press. https://www.ipcc.ch/site/assets/uploads/2018/02/ar4-wg1-spm-1.pdf

Solt, L. A., Wang, Y., Banerjee, S., Hughes, T., Kojetin, D. J., Lundasen, T., Shin, Y., Liu, J., Cameron, M. D., Noel, R., Yoo, S. H., Takahashi, J. S., Butler, A. A., Kamenecka, T. M., & Burris, T. P. (2012). Regulation of circadian behaviour and metabolism by synthetic REV-ERB agonists. *Nature*, *485*(7396), 62–68. doi:10.1038/nature11030 PMID:22460951

Soni, P., Sharma, S., Sharma, S., Kumar, S., & Sharma, K. P. (2006). A comparative study on the toxic effects of textile dye waste waters (untreated and treated) on mortality and RBC of fresh water fish *Gambusia affinis*. *Journal of Environmental Biology*, *27*, 623–628. PMID:17405321

Sorensen, C., Saunik, S., Sehgal, M., Tewary, A., Govindan, M., Lemery, J., & Balbus, J. (2018, October 17). Climate Change and Women's Health: Impacts and Opportunities in India. *Geohealth.*, *2*(10), 283–297. doi:10.1029/2018GH000163 PMID:32159002

Soto, P., Natzke, R., & Hansen, P. (2003). Identification of possible mediators of embryonic mortality caused by mastitis: Actions oflipopolysaccharide, prostaglandin F2alpha, and the nitric oxidegenerator, sodium nitroprusside dihydrate, on oocyte maturationand embryonic development in cattle. *American Journal of Reproductive Immunology*, *50*(3), 263–272. doi:10.1034/j.1600-0897.2003.00085.x PMID:14629032

Soto, P., & Smith, L. C. (2009). BH4 peptide derived from Bel –Xl and Bax inhibitor peptide suppresses apoptotic mitochondrial changes in heat stressed bovine oocytes. *Molecular Reproduction and Development*, *76*(7), 637–646. doi:10.1002/mrd.20986 PMID:19062170

Sousa, M. L. D., Moraes, P., Lopes, P. R. M., & Montagnolli, R. (2012). Contamination by Remazol Red Brilliant dye and its impact in aquatic photosynthetic microbiota. *Journal of Environmental Management and Sustainable Development*, *1*, 129–138.

Souza, A., Narciso, C., Batista, E., Carvalho, P., & Wiltbank, M. (2018). Effect of uterine environment on embryo production and fertility in cows. *Animal Reproduction (AR)*, *11*, 159–167.

Soyinka, O. O., Oritogun, K. S., Amballi, A. A., Olorundami, B. O., & Ogundahunsi, O. A. (2020). Effects of occupational exposure to VAT dye on fertility and pregnancy outcome of women dye users. *Arch Basic Applied Medicine*, *8*, 21–25.

Specht, I. O., Bonde, J. P., Toft, G., Lindh, C. H., Jonsson, B. A., & Jorgensen, K. T. (2015). Serum phthalate levels and time to pregnancy in couples from Greenland, Poland and Ukraine. *PLoS One*, *10*(3), e0120070. doi:10.1371/journal.pone.0120070

Speroff, L., & Fritz, M. A. (Eds.). (2005). Clinical Gynecologic Endocrinology and Infertility. Lippincott Williams and Wilkins.

Spicer, L., & Echternkamp, S. (1986). Ovarian follicular growth, function and turnover in cattle: A review. *Journal of Animal Science*, *62*(2), 428–451. doi:10.2527/jas1986.622428x

Spira, A., & Multigner, L. (1998). Environmental factors and male infertility. *Human Reproduction (Oxford, England)*, *13*, 2041–2042. doi:10.1093/humrep/13.8.2041 PMID:9756262

Spitz, I. M., Croxatto, H. B., & Robbins, A. (1996). Antiprogestins: Mechanism of Action and Contraceptive. *Annual Review of Pharmacology and Toxicology*, *36*(1), 47–81. doi:10.1146/annurev.pa.36.040196.000403

Srinivasan, A. K., Freyle, J., Gitlin, J. S., & Palmer, L. S. (2007). Climatic conditions and the risk of testicular torsion in adolescent males. *The Journal of Urology*, *178*(6), 2585–2588. doi:10.1016/j.juro.2007.08.049 PMID:17945301

Sripriya, L., Vijayalakshmi, M., Sumathy, R., & Sharmila, J. (2014). The impact of textile dyes on the biochemistry and histology of liver, a fresh water fish, tilapia, *Oreochromis mossambicus. Journal of Pharmaceutical and Biosciences*, *5*, 271–298.

Staples, C., Burke, J., & Thatcher, W. (1998). Influence of supplemental fats on reproductive tissues and performance of lactating cows. *Journal of Dairy Science*, *81*(3), 856–871. doi:10.3168/jds.S0022-0302(98)75644-9

Statista. (2019). https://www.statista.com/statistics/268184/leading-import-countries-worldwide/

Stavreus-Evers, A., Nikas, G., Sahlin, L., Eriksson, H., & Landgren, B. M. (2001). Formation of pinopodes in human endometrium is associated with the concentrations of progesterone and progesterone receptors. *Fertility and Sterility*, *76*(4), 782–791. doi:10.1016/S0015-0282(01)01993-8 PMID:11591414

Steffan, J. J., Brevik, E. C., Burgess, L. C., & Cerdà, A. (2017). The effect of soil on human health: An overview. *European Journal of Soil Science*, *69*(1), 159–171. doi:10.1111/ejss.12451 PMID:29430209

Stem, E., Mertz, G. A., Stryker, J. D., & Huppi, M. (1989). Changing animal disease patterns induced by the greenhouse effect. In The potential effects of global change on the United States. US Environmental Protection Agency.

Stensgaard, A. S., Vounatsou, P., Sengupta, M. E., & Utzinger, J. (2019, February). Schistosomes, snails and climate change: Current trends and future expectations. *Acta Tropica*, *190*, 257–268. doi:10.1016/j.actatropica.2018.09.013 PMID:30261186

Stevens, R. G. (1987). Electric power use and breast cancer: A hypothesis. *American Journal of Epidemiology*, *125*(4), 556–561. doi:10.1093/oxfordjournals.aje.a114569 PMID:3548332

Stewart, B. M., Block, J., Morelli, P., Navarette, A. E., Amstalden, M., Bonilla, L., Hansen, P. J., & Bilby, T. R. (2011). Efficacy of embryo transfer in lactating dairy cows during summer using fresh or vitrified embryos produced in vitro with sexsorted semen. *Journal of Dairy Science*, *94*(7), 3437–3445. doi:10.3168/jds.2010-4008 PMID:21700029

Stewart, Z. P., Pierzynski, G. M., Middendorf, B. J., & Vara Prasad, P. V. (2020). Approaches to improve soil fertility in sub-Saharan Africa. *Journal of Experimental Botany*, *71*(2), 632–641. doi:10.1093/jxb/erz446 PMID:31586430

Stiborova, M., Frei, E., & Schmeiser, H. H. (1992). Comparison of cytochrome P-450- and peroxidase-mediated activations of carcinogenic azo dyes and N-nitrosamines. *General Physiology and Biophysics*, *11*, 489–498. PMID:1291449

Stone, B. A. (1982). Heat induced infertility of boars: The interrelationshipbetween depressed sperm output and fertility andan estimation of the critical air temperature above which spermoutput is impaired. *Animal Reproduction Science*, *4*(4), 283–299. doi:10.1016/0378-4320(82)90043-4

Stoss, J. (1983). Fish gamete preservation and spermatozoan physiology. In Fish physiology (pp. 305-50). Academic Press. doi:10.1016/S1546-5098(08)60306-4

Strand, L. B., Barnett, A. G., & Tong, S. (2011). The influence of season andambient temperature on birth outcomes: A review of theepidemiological literature. *Environmental Research*, *111*(3), 451–462. doi:10.1016/j.envres.2011.01.023 PMID:21333980

Strand, L. B., Barnett, A. G., & Tong, S. (2012). Maternal exposure to ambient temperature and the risks of preterm birth and stillbirth in Brisbane, Australia. *American Journal of Epidemiology*, *175*(2), 99–107. doi:10.1093/aje/kwr404 PMID:22167749

Stumpf, P. G., Garry, F., & Lloyd, T. (1984). Effects of temperature on*in vitro* penetration of human sperm into bovine cervicalmucus. *The Journal of Reproductive Medicine*, *29*, 833–836. PMID:6520822

Suarez, S. S. (2008). Control of hyperactivation in sperm. *Human Reproduction Update*, *14*(6), 647–657. doi:10.1093/humupd/dmn029 PMID:18653675

Subedi, R., Bhatta, L. D., Udas, E., Agrawal, N. K., Joshi, K. D., & Panday, D. (2019). Climate-smart practices for improvement of crop yields in mid-hills of Nepal. *Cogent Food & Agriculture*, *5*(1), 1–15. doi:10.1080/23311932.2019.1631026

Su, J., Wang, Y., Xing, X., Zhang, L., Sun, H., & Zhang, Y. (2015). Melatonin significantly improves the developmental competence of bovine somatic cell nuclear transfer embryos. *Journal of Pineal Research*, *59*(4), 455–468. doi:10.1111/jpi.12275 PMID:26331949

Su, R. W., & Fazleabas, A. T. (2015). Implantation and establishment of pregnancy in human and nonhuman primates. In *Regulation of Implantation and Establishment of Pregnancy in Mammals* (pp. 189–213). Springer. doi:10.1007/978-3-319-15856-3_10

Suriyasomboon, A., Kunavongkrit, A., Lundeheim, N., & Einarsson, S. (2004). Effect of temperature and humidity onsperm production in Duroc boars under different housing systemsin Thailand. *Livestock Production Science*, *89*(1), 19–31. doi:10.1016/j.livprodsci.2003.12.008

Suryavathia, S., Sharma, S., Sharma, S., Saxena, P., Pandey, S., Grover, R., & ... (2005). Acute toxicity of textile dye wastewaters (untreated and treated) of Sanganer on male reproductive systems of albino rats and mice. *Reproductive Toxicology (Elmsford, N.Y.)*, *19*(4), 547–556. doi:10.1016/j.reprotox.2004.09.011 PMID:15749270

Suteu, D., Zaharia, C., & Malutan, T. (2011). Removal of orange 16 reactive dye from aqueous solutions by waste sunflower seed shells. *Journal of the Serbian Chemical Society*, *76*(4), 607–624. doi:10.2298/JSC100721051S

Sutton-McDowall, M. L., Wu, L. L., Purdey, M., Abell, A. D., Goldys, E. M., MacMillan, K. L., Thompson, J. G., & Robker, R. L. (2016). Nonesterified fatty acid-induced endoplasmic reticulum stress in cattle cumulus oocyte complexes alters cell metabolism and developmental competence. *Biology of Reproduction*, *94*(23), 21–29.

Sweeney, T., Nicol, L., Roche, J. F., & Brooks, A. N. (2000). Maternal exposure to octylphenol suppresses ovine fetal follicle-stimulating hormone secretion, testis size, and sertoli cell number. *Endocrinology*, *141*(7), 2667–2673. doi:10.1210/endo.141.7.7552 PMID:10875272

Szymanski, L., Schneider, J., Friedman, M., Ji, H., Kurose, Y., Blache, D., Rao, A., Dunshea, F. R., & Clarke, I. J. (2007). Changes in Insulin, Glucose and Ketone Bodies, But Not Leptin or Body Fat Content Precede Restoration of Luteinising Hormone Secretion in Ewes. *Journal of Neuroendocrinology*, *19*(6), 449–460. doi:10.1111/j.1365-2826.2007.01551.x

Takahashi, M. (2011). Heat stress on reproductive function and fertility in mammals. *Reproductive Medicine and Biology*, *11*(1), 37–47. doi:10.100712522-011-0105-6 PMID:29699104

Takahashi, M. (2012). Oxidative stress and redox regulation on in vitro development of mammalian embryos. *The Journal of Reproduction and Development*, *58*(1), 1–9. doi:10.1262/jrd.11-138N PMID:22450278

Takata, M., Burioka, N., Ohdo, S., Takane, H., Terazono, H., Miyata, M., Takanori, S., Hisashi, S., Yasushi, F., Katsuyuki, T., & Eiji, S. (2002). Daily expression of mRNAs for the mammalian Clock genes Per2 and clock in mouse suprachiasmatic nuclei and liver and human peripheral blood mononuclear cells. *Japanese Journal of Pharmacology*, *90*(3), 263–269. doi:10.1254/jjp.90.263 PMID:12499581

Taketani, T., Tamura, H., Takasaki, A., Lee, L., Kizuka, F., Tamura, I., Taniguchi, K., Maekawa, R., Asada, H., Shimamura, K., Reiter, R. J., & Sugino, N. (2011). Protective role of melatonin in progesterone production by human luteal cells. *Journal of Pineal Research*, *51*(2), 207–213. doi:10.1111/j.1600-079X.2011.00878.x PMID:21585519

Talukdar, D., Talukdar, P., & Ahmed, K. (2016). Minerals and its impact on fertility of livestock: A review. *Agricultural Reviews (Karnal)*, *37*(4), 333–337. doi:10.18805/ag.v37i4.6464

Tamarkin, L., Cohen, M., Reichert, C., Reichert, C., Lippman, M., & Chabner, B. (1981). Melatonin Inhibition and Pinealectomy Enhancement of 7,12-Dimethylbenz(a)anthracene-induced Mammary Tumors in the Rat. *Cancer Research*, *41*, 4432–4436. PMID:6796259

Tamura, H., Nakamura, Y., Takiguchi, S., Kashida, S., Yamagata, Y., Sugino, N., & Kato, H. (1998). Melatonin directly suppresses steroid production by preovulatory follicles in the cyclic hamster. *Journal of Pineal Research*, *25*(3), 135–141. doi:10.1111/j.1600-079X.1998.tb00551.x PMID:9745981

Tamura, H., Takasaki, A., Miwa, I., Taniguchi, K., Maekawa, R., Asada, H., Taketani, T., Matsuoka, A., Yamagata, Y., Shimamura, K., Morioka, H., Ishikawa, H., Reiter, R. J., & Sugino, N. (2008). Oxidative stress impairs oocyte quality and melatonin protects oocytes from free radical damage and improves fertilization rate. *Journal of Pineal Research*, *44*(3), 280–287. doi:10.1111/j.1600-079X.2007.00524.x PMID:18339123

Tamura, H., Takasaki, A., Taketani, T., Tanabe, M., Kizuka, F., Lee, L., Tamura, I., Maekawa, R., Asada, H., Yamagata, Y., & Sugino, N. (2013). Melatonin as a free radical scavenger in the ovarian follicle. *Endocrine Journal*, *60*(1), 1–13. doi:10.1507/endocrj.EJ12-0263 PMID:23171705

Tan, D. X., Manchester, L. C., Esteban-Zubero, E., Zhou, Z., & Reiter, R. J. (2015). Melatonin as a potent and inducible endogenous antioxidant: Synthesis and metabolism. *Molecules (Basel, Switzerland)*, *20*(10), 18886–18906. doi:10.3390/molecules201018886 PMID:26501252

Tan, D. X., Manchester, L. C., Hardeland, R., Lopez-Burillo, S., Mayo, J. C., Sainz, R. M., & Reiter, R. J. (2003). Melatonin: A hormone, a tissue factor, an autocoid, a paracoid, and an antioxidant vitamin. *Journal of Pineal Research*, *34*(1), 75–78. doi:10.1034/j.1600-079X.2003.02111.x PMID:12485375

Tan, O., & Bukulmez, O. (2011). Biochemistry, molecular biology and cell biology of gonadotropin-releasing hormone antagonists. *Current Opinion in Obstetrics & Gynecology*, *23*(4), 238–244. doi:10.1097/GCO.0b013e328348a3ce PMID:21666463

Tao, S., & Dahl, G. E. (2013). Invited review: Heat stress effects during late gestation on dry cows and their calves. *Journal of Dairy Science*, *96*(7), 4079–4093. doi:10.3168/jds.2012-6278 PMID:23664343

Tao, S., Monteiro, A. P., Thompson, I. M., Hayen, M. J., & Dahl, G. E. (2012). Effect of late gestation maternal heat stress on growth and immune function of dairy calves. *Journal of Dairy Science*, *95*(12), 7128–7136. doi:10.3168/jds.2012-5697

Tarara, R., Enders, A. C., Hendrickx, A. G., Gulamhusein, N., Hodges, J. K., Hearn, J. P., Eley, R. B., & Else, J. G. (1987). Early implantation and embryonic development of the baboon: Stages 5, 6 and 7. *Anatomy and Embryology*, *176*(3), 267–275. doi:10.1007/BF00310182 PMID:3631530

Taskinen, H., Anttila, A., Lindbohm, M. L., Sallme'n, M., & Hemminki, K. (1989). Spontaneous abortions and congenital malformations among the wivesof men occupationally exposed to organic solvents. *Scandinavian Journal of Work, Environment & Health*, *15*(5), 345–352. doi:10.5271jweh.1839 PMID:2799322

Taskinen, H., Kyyronen, P., Hemminki, K., Hoikkala, M., Lajunen, K., & Lindbohm, M. L. (1994). Laboratory work and pregnancy outcome. *Journal of Occupational and Environmental Medicine*, *36*, 311–319. doi:10.1097/00043764-199403000-00008 PMID:8195901

Taskinen, H., Lindbohm, M. L., & Hemminki, K. (1986). Spontaneous abortionsamong women working in the pharmaceutical industry. *British Journal of Industrial Medicine*, *43*, 199–205. PMID:3947584

Taylor, H. S. (2000). The role of HOX genes in the development and function of the female reproductive tract. In Seminars in reproductive medicine (Vol. 18, No. 01, pp. 081-090). Thieme Medical Publishers, Inc. doi:10.1055-2000-13478

Taylor, H. S., Arici, A., Olive, D., & Igarashi, P. (1998). HOXA10 is expressed in response to sex steroids at the time of implantation in the human endometrium. *The Journal of Clinical Investigation*, *101*(7), 1379–1384. doi:10.1172/JCI1597 PMID:9525980

Taylor, H. S., Igarashi, P., Olive, D. L., & Arici, A. (1999). Sex steroids mediate HOXA11 expression in the human peri-implantation endometrium. *The Journal of Clinical Endocrinology and Metabolism*, *84*(3), 1129–1135. doi:10.1210/jc.84.3.1129 PMID:10084606

Taylor, H. S., VandenHeuvel, G. B., & Igarashi, P. (1997). A conserved Hox axis in the mouse and human female reproductive system: Late establishment and persistent adult expression of the Hoxa cluster genes. *Biology of Reproduction*, *57*(6), 1338–1345. doi:10.1095/biolreprod57.6.1338 PMID:9408238

Teboul, M., Barrat-Petit, M. A., Li, X. M., Claustrat, B., Formento, J. L., Delaunay, F., Lévi, F., & Milano, G. (2005). Atypical patterns of circadian clock gene expression in human peripheral blood mono nuclear cells. *Journal of Molecular Medicine*, *83*(9), 693–699. doi:10.100700109-005-0697-6 PMID:16096739

Tei, C., Maruyama, T., Kuji, N., Miyazaki, T., Mikami, M., & Yoshimura, Y. (2003). Reduced expression of αvβ3 integrin in the endometrium of unexplained infertility patients with recurrent IVF-ET failures: Improvement by danazol treatment. *Journal of Assisted Reproduction and Genetics*, *20*(1), 13–20. doi:10.1023/A:1021254620888 PMID:12645863

Thimon, V., Calvo, E., Koukoui, O., Légaré, C., & Sullivan, R. (2008). Effects of vasectomy on gene expression profiling along the human epididymis. *Biology of Reproduction*, *79*(2), 262–273. doi:10.1095/biolreprod.107.066449 PMID:18434627

Thomas, C. D., Cameron, A., Green, R. E., Bakkenes, M., Beaumont, L. J., Collingham, Y. C., & Hughes, L. (2004). Extinction risk from climate change. *Nature*, *427*(6970), 145–148. doi:10.1038/nature02121 PMID:14712274

Thompson, C. L., Larkin, E. K., Patel, S., Berger, N. A., Redline, S., & Li, L. (2011). Short duration of sleep increases risk of colorectal adenoma. *Cancer*, *117*(4), 841–847. doi:10.1002/cncr.25507 PMID:20936662

Thonneau, P., Abell, A., Larsen, S. B., Bonde, J. P., Joffe, M., Clavert, A., Ducot, B., Multigner, L., & Danscher, G. (1999a). Effects of pesticide exposure ontime to pregnancy: Results of a multicenter study in France andDenmark. ASCLEPIOS Study Group. *American Journal of Epidemiology*, *150*(2), 157–163. doi:10.1093/oxfordjournals.aje.a009975 PMID:10412960

Thonneau, P., Bujan, L., Multigner, L., & Mieusset, R. (1998, August). Occupational heat exposure and male fertility: A review. *Human Reproduction (Oxford, England)*, *13*(8), 2122–2125. doi:10.1093/humrep/13.8.2122 PMID:9756281

Thonneau, P., Ducot, B., Bujan, L., Mieusset, R., & Spira, A. (1997). Effect of maleoccupational heat exposure on time to pregnancy. *International Journal of Andrology*, *20*(5), 774–778. doi:10.1046/j.1365-2605.1997.d01-303.x PMID:16130271

Thonneau, P., Larsen, S. B., Abell, A., Clavert, A., Bonde, J. P., Ducot, B., & Multinger, L. (1999b). Time to pregnancy and paternal exposure to pesticidesin preliminary results from Danish and French studies. Asclepios. *Scandinavian Journal of Work, Environment & Health*, *25*(Suppl 1), 49–61. PMID:10235412

Thornton, P., & Gerber, P. (2010). Climate change and the growth of the livestock sector in developing countries. *Mitigation and Adaptation Strategies for Global Change*, *15*(2), 169–184. doi:10.100711027-009-9210-9

Thornton, P., Herrero, M., Freeman, A., Mwai, O., Rege, E., Jones, P., & McDermott, J. (2007). Vulnerability, climate change and livestock research opportunities and challenges for poverty alleviation. *Journal of SAT Agricultural Research*, *4*(1), 1–23.

Thrall, P. H., Oakeshott, J. G., Fitt, G., Southerton, S., Burdon, J. J., Sheppard, A., & Ford Denison, R. (2011). Evolution in agriculture: The application of evolutionary approaches to the management of biotic interactions in agro-ecosystems. *Evolutionary Applications*, *4*(2), 200–215. doi:10.1111/j.1752-4571.2010.00179.x PMID:25567968

Thulstrup, A. M., & Bonde, J. P. (2006). Maternal occupational exposure and risk of specific birth defects. *Occupational Medicine*, *56*(8), 532–543. doi:10.1093/occmed/kql115

Thureen, P.J., Trembler, K.A., Meschia, G., Makowski, E.L., & Wilkening, R.B. (1992). Placental glucose transport in heat induced fetal growth retardation. *American Journal of Physiology, 263*, R578 – R585.

Tielemans, E., Van Kooij, R., te Velde, E.R., Burdorf, A., & Heederik, D. (1999a). Pesticide exposure and decreased fertilisation rates in vitro. *Lancet, 354*, 484–485.

Tielemans, E., Burdorf, A., te Velde, E. R., Weber, R. F. A., Van Kooji, R. J., Veulemans, H., & Heederik, D. J. J. (1999b). Occupationally relatedexposures and reduced semen quality: A case-control study. *Fertility and Sterility*, *71*(4), 690–696. doi:10.1016/S0015-0282(98)00542-1 PMID:10202880

Tielemans, E., Van Kooij, R., Looman, C., Burdorf, A., te Velde, E., & Heederik, D. (2000). Paternal occupational exposures and embryo implantation ratesafter IVF. *Fertility and Sterility*, *74*(4), 690–695. doi:10.1016/S0015-0282(00)00720-2 PMID:11020508

Tompkins, E. C., Heidenreich, C. J., & Stob, M. (1967). Effect of post-breeding thermal stress on embryonic mortality in swine. *Journal of Animal Science*, *26*(2), 377–380. doi:10.2527/jas1967.262377x

Torres-Junior, J. R. D., Pires, M. D. A., De Sa, W. F., Feffeira, A. D., Viana, J. H. M., Camargo, L. S. A., Ramos, A. A., Folhadella, I. M., Polisseni, J., de Freitas, C., Clemente, C. A. A., De Sa, M. F., Paula-Lopes, F. F., & Baruselli, P. S. (2008). Effect of maternal heat stress on follicular growth and oocyte competence in Bosindicus cattle. *Theriogenology, 69*(2), 155–166. doi:10.1016/j.theriogenology.2007.06.023 PMID:17953981

Tosini, G., Pozdeyev, N., Sakamoto, K., & Iuvone, P. M. (2008). The circadian clock system inthe mammalian retina. *BioEssays, 30*(7), 624–633. doi:10.1002/bies.20777 PMID:18536031

Tremellen, K. (2008). Oxidative stress and male infertility-A clinical perspective. *Human Reproduction Update, 14*(3), 243–258. doi:10.1093/humupd/dmn004 PMID:18281241

Trentini, G. P., Mess, B., De Gaetani, C. F., & Ruzsás, C. (1978). Effect of melatonin on induction of ovulation in the light- induced constant estrous-anovulatory syndrome and possible role of the brain serotoninergic system. *Journal of Endocrinological Investigation: Official Journal of the Italian Society of Endocrinology, 1*(4), 305–310. doi:10.1007/BF03350974 PMID:574524

Tricot, G. (2000). New insights into role of microenvironment in multiple myeloma. *TheLancet, 355*(9200), 248–250. doi:10.1016/S0140-6736(00)00019-2 PMID:10675068

Troy, P. J., Daftary, G. S., Bagot, C. N., & Taylor, H. S. (2003). Transcriptional repression of peri-implantation EMX2 expression in mammalian reproduction by HOXA10. *Molecular and Cellular Biology, 23*(1), 1–13. doi:10.1128/MCB.23.1.1-13.2003 PMID:12482956

Tsuboi, T., Osawa, T., Kimura, K., Kubo, M., & Haritani, M. (2011). Experimental infection of early pregnant cows with bovine viral diarrhea virus: Transmission of virus to the reproductive tract and conceptus. *Research in Veterinary Science, 90*(1), 174–178. doi:10.1016/j.rvsc.2010.04.024 PMID:20553701

Tsutsui, K., Saigoh, E., Ukena, K., Teranishi, H., Fujisawa, Y., Kikuchi, M., Ishii, S., & Sharp, P. J. (2000). A novel avian hypothalamic peptide inhibiting gonadotropin release. *Biochemical and Biophysical Research Communications, 275*(2), 661–667. doi:10.1006/bbrc.2000.3350

Tuc, V. P., Wangsuphachart, V., Tasanapradit, P., Fungladda, W., Van Trong, P., & Nhung, N. T. (2007). Impacts of pesticide use on semen characteristics among rice famers in Kienxuong District, Thaibinh Province, Vietnam. *The Southeast Asian Journal of Tropical Medicine and Public Health, 38*(3), 569.

Tufekci, N., Sivri, N., & Toroz, I. (2007). Pollutants of textile industry wastewater and assessment of its discharge limits by water quality standards. *Turkish Journal of Fisheries and Aquatic Sciences, 7*, 97–103.

Turek, F. W., Joshu, C., Kohsaka, A., Lin, E., Ivanova, G., McDearmon, E., Laposky, A., Losee-Olson, S., Easton, A., Jensen, D. R., Eckel, R. H., Takahashi, J. S., & Bass, J. (2005). Obesity and metabolic syndrome in circadian Clock mutant mice. *Science, 308*(5724), 1043–1045. doi:10.1126cience.1108750 PMID:15845877

Tusell, L., David, I., Bodin, L., Legarra, A., Rafel, O., López-Bejar, M., & Piles, M. (2011). Using the product threshold model for estimating separately the effect of temperature on male and female fertility. *Journal of Animal Science, 89*(12), 3983–3995. doi:10.2527/jas.2011-3924 PMID:21764834

Twig, G., Shina, A., Amital, H., & Shoenfeld, Y. (2012). Pathogenesis of infertility and recurrent pregnancy loss in thyroid autoimmunity. *Journal of Autoimmunity, 38*(2-3), J275–JJ28. doi:10.1016/j.jaut.2011.11.014 PMID:22218218

Twombly, R. (1995). Assault on the male. *Environmental Health Perspectives, 103*(9), 802–805. doi:10.1289/ehp.95103802 PMID:7498090

Uehata, T., & Sasakawa, N. (1982). The fatigue and maternity disturbances of night workwomen. *Journal of Human Ergology*, *11*, 465–474. PMID:7188479

Ulberg, L. C., & Burfening, P. J. (1967). Embryo death resulting from adverse environment on spermatozoa or ova. *Journal of Animal Science*, *26*(3), 571–577. doi:10.2527/jas1967.263571x PMID:6068502

Umbuzeiro, G. A., Freeman, H. S., Warren, S. H., Oliveira, D. P., Terao, Y., Watanabe, T., & ... (2005). The contribution of azo dyes to the mutagenic activity of the Cristais river. *Chemosphere*, *60*(1), 55–64. doi:10.1016/j.chemosphere.2004.11.100 PMID:15910902

UN. (2017). *United Nations, Department of Economic and Social Affairs, Population Division 2017. World population prospects: the 2017 revision, key findings and advance tables*. Working Paper No. ESA/P/WP/248. Available from https://esa.un.org/unpd/wpp/publications/files/wpp2017_keyfindings.pdf

UNSCEAR. (1986). Biological effects of pre-natal irradiation. Annex C. In *United Nations Scientiðc Committee on the Effects of Atomic Radiation, Genetic and Somatic Effects of Ionising Radiation. UNSCEAR Report to the General Assembly with Scientific Annexes* (pp. 263–366). United Nations.

Upadhyay, R. C., & Ashutosh, S.V. (2009). Impact of climate change on reproductive functions of cattle and buffalo. In Global Climate Change and Indian Agriculture. New Delhi: ICAR.

Upfold, J., Smith, M. S. R., & Edwards, M. J. (1989). Quantitative study of the effects of maternal hyperthermia on cell death and proliferation in the guinea pig brain on day 21 of pregnancy. *Teratology*, *39*(2), 173–179. doi:10.1002/tera.1420390209 PMID:2928964

US Congress. (1985). Evidence for workplace hazards to reproductive function. In *Reproductive health hazards in the workplace* (p. 93). Office of Technological Assessment.

Vajnhandl, S., & Le Marechal, A. M. (2005). Ultrasound in textile dyeing and the decolouration/mineralization of textile dyes. *Dyes and Pigments*, *65*(2), 89–101. doi:10.1016/j.dyepig.2004.06.012

Valanis, B., Vollmer, W. M., & Steele, P. (1999). Occupational exposure to antineoplasticagents: Self-reported miscarriages and stillbirths among nursesand pharmacists. *Journal of Occupational and Environmental Medicine*, *41*(8), 632–638. doi:10.1097/00043764-199908000-00004 PMID:10457505

Van Bergenen, H. (1985). Studies on a possible relationship between alterations in the cytoskeleton and induction of heat shock protein synthesis in mammalian cells. *International Journal of Hyperthermia*, *1*, 59–83.

Van den Boogaard, E., Vissenberg, R., Land, J. A., van Wely, M., van der Post, J. A., Goddijn, M., & Bisschop, P. H. (2011). Significance of (sub) clinical thyroid dysfunction and thyroid autoimmunity before conception and in early pregnancy: A systematic review. *Human Reproduction Update*, *17*(5), 605–619. doi:10.1093/humupd/dmr024 PMID:21622978

van der Horst, G. T., Muijtjens, M., Kobayashi, K., Takano, R., Kanno, S., Takao, M., De Wit, J., Verkerk, A., Eker, A. P. M., Van, L. D., Buijs, R., Bootsma, D., Hoeijmakers, J. H. J., & Yasui, A. (1999). Mammalian Cry1 and Cry2 are essential for maintenance of circadian rhythms. *Nature*, *398*(6728), 627–630. doi:10.1038/19323 PMID:10217146

van der Horst, G., & Maree, L. (2014). Sperm form and function in the absence of sperm competition. *Molecular Reproduction and Development*, *81*(3), 204–216. doi:10.1002/mrd.22277 PMID:24273026

Van der Linden, M., Buckingham, K., Farquhar, C., Kremer, J. A., & Metwally, M. (2011). Luteal phase support for assisted reproduction cycles. *Cochrane Database of Systematic Reviews*, 10. PMID:21975790

Van Langendonckt, A., Casanas-Roux, F., & Donnez, J. (2002). Oxidative stress and peritoneal endometriosis. *Fertility and Sterility*, *77*(5), 861–870. doi:10.1016/S0015-0282(02)02959-X PMID:12009336

Van Oostdam, J., Gilman, A., Dewailly, E., Usher, P., Wheatley, B., Kuhnlein, S., Walker, J., Tracy, B., Feeley, M., & Jerome, V. (1999). Human healthimplications of environmental contaminants in Canada: A review. *The Science of the Total Environment*, *230*(1-3), 1–82. doi:10.1016/S0048-9697(99)00036-4 PMID:10466227

Van Zutphen, A. R., Fletcher, B. A., & Hwang, S. A. (2012). A population-based case–control study of extreme summer temperature and birth defects. National Institute of Environmental Health Sciences. *Environmental Health Perspectives*, *120*(10), 1443–1449. doi:10.1289/ehp.1104671 PMID:23031822

Vandaele, L., & Van Soom, A. (2011). Intrinsic factors affecting apoptosis in bovine in vitro produced embryos. *Verhandelingen - Koninklijke Academie voor Geneeskunde van Belgie*, *73*(1-2), 79–104.

Vandenberg, L. N., Colborn, T., Hayes, T. B., Heindel, J. J., Jacobs, D. R. Jr, Lee, D. H., Shioda, T., Soto, A. M., vom Saal, F. S., Welshons, W. V., Zoeller, R. T., & Myers, J. P. (2012). Hormones and endocrine-disrupting chemicals: Low-dose effects and nonmonotonic dose responses. *Endocrine Reviews*, *33*(3), 378–455. doi:10.1210/er.2011-1050

vanVliet, M. T. H., Ludwig, F., Zwolsman, J. J. G., Weedon, G. P., & Kabat, P. (2011). Global river temperatures and sensitivity to atmospheric warming andchanges in river flow. *Water Resources Research*, *47*, W02544.

Várallyay, G. (2010). The impact of climate change on soils and on their water management. *Agronomy Research*, *8*, 385–396.

Varcoe, T. J., Gatford, K. L., & Kennaway, D. J. (2018). Maternal circadian rhythms and the programming of adult health and disease. *American Journal of Physiology. Regulatory, Integrative and Comparative Physiology*, *314*(2), R231–R241. doi:10.1152/ajpregu.00248.2017 PMID:29141950

Varcoe, T. J., Wight, N., Voultsios, A., Salkeld, M. D., & Kennaway, D. J. (2011). Chronic phase shifts of the photoperiod throughout pregnancy programs glucose intolerance and insulin resistance in the rat. *PLoS One*, *6*(4), e18504. doi:10.1371/journal.pone.0018504 PMID:21494686

Vasconcelos, J. L. M., Silcox, R. W., Lacerda, J. A., Pursley, G. R., & Wiltbank, M. C. (1998). Pregnancy rate, pregnancy loss, and response to heat stress after AI at 2 different times from ovulation in dairy cows. *Biology of Reproduction*, *56*(Suppl 1), 140.

Velazquez, M. (2015). Impact of maternal malnutrition during the peri-conceptional period on mammalian preimplantation embryo development. *Domestic Animal Endocrinology*, *51*, 27–45. doi:10.1016/j.domaniend.2014.10.003

Verga, M., Luzi, F., & Carenzi, C. (2007). Effects of husbandry and management systems on physiology and behaviour of farmed and laboratory rabbits. *Hormones and Behavior*, *52*(1), 122–129. doi:10.1016/j.yhbeh.2007.03.024 PMID:17482618

Vergetaki, A., Jeschke, U., Vrekoussis, T., Taliouri, E., Sabatini, L., Papakonstanti, E. A., & Makrigiannakis, A. (2014). Galectin-1 overexpression in endometriosis and its regulation by neuropeptides (CRH, UCN) indicating its important role in reproduction and inflammation. *PLoS One*, *9*(12), e114229. doi:10.1371/journal.pone.0114229 PMID:25473847

Verma, S. K., Calero-Bernal, R., Cerqueira-Cézar, C. K., Kwok, O. C., Dudley, M., Jiang, T., Su, C., Hill, D., & Dubey, J. P. (2016, May). Toxoplasmosis in geese and detection of two new atypical Toxoplasma gondii strains from naturally infected Canada geese (Branta canadensis). *Parasitology Research*, *115*(5), 1767–1772. doi:10.100700436-016-4914-8 PMID:26796021

Vermunt, J. J., & Tranter, B. P. (2011). *Heat stress in dairy cattle–a review, and some of the potential risks associated with the nutritional management of this condition*. Paper presented at the review of AVA QLD Division Conference.

Veulemans, H., Steeno, O., Masschelein, R., & Groeseneken, D. (1993). Exposure to ethylene glycol ethers and spermatogenic disorders in man: A case-control study. *British Journal of Industrial Medicine*, *50*(1), 71–78. doi:10.1136/oem.50.1.71 PMID:8431394

Vicedo-Cabrera, A. M., Iñíguez, C., Barona, C., & Ballester, F. (2014). Exposure to elevated temperatures and risk of preterm birth in Valencia, Spain. *Environmental Research*, *134*, 210–217. doi:10.1016/j.envres.2014.07.021 PMID:25173054

Vicedo-Cabrera, A. M., Olsson, D., & Forsberg, B. (2015). Exposure to seasonal temperatures during the last month of gestation and the risk of preterm birth in Stockholm. *International Journal of Environmental Research and Public Health*, *12*(4), 3962–3978. doi:10.3390/ijerph120403962 PMID:25867199

Vidal-Vanaclocha, F., Fantuzzi, G., Mendoza, L., Fuentes, A. M., Anasagasti, M. J., Martin, J., Carrascal, T., Walsh, P., Reznikov, L. L., Kim, S. H., Novick, D., Rubinstein, M., & Dinarello, C. A. (2000). IL-18 regulates IL-1β-dependent hepatic melanoma metastasis via vascular cell adhesion molecule-1. *Proceedings of the National Academy of Sciences of the United States of America*, *97*(2), 734–739. doi:10.1073/pnas.97.2.734 PMID:10639148

Visconti, P. E., Krapf, D., De La Vega-beltrán, J. L., Acevedo, J. J., & Darszon, A. (2011). Ion channels, phosphorylation and mammalian sperm capacitation. *Asian Journal of Andrology*, *13*(3), 395–405. doi:10.1038/aja.2010.69 PMID:21540868

Viswanathan, A. N., Hankinson, S. E., & Schernhammer, E. S. (2007). *Night Shift Work and the Risk of Endometrial Cancer*. Academic Press.

Vitaterna, M. H., Selby, C. P., Todo, T., Niwa, H., Thompson, C., & Fruechte, E. M. (1999). Differential regulation of mammalian period genes and circadian rhythmicity by cryptochromes 1 and 2. *Proceedings of the National Academy of Sciences of the United States of America*, *96*, 12114-12119. 10.1073/pnas.96.21.12114

Vladic, T., & Jarvi, T. (1997). Sperm motility and fertilization time span in Atlanticsalmon and brown trout - The effect of water temperature. *Journal of Fish Biology*, *50*, 1088–1093.

Voglera, C. J., Bamea, J. H., DeJarnettea, J. M., McGilliarda, M. L., & Saacke, R. G. (1993). Effects of elevated testicular temperature onmorphology characteristics of ejaculated spermatozoa in the bovine. *Theriogenology*, *40*(6), 1207–1219. doi:10.1016/0093-691X(93)90291-C

Volt, H., García, J. A., Doerrier, C., Díaz-Casado, M. E., Guerra-Librero, A., Lõpez, L. C., Escames, G., Tresguerres, J. A., & Acuña-Castroviejo, D. (2016). Same molecule but different expression: Aging and sepsis trigger NLRP3 inflammasome activation, a target of melatonin. *Journal of Pineal Research*, *60*(2), 193–205. doi:10.1111/jpi.12303 PMID:26681113

Vries, F. (2018). Impacts of Climate Change on Soil Microbial Communities and Their Functioning. *Developments in Soil Science*, *35*, 111–129. doi:10.1016/B978-0-444-63865-6.00005-3

Wahl, R. L., & Reif, J. S. (2009). Temporal trends in bull semen quality: A comparative model for human health. *Environmental Research*, *109*(3), 273–280. doi:10.1016/j.envres.2008.10.012

Wale, P. L., & Gardner, D. K. (2016). The effects of chemical and physical factors on mammalian embryo culture and their importance for the practice of assisted human reproduction. *Human Reproduction Update*, *22*(1), 2–22. doi:10.1093/humupd/dmv034

Wallace, J. M., Regnault, T. R., Limesand, S. W., Hay, W. W. Jr, & Anthony, R. V. (2005). Investigating the causes of low birth weight in contrasting ovine paradigms. *The Journal of Physiology*, *565*(1), 19–26. doi:10.1113/jphysiol.2004.082032 PMID:15774527

Walsh, D., Li, K., Wass, J., Dolnikov, A., Zeng, F., & Edwards, M. (1993). Heat-shock gene expression and cell cycle changes during mammalian embryonic development. *Development & Genetics, 14*, 127–36.

Walsh, D., Zhe, L., Zeng, F., Yan, W., & Edwards, M. (1994). Heat shock genes and cell cycle regulation in early mammalian development. *Environmental Medicine*, *38*, 1–6.

Walsh, D., Zhe, L., Zeng, F., Yan, W., & Edwards, M. (1994). Heat shocks genes and cell cycle regulation in early mammalian development. *Environmental Medicine*, *38*, 1–6.

Walsh, F. C. (2001). Electrochemical technology for environmental treatment and clean energy conversion. *Pure and Applied Chemistry*, *73*(12), 1819–1837. doi:10.1351/pac200173121819

Wan, G. H., & Chung, F. F. (2012). Working conditions associated with ovarian cycle in a medical center nurses: A Taiwan study. *Japan Journal of Nursing Science*, *9*(1), 112–118. doi:10.1111/j.1742-7924.2011.00191.x PMID:22583945

Wang, H., Critchley, H. O., Kelly, R. W., Shen, D., & Baird, D. T. (1998). Progesterone receptor subtype B is differentially regulated in human endometrial stroma. *Molecular Human Reproduction*, *4*(4), 407–412. doi:10.1093/molehr/4.4.407 PMID:9620842

Wang, H., & Dey, S. K. (2006). Roadmap to embryo implantation: Clues from mouse models. *Nature Reviews. Genetics*, *7*(3), 185–199. doi:10.1038/nrg1808 PMID:16485018

Wang, J. Z., Sui, H. S., Miao, D. Q., Liu, N., Zhou, P., Ge, L., & Tan, J. (2009). Effects of heat stress during in vitro maturation on cytoplasmic versus nuclear components of mouse oocytes. *Reproduction (Cambridge, England)*, *137*(2), 181–189. doi:10.1530/REP-08-0339 PMID:19029342

Wang, J., Williams, G., Guo, Y., Pan, X., & Tong, S. (2013). Maternal exposure to heatwave and preterm birth in Brisbane, Australia. *BJOG*, *120*(13), 1631–1641. doi:10.1111/1471-0528.12397 PMID:24034563

Wang, L., Barrington, S., & Kim, J. W. (2007). Biodegradation of pentylamine and aniline from petrochemical wastewater. *Journal of Environmental Management*, *83*(2), 191–197. doi:10.1016/j.jenvman.2006.02.009 PMID:16678965

Wang, X. X., Ying, P., Diao, F., Wang, Q., Ye, D., Jiang, C., Shen, N., Xu, N., Chen, W. B., Lai, S. S., Jiang, S., Miao, X. L., Feng, J., Tao, W. W., Zhao, N. W., Yao, B., Xu, Z. P., Sun, H. X., Li, J. M., ... Li, C. J. (2013, July 29). Altered protein prenylation in Sertoli cells is associated with adult infertility resulting from childhood mumps infection. *The Journal of Experimental Medicine*, *210*(8), 1559–1574. doi:10.1084/jem.20121806 PMID:23825187

Wang, Y. Y., Tawfik, O., & Wood, G. W. (1998). Endotoxin-induced abortion in mice is mediated by activated fetal macrophages. *Journal of Leukocyte Biology*, *63*(1), 40–50. doi:10.1002/jlb.63.1.40 PMID:9469471

Wang, Y., Goldberg, J., Sharma, R. K., Agarwal, A., & Falcone, T. (1997). Importance of reactive oxygen species in the peritoneal fluid of women with endometriosis or idiopathic infertility. *Fertility and Sterility*, *68*(5), 826–830. doi:10.1016/S0015-0282(97)00343-9 PMID:9389810

Wang, Y., Huang, J., Xia, P., He, J., Wang, C., Ju, Z., Li, J., Li, R., Zhong, J., & Li, Q. (2013). Genetic variations of HSBP1 gene and its effect on thermal performance traits in Chinese Holstein cattle. *Molecular Biology Reports*, *40*(6), 3877–3882. doi:10.100711033-012-1977-1

Wanner, R. A., Edwards, M. J., & Wright, R. G. (1976). The effect of hyperthermia on the neuroepithelium of the 21-day guinea-pig foetus: Histologic and ultrastructural study. *The Journal of Pathology*, *118*(4), 235–244. doi:10.1002/path.1711180406 PMID:1271135

Wan, X., He, X., Liu, Q., Duan, P., & Li, H. (2020). Frequent and mild scrotal heat stress impairs embryo development, implantation, and offspring sex ratio in mice. *Reproductive Biomedicine Online*, *40*(5), 617–626. doi:10.1016/j.rbmo.2020.01.020 PMID:32276888

Watanabe, H., Vriens, J., Suh, S. H., Benham, C. D., Droogmans, G., & Nilius, B. (2002). Heat-evoked activation of TRPV4 channels in a HEK293 cell expression system and in native mouse aorta endothelial cells. *The Journal of Biological Chemistry*, *277*(49), 47044–47051. doi:10.1074/jbc.M208277200 PMID:12354759

Wathes, D. C., Fenwick, M., Cheng, Z., Bourne, N., Liewellyn, S., & Morris, D. G. (2007). Influence of negative energy balance on cyclicity and fertility in the high producing dairy cow. *Theriogenology*, *68*(Suppl 1), S232–S241. doi:10.1016/j.theriogenology.2007.04.006 PMID:17475319

Watson, R. T. (2001). A Contribution of Working Groups I, II, and III to the Third Assessment Report of the Intergovernmental Panel on Climate Change. Intergovernmental Panel on Climate Change (IPCC). In *Climate Change 2001: Synthesis Report*. Cambridge University Press. https://www.ipcc.ch/site/assets/uploads/2018/05/SYR_TAR_full_report.pdf

Weber, A. L., & Adler, N. T. (1979). Delay of constant light-induced persistent vaginal estrus by 24-hour time cues in rats. *Science*, *204*(4390), 323–325. doi:10.1126cience.571146 PMID:571146

Webster, W. S., & Edwards, M. J. (1984). Hyperthermia and the induction of neural tube defects in mice. *Teratology*, *29*(3), 417–425. doi:10.1002/tera.1420290313 PMID:6463904

Wegierski, T., Lewandrowski, U., Muller, B., Sickmann, A., & Walz, G. (2009). Tyrosine phosphorylation modulates the-activity of TRPV4 in response to defined stimuli. *The Journal of Biological Chemistry*, *284*(5), 2923–2933. doi:10.1074/jbc.M805357200 PMID:19033444

Weibel, L., Spiegel, K., Gronfier, C., Follenius, M., & Brandenberger, G. (1997). Twenty-four-hour melatonin and core body temperature rhythms: Their adaptation in night workers. *The American Journal of Physiology*, *272*, R948–R954. PMID:9087659

Wei, D., Zhang, C., Xie, J., Song, X., Yin, B., Liu, Q., Hu, L., Hao, H., Geng, J., & Wang, P. (2013). Supplementation with low concentrations of melatonin improves nuclear maturation of human oocytes in vitro. *Journal of Assisted Reproduction and Genetics*, *30*(7), 933–938. doi:10.100710815-013-0021-2 PMID:23737216

Weigert, M., Feichtinger, W., Kulin, S., Kaali, S. G., Dorau, P., & Bauer, P. (2001). Seasonal influences on in vitro fertilization and embryo transfer. *Journal of Assisted Reproduction and Genetics*, *18*(11), 598–602. doi:10.1023/A:1013160905497 PMID:11804428

Weihua, Z., Saji, S., Mäkinen, S., Cheng, G., Jensen, E. V., Warner, M., & Gustafsson, J. Å. (2000). Estrogen receptor (ER) β, a modulator of ERα in the uterus. *Proceedings of the National Academy of Sciences of the United States of America*, *97*(11), 5936–5941. doi:10.1073/pnas.97.11.5936 PMID:10823946

Welch, W. J., & Suhan, J. P. (1985). Morphological study of the mammalian stress response: Characterization of changes in cytoplasmic organelles, cytoskeleton, and nucleoli, and appearance of intranuclear actin filaments in rat fibroblasts after heat-shock treatment. *The Journal of Cell Biology*, *101*(4), 1198–1211. doi:10.1083/jcb.101.4.1198 PMID:3900086

Wennborg, H., Bodin, L., Vainio, H., & Axelsson, G. (2000). Pregnancy outcomeof personnel in Swedish biomedical research laboratories. *Journal of Occupational and Environmental Medicine*, *42*(4), 438–446. doi:10.1097/00043764-200004000-00022 PMID:10774513

Wennborg, H., Bonde, J. P., Stenbeck, M., & Olsen, J. (2002). Adverse reproduction outcomes among employees working in biomedical research laboratories. *Scandinavian Journal of Work, Environment & Health*, *28*(1), 5–11. doi:10.5271jweh.640 PMID:11871853

West, J. W. (2003). Effects of heat stress on production in dairy cattle. *Journal of Dairy Science*, *86*(6), 2131–2144. doi:10.3168/jds.S0022-0302(03)73803-X PMID:12836950

Wettemann, R. P., & Bazer, F. W. (1985). Influence of environmental temperature on prolificacy of pigs. *Journal of Reproduction and Fertility. Supplement, 1985*(33), 199–208. PMID:3910825

Wettemann, R. P., & Desjardins, C. (1979). Testicular function in boars exposed to elevated ambient temperature. *Biology of Reproduction, 20*(2), 235–241. doi:10.1095/biolreprod20.2.235 PMID:454734

Wettemann, R. P., Wells, M. E., Omtvedt, I. T., Pope, C. E., & Turman, E. J. (1976). Influence of elevated ambient temperatureon reproductive performance of boars. *Journal of Animal Science, 42*(3), 664–669. doi:10.2527/jas1976.423664x PMID:1262278

Wettemann, R., Bazer, F., Thatcher, W., Caton, D., & Roberts, R. (1988). Conceptus development, uterine response, blood gases and endocrine function of gilts exposed to increased ambient temperature during early pregnancy. *Theriogenology, 30*(1), 57–74. doi:10.1016/0093-691X(88)90263-4 PMID:16726449

Wheelock, J. B., Rhoads, R. P., Vanbaala, M. J., Sanders, S. R., & Baumgard, I. H. (2010). Effects of heat stress on energetic metabolism in lactating Holstein cows. *Journal of Dairy Science, 93*(2), 644–655. doi:10.3168/jds.2009-2295 PMID:20105536

Whirledge, S. D., Oakley, R. H., Myers, P. H., Lydon, J. P., DeMayo, F., & Cidlowski, J. A. (2015). Uterine glucocorticoid receptors are critical for fertility in mice through control of embryo implantation and decidualization. *Proceedings of the National Academy of Sciences of the United States of America, 112*(49), 15166–15171. doi:10.1073/pnas.1508056112 PMID:26598666

Whirledge, S., & Cidlowski, J. A. (2013). A role for glucocorticoids in stress-impaired reproduction: Beyond the hypothalamus and pituitary. *Endocrinology, 154*(12), 4450–4468. doi:10.1210/en.2013-1652 PMID:24064362

Wilcox, A. J., Baird, D. D., & Weinberg, C. R. (1999). Time of implantation of the conceptus and loss of pregnancy. *The New England Journal of Medicine, 340*(23), 1796–1799. doi:10.1056/NEJM199906103402304 PMID:10362823

Wilking, M., Ndiaye, M., Mukhtar, H., & Ahmad, N. (2013). Circadian rhythm connections to oxidative stress: Implications for human health. Antioxidants and Redox Signaling, 19, 192-208.

Williams, S. J. (2013). Sea-Level Rise Implications for Coastal Regions. *Journal of Coastal Research, 63*(Special Issue), 184–196. doi:10.2112/SI63-015.1

Williot, P., Kopeika, E. F., & Goncharov, B. F. (2000). Influence of testis state,temperature and delay in semen collection on spermatozoa motilityin the cultured Siberian sturgeon (*Acipenserbaeri* Brandt). *Aquaculture (Amsterdam, Netherlands), 189*(1-2), 53–61. doi:10.1016/S0044-8486(00)00358-6

Willy, D. K., Muyanga, M., Mbuvi, J., & Jayned, T. (2019). The effect of land use change on soil fertility parameters in densely populated areas of Kenya. *Geoderma, 343*, 254–262. doi:10.1016/j.geoderma.2019.02.033 PMID:31160827

Wilson, S. J., Marion, R. S., Spain, J. N., Spiers, D. E., Keisler, D. H., & Lucy, M. C. (1998). Effects of controlled heat stress on ovarian function of dairy cattle. *Journal of Dairy Science, 81*, 2139–2144.

Wiltbank, M. C., Baez, G. M., Garcia-Guerra, A., Toledo, M. Z., Monteiro, P. L. J., Melo, L. F., & Sartori, R. (2016). Pivotal periods for pregnancy loss during the first trimester of gestation in lactating dairy cows. *Theriogenology, 86*(1), 239–253. doi:10.1016/j.theriogenology.2016.04.037 PMID:27238438

Windham, G. C., Lee, D., Mitchell, P., Anderson, M., Petreas, M., & Lasley, B. (2005). Exposure to organochlorine compounds and effects on ovarian function. *Epidemiology (Cambridge, Mass.), 16*(2), 182–190. doi:10.1097/01.ede.0000152527.24339.17

Winfield, C. G., Hemsworth, P. H., Galloway, D. B., & Makin, A. W. (1981). Sexual behaviour and semen characteristics of boars: Effects of high temperature. *Australian Journal of Experimental Agriculture and Animal Husbandry*, *21*(108), 39–45. doi:10.1071/EA9810039

Wirbisky, S. E., & Freeman, J. L. (2015). Atrazine exposure and reproductive dysfunction through the hypothalamus-pituitary-gonadal (HPG) axis. *Toxics*, *3*(4), 414–450. doi:10.3390/toxics3040414

Wirth, J. J., & Mijal, R. S. (2010). Adverse effects of low level heavy metal exposure on male reproductive function. *Systems Biology in Reproductive Medicine*, *56*(2), 147–167. doi:10.3109/19396360903582216

Wise, M. E., Armstrong, D. V., Huber, J. T., Hunter, R., & Wiersma, F. (1988). Hormonal alterations in the lactating dairy cow in response to thermal stress. *Journal of Dairy Science*, *71*(9), 2480–2485. doi:10.3168/jds.S0022-0302(88)79834-3 PMID:3183143

Wisniewski, P., Romano, R. M., Kizys, M. M., Oliveira, K. C., Kasamatsu, T., Giannocco, G., Chiamolera, M. I., Dias-da-Silva, M. R., & Romano, M. A. (2015). Adult exposure to bisphenol A (BPA) in Wistar rats reduces sperm quality with disruption of the hypothalamic–pituitary–testicular axis. *Toxicology*, *329*, 1–9. doi:10.1016/j.tox.2015.01.002

Wolfensen, D., Roth, Z., & Meidan, R. (2000). Impaired reproduction in heat stressed cattle: basic and applied aspects. *Animal Reproduction Science*, *60-61*, 535-547.

Wolfenson, D., Flamenbaum, I., & Berman, A. (1988). Hyperthermia and body energy store effects on estrous behavior, conception rate, and corpus luteum function in dairy cows. *Journal of Dairy Science*, *71*(12), 3497–3504. doi:10.3168/jds.S0022-0302(88)79956-7

Wolfenson, D., Luft, O., Berman, A., & Meidan, R. (1993). Effects of season, incubation temperature and cell age on progesterone and prostaglandin F2α production in bovine luteal cells. *Animal Reproduction Science*, *32*(1-2), 27–40. doi:10.1016/0378-4320(93)90055-V

Wolfenson, D., & Roth, Z. (2019). Impact of heat stress on cow reproduction and fertility. *Animal Frontiers*, *9*(1), 32–39. doi:10.1093/af/vfy027 PMID:32002237

Wolfenson, D., Roth, Z., & Meidan, R. (2000). Impaired reproduction in heat-stressed cattle: Basic and applied aspects. *Animal Reproduction Science*, *60*, 535–547. doi:10.1016/S0378-4320(00)00102-0 PMID:10844222

Wolfenson, D., Sonego, H., Bloch, A., Shaham-Albalancy, A., Kaim, M., Folman, Y., & Meidan, R. (2002). Seasonal differences in progesterone production by luteinized bovine thecal and granulosa cells. *Domestic Animal Endocrinology*, *22*(2), 81–90. doi:10.1016/S0739-7240(01)00127-8 PMID:11900966

Wolfenson, D., Thatcher, W., Badinga, L., Savi, J., Meidan, R., Lew, B., & Berman, A. (1995). Effect of heat stress on follicular development during the estrous cycle in lactating dairy cattle. *Biology of Reproduction*, *52*(5), 1106–1113. doi:10.1095/biolreprod52.5.1106

Wolf, J., & Armstrong, B. (2012). The association of season and temperature with adverse pregnancy outcome in two German states, a time-series analysis. *PLoS One*, *7*(7), e40228. doi:10.1371/journal.pone.0040228 PMID:22792247

Wong, M. (2008). *Development and Trends in the Chinese Garment and Textile Industry. MFA+3 Labour rights in a changing garment industry.* https://en.archive.maquilasolidarity.org/ sites/maquilasolidarity.org/files/GM-MFA3-TrendsofChineseTextileIndustry-2008-09-28.pdf

Wong, E. Y., Ray, R. M., Gao, D. L., Wernli, K. J., Li, W., Fitzgibbons, E. D., & et al., (2009). Dust and chemical exposures, and miscarriage risk among women textile workers in Shanghai, China. *Occupational and Environmental Medicine*, *66*(3), 161–168. doi:10.1136/oem.2008.039065 PMID:18805889

Woodall, P. A., & Kramer, M. R. (2018, April). Schistosomiasis and Infertility in East Africa. *The American Journal of Tropical Medicine and Hygiene*, *98*(4), 1137–1144. doi:10.4269/ajtmh.17-0280 PMID:29313478

Wood, G. W., Hausmann, E., & Choudhuri, R. (1997). Relative role of CSF-1, MCP-1/JE, and RANTES in macrophage recruitment during successful pregnancy. *Molecular Reproduction and Development: Incorporating Gamete Research*, *46*(1), 62–70. doi:10.1002/(SICI)1098-2795(199701)46:1<62::AID-MRD10>3.0.CO;2-5 PMID:8981365

Woodruff, T. J., Zota, A. R., & Schwartz, J. M. (2011). Environmental chemicals in pregnant women in the United States: NHANES 2003–2004. *Environmental Health Perspectives*, *119*(6), 878–885. doi:10.1289/ehp.1002727

World Health Organisation. (1999). *WHO laboratory manual for the examination of human semen and sperm-cervical mucus interaction*. Cambridge University Press.

World Health Organization (WHO). (2018). *Global Prevalence of Infertility, Infecundity and Childlessness*. http://www.who.int/reproductivehealth/ topics/infertility/burden/en

World Health Organization. (n.d.). *Schistosomiasis*. https://www.who.int/mediacentre/factsheets/fs115/en/)

Wu, H. M., Huang, H. Y., Lee, C. L., Soong, Y. K., Leung, P. C., & Wang, H. S. (2015). Gonadotropin-releasing hormone type II (GnRH-II) agonist regulates the motility of human decidual endometrial stromal cells: Possible effect on embryo implantation and pregnancy. *Biology of Reproduction*, *92*(4), 98–1. doi:10.1095/biolreprod.114.127324 PMID:25761596

Wu, X., Lu, Y., Zhou, S., Chen, L., & Xu, B. (2016). Review article: impact of climate change on human infectious diseases: empirical evidence and human adaptation. *Environment International*, *86*, 14–23. doi:10.1016/j.envint.2015.09.007 PMID:26479830

Yaeram, J., Setchell, B. P., & Maddocks, S. (2006). Effect of heat stress on the fertility of male mice in vivo and in vitro. *Reproduction, Fertility, and Development*, *18*(6), 647–653. doi:10.1071/RD05022 PMID:16930511

Yalcin, M., Dyskin, E., Lansing, L., Bharali, D. J., Mousa, S. S., Bridoux, A., Hercbergs, A. H., Lin, H. Y., Davis, F. B., Glinsky, G. V., Glinskii, A., Ma, J., Davis, P. J., & Mousa, S. A. (2010). Tetraiodothyroacetic acid (tetrac) and nanoparticulatetetrac arrest growth of medullary carcinoma of the thyroid. *The Journal of Clinical Endocrinology and Metabolism*, *95*(4), 1972–1980. doi:10.1210/jc.2009-1926 PMID:20133461

Yamada, M., & Isaji, Y. (2011). Structural and functional changes linked to, and factors promoting, cytoplasmic maturation in mammalian oocytes. *Reproductive Medicine and Biology*, *10*(2), 69–79. doi:10.100712522-011-0079-4

Yamanaka, A., Beuckmann, C. T., Willie, J. T., Hara, J., Tsujino, N., Mieda, M., Tominaga, M., Yagami, K. I., Sugiyama, F., Goto, K., Yanagisawa, M., & Sakurai, T. (2003). Hypothalamic orexin neurons regulate arousal according to energy balance in mice. *Neuron*, *38*(5), 701–713. doi:10.1016/S0896-6273(03)00331-3 PMID:12797956

Yangli, P., Yingjie, W., Jie, C., & Yinghe, Q. (2012). Effects of chronic heat stress on the reproductive capacity of male Rex rabbits. *Livestock Science*, *146*(1), 13–21. doi:10.1016/j.livsci.2012.02.015

Yang, M. Y., & Rajamahendran, R. (2002). Expression of Bcl-2 and Bax proteins in relation to quality of bovine oocytes and embryos produced in vitro. *Animal Reproduction Science*, *70*(3-4), 159–169. doi:10.1016/S0378-4320(01)00186-5

Yang, Q., Yang, Z., Ding, H., Zhang, X., Dong, Z., Hu, W., Liu, X., Wang, M., Hu, G., & Fu, C. (2014). The relationship between meteorological factors and mumps incidence in Guangzhou, China, 2005-2012. *Human Vaccines & Immunotherapeutics*, *10*(8), 2421–2432. doi:10.4161/hv.29286 PMID:25424950

Yang, S., Liu, A., Weidenhammer, A., Cooksey, R. C., McClain, D., Kim, M. K., Aguilera, G., Abel, E. D., & Chung, J. H. (2009). The role of mPer2 clock gene in glucocorticoid and feeding rhythms. *Endocrinology*, *50*(5), 2153–2160. doi:10.1210/en.2008-0705 PMID:19179447

Yang, T., Siddique, K. H. M., & Liu, K. (2020). Cropping systems in agriculture and their impact on soil health-A review. *Global Ecology and Conservation*, *23*, 1–13. doi:10.1016/j.gecco.2020.e01118

Yang, X., Zhang, S., Ju, M., & Liu, L. (2019). Preparation and Modification of Biochar Materials and their Application in Soil Remediation. *Applied Sciences (Basel, Switzerland)*, *9*(7), 1365–1390. doi:10.3390/app9071365

Yang, Y. T., & Sarfaty, M. (2016, August 12). Zika virus: A call to action for physicians in the era of climate change. *Preventive Medicine Reports*, *4*, 444–446. doi:10.1016/j.pmedr.2016.07.011 PMID:27617189

Yao, Z., Li, J., Xie, H., & Yu, C. (2012). Review on Remediation Technologies of Soil Contaminated by Heavy Metals. *Procedia Environmental Sciences*, *16*, 722–729. doi:10.1016/j.proenv.2012.10.099

Yarsley, V. E., & Couzens, E. G. (1945). *Plastics Middlesex*. Penguin Books Limited.

Ye, L., Zhao, X., Bao, E., Li, J., Zou, Z., & Cao, K. (2020). Bio-organic fertilizer with reduced rates of chemical fertilization improves soil fertility and enhances tomato yield and quality. *Scientific Reports*, *10*(1), 177–188. doi:10.103841598-019-56954-2 PMID:31932626

Ye, M., Beach, J., Martin, J. W., & Senthilselvan, A. (2017). Pesticide exposures and respiratory health in general populations. *Journal of Environmental Sciences (China)*, *51*, 361–370. doi:10.1016/j.jes.2016.11.012

Yie, S.-M., Brown, G. M., Liu, G.-Y., Collins, J. A., Daya, S., Hughes, E. G., Foster, W., & Gand Younglai, E. V. (1995). Melatonin and steroids in human pre-ovulatoryfollicular fluid: Seasonal variations, granulosa cell steroid production. *Human Reproduction (Oxford, England)*, *10*(1), 50–55. doi:10.1093/humrep/10.1.50 PMID:7745070

Yoo, H. J., Barlow, D. H., & Mardon, H. J. (1997). Temporal and spatial regulation of expression of heparin-binding epidermal growth factor-like growth factor in the human endometrium: A possible role in blastocyst implantation. *Developmental Genetics*, *21*(1), 102–108. doi:10.1002/(SICI)1520-6408(1997)21:1<102::AID-DVG12>3.0.CO;2-C PMID:9291586

Yoshikawa, T., Sellix, M., Pezuk, P., & Menaker, M. (2009). Timing of the ovarian circadian clock is regulated by gonadotropins. *Endocrinology*, *150*(9), 4338–4347. doi:10.1210/en.2008-1280 PMID:19520783

Yoshinaga, K. (1989). Receptor concept in implantation research. *Progress in Clinical and Biological Research*, *294*, 379–387. PMID:2542989

Yoshinaga, K. (2014). Progesterone and its downstream molecules as blastocyst implantation essential factors. *American Journal of Reproductive Immunology*, *72*(2), 117–128. doi:10.1111/aji.12253 PMID:24754263

Younas, M., Ishaq, M., & Ali, I. (2012). *Effect of Climate Change on Livestock Production in Pakistan*. Academic Press.

Youness, E. R., Mohammed, N. A., & Morsy, F. A. (2012). Cadmium impact and osteoporosis: Mechanism of action. *Toxicology Mechanisms and Methods*, *22*(7), 560–567. doi:10.3109/15376516.2012.702796

Younglai, E. V., Foster, W. G., Hughes, E. G., Trim, K., & Jarrell, J. F. (2002). Levelsof environmental contaminants in human follicular fluid, serum andseminal plasma of couples undergoing in vitro fertilization. *Archives of Environmental Contamination and Toxicology*, *43*(1), 121–126. doi:10.100700244-001-0048-8 PMID:12045882

Young, S. L. (2013). Eestrogen and progesterone action on endometrium: A translational approach to understanding endometrial receptivity. *Reproductive Biomedicine Online*, *27*(5), 497–505. doi:10.1016/j.rbmo.2013.06.010 PMID:23933037

Yousef, M. K. (1985). *Stress physiology in livestock*. CRC Press.

Yu, B., Ruman, J., & Christman, G. (2011). The role of peripheral gonadotropin-releasing hormone receptors in female reproduction. *Fertility and Sterility*, *95*(2), 465–473. doi:10.1016/j.fertnstert.2010.08.045 PMID:20888559

Yucra, S., Rubio, J., Gasco, M., Gonzales, C., Steenland, K., & Gonzales, G. F. (2006). Semen quality and reproductive sex hormone levels in Peruvian pesticide sprayers. *International Journal of Occupational and Environmental Health, 12*(4), 355–361. doi:10.1179/oeh.2006.12.4.355

Yu, M., Wang, Y., Jiang, J., Wang, C., Zhou, G., & Yan, J. (2019). Soil Organic Carbon Stabilization in the Three Subtropical Forests: Importance of Clay and Metal Oxides. *Journal of Geophysical Research. Biogeosciences, 124*(10), 2976–2990. doi:10.1029/2018JG004995

Yu, X., Feric, Z., Cordero, J. F., Meeker, J. D., & Alshawabkeh, A. (2018). Potential influence of temperature and precipitation on preterm birth rate in Puerto Rico. *Scientific Reports, 8*(1), 16106. doi:10.103841598-018-34179-z PMID:30382121

Zafar, A. H., Ahmed, N., & Shah, S. K. (1988). Effect of seasonal variation on semen production of Nili-Ravi buffalo bulls. *Buffalo Journal, 4*(1), 61–68.

Zaharia, C., Suteu, D., Muresan, A., Muresan, R., & Popescu, A. (2009). Textile wastewater treatment by homogeneous oxidation with hydrogen peroxide. *Environmental Engineering and Management Journal, 8*(6), 1359–1369. doi:10.30638/eemj.2009.199

Zegers-Hochschild, F., Adamson, G. D., Dyer, S., Racowsky, C., De Mouzon, J., Sokol, R., Rienzi, L., Sunde, A., Schmidt, L., Cooke, I. D., Simpson, J. L., & van der Poel, S. (2017). The international glossary on infertility and fertility care, 2017. *Human Reproduction (Oxford, England), 32*(9), 1786–1801. doi:10.1093/humrep/dex234

Zeidan, A. E. B., Marai, I. F. M., & Abd El-Kariem, Z. A. (1997). Effects of intratesticular injection of gonadotropin-releasing hormone on reproductive performance of low fertile male rabbits under Egyptian summer conditions. *Proceedings of 1st International Conference on Animal Production and Health*, 557–566.

Zeller, J. M., Henig, I., Radwanska, E., & Dmowski, W. P. (1987). Enhancement of Human Monocyte and Peritoneal Macrophage Chemiluminescence Activities in Women With Endometriosis. *American Journal of Reproductive Immunology and Microbiology, 13*(3), 78–82. doi:10.1111/j.1600-0897.1987.tb00097.x PMID:3605484

Zeng, X., Jin, T., Zhou, Y., & Kong, Q. (2002). Alterations of serum hormone levels in male workers occupationally exposed to cadmium. *Journal of Toxicology and Environmental Health. Part A., 65*(7), 513–521. doi:10.1080/15287390252807975

Zeron, Y., Ocheretny, A., Kedar, O., Borochov, A., Sklan, D., & Arav, A. (2001). Seasonal changes in bovine fertility: Relation to developmental competence of oocytes, membrane properties and fatty acid composition of follicles. *Reproduction (Cambridge, England)*, 447–454. doi:10.1530/rep.0.1210447

Zhang, L., Zhang, Z., Wang, F., Tian, X., Ji, P., & Liu, G. (2017). Effects of melatonin administration on embryo implantation and offspring growth in mice under different schedules of photoperiodic exposure. *Reproductive Biology and Endocrinology, 15*(1), 78. doi:10.118612958-017-0297-7 PMID:28969693

Zhang, S., Zeitzer, J. M., Yoshida, Y., Wisor, J. P., Nishino, S., Edgar, D. M., & Mignot, E. (2004). Lesions of the suprachiasmatic nucleus eliminate the daily rhythm of hypocretin-1 release. *Sleep, 27*(4), 619–627. doi:10.1093leep/27.4.619 PMID:15282996

Zhang, W., Du, Z., Huang, S., Chen, L., Tang, W., Zheng, H., Yang, B., & Hao, Y. (2017, September 1). The association between human perceived heat and early-stage syphilis and its variance: Results from a case-report system. *The Science of the Total Environment, 593-594*, 593–594. doi:10.1016/j.scitotenv.2017.03.194 PMID:28364611

Zhao, M., Wan, J., Zeng, K., Tong, M., Lee, A. C., Ding, J., & Chen, Q. (2016). The reduction in circulating melatonin level may contribute to the pathogenesis of ovarian cancer: A retrospective study. *Journal of Cancer, 7*(7), 831–836. doi:10.7150/jca.14573 PMID:27162542

Zha, W. T., Li, W. T., Zhou, N., Zhu, J. J., Feng, R., Li, T., Du, Y. B., Liu, Y., Hong, X. Q., & Lv, Y. (2020). Effects of meteorological factors on the incidence of mumps and models for prediction, China. *BMC Infectious Diseases*, *20*(1), 468. doi:10.118612879-020-05180-7 PMID:32615923

Zhou, J., & Bondy, C. (1992). Insulin-like growth factor-II and its binding proteins in placental development. *Endocrinology*, *131*(3), 1230–1240. doi:10.1210/endo.131.3.1380437 PMID:1380437

Zhou, S. Y., Liang, Y. X., Chen, Z. Q., & Wang, Y. L. (1988). Effects of occupational exposure to low-level carbon disulfide (CS2) on menstruation and pregnancy. *Industrial Health*, *26*(4), 203–214. doi:10.2486/indhealth.26.203 PMID:3230053

Zhou, W., Fang, F., Zhu, W., Chen, Z. J., Du, Y., & Zhang, J. (2017). Bisphenol A and ovarian reserve among infertile women with polycystic ovarian syndrome. *International Journal of Environmental Research and Public Health*, *14*(1), 18. doi:10.3390/ijerph14010018

Zhou, Y., Fu, X. M., He, D. L., Zou, X. M., Wu, C. Q., Guo, W. Z., & Feng, W. (2016). Evaluation of urinary metal concentrations and sperm DNA damage in infertile men from an infertility clinic. *Environmental Toxicology and Pharmacology*, *45*, 68–73. doi:10.1016/j.etap.2016.05.020

Ziskin, M. C., & Morrissey, J. (2011). Thermal thresholds for teratogeni- city, reproduction, and development. *International Journal of Hyperthermia*, *27*(4), 374–387. doi:10.3109/02656736.2011.553769 PMID:21591900

Ziv-Gal, A., & Flaws, J. A. (2016). Evidence for bisphenol A-induced female infertility: A review (2007–2016). *Fertility and Sterility*, *106*(4), 827–856. doi:10.1016/j.fertnstert.2016.06.027

Zi, X.-D., Yin, R.-H., Chen, S.-W., Liang, G.-N., Zhang, D.-W., & Guo, C.-H. (2009). Developmental competence of embryos derived from reciprocal in vitro fertilization between yak (Bosgrunniens) and cattle (Bostaurus). *The Journal of Reproduction and Development*, 0906090175–090609017.

Zvonic, S., Ptitsyn, A. A., Conrad, S. A., Scott, L. K., Floyd, Z. E., Kilroy, G., Wu, X., Goh, B. C., Mynatt, R. L., & Gimble, J. M. (2006). Characterization of peripheral circadian clocks in adipose tissues. *Diabetes*, *55*(4), 962–970. doi:10.2337/diabetes.55.04.06.db05-0873 PMID:16567517

Zwolak, A., Sarzyńska, M., Szpyrka, E., & Stawarczyk, K. (2019). Sources of Soil Pollution by Heavy Metals and Their Accumulation in Vegetables: A Review. *Water, Air, and Soil Pollution*, *230*(7), 164–173. doi:10.100711270-019-4221-y

About the Contributors

Khursheed Ahmad Wani is Lecturer in the Department of Environmental Science, Government Degree College Bijbehara, J&K. Dr. Wani served at ITM University Gwalior as Assistant Professor in the Department of Environmental Science from 2011 to 2017. He has completed his M. Sc. and Ph. D. in Environmental Science from Jiwaji University, Gwalior, Madhya Pradesh. He has also qualified Jammu and Kashmir State Level Eligibility Test (JKSET). His areas of interest include Environmental management, Environmental Impact Assessment and Occupational Health. He has published research papers in various national and international journals and is the author of three books on Environmental Science. He is a member of several national and international organizations and is in the editorial board of various journals. He has organized different programmes on Environmental Awareness and has completed research project sponsored by Madhya Pradesh Council of Science and Technology, Bhopal, India. Dr. Wani has organized various national and international conferences/seminars/ workshops as well.

Nibedita Naha, PhD, is Scientist 'D' and Head of the Biochemistry Division, ICMR-National Institute of Occupational Health (Ministry of Health and Family Welfare, Department of Health Research, Govt. of India). She did PhD work on reproductive health of the lead factory workers from Jadavpur University, India; and Postdoctorate on prostate cancer cell signalling, tumour xenograft, and neurobiochemistry of substance abuse from Gyeongsang National University, South Korea. She also served as a faculty of Govt. Degree College, Medical College, and Universities in India and abroad. She has vast experience on mechanism-based approach of different techniques using pre-clinical animal models, cell lines and human subjects (patient's samples) with respect to occupational and environmental exposure since 2000. Dr. Naha is a recognized PhD Guide of Gujarat University, India; and a member of several PhD Research Committees (RDC) and Board of Studies (BoS) of different universities, CPCSEA nominee of Animal Ethics Committee (Ministry of Fisheries, Animal Husbandry and Dairying, Govt. of India) of several research organizations and pharmaceutical companies; Chairperson of Human Ethics Committee (Ministry of Health and Family Welfare, Govt. of India) of pharmaceutical colleges; and Department of Biotechnology nominee (Ministry of Science and Technology, Govt. of India) for the Institute Biosafety Committee. Apart from that, Dr. Naha is also a reviewer and editorial member of several international and national peer-reviewed journals; and invited speakers of different national and international conferences. She is also a trained NABL internal auditor of ISO/IEC17025:2017 and ISO/IEC15189:2012. She has authored more than 30 research articles, invited reviews and book chapters along with patents on her research work. As per recent Google scholar, her current citation is nearly 300 with 9 as both h-index and i10-index excluding self-citation. Dr. Naha is an elected member of the Physiological Society of India (PSI) since 2013, and the advisory board member of some national conferences sponsored by

Govt. of India. She has also honoured as a life member of several scientific societies in India and US such as, SRBCE, ISSRF, NESA, ISNA, ISCA, SfN and AAAS etc.; and awardees of many national and international scientific reputes including SfN (USA), Brain Korea 21 (South Korea), PB Sen Gold Medal (Physiological Society of India), Young Scientist Award (Indian Science Congress), National Scholarship (Govt. of India) etc. Dr. Naha considers herself extremely fortunate to contribute in improving the health and wellbeing through her research findings in the respective fields.

* * *

Ranjana Choudhari is a qualified medical professional, currently employed as a senior medical scientist (Deputy director grade) in the Department of Clinical Epidemiology at ICMR- National Institute of Occupational Health, Ahmedabad, under the aegis of Indian Council of Medical Research, Department of health research, Ministry of Health and Family Welfare, Government of India. She has more than 16 years' research experience in planning and executing research on various occupational and environmental health issues of public health significance at the national level. Additionally, she has rich experience in public health care delivery in rural areas of India and clinical practice in urban areas of India. She has also served for a period about three years as a Medical officer for Government of Maharashtra State (India), delivering health care services at the grass-root level, including curative medical services, Child immunization, Epidemics investigations, Planning & implementation of prevention and control strategies for epidemics, Antenatal and perinatal health care, Execution of national programs on prevention and control of various diseases of public health importance, etc. She is an alumna of Government Medical College, Aurangabad during her professional courses on medicine studies and an awardee of WHO fellowship on "Epidemiology of non-communicable diseases" at All India Institute of Hygiene and Public health, Kolkata. Her areas of interest include Human biomonitoring, Heavy metal toxicity, Air pollution, Environmental and health risk assessment, Dust exposure and Occupational diseases, etc. She has credit for publishing research papers in various national and international journals and is a member of several national and international organizations. As a lead researcher, she has successfully completed research projects on occupational and environmental issues, including Hon'ble Supreme court and High court of Gujarat directive research projects, also projects sponsored by the Government of Gujarat and State Pollution control boards.

Uiase Bin Farooq is a Ph.D. Scholar in the Division of Veterinary Surgery and Radiology, SKUAST – Kashmir.

Younis Ahmad Hajam holds a doctorate in Zoology from Guru Ghasidas Central University, Bilaspur, Chhattisgarh, India. Currently, he is working as assistant Professor at Career Point University, Hamirpur, Himachal Pradesh, India. He also served as an Junior Research Fellow (JRF) at Zoology, Guru Ghasidas Central University, Bilaspur, Chhattisgarh, India. Dr. Younis Ahmad has published 20 research papers in leading national and international journals (such as Life Sciences, Impact factor 3.4), two book chapters and one national proceedings. Dr. Younis Ahmad has received national award for his contributions in biological science. Dr. Younis Ahmad's current research focus is therapeutic management for diabetes/Polycystic Ovarian Cancer, melatonin physiology, antioxidants and natural therapies for other metabolic and endocrine disorders.

Pir Ishfaq is a PhD in Zoology at the School of Biological Sciences.

Irfan Jamila is senior Assistant Professor in Department of Zoology, Government Womens College Srinagar.

Ashiq Khanday is a research student.

Rajesh Kumar holds a doctorate in Zoology from Jiwaji University, Gwalior, Madhya Pradesh, India. Currently, he is working as assistant Professor at Career Point University, Hamirpur, Himachal Pradesh, India. He also served as an Assistant Professor at the Arni School of Basic Science, Arni University, Indora, Himachal Pradesh, India. Dr. Kumar has published 24 research papers in leading international journals, three books, one book chapter and seven national proceedings. He is also a reviewer for numerous SCI journals. Dr. Kumar has received several national & international awards for his contributions in biological science. Dr. Kumar's current research focus is Bee nutrition, natural products, therapeutic and biomedicines.

Javid Malik completed his Ph.D from Guru Ghasidas Vishwavidyalya (A Central University), Bilaspur, Chhattisgarh, India. Dr Malik has published many research papers in various national and international journals, Dr Malik has been awarded at national as well as international platforms.

Junaid Ahmad Malik has completed his Ph.D in Zoology (Wildlife) in 2015. He is working as a Lecturer in Zoology at Govt. Degree College, Bijbehara, Anantnag. Dr. Malik has published many research articles and technical papers in many International and National reputed Journals like Springer, Elsevier, Taylor and Francis etc. He has also authored several books and book chapters of reputed publishers. He is in the editorial board and a regular reviewer of several journals of reasonable reputation. He is a lifetime member of several organizations and is in the editorial board of various journals. Dr. Malik has participated in various national and international conferences/seminars/symposiums/workshops as well.

Gokul Manickavachagam is a Ph.D. Fellow in the Biochemistry Division. His research interests are in cognitive and behavioral neuroscience and revolves around how nicotine affects the neurobehavior of adult rats. For his P.h.D thesis, he wrote about the role on nicotine and cigarette smoking in the upregulation of BDNF-Dopaminergic system.

Javid Manzoor is a Ph.D. student in the Department of Environmental science, Jiwaji University Gwalior (M.P) India. He has published research papers in international journals and has presented his work in various conferences. His area of research is environmental toxicology, pollution and its control.

Urfeya Mirza is a Ph.D. Scholar in the Division of Veterinary Surgery and Radiology, SKUAST – Kashmir.

Sukanta Mondal is currently working as Principal Scientist, Physiology Division, ICAR-National Institute of Animal Nutrition and Physiology, Bangalore, India. His major research interests involve cloning, characterization and expression of genes regulating early embryonic loss, molecular characterization and expression of hormone receptors, impact of stress on maternal recognition of pregnancy etc. He is a noted environmental physiologist who has deciphered the molecular basis of climate change impact on maternal recognition of pregnancy and embryo survival in sheep. He was instrumental in developing ideal heat stress and nutritional model in sheep. He has published over 130 research publications in various National/International journals and presented more than 30 papers at various National/International Conferences. Dr. Mondal's pioneering and path breaking contribution in hormone receptor as well as sheep and buffalo genomics has been recognized nationally and has received several prestigious awards like Prof. G. P. Talwar Midcareer Scientist award (2014), Rashtriya Gaurav award (2014), Prof. G. K. Pal award (2013), Fellow of Indian Chemical Society (2013), Fellow of Society for Applied Biotechnology (2012), Fellow of Indian Association of BioMedical Scientitsts (2010), Dr. K. Anji Reddy award (2011), Prof. P. B. Sen Memorial Oration award (2008), Biotechnology Overseas Associateship award (Long term, 2007-08), ISSRF Young Scientist Award (2006), 2000 Outstanding Scientists of the 21st Century and 21st Century Award by the International Biographical Centre, England, UK as well as ABI's Research Board of Advisors by American Biographical Institute, Inc, USA. Dr. Mondal has been bestowed with several fellowships including National Scholarship (Govt. of India), S J Jindal Trust fellowship, Junior Research Fellowship (NDRI), Senior Research Fellowship (IVRI) and CRBR Postdoctoral fellowship (Canada). He is serving as Joint Secretary, National Environmental Science Academy, Executive member, Physiological Society of India and Editorial board member of Indian Association for Animal Production. He is Associate Editor of journals viz., American Journal of Biochemistry and Molecular Biology, Asian Journal of Biotechnology, Asian Journal of Cell Biology, Current Research in Poultry Science and Biotechnology etc.

Joan Nyika is an Environmental Science PhD student at the University of South Africa and a Tutorial Fellow at the Technical University of Kenya, Geoscience and the Environment Department. She obtained Msc and Bsc degrees in Land and Water Management and Biochemistry from the University of Nairobi and Jomo Kenyatta University of Agriculture and Technology, Kenya in 2017 and 2011, respectively. Her PhD work revolves around solid waste management and the effects of landfill leachate on soil and groundwater. She has interests in sustainable environmental management, water management and ecosystem management. Joan has published more than 15 articles and 5 book chapters in peer reviewed journals and books, respectively. She has and is mentoring both undergraduate and diploma students and has more than 5 years experience teaching and training in environmental science.

Jayita Pal Chowdhury is a Research Fellow working in the field of melatonin physiology and female reproduction.

Mohd Wani has completed his Ph.D from Department of Forestry, Wild Life and Environmental Sciences, Guru Ghasidas Vishwavidyalaya, Bilaspur, Chhattisgarh-495001, India. Dr Wani has published various research papers in various journals of repute. Dr Wani is presently working as forestry officer in the department of wild and environment, Govt. of India.

Index

F

H

I

L

M

N

O

P

R

S

T

V

X

Purchase Print, E-Book, or Print + E-Book

IGI Global books are available in three unique pricing formats:
Print Only, E-Book Only, or Print + E-Book. Shipping fees apply.

www.igi-global.com

Recommended Reference Books

ISBN: 978-1-5225-7635-8
© 2019; 449 pp.
List Price: $255

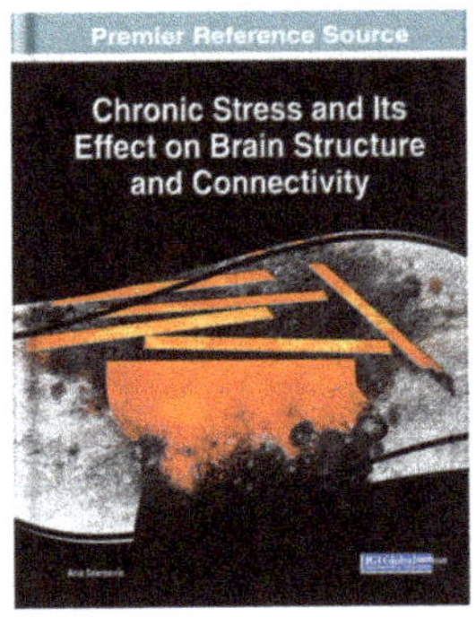

ISBN: 978-1-5225-7513-9
© 2019; 318 pp.
List Price: $225

ISBN: 978-1-5225-7576-4
© 2019; 245 pp.
List Price: $245

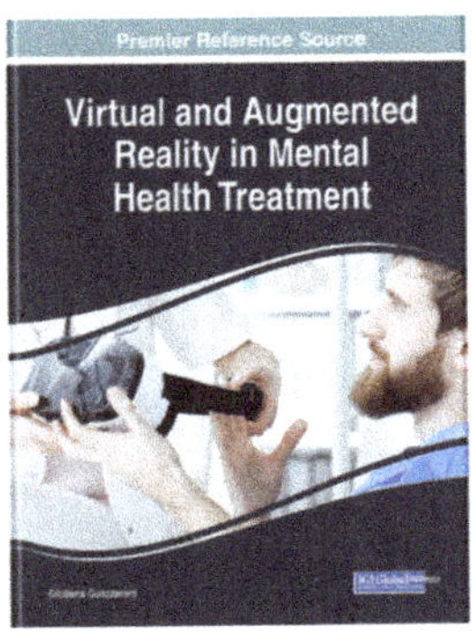

ISBN: 978-1-5225-7168-1
© 2019; 329 pp.
List Price: $265

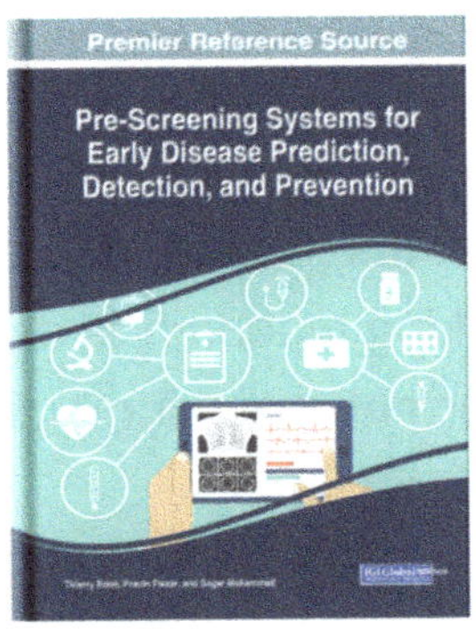

ISBN: 978-1-5225-7131-5
© 2019; 395 pp.
List Price: $245

ISBN: 978-1-5225-7402-6
© 2019; 323 pp.
List Price: $245

Do you want to stay current on the latest research trends, product announcements, news and special offers?
Join IGI Global's mailing list today and start enjoying exclusive perks sent only to IGI Global members.
Add your name to the list at **www.igi-global.com/newsletters.**

Publisher of Peer-Reviewed, Timely, and Innovative Academic Research

IGI Global
DISSEMINATOR OF KNOWLEDGE

www.igi-global.com Sign up at www.igi-global.com/newsletters facebook.com/igiglobal twitter.com/igiglobal linkedin.com/igiglobal

www.ingramcontent.com/pod-product-compliance
Lightning Source LLC
LaVergne TN
LVHW060300011026
849161LV00011B/2740